Mobile Platforms, Design, and Apps for Social Commerce

Jean-Éric Pelet
ESCE International Business School, Paris, France

A volume in the Advances in E-Business Research (AEBR) Book Series

www.igi-global.com

Published in the United States of America by
 IGI Global
 Business Science Reference (an imprint of IGI Global)
 701 E. Chocolate Avenue
 Hershey PA, USA 17033
 Tel: 717-533-8845
 Fax: 717-533-8661
 E-mail: cust@igi-global.com
 Web site: http://www.igi-global.com

Copyright © 2017 by IGI Global. All rights reserved. No part of this publication may be reproduced, stored or distributed in any form or by any means, electronic or mechanical, including photocopying, without written permission from the publisher. Product or company names used in this set are for identification purposes only. Inclusion of the names of the products or companies does not indicate a claim of ownership by IGI Global of the trademark or registered trademark.
      Library of Congress Cataloging-in-Publication Data

Names: Pelet, Jean-Eric, 1976- editor.
Title: Mobile platforms, design, and apps for social commerce / Jean-Eric
 Pelet, editor.
Description: Hershey, PA : Business Science Reference, [2017] | Includes
 bibliographical references.
Identifiers: LCCN 2017003171| ISBN 9781522524694 (h/c) | ISBN 9781522524700
 (eISBN)
Subjects: LCSH: Mobile apps--Development. | Business--Computer programs.
Classification: LCC HF5548.38.M63 M63 2017 | DDC 650.0285/535--dc23 LC record available at https://lccn.loc.
gov/2017003171

This book is published in the IGI Global book series Advances in E-Business Research (AEBR) (ISSN: 1935-2700; eISSN: 1935-2719)

British Cataloguing in Publication Data
A Cataloguing in Publication record for this book is available from the British Library.

All work contributed to this book is new, previously-unpublished material. The views expressed in this book are those of the authors, but not necessarily of the publisher.

For electronic access to this publication, please contact: eresources@igi-global.com.

Advances in E-Business Research (AEBR) Book Series

In Lee
Western Illinois University, USA

ISSN:1935-2700
EISSN:1935-2719

Mission

Technology has played a vital role in the emergence of e-business and its applications incorporate strategies. These processes have aided in the use of electronic transactions via telecommunications networks for collaborating with business partners, buying and selling of goods and services, and customer service. Research in this field continues to develop into a wide range of topics, including marketing, psychology, information systems, accounting, economics, and computer science.

The **Advances in E-Business Research (AEBR) Book Series** provides multidisciplinary references for researchers and practitioners in this area. Instructors, researchers, and professionals interested in the most up-to-date research on the concepts, issues, applications, and trends in the e-business field will find this collection, or individual books, extremely useful. This collection contains the highest quality academic books that advance understanding of e-business and addresses the challenges faced by researchers and practitioners.

Coverage

- Mobile Business Models
- Collaborative commerce
- Electronic Supply Chain Management
- Social network
- Trust, Security, and Privacy
- Valuing e-business assets
- Virtual Collaboration
- Electronic communications
- Evaluation methodologies for e-business systems
- E-Business Management

IGI Global is currently accepting manuscripts for publication within this series. To submit a proposal for a volume in this series, please contact our Acquisition Editors at Acquisitions@igi-global.com or visit: http://www.igi-global.com/publish/.

The Advances in E-Business Research (AEBR) Book Series (ISSN 1935-2700) is published by IGI Global, 701 E. Chocolate Avenue, Hershey, PA 17033-1240, USA, www.igi-global.com. This series is composed of titles available for purchase individually; each title is edited to be contextually exclusive from any other title within the series. For pricing and ordering information please visit http://www.igi-global.com/book-series/advances-business-research/37144. Postmaster: Send all address changes to above address. Copyright © 2017 IGI Global. All rights, including translation in other languages reserved by the publisher. No part of this series may be reproduced or used in any form or by any means – graphics, electronic, or mechanical, including photocopying, recording, taping, or information and retrieval systems – without written permission from the publisher, except for non commercial, educational use, including classroom teaching purposes. The views expressed in this series are those of the authors, but not necessarily of IGI Global.

Titles in this Series

For a list of additional titles in this series, please visit: www.igi-global.com/book-series

Driving Innovation and Business Success in the Digital Economy
Ionica Oncioiu (Titu Maiorescu University, Romania)
Business Science Reference • copyright 2017 • 321pp • H/C (ISBN: 9781522517795) • US $200.00 (our price)

Social Media Listening and Monitoring for Business Applications
N. Raghavendra Rao (FINAIT Consultancy Services, India)
Business Science Reference • copyright 2017 • 470pp • H/C (ISBN: 9781522508465) • US $205.00 (our price)

Analyzing the Strategic Role of Social Networking in Firm Growth and Productivity
Vladlena Benson (Kingston University, UK) Ronald Tuninga (Kingston Business School, UK) and George Saridakis (Kingston University, UK)
Business Science Reference • copyright 2017 • 525pp • H/C (ISBN: 9781522505594) • US $220.00 (our price)

Securing Transactions and Payment Systems for M-Commerce
Sushila Madan (University of Delhi, India) and Jyoti Batra Arora (Banasthali Vidyapeeth University, India)
Business Science Reference • copyright 2016 • 349pp • H/C (ISBN: 9781522502364) • US $205.00 (our price)

E-Retailing Challenges and Opportunities in the Global Marketplace
Shailja Dixit (Amity University, India) and Amit Kumar Sinha (Amity University, India)
Business Science Reference • copyright 2016 • 358pp • H/C (ISBN: 9781466699212) • US $215.00 (our price)

Successful Technological Integration for Competitive Advantage in Retail Settings
Eleonora Pantano (Middlesex University London, UK)
Business Science Reference • copyright 2015 • 405pp • H/C (ISBN: 9781466682979) • US $200.00 (our price)

Strategic E-Commerce Systems and Tools for Competing in the Digital Marketplace
Mehdi Khosrow-Pour (Information Resources Management Association, USA)
Business Science Reference • copyright 2015 • 315pp • H/C (ISBN: 9781466681330) • US $185.00 (our price)

The Evolution of the Internet in the Business Sector Web 1.0 to Web 3.0
Pedro Isaías (Universidade Aberta (Portuguese Open University), Portugal) Piet Kommers (University of Twente, The Netherlands) and Tomayess Issa (Curtin University, Australia)
Business Science Reference • copyright 2015 • 407pp • H/C (ISBN: 9781466672628) • US $235.00 (our price)

www.igi-global.com

701 East Chocolate Avenue, Hershey, PA 17033, USA
Tel: 717-533-8845 x100 • Fax: 717-533-8661
E-Mail: cust@igi-global.com • www.igi-global.com

Table of Contents

Preface xv

Chapter 1
MOOCs on Mobiles: Curating the Web and Using Social Media to Enhance E-Learning 1
    Jean-Éric Pelet, ESCE International Business School, Paris, France
    Marlene A. Pratt, Griffith University, Australia
    Stéphane Fauvy, ESSCA Ecole de Management, France

Chapter 2
College Students' Perception on the Use of Social Network Tool for Education Learning in USA 27
    Mohammed Alfadil, University of Northern Colorado, USA
    Hamzah Alhababi, University of Northern Colorado, USA
    Ali Buhamad, University of Northern Colorado, USA

Chapter 3
Heuristic Evaluation on M-Learning Applications: A Comparative Analysis of Two Heuristic
Sets 38
    Christofer Ramos, Universidade do Estado de Santa Catarina, Brazil
    Flávio Anthero Nunes Vianna dos Santos, Universidade do Estado de Santa Catarina, Brazil
    Monique Vandresen, Universidade do Estado de Santa Catarina, Brazil

Chapter 4
Mobile News Apps in India: Relocating News in the Mobile Platform 56
    Saayan Chattopadhyay, University of Calcutta, India

Chapter 5
Are We Ready to App? A Study on mHealth Apps, Its Future, and Trends in Malaysia Context 69
    Sharidatul Akma Abu Seman, Universiti Teknologi MARA, Malaysia
    Ramayah T., Universiti Sains Malaysia, Malaysia

Chapter 6
Models of Privacy and Security Issues on Mobile Applications 84
    Lili Nemec Zlatolas, University of Maribor, Slovenia
    Tatjana Welzer, University of Maribor, Slovenia
    Marjan Heričko, University of Maribor, Slovenia
    Marko Hölbl, University of Maribor, Slovenia

Chapter 7
Virtualization in Mobile Cloud Computing (VMCC) Environments 106
    Raghvendra Kumar, LNCT College, India
    Prasant Kumar Pattnaik, KIIT University, India
    Priyanka Pandey, Lakshmi Narain College of Technology, India

Chapter 8
The SMAC Opportunity Contracts: Generating Value From Responsive Agile Risk-Oriented Techniques 115
    Mohammad Ali Shalan, Jordan Engineers Association, Jordan
    Nebal Abdulrazzak Anaim, SJ Group, Saudi Arabia

Chapter 9
Irritating Factors While Navigating on Websites and Facebook and Its Reactions Using Different Devices 135
    Sana El Mouldi, IAE Université Bordeaux IV, France & ISG Tunisia, Tunisia
    Norchene Ben Dahmane Mouelhi, Université de Carthage, Tunisia

Chapter 10
Mobile Augmented Reality: Evolving Human-Computer Interaction 153
    Miguel A. Sánchez-Acevedo, Universidad de la Cañada, Mexico
    Beatriz A. Sabino-Moxo, Universidad de la Cañada, Mexico
    José A. Márquez-Domínguez, Universidad de la Cañada, Mexico

Chapter 11
Using Cognitive Psychology to Understand Anticipated User Experience in Computing Products 175
    Emmanuel Eilu, Makerere University Kampala, Uganda

Chapter 12
Mobile Commerce Technologies and Management 197
    Kijpokin Kasemsap, Suan Sunandha Rajabhat University, Thailand

Chapter 13
mMarketing Opportunities for User Collaborative Environments in Smart Cities 219
    Artemis D. Avgerou, Imperial College, UK
    Despina A. Karayanni, University of Patras, Greece
    Yannis C. Stamatiou, University of Patras, Greece

Chapter 14
The Effect of Cultural Values in Mobile Payment Preference 248
    Jashim Khan, University of Surrey, UK
    Jean-Éric Pelet, ESCE International Business School, Paris, France
    Gary James Rivers, University of Surrey, UK
    Na Zuo, Digital Economy Consulting Centre, New Zealand

Chapter 15
Consumer Behavior in M-Commerce: Literature Review and Research Agenda 264
   Saïd Aboubaker Ettis, Gulf College – Muscat, Oman
   Afef Ben Zin El Abidine, ISET Kairouan, Tunisia

Chapter 16
Designing Website Interfaces for M-Commerce With Consideration for Adult Consumers.............. 288
   Jean-Éric Pelet, ESCE International Business School, France
   Basma Taieb, Cergy Pontoise, France

Chapter 17
Mobile Customer Relationship Management: An Overview... 309
   Tolga Dursun, Abant İzzet Baysal Üniversity, Turkey
   Süleyman Çelik, Abant İzzet Baysal Üniversity, Turkey

Chapter 18
Benefits of Using Social Media Commerce Applications for Businesses 322
   Ardian Hyseni, South East European University, Kosovo

Compilation of References 347

About the Contributors 401

Index 409

Detailed Table of Contents

Preface xv

Chapter 1
MOOCs on Mobiles: Curating the Web and Using Social Media to Enhance E-Learning 1
    Jean-Éric Pelet, ESCE International Business School, Paris, France
    Marlene A. Pratt, Griffith University, Australia
    Stéphane Fauvy, ESSCA Ecole de Management, France

MOOCs (Massive Open Online Courses) have gained popularity for e-learning purposes. Effectiveness of a MOOC depends on the platform interface design and management, which should create student cohesiveness and optimize collaboration. A MOOC prototype was developed and e-learning applications were pilot-tested for one semester with a total of 160 students from graduate courses in a French business school. Students used a mobile supported e-learning environment and reported their experiences through the writing of a synthesis, the building of a CMS (Content Management System) and the elaboration of a content curation system.

Chapter 2
College Students' Perception on the Use of Social Network Tool for Education Learning in USA 27
    Mohammed Alfadil, University of Northern Colorado, USA
    Hamzah Alhababi, University of Northern Colorado, USA
    Ali Buhamad, University of Northern Colorado, USA

This study presents a survey research design on college students' perception on the use of social network tools for education in USA. The survey research design is a very valuable tool for assessing opinions and trends and it includes both graduate and undergraduate students, as well as male and female respondents. The data collection was via University of Northern Colorado (UNC) website (list-serves and Facebook groups). Composite 2 of the survey was eliminated because the responses orders of the items were different from composite 1 as the responses in composite 1 went from strongly disagree to strongly agree while in composite 2 the responses went the other direction and we were afraid the participants did not pay attention to that.

Chapter 3
Heuristic Evaluation on M-Learning Applications: A Comparative Analysis of Two Heuristic
Sets 38
    Christofer Ramos, Universidade do Estado de Santa Catarina, Brazil
    Flávio Anthero Nunes Vianna dos Santos, Universidade do Estado de Santa Catarina, Brazil
    Monique Vandresen, Universidade do Estado de Santa Catarina, Brazil

Heuristic evaluation stands out among the usability evaluation methods regarding its benefits related to time and costs. Nevertheless, generic heuristic sets require improvements when it comes to specific interfaces as seen on m-learning applications that have acquired considerable evidence within the current technologic context. Regarding the lack of studies aimed at interfaces of this sort, the authors propose, through a systematic methodology, the comparative study between a heuristic set specific to the assessment on e-learning interfaces and other, on mobile. The identified usability problems were matched with the aspects of coverage, distribution, redundancy, context and severity, in a way that it was possible to understand the efficiency of each set in covering m-learning issues. Among the findings, e-learning's heuristic set could detect a larger number of usability problems not found by mobile's.

Chapter 4
Mobile News Apps in India: Relocating News in the Mobile Platform 56
    Saayan Chattopadhyay, University of Calcutta, India

This chapter on the consumption of mobile news in developing country draws on the limited but growing scholarship on journalism and mobile media. India, becomes an emblematic instance, as India's mobile phone subscriber base peaked to more than 1 billion users in late 2015, making India the second largest mobile phone user base. In recent years, mobile media in India have also penetrated individuals' news consumption and sharing behaviors. These emerging practices can be posited in relation to the ubiquitous presence of mobile devices and a steadily expanding digital ecosystem. Utilizing both quantitative and qualitative methods, this chapter seeks to explore how mobile apps position themselves into wider news media assemblages in a developing country like India and what are the factors that influence mobile apps usage for news consumption in India? Hence, broadly, the article aims to explore, how these emerging practices are transforming not only the dominant ways of distributing the news but also the very nature of the relationship between news media and its audience.

Chapter 5
Are We Ready to App? A Study on mHealth Apps, Its Future, and Trends in Malaysia Context 69
    Sharidatul Akma Abu Seman, Universiti Teknologi MARA, Malaysia
    Ramayah T., Universiti Sains Malaysia, Malaysia

In Malaysia, adoption of the mobile application for smartphones and tablet computers are growing in number and are actively applied in healthcare. However, limited studies were found looking at mHealth app that is focusing on Malaysia context. This study aims to examine the current mHealth app that is available in Malaysia. This study also seeks to rank the pricing of top paid apps from two major platforms, Apple iOS, and Android PlayStore. In mid-2016, the authors overviewed the Medical app and Health and Fitness category from two dominant platforms; Apple iOS and Android Play Store. The only app that was related to human healthcare, described in Bahasa Malaysia or English, was examined. Most app that is designed specifically for Malaysia is informational apps, which provide information on healthcare and medical information. The study also reveals that most consumers in Malaysia are ready and are willing to pay for mHealth app. Majority of app price is between RM10.01 to RM25.

Chapter 6
Models of Privacy and Security Issues on Mobile Applications 84

    Lili Nemec Zlatolas, University of Maribor, Slovenia
    Tatjana Welzer, University of Maribor, Slovenia
    Marjan Heričko, University of Maribor, Slovenia
    Marko Hölbl, University of Maribor, Slovenia

The development of smart phones and other smart devices has led to the development of mobile applications, which are in use frequently by the users. It is also anticipated that the number of mobile applications will grow rapidly in the next years. This topic has, therefore, been researched highly in the past years. Mobile applications gather user data and that is why privacy and security in mobile applications is a very important research topic. In this chapter we give an overview of the current research on privacy and security issues of mobile applications.

Chapter 7
Virtualization in Mobile Cloud Computing (VMCC) Environments 106

    Raghvendra Kumar, LNCT College, India
    Prasant Kumar Pattnaik, KIIT University, India
    Priyanka Pandey, Lakshmi Narain College of Technology, India

Unfortunately, most of the widely used protocols for remote desktop access on mobile devices have been designed for scenarios involving personal computers. Furthermore, their energy consumption at the mobile device has not been fully characterized. In this chapter, we specially address energy consumption of mobile cloud networking realized through remote desktop technologies. In order to produce repeatable experiments with comparable results, we design a methodology to automate experiments with a mobile device. Furthermore, we develop an application that allows recording touch events and replaying them for a certain number of times. Moreover, we analyze the performance of widely used remote desktop protocols through extensive experiments involving different classes of mobile devices and realistic usage scenarios. We also relate the energy consumption to the different components involved and to the protocol features. Finally, we provide some considerations on aspects related to usability and user experience.

Chapter 8
The SMAC Opportunity Contracts: Generating Value From Responsive Agile Risk-Oriented
Techniques 115

    Mohammad Ali Shalan, Jordan Engineers Association, Jordan
    Nebal Abdulrazzak Anaim, SJ Group, Saudi Arabia

The concept of Social, Mobile, Analytics and Cloud (SMAC) is increasingly asserted as the phenomena with the potential to change technology and business relationships. In this SMAC era plenty of Middle Circle Contractors (MCCs) are being introduced as principal suppliers, integrators or outsourced contractors. This is reducing the Client Enterprise (CE) controls over their technology assets. Because it is not mature yet, SMAC nature is very disruptive and agile, thus rapidly changing the landscape of contracting, and ultimately turning the long-held promise of utility based computing into a reality. Such changes necessitate contracting transformation with innovated approaches to get targeted benefits, reduce risks and enhance operational controls. The main objective of this chapter is to provide guidelines to generate SMAC Opportunity Contracts (OCs) that are responsive and agile to provide the maximum business value, enhance risk governance and re-invent the roles and obligations in an ever-changing environment.

Chapter 9
Irritating Factors While Navigating on Websites and Facebook and Its Reactions Using Different Devices 135
    Sana El Mouldi, IAE Université Bordeaux IV, France & ISG Tunisia, Tunisia
    Norchene Ben Dahmane Mouelhi, Université de Carthage, Tunisia

The research presented in this chapter identifies sources of the irritation felt by internet users while browsing websites and Facebook. A qualitative approach was taken, including 40 individual interviews, enabled the authors to determine the irritating factors and user reactions when using different devices such as smartphones, computers and tablets to navigate websites and Facebook. The implications of this research will help marketers and web developers to reduce internet user irritation and better understand their behavior to better meet their expectations.

Chapter 10
Mobile Augmented Reality: Evolving Human-Computer Interaction 153
    Miguel A. Sánchez-Acevedo, Universidad de la Cañada, Mexico
    Beatriz A. Sabino-Moxo, Universidad de la Cañada, Mexico
    José A. Márquez-Domínguez, Universidad de la Cañada, Mexico

Users who have access to a mobile device have increased in recent years. Therefore, it is possible to use a mobile device as a tool which helps to users in their daily life activities, not only for communication. On the other hand, augmented reality is a growing technology which allows the interaction with real and virtual information at the same time. Mixing mobile devices and augmented reality open the possibility to develop useful applications that users can carry with them all the time. This chapter describes recent advances in the application of mobile augmented reality in automotive industry, commerce, education, entertainment, and medicine; also identifies the different devices used to generate augmented reality, highlights factors to be taken into account for developing mobile augmented applications, introduces challenges to be addressed, and discusses future trends.

Chapter 11
Using Cognitive Psychology to Understand Anticipated User Experience in Computing Products... 175
    Emmanuel Eilu, Makerere University Kampala, Uganda

User Experience assessment is an evaluation of user's experience with the product, system or service during 'use' (i.e., actual interaction experience) as well as 'anticipated or before use' (i.e., pre-interaction experience).Whereas many user experience researchers may be conversant with explaining a person's experience during use of a product, system or service, they find it difficult to explain experience before a product or service is used (Anticipated Use), which in this chapter is referred to as Anticipated User Experience (AUX). This chapter applies the theory of cognitive psychology and its principles to best explain how Anticipated User Experience occurs and how this experience can be achieved. This chapter goes a long way in informing user experience researchers and practitioners on the relevance of attaining AUX in a computing product and how it can be achieved.

Chapter 12
Mobile Commerce Technologies and Management.. 197
    Kijpokin Kasemsap, Suan Sunandha Rajabhat University, Thailand

This chapter reveals the prospect of mobile commerce (m-commerce); m-commerce and trust; m-commerce, privacy, and security issues; m-commerce adoption and technology acceptance model (TAM); and the significant perspectives on m-commerce. M-commerce is used for business transactions conducted by mobile phones for the promotional and financial activities using the wireless Internet connectivity. M-commerce is the important way to purchase the online items through online services. The main goal of m-commerce is to ensure that customers' shopping experience is well-suited to the smaller screen sizes that they can see on smartphones and tablets. Computer-mediated networks enable these transaction processes through electronic store searches and electronic point-of-sale capabilities. M-commerce brings the new possibility for businesses to sell and promote their products and services toward gaining improved productivity and business growth.

Chapter 13
mMarketing Opportunities for User Collaborative Environments in Smart Cities............................ 219
    Artemis D. Avgerou, Imperial College, UK
    Despina A. Karayanni, University of Patras, Greece
    Yannis C. Stamatiou, University of Patras, Greece

Smart City infrastructures connect people with their devices through wireless communications networks while they offer sensor-based information about the city's status and needs. Connecting people carrying mobile devices equipped with sensors through such an infrastructure leads to the "collective intelligence" or "crowdsourcing" paradigm. This paradigm has been deployed in numerous contexts such as performing large-scale experiments (e.g., monitoring the pollution levels or analyzing mobility patterns of people to derive useful information about rush hours in cities) or gathering and sharing user collected experiences in efforts to increase privacy awareness and personal information protection levels. In this chapter, we will focus on employing this paradigm in the mMarketing/mCommerce domain and discuss how crowdsourcing can create new opportunities for commercial activities as well as expansion of existing ones.

Chapter 14
The Effect of Cultural Values in Mobile Payment Preference 248
    Jashim Khan, University of Surrey, UK
    Jean-Éric Pelet, ESCE International Business School, Paris, France
    Gary James Rivers, University of Surrey, UK
    Na Zuo, Digital Economy Consulting Centre, New Zealand

The purpose of this study is to compare French and New Zealand consumers' perceptions of mobile payments (m-payments) relative to other options to identify the preferred mode of payment and related spending behaviour. Evidence suggests that payment modes can influence spending behaviours and therefore this is important to commerce to promote payment modes that facilitate transactions. Using the Perceptions of Payment Mode (PPM) scale, this study was able to identify cultural differences on perceptions of cash payments, though both countries' consumers held negative perceptions of, and emotions towards, m-payments relative to other options. The empirical results are useful in understanding cultural aspects of payment modes and for companies to recognise consumers' associations with these modes to enhance relations, services and the use of m-payments.

Chapter 15
Consumer Behavior in M-Commerce: Literature Review and Research Agenda 264
    Saïd Aboubaker Ettis, Gulf College – Muscat, Oman
    Afef Ben Zin El Abidine, ISET Kairouan, Tunisia

Our work has a two-fold general objectives: on the one hand, we wish to describe the mobile commerce environment and, on the other hand, to establish the determinants of the mobile consumer behavior. To achieve our objectives, first, we describe the concept of mobile commerce, constraints and benefits. Second, we study the determinants of mobile commerce adoption. Third, we focus on the determinants of mobile consumer satisfaction and loyalty. Finally, we summarize future avenues of investigation in a research agenda.

Chapter 16
Designing Website Interfaces for M-Commerce With Consideration for Adult Consumers.............. 288
    Jean-Éric Pelet, ESCE International Business School, France
    Basma Taieb, Cergy Pontoise, France

This chapter analyzes the interaction effects between the principal design cues of a mobile commerce website, such as background/foreground colors, font text and layout. Three experiments have been conducted based on visits to a fictitious m-commerce website. Experiment 1 manipulates the levels of color contrast: positive contrast (light text on a dark background) versus negative contrast (dark text on a light background). In experiment 2, contrast and font have been manipulated with a complete factorial plan: 2 x 2 (negative vs positive contrast x serif font vs sans serif font). Finally, contrast and layout have been manipulated in a third experimental 2 x 2 plan (negative vs positive contrast x dense vs airy layout). This research involved 219 French participants. Results show significant effects of the positive contrast (light text on a dark background) of the mobile website design on the purchase and revisit intentions of adults. Discussions about the interaction effects of design elements, limitations and directions for future research follow.

Chapter 17
Mobile Customer Relationship Management: An Overview.. 309
    Tolga Dursun, Abant İzzet Baysal Üniversity, Turkey
    Süleyman Çelik, Abant İzzet Baysal Üniversity, Turkey

Electronic platforms provide many advantages both customers and companies due to development of communication technology. Today almost every people have smartphones and tablets. Thus mobile customer relationship management became an significant concept for generating long-term relationships and increasing customer satisfaction, retention and loyalty. In addition companies use mobile CRM to facilitate salespeople for better performance in marketing activities. M-CRM offers interactive relationships between firms and companies. In this study, we define what is customer relationship management and origins of CRM. After that we stated electronic customer relationship management concept and finally we mentioned about mobile CRM especially benefits and characteristics of it.

Chapter 18
Benefits of Using Social Media Commerce Applications for Businesses 322
    Ardian Hyseni, South East European University, Kosovo

Social media commerce has changed the way of commerce globally; customers are affected more and more by social media, in decision making for buying a product or a service. While in the past people were affected by traditional marketing ways like newspapers, televisions and radios for buying a product, nowadays, through social media customers can find feedbacks and reviews on social media and can see thousands of photos of a single product with less a minute of searching in a social networking sites like. With the growth of social media's impact on businesses, social commerce has become a trending way of making commerce. In this paper it demonstrated a platform for businesses to make commerce through Facebook which is called Facebook commerce.

Compilation of References 347

About the Contributors 401

Index 409

Preface

MOBILE IS CHANGING OUR LIFE IN THE REAL WORLD

Mobile is not only reinventing our digital experiences, it is changing our life in the real world: in the streets, public transportation, at school, at home, in retail stores, etc.

These changes are due to a large extent to designers who invented the new interfaces of our smartphones and touch screen devices. Ramos et al. have investigated this area of interfaces, in particular in an m-learning context, focusing on heuristic evaluation. The usability problems of m-learning they identified were matched with aspects of coverage, distribution, redundancy, context and severity, in a way that it was possible to understand the efficiency of each set in covering m-learning issues. One of their most important findings is that the heuristic set of e-learning issues detect a large number of usability problems not found by mobiles.

Doueihi (2011) stated that

The greater influence of Steve Jobs has been the return of the body in our daily digital, return which has transformed our habitus by changing the working space, public space and intimate space. A return which also means the emergence of new digital culture. A culture of the bureau and the Chair turns with the iPhone in a walking culture.

Referring to "walking culture", no one can ignore people in trains, on buses or in the street using their smartphone to play, to discuss, to visit, to read or to work. With smartphones the frontier between the real and the virtual is blurring. In this vein, Alfadil presents college students' perception of the use of social network tools for learning in USA.

Mobile devices have recently become so pervasive that they are increasingly replacing personal computers in everyday activities. However, due to their limited resources, mobile devices cannot or the same performance of personal computers and workstations. One option to overcome such limitations is given by remote desktop access, wherein the mobile device uses thin client software to connect to a remote desktop host. When such host is virtualized, remote desktop access becomes a form of mobile cloud networking. Unfortunately, most of the widely used protocols for remote desktop access on mobile devices have been designed for scenarios involving personal computers. This is the chapter written by Kumar et al.

When using maps and city guides to gather information on restaurants, museums or theaters, when communicating with friends and family, in education and health, in shopping and retail store experiences, smartphones enrich our life as well as increase and improve our experiences. Both connecting people, and carrying mobile devices equipped with sensors through such an infrastructure can lead to the development of the "collective intelligence" or "crowdsourcing" paradigm, according to Avgerou et al. Their chapter is about Smart City infrastructures which connect people with their devices through wireless communication networks while they offer sensor-based information about the city's status and needs. They discuss how crowdsourcing can create new opportunities for commercial activities as well as how to expand existing ones in the marketing/m-commerce domain. Mobile therefore introduces digital in a new Walking Culture.

FROM "MOBILE FIRST" TO "APP FIRST"

With the proliferation of touch screens over the last couple of years the mobile surpassed desktops in terms of total digital engagement. One has noticed this change step by step with respect to social networking, reading e-mails and watching videos. For example, in his chapter, Saayan shows that the consumption of mobile news in developing countries draws on the limited but growing scholarship on journalism and mobile media. News audiences are increasingly demanding improved access and are trying to navigate through the array of news messages which emanate from print media, broadcasting, web, and mobile media platforms. India is becoming an emblematic instance in this process, as India's mobile phone subscriber base peaked at more than 1 billion users in late 2015, making India the second largest mobile phone user base after China. The authors explore how mobile apps position themselves into wider news media assemblages in a developing country like India, and how these mobile news apps are linked to one another, in addition to the legacy news media in the wider field of using and managing news media in India.

Finally, now the majority of all digital media time spent occurs on mobile. Recently comScore announced in their U.S. Mobile App Report 2016 that "Mobile now represents almost 2 out of 3 digital media minutes." This is also true in the health market. In Malaysia, the number of applications adopted for smartphones has grown substantially and are actively applied in healthcare. In their chapter, Abu-Seman and Ramayah examine the current mHealth app (see chapter for the permalink if you want to download the apps) that is available in Malaysia. This study also seeks to rank the pricing of top paid apps from two major platforms, Apple iOS, and Android PlayStore. The study reveals that most consumers in Malaysia are willing to pay for the mHealth app. The prices of this app range between RM[1] 10.01 (2.24 US Dollars) to RM25 (5.59 US Dollars).

The convenience of smartphones and tablet devices have completely shifted the digital media landscape in favor of mobile. As stated by Kasemsap, the main goal of m-commerce is to ensure that customers' shopping experiences are well-suited to the smaller screen sizes that they can see on their smartphones and tablets. Computer-mediated networks enable these transaction processes through electronic store searches and electronic point-of-sale capabilities. M-commerce opens up the new opportunity for businesses to sell and promote their products and services toward gaining improved productivity and business growth. In 10 years (the first iPhone was presented in Steve Jobs' Key note in January 2007) we switched from a large horizontal screen and a mouse to a vertical touch screen. Mobile's screens have become the number one access to the internet and world's internet population is now "Mobile First".

"Mobile First" does not only mean that we spend more than half of our digital time on our smartphones; it also means that publishers, brands and retailers' digital interfaces should be first thought for mobile

Preface

users before being built for desktop users. We should think rather vertically than horizontally, prefer touch screens over mice as well as cursors and mobile application over web browsing. The smartphone reinvents our digital experiences.

APPS ARE EATING THE INTERNET

In 2016, mobile applications have become the primary access vehicle to the internet, representing half of total digital media time spent, according to comScore (2016). Apps are the fuel that is driving mobile's growth and utility. They allow individuals to perform new digital tasks such as playing games, posting a Facebook status, engaging messaging conversation, listening to music, watching videos, hailing a cab, checking the weather, reading news, shopping with On-line and Off-line retailers. In their research, El Mouldi and Ben Dahmane Mouelhi present identified sources of the irritation felt by Internet users while browsing websites and Facebook. A qualitative approach was taken, including 40 individual interviews. It enables the authors to determine the irritating factors and user reactions when using different devices such as smartphones, computers and tablets to navigate websites and Facebook.

Most of the time, Smartphones and Tablets have limited usefulness without apps. Social Networking (including the fast-growing activity of messaging conversation), Games and Music contribute for nearly half of the total time spent on mobile apps. The strength of these categories highlights that mobile devices are more heavily used for entertainment and communication than their desktop counterparts. Cognitive psychology can be used to understand anticipated user experience (AUX) in computing products. In his chapter, Eilu explains that user experience assessment is an evaluation of user's experience with the product, system or service during 'use' (i.e., actual interaction experience) as well as 'anticipated or before use' (i.e., pre-interaction experience). Whereas many user experience researchers may be conversant with explaining a person's experience during use of a product, system or service, they find it difficult to explain experience before a product or service is used (Anticipated Use), which in this chapter is referred to as Anticipated User Experience (AUX). This chapter applies the theory of cognitive psychology and its principles to best explain how Anticipated User Experience occurs and how this experience can be achieved. This chapter goes a long way in informing user experience researchers and practitioners on the relevance of attaining AUX in a computing product and how it can be achieved.

The ranking of the top Apps is strongly dominated by App constellations of some of the largest digital media brands: Facebook, Google, Apple, Yahoo, Amazon and eBay. Facebook's mobile App especially shows an immense influence in the app ecosystem in terms of engagement and audience reach. In order to better apprehend this ecosystem, Ettis and Ben Zin El Abidine provide readers with a chapter with a two-fold general objective: on the one hand, authors wish to describe the mobile commerce environment and, on the other hand, to establish the determinants of the mobile consumer's behavior. To achieve this objective, first, authors describe the concept of mobile commerce, constraints, and benefits. Second, they study the determinants of mobile commerce adoption. Third, they focus on the determinants of mobile consumer satisfaction and loyalty.

App is the new battle of the Internet giants. The development of smartphones and other smart devices has led to the development of mobile applications, which are in use frequently by the users as stated by Nemec Zlatolas et al. It is also anticipated that the number of mobile applications will grow rapidly in the next years. This topic has, therefore, been researched highly in the past years. Mobile applications gather user data and that is why privacy and security in mobile applications is a very important research topic. In this chapter authors give an overview of the current research on privacy and security issues of mobile applications.

APP USERS ARE DEMANDING BUT LOYALTY

Mobile web is the primary vehicle for expanding audience reach, but Apps are where heavy engagement happens. After decades working on web sites optimization as digital marketers or web designers trying to crack Google's secret ranking algorithm, it is even more difficult today to build a large audience on Apps. This audience, as for the whole mobile industry, must take into account the growing age of mobile users, and the relative deficiency occurred by their vision. To better understand this change in the range of consumers, Pelet and Taieb present a study that investigates the effects of mobile website design on the behavioral intentions of adult consumers. More specifically, the authors analyze the interaction effects between the principal design cues of a mobile commerce (m-commerce) website, such as background/foreground colors, font text and layout. Results show significant effects of the positive contrast (light text on a dark background) of the mobile website design on the purchase and revisit intentions of adults. Discussions about the interaction effects of design elements, limitations and directions for future research follow.

The competition is much more challenging on App when it reveals crucial to find attention on Smartphone's screen: promote your Application, find your Application (App store Optimization), face user's reviews and notes, manage downloads, develop usage, optimize service level and revenue in order to leverage the Customer Relationship Management. Because today, almost everyone has smartphones and tablets, Tolga and Süleyman explain how mobile customer relationship management becomes a significant concept for generating long-term relationships and increasing customer satisfaction, retention and loyalty. In addition, companies use mobile CRM (M-CRM) to facilitate salespeople for better performance in marketing activities. M-CRM offers interactive relationships between firms and companies. In this study, authors define what is customer relationship management and origins of CRM. They then define electronic customer relationship management concept and finally mention benefits and characteristics of M-CRM.

Having a new kind of relationship with clients and users within application is a new challenge. Most of that time spent on application is concentrated in the highest engagement Apps owned by a few of the largest internet companies. Those large Internet companies command the majority of app time spent on smartphone (Husson, 2016).

It will be very challenging for lesser-known brands and offline brands to create a relationship with users and clients within an application even if the real value of Apps is loyalty from user.

Retailers, brands and digital publishers must have a well-developed platform strategy for converting their large mobile web audiences into highly engaged, loyal app users.

Without strong Apps the transition to mobile is not effective. In this direction, the concept of Social, Mobile, Analytics and Cloud (SMAC) is increasingly asserted as the phenomena with the potential to change technology and business relationships according to Shalan and Anaim. Because it is not mature yet, SMAC nature is very disruptive and agile, thus rapidly changing the landscape, and ultimately turning the long-held promise of utility based computing into a reality. Such changes necessitate contracting transformation with innovated approaches to get targeted benefits, reduce risks and enhance operational

Preface

controls. The main objective of this chapter is to provide guidelines to generate SMAC Opportunity Contracts (OCs) that are responsive and agile to provide the maximum business value, enhance risk governance and re-invent the roles and obligations in an ever-changing environment.

HAVE WE REACHED THE POINT OF "PEAK APP"?

A new challenge for marketers is related to "Peak App" as App downloads are beginning to plummet, and augmented reality to grow. Mixing mobile devices and augmented reality opens the possibility to develop useful applications that users can carry with them all the time. In their chapter, Sánchez-Acevedo et al. describe recent advances in the application of mobile augmented reality in automotive industry, commerce, education, entertainment, and medicine; also identifies the different devices used to generate augmented reality, highlights factors to be taken into account for developing mobile augmented applications, introduces challenges to be addressed, and discusses future trends.

Nearly half of smartphone users in the US market don't download any apps on a monthly basis and the average user downloads two applications (Comscore, 2016). The App market is definitely tightening and App publishers need to rethink new way of breaking through the consumer's screen.

"Mobile app growth is in the later innings, especially in developed markets" declare a report from Nomura which used data from App analytics company Sensor Tower (2016).

It will probably be more and more difficult for marketers to create a true and valuable relationship with customers in the application ecosystem.

App acquisition will certainly be moving from "pull marketing" to "push marketing". App stores remain the most important method to increase downloads, but they are no longer growing in terms of relevancy. More users are now discovering apps from websites, digital ads and traditional media ads, highlighting the increasing importance of traditional push marketing for user acquisition. App management is a new frontier for marketers.

MESSAGING APPS INTRODUCE A NEW PARADIGM

Messaging Apps appears to be the fastest growing activity for users and we expect they could play a key role throughout the customer life cycle, especially to enable brands and retailers to deepen conversations with their customers (Husson, 2016).

Messaging apps combine three strong digital assets: frequency of use, convenience and emotion. Those new assets will drive deeper relationships between consumers and brands. Messaging Apps will introduce a paradigm shift for marketers where interactive and contextual conversations will replace ad broadcasting. These new conversation interfaces will capture and mediate consumers' digital moments and change the relationship that marketers have with customers.

While Asian messaging apps such as WeChat or Line are the most advanced, Facebook is really the only one that competes at scale, due to its combined reach of WhatsApp and Messenger (Husson, 2016).

Messaging Apps will introduce a new kind of conversations with customers. This also exists in the e-learning industry, as depicted by Pelet, Pratt and Fauvy in their chapter about MOOCs and mobiles. MOOCs (Massive Open Online Courses) have gained popularity for e-learning purposes. Effectiveness of a MOOC depends on the platform interface design and management, which should create student cohesiveness and optimize collaboration. For the purpose of the chapter, a MOOC prototype was developed and e-learning applications were pilot-tested. Students used a mobile supported e-learning environment and reported their experiences through the writing of a synthesis, the building of a CMS (Content Management System) and the elaboration of a content curation system.

ON- AND OFF-LINE MOBILE COMMERCE

Due to the native environment and additional control of the purchase path, e-retailers see a higher conversion rate on their apps than both mobile web and desktop according to the latest study conducted by Criteo (http://www.criteo.com/).

"E-Retailers whose apps focus on providing shoppers with relevant and useful products and remove barriers to purchase drive a higher share of transactions than mobile web. Apps drive more transactions and value by removing barriers to purchase" (Critéo, 2016). In any case, payment will be a differential asset. This is the topic of Khan et al. The purpose of their chapter is to compare French and New Zealand consumers' perceptions of mobile payments (m-payments) relative to other options to identify the preferred mode of payment and related spending behavior. Evidence suggests that payment modes can influence spending behaviors and therefore this is important to commerce to promote payment modes that facilitate transactions. The empirical results are useful in understanding cultural aspects of payment modes and for companies to recognize consumers' associations with these modes to enhance relations, services and the use of m-payments.

The app conversion funnel is wider at every point of contact than mobile web, indicating that app users are your most loyal and dedicated shoppers. Per transaction, app buyers spend more than both mobile web and desktop.

For In-store retailers and brands, the new "walking culture" driven by Smartphone and Application is the next opportunity. "Mobile to store" and "in-store" new Digital customer relationship using Apps is one of the next mobile challenge.

As frontier between in-store and virtual store is blurring, Apps will help to develop "conversational and social commerce".

Social media commerce has changed the way of commerce globally; customers are affected more and more by social media, in decision making for buying a products or a service. While in the past people were affected by traditional marketing ways like newspapers, televisions and radios for buying a product, nowadays, through social media customers can find feedbacks and reviews on social media and can see thousands of photos of a single product with less a minute of searching in a social networking sites like. With the growth of social media's impact on businesses, social commerce has become a trending way of making commerce. In his chapter, Hyseni presents a platform for businesses to make commerce through Facebook which is called "Facebook commerce."

Bertrand Jonquois
Atsukè, France

Jean-Éric Pelet
ESCE International Business School, Paris, France

REFERENCES

Comscore. (2016). *The 2016 U.S. Mobile App Report*. Comscore.

Critéo. (2016). *State of Mobile Commerce 2016*. Author.

Doueihi, M. (2011). L'esthète du numérique. *Le Monde*. Retrieved from www.lemonde.fr/idees/article/2011/10/07/l-esthete-du-numerique_1583940_3232.html

Husson, T. (2016). *The Future of Messaging Apps*. Comscore.

Sensor Tower. (2016). *Nomura research*. Author.

ENDNOTE

[1] Malaysian Ringgit.

Chapter 1
MOOCs on Mobiles:
Curating the Web and Using Social Media to Enhance E-Learning

Jean-Éric Pelet
ESCE International Business School, Paris, France

Marlene A. Pratt
Griffith University, Australia

Stéphane Fauvy
ESSCA Ecole de Management, France

ABSTRACT

MOOCs (Massive Open Online Courses) have gained popularity for e-learning purposes. Effectiveness of a MOOC depends on the platform interface design and management, which should create student cohesiveness and optimize collaboration. A MOOC prototype was developed and e-learning applications were pilot-tested for one semester with a total of 160 students from graduate courses in a French business school. Students used a mobile supported e-learning environment and reported their experiences through the writing of a synthesis, the building of a CMS (Content Management System) and the elaboration of a content curation system.

INTRODUCTION

The "Learning for all" movement is stimulating active debates in the education space around the world. These debates combined with the emergence of new forms of blended learning as well as the arrival of Massive Open Online Courses (MOOCs) and other forms of open educational resources (OERs) have made e-learning front page news across all continents and societies.

Collaborative learning is one of the key instructional strategies that are being adopted and has gained an increasing role in educational research and practices in recent years. Computer-supported collaborative learning (CSCL) is a pedagogical approach wherein learning takes place via social interaction using a computer or through the Internet (Zheng, Junfeng, Wei, & Ronghuai, 2014). This is possible thanks

DOI: 10.4018/978-1-5225-2469-4.ch001

to the use of social media, enabling students to correspond, chat and comment on content related to a course. Many new technologies are emerging which offer new ways of teaching and learning, such as ubiquitous learning technologies, gesture-based computing, augmented reality technology, and learning analytics. Students who have grown up amidst new technologies are keen to use and adopt new devices, apps and various kinds of new ICT. Indeed, collaborative learning aims to promote students' individual cognition, group cognition and community cognition through the use of appealing, fun, easy-to-use and instantaneous tools. These tools enable students to communicate with each other, as well as sharing documents and ideas, as if they were in the same classroom or spaces. The new generation of students are experiential, interactive and social learners, multi-taskers, structured and relevant learners, and technology immersed learners (Zheng et al., 2014).

The CSCL setting is characterized by the sharing and construction of knowledge among participants using technology as their primary means of communication or as a common resource (Stahl, Koschmann, & Suthers, 2006). The latter can be implemented online and in classroom learning environments, which can take place synchronously or asynchronously. The appropriate processes, assessment and interaction methods can provide insight into effectiveness of collaborative learning in face-to-face and online contexts. Accompanying CSCL, ubiquitous e-learning is a notion that is becoming a pertinent factor in education (Pelet & Papadopoulou, 2013; Stahl et al., 2006). Many universities are starting to experiment with hybrid educational models mixing digital technologies and social media with traditional teaching approaches. This has led to an increased rate of learning outcomes as a result of applying traditional and e-learning hybrid models (Bowen, 2012). The proliferation of social media and the development of learning lead to a sharp increase in the use of social media as a support in the learning process because they participate in the permanent construction of new knowledge (Zaina, Ameida, & Torres, 2014). Among the best known micro-blogging tools, Twitter (20 million users, 50 million tweets every day) (Cormode, Krishnamurthy, & Willinger, 2010) allows users to send free short messages (called tweets limited to 140 characters) and so communicate, share ideas, links or photos to other users (Dunlap & Lowenthal, 2009). The high cost of higher education is considered one of the principal problems of today's educational system (Bowen, 2012), where the technological shift towards digital learning environments is a partial solution. Universities recognize a more digitally coherent system of operation is less expensive than the traditional model of education, and has led to universities reviewing their investment strategies (Bourcieu & Léon, 2013).

MOOCs may be a catalyst in the process of re-imagining higher education or re-enchanting e-learning, due to the powerful elements constituting the MOOC architecture. Whether MOOCs are part of a global open education initiative or a for-profit education model, today there is certainly growing R&D interest, as well as entrepreneurial attention to this form of learning. There is, however, substantial criticism and typical bystander scepticism about MOOCs, which typically include low completion rates.

This chapter provides an overview of the development and application of a MOOC. It integrates social media and curation tools as a hot topic in e-learning and presents concrete ideas on how to enable and support learning in higher education with the use of electronic devices and free Internet tools. The chapter will focus on learning as a collaborative process in which students developed their own functional knowledge management tools and actively participated in an expansive learning experience. Interaction between students and lecturers were formed by a self-regulated group of students, embracing one of the primary characteristics of a MOOC: collaborative development and constructivist learning situations.

REVIEW OF LITERATURE

MOOCs

Created in 2007, MOOCs are available worldwide and diffused by universities or by private platforms (Karsenti, 2013). This market faces strong challenges which will be discussed throughout this chapter, which include the fees and costs of creation, the quality of teaching, participation of students and the evaluation of large courses.

MOOC can be defined as aggregate classes from multiple organizations, universities and schools, offered on a single digital platform. They are designed in a way to enable the delivery of specific courses to thousands of recipients simultaneously. There are many courses on a wide array of themes and topics available on MOOCs, many of which are free or at a very low cost. Gaebel (2013) defined MOOCs as free, credit-less online courses where people can participate without limits on the amount of classes they can enrol in. De Waar (2011) reinforced this definition by describing them as "time and cost efficient". MOOCs global success is based on their potential to "MacDonalise higher education" (Lane & Kinser, 2015), of reducing tuition fees, democratize courses from prestigious universities, beyond geographical barriers and at no charge (Lewin, 2012).

There are free tools available for building these courses, which enable functions where languages can be chosen, tools tailored to the preferences of the participants and courses set up quickly. A small proportion of MOOCs are financed by examination and diploma activities, however, new business models are emerging regularly. MOOCs can be beneficial to students as an informal means of supplementing their knowledge base and enhancing their productivity. While the high number of registrations in MOOCs (over 20 million from 203 countries) was unthinkable a few years ago (Karsenti, 2013), MOOCs are based on the same teaching methods as distance learning. Issues raised highlight the views between pro-technological (in favor to the development of MOOC in education because they provide a balance between work-family-school democratization of education, free tuition, development of autonomy, creating communities of learners, student satisfaction) and skeptics (more cautious about using MOOC because their benefits are the same as distance learning) (Karsenti, 2013).

The current challenge is to reduce the dropout rate, maximize engagement and participant satisfaction. For their novelty and popularity, MOOC revolutionized learning but face a high dropout rate (Clow, 2013; Knowledge@Wharton, 2013). To increase student engagement and reduce the number of dropouts, ideally a MOOC should be of medium difficulty (Willging & Johnson, 2004). It requires a moderate amount of work of approximately six hours a week, over a short period (less than 2 months) on a fixed schedule. It contains a draft or final exam, a free manual open access, a forum to facilitate the construction of knowledge and communication between students and professors, an evaluation system by peers (Piech et al., 2013), and delivers a recognized certificate or equivalent. Adamopoulos (2013) classified in order of importance the determinants that affect the retention of online courses. Research show students completion depends on factors, which have a positive or neutral impact. These include: course evaluation by the students including the professor (Cormier & Siemens, 2010; Masters, 2011), course characteristics (difficulty, discipline, time, load and work rate), characteristics of the university (ranking, number of proposed MOOCs), the platform (usability, access to the manual, discussion forum) and students specificities (gender, degree). The results lead to the ideal MOOC design and recommendations on the use of technology in learning. According to Adamopoulos (2013), if students enrolled in a MOOC feel engaged and satisfied with the course, interactions, educational tools and evaluation system, they will tend to fully complete the MOOC. It therefore appears necessary to study the factors influencing positively the commitment and satisfaction, which could potentially increase the self-determination of participants.

The MOOC term connotes Open Access, which means that learners don't need to be registered at any particular college, university or campus as a prerequisite to enrolment. One teacher can be responsible for hundreds or thousands of students. The high number of registered participants, however, makes it difficult and sometimes impossible to link learners and teachers (Khalil & Ebner, 2013). The large number of enrolees and courses allows MOOCs to offer two approaches to instructional design: 1) peer-review, group collaborations through "crowd sourcing" or 2) Automated feedback and self-assessments (Kop, 2011).

MOOCs generally consist of pre-recorded course videos, mini- presentations of experts, MCQ (multichoice questions), forums and sometimes peer review, and are broadcast to registered participants on platforms. The primary goal remains the transmission of knowledge (Vaufrey, 2015). Activities are differentiated by the public. Here are five categories of pedagogical approaches (mandatory for students and recommended for external):

- **Annotated Reading Document:** These annotations visible to all participants to allow further discussion and analysis by the contributions of each participant;
- **Document Analysis:** For this, the use of a pad enables the writing of a collaborative work where everyone can add a differential element with a different colour;
- **Development of a Topical Question:** Chosen from a pre-set list, it results in writing an online article where the other participants are required to read, comment, and mark;
- **Individual Courses Summary:** Development of one or more strong ideas, outstanding references, and potential areas for improvement of the course. A final e-book consists of all the collected summaries, is given to each participant. External participants can also perform these syntheses but as a collaborative work.
- **Participation in the Webinars:** These interactive meetings encourage collaborative work.

Often, MOOC students watch short videos (blended learning) which are graded either by computers or by other students. They can also be asked to participate in forums. Forums develop the autonomy of participants through the building of communities of learners, experience sharing and collaborative dialogue between peers and experts (Luca & Mcloughlin, 2004). The abundance of messages and the poor attitude of some students mean that many forums are underutilized and unproductive. In return, reputation systems are sometimes set up where staff reward by scoring and ranking the most active students on the forum (Resnick, Kuwabara, Zeckhauser, & Friedman, 2000). Coetzee et al. (2014) examine the usefulness and effectiveness of forums and reputation systems to promote learning and a sense of community. The results show that participation in the forum improves the ratings and the persistence of students. Reputation systems improve participation indicators such as the number and response times and decrease the number of calls for help; but do not impact the notes, perseverance and the feeling of belonging to the community of participants. The research they conducted explains the massive use of forums in MOOCs, characterized by the influence of some very active members. Reputation systems have the ability to enhance the experience of the forums, but do not contribute to the formation of communities of learners. They are perceived as fair and helpful but would be more motivating if they impacted other courses or the final grade.

As students seek autonomy and computer skills, it is essential to quickly identify participants who experience difficulty. Hence the question: how to help and encourage students to continue learning? Sinclair et al. (2014) found that good marks obtained in MCQ shows that participants were serious. The experience was found to be positive for 98% of participants who believe that training was at the right level, well built with a motivating introduction, simple examples and useful exercises. Participants however want the duration of videos and sound recordings improved and a card with the topics to be created easily and quickly to find the necessary information. Due to the short timeframe, it was assumed that participants would have great motivation and enough autonomy to take a course, however, this did not relate to a good success rate. To remedy this, the creation of a mini MOOC with minimum content, lasting a week was contemplated. Sinclair et al. (2014) highlight the dual movement of dropouts to which MOOCs are confronted: there was first a strong difference between enrolment and the number of participants and a participation rate continuously declining and measuring the progress of the course.

Peer relationships and student engagement to complete a MOOC is a topic of paramount importance (D. Yang, Wen, & Rose, 2014). Previous research has shown the impact of participation on the abandonments of many MOOCs (Diyi Yang, Sinha, Adamson, & Rose, 2013). The authors were interested in explanatory factors, especially the links that are formed during the course, questioning whether these relations influence the commitment created. Five steps exist to measure the influence of peer relationships on student commitment to the MOOC according to Yang et al. (2014):

- **Response Interactions:** Response interactions are an indicator of the willingness of students to socialize.
- **Proof of Co-Occurrence:** The repetition of several identical words in the same extract which indicates that students participate in the same conversation and share a common interest.
- **Community Connections:** The student participation model can be summarized by a set of graphs and sub-graphs of social relations.
- **Modeling Topics:** Sharing the same interest, the same topic of conversation can lead to the creation of several sub-communities.
- **Cohort:** Can occur when individuals began their participation in the same time and therefore their level of commitment is similar.

Results indicate that students are four times less likely to continue MOOCs when the drop of students leaving the course of the current week and the previous week mainly concern close connections with other students. When friendly relationships are important, students are motivated to pursue the course and therefore their commitment is stronger (D. Yang et al., 2014). This experience emphasizes the influence of peer behaviour on the motivation of everyone to take the course. The more a student sees his close friend/colleague abandon the MOOC, the less he becomes motivated to participate in forum discussions and he will even be tempted to do the same. Peer influence on the high attrition in the MOOC is even stronger when individuals lose their friendly relationships built up over the course. This highlights the importance of creating a virtual environment that promotes social relationships and encourage exchanges in the forum. Peer influence impacts the motivation of participants to continue or abandon the course, and therefore weighs on the self-determination of MOOC users (D. Yang et al., 2014).

One of the problems encountered by students is the rather limited possibility of interacting with other students (Rivard, 2013). Heutte et al. (2014) analyse perseverance of students enrolled in a MOOC. The results show that through the progression of the course fewer participants remain active. Links have been established with the place of residence, affect and feelings of wellbeing. Additionally, Xu and Jaggars (2014) explain that learners may receive inferior educational experiences when receiving their education through MOOCs due to the lack of a teacher-student relationship. MOOCs make higher education more affordable and could benefit the global economy by helping students and workers become lifelong learners. According to Kolowich (2013), the motivating factors fuelling support for MOOCs include:

1. An altruistic initiative to increase access to higher education worldwide,
2. The desire to stay up-to-date with new pedagogical approaches without being forced into using online techniques, regardless if the emergent techniques takes a different form than MOOCs over time,
3. A desire to broadly increase their personal visibility in Academia.

Also, learners can make use of the wide range of technology-based multimedia activities in order to:

- Manage and reflect their learning process.
- Create content for collaborating and communicating with others.
- Grade their peers and receive peer evaluation.
- Read and curate content and share it with their peers.

MOOCs and Self-Determination Theory

According to Self-Determination Theory (SDT), individuals basically need to see themselves as the main cause of their behaviour (Ryan & Deci, 2006), and be able to make choices, decisions, and to have control over their activity (the need of autonomy). They also need to feel effective in the activities they want to achieve and in dealing with their environment (the need of competence). Perceptions of efficiency during an activity provide a level of satisfaction that will increase motivation to continue the activity. Finally people have a need for social belonging, to feel connected to people (the need of relatedness), both on a cognitive side, which means to feel understood, accepted, respected (recognition of competence) and on the emotional side. Satisfaction of these three requirements helps to determine the emergence of different types of motivation. These motivations are positioned on an axis of self-determination (Ryan & Deci, 2006).

Using SDT, Deci and Ryan (1985) distinguish between different types of motivation based on the different reasons or goals that give rise to an action. The most basic distinction is between intrinsic motivation, which refers to doing something because it is inherently interesting or enjoyable, and extrinsic motivation, which refers to doing something because it leads to a separable outcome. Intrinsic motivation has emerged as an important phenomenon for educators because it results in high-quality learning and creativity; and it is especially important to detail the factors and forces that engender versus undermine it. The latest developments about extrinsic motivation highlight that extrinsic motivation is argued to vary considerably in its relative autonomy and thus can either reflect external control or true self-regulation. Understanding these different types of extrinsic motivation, and what fosters each of them, is an important issue for educators who cannot always rely on intrinsic motivation to foster learning (Ryan & Deci, 2000).

Furthermore, with the self-determination theory framework, Hartnett, St. George and Dron (2011) highlight that the perceived importance, relevance, and the utility value of the activity (associated with identified regulation) are just as important as the interest or enjoyment of the task (associated with intrinsic motivation). As a result, the relevance and value of the task (e.g., online discussions) need to be clearly identified and linked to learning objectives to help learners understand how the activity can aid in the realisation of personal goals, aspirations, and interests, both in the short and longer term. By offering meaningful choices (i.e., not just option choices) to learners that allow them to pursue topics that are of interest to them, the perceived value of the activity is further enhanced. Finally, by establishing frequent, ongoing communication with learners, where they feel able to discuss issues in an open and honest manner, practitioners are in a better position to accurately monitor and respond to situational factors that could potentially undermine learner motivation.

In his work, Kaplan (2014) focuses on the different strategies used by learners to regulate their learning in a technological environment. The author distinguishes the individual regulation (self-regulation) related to knowledge, motivation and emotions, from the collective regulation (co-regulation) related to peer interactions. With the deployment of Internet, technology has created a virtuous circle: each new technology changes the perception of the environment, which results in the creation of technological innovations to be tested (McLuhan, 1994). MOOCs embody this process, classified between "c-MOOC" (knowledge is constructed through interactions with peers, which thus form communities) and "x-MOOC" (knowledge is transmitted from the legitimate authority learners) (Karsenti, 2013; Rodriguez, 2013). While c-MOOC corresponds to the educational ideal where each learner defined its own objectives, x-MOOC reproduce traditional teaching that was done without the support of technology (Kaplan, 2014). The main argument in favour of the effectiveness of technology is rarely a MOOC (Bady, 2013) but the lower cost. Kaplan (2014) therefore underlines the role of regulation as a guarantor of the quality of learning where individual and collective regulations coexist. It is the learning environment (design, objectives, activities, subject, and personal variables) that determines how participants will regulate their learning. In the literature, self-regulation (individual strategies for maximizing performance) is distinguished from co-regulation (due to the influence of peer interactions). MOOC persistence of low rates is caused by the inability of tutors to help each participant, given the high number of registered students (Heutte et al., 2014). This weakness can be remedied by providing each learner individual regulatory tools, (e.g. present the objectives of each module, keep old activities, assess its progress by creating a reference standard, free choice of the type of exercise) and/or collective tools (e.g. multiply the communication channels, archive conversations, create a FAQ, favour the sharing of documents) (Kaplan, 2014). The new virtual learning environments facilitate the construction of knowledge in the learner community. For this, the latter must develop autonomy, learner control authority and cooperation between all participants to maximize commitment, persistence and performance of all. The design of the MOOC must especially emphasize ergonomics and service in favour of co-regulation, as learners seek cooperation to optimize learning.

E-Learning and Mobile Learning (M-Learning)

Arrival of the Internet and advances in technology has increased the usability of mobile phones as a training tool (J.-E. Pelet, Khan, Papadopoulou, & Bernardin, 2014). M-learning refers to wireless handheld devices such as personal digital assistants (PDAs), smartphones and tablets. Often these systems operate with wireless access protocols (WAPs) and wireless markup language (WML). The lightweight architecture of these protocols makes accessibility possible with a wide range of affordable devices. The main advantage of m-learning is its flexibility: this ubiquitous technology itself allows the user to access a wealth of information anywhere and at any time. For this, the design of mobile applications should be adapted to facilitate readability and storing content, and thus the usability of a mobile learning platform (Pelet, 2013). Websites that have responsive design is suitable for any screen size, and therefore consulted on any type of computer (e.g. desktop, laptop, tablet, smartphone). In order to effectively adapt interfaces to mobile phones, designers need to consider the following challenges: small size screens, little space between the keys, no visual comfort to track interactions and non-expertise of users. Numerous research conducted on the use and effectiveness of websites (Hall & Hanna, 2003; Kiritani & Shirai, 2003) found hue, saturation, colour brightness and contrasts between the background and the foreground influence emotions, attitudes and user confidence. In their work, Pelet and Uden (2014) were interested in the effectiveness of the application of mobile phones as a training support. The study found the colours of the interface influence the ease of use experienced by users, and their ability to memorize its contents. Pelet (2013) deals specifically with the effects of colour mobile applications on the effectiveness of learning and learner satisfaction. In order to analyse the effects of colour on mobile phones, Pelet and Uden (2014) conducted an experiment on 160 students during a course of e-marketing. The results showed that students perceived the interface created as useful because its content was accessible in an ubiquitous manner. The attractive design, the contrast between the colours, the right size and line spacing the right length of paragraphs facilitated the readability and memorization of information. Minimal information has also been promoted through sleek design of the site (some photos) and through its ability to access different types of content (e.g. website, power point). Furthermore, Pelet and Uden (2014) point out that certain colours enhance the ease of learning on mobile platforms.

It would be interesting to study respectively motivation to first enrol in a MOOC, and then complete it entirely. In the situational case of self-directed e-learning (SDEL), referring to electronic learning environments where there are often no peer learners or instructors regularly available, Kim and Frick's (2011) analysis indicated that the best predictors of motivation to begin e-learning were perceived relevance, reported technology competence, and age. In addition, motivation during the e-learning process was perceived quality of instruction and learning, and motivation to begin.

The design of an e-learning platform is of paramount importance for influencing learner interaction and behaviour as well as the overall success of the learning experience. Learners can benefit from the socialization of their platform which fosters the multiplication of social links, facilitating the curation of content to read, and to learn and share (J.-É. Pelet & Uden, 2014). As pointed out by Traxler (2007), with increased popular access to information and knowledge anywhere, anytime, the role of education is challenging and the relationships between education, society, and technology are now more dynamic than ever.

One of the most useful aspects of m-learning is that users have the capacity to make documentations while they are in the field; thus bridging the gap between theoretical and practical knowledge (Setaro, 2001; Stone, Briggs, & Smith, 2002). Although learning on mobile devices may never completely replace traditional in class teaching, it is widely accepted that if used correctly, this technology offers a significant complement to the learning environment (Chatti, Jarke, & Frosch-Wilke, 2007). Wireless handheld devices can be individualized to meet the needs and desires of its user, enhancing the collaborative process with automated information such as real-time course updates, deadlines and notifications. The learning sphere has become ubiquitous, centralized around the learner and increasingly oriented towards creating flexibility and optimizing content delivery (Corbeil & Valdes-Corbeil, 2007). Students enjoy using wireless handheld devices and appreciate the new age interactive and ubiquitous learning environment. These types of interactive social tools have broken the barrier between the academic and private spheres, and foster a sense of pleasure in taking part in the online learning game (Pelet, 2013). Learners are more successful and have higher retention rates when they enjoy the learning process. As wireless handheld devices become more affordable, the potential for integrating this technology into learning environments becomes more considerable (Motiwalla, 2007).

Social Media and E-Learning

In 2013, 89% of European Internet users were 16 to 24 year olds who participated in some form of social networking (Eurostat, 2014). As a part of modernizing the traditional approach to education, many higher education institutions (and educators) find themselves in a situation where they must adapt to the heightened use of social media and create a link to educational engagement (Selwyn, 2012). The majority of university students have mobile technology and use social media regularly; which is all the more reason why these elements should be integrated into tertiary level education (Lewis, Pea, & Rosen, 2010). As the technological framework is already in place, it is just a question of creating structured learning environments with the integration of these tools. Social networks such as Facebook have potentially positive benefits to teaching and learning, particularly with the development of educational micro-communities (Bosch, 2009). These micro-communities can be complemented with the use of other Web 2.0 applications that permit blogging, collaborative content sharing, podcasting and multimedia sharing. Structured learning environments can be created with simple collaborative features such as "Facebook groups" which can act as collaborative discussion boards in synchronous and asynchronous settings. Once the micro-community is established through the development of a group, other social media applications such as collaborative WIKIs can be integrated in order to add structural consistency.

Students are more likely to be connected simultaneously on their Facebook network than on any formal University Web portal and this enhances the potential for collaborative development between learning community members. Some universities have integrated micro-blogging on Twitter into the context of lecture hall discussions as students communicate synchronously with each other and the professor during the course. Certain studies show that the integration of micro-blogging into the educative experience successfully promoted active and continual feedback from the students (Pelet, 2013). Social media supports various innovations including: content creation, enhanced learner connectedness and collaboration (Redecker, Ala-Mutka, & Punie, 2010). Social media applications provide capacities which face-to-face instructions do not such as individualized tools permitting knowledge exchange and consultation without temporal or spatial barriers. In terms of education, social media is predominantly used by youth as a means of informal learning (Ito et al., 2008). However, the gap between informal and formal learning can be filled with the implementation of structured learning spaces such as micro-communities and interactive videos that contain integrated quizzes, similar to social media.

Cernezel et al. (2014) focused on the ability to predict the final scores of students by the number of activities they carried out and by their intermediary note. This prediction is reliable over 90% of the time. Online course management systems allow teachers to communicate and interact with their students and provide feedback on their work. They also record the student activity data (number of connections, viewed course and carried out evaluations), to identify the student preferences and performance (Pahl & Donnellan, 2002; Pritchard & Warnakulasooriya, 2005). This tool then fills the need for physical contact. Cernezel et al.'s work (2014) aims to see if it could also help to predict the final scores. Two predictive variables of final grades have been tested and released in their experiment:

- The number of activities carried out by students (number of views examinations and teaching resources), the higher the final score will be.
- With intermediate scores, the scores in the first and second experiment are similar. So there is a real correlation between the midterm scores and the final grades. The final scores of learners can be predicted from their intermediate note of the learning environment: the higher the mark will be during the intermediate rating, the higher the final score will be.

Other hypotheses were tested: the more students complete and send in their assessment, the sooner the notes will be available. But like all examinations completed at the end of the course, the predictor variable is not verifiable. The two variables that are identified comprise grades and student activities. When these two are combined, it shows a model able to predict marks with a reliability rating of 91.7%. These two variables are reliable and easy to use which are available through the online course management system. Cernezel et al. (2014) state that the activities and intermediate scores of students are correlated to their final grades. This underlying meaning that students satisfied with the activities, with the variety of the activities, and their intermediate score, are more likely to complete the MOOC. Driven by the development of MOOCs in the US and by the French government, French universities are no exception to the trend of the development of MOOC (Karsenti, 2013). After differentiating the c-MOOC (knowledge co-constructed by peers through online discussions and interactions) from x-MOOC (knowledge transmitted by the teacher to participants) (Boyatt, Joy, Rocks, & Sinclair, 2014; Rodriguez, 2013), Heutte et al. (2014) explain that MOOC have the common goal to preserve the largest number of active participants until the end of the course. Now, as the learning environment and learners provisions (motivation and participation) affect each other, dropouts have grown.

METHODOLOGY

The purpose of this research was to develop student proficiency with creating a landing page linked to a CMS and search engine optimization, as well as effective team interaction skills in a course. The main learning objectives of the course were to provide students with an experiential learning process using social media embedded on mobile devices.

During an e-marketing course, 4 lectures were given to masters students in a Business School (BS) between September 2013 and January 2014. These 4 lectures included:

- Web 2.0 Strategy,
- Fundamentals of e-business and e-marketing,
- Communitarian and Sensorial Marketing,
- New Marketing.

Approximately 160 students from the business school used the main website (www.kmcms.net - Knowledge & Management System/Content Management System) to follow the course and prepare for their exam. This platform provided students with up-to-date lectures and theoretical content (books and articles). The platform also included roughly 1,700 posts ranging from one to several pages of content depending on the source. The platform was accessible to students, after registering and choosing the course they wanted to attend. Four "image links" were positioned on the homepage of www.kmcms.net

redirecting students to 4 CMS (Content Management Systems). Websites on *e-business and e-marketing fundamentals* were available to the students, see Figures 1 and 2.

These 2 CMS used responsive templates enabled students to read, comment, grade, and write. They provided;

Figure 1. E-marketing homepage

Figure 2. Strategy 2.0 homepage

MOOCs on Mobiles

- Lectures on the two evoked topics.
- Explanations regarding the content and revisions for the exam.
- Explanations about their assessment during the course.

Two other curation platforms were available for the purpose of concatenating and curating content from the Web, such as blogs, organizational/business websites and management websites[1]. These curating sites were used in order to prepare topics on New Marketing as well as Communitarian and Sensorial Marketing, see Figures 3 and 4.

Figure 3. KMCMS homepage with 4 links

Figure 4. Image of a synthesis on Tumblr (x3 print screens)

Students were assuming an evolving role as the principal players in their educational endeavours. Within the course students were assigned a role as autonomous researchers and had the responsibility of curating content with a unique knowledge management tool that they themselves created. Content curators are individuals who continually find, organize and share the best and most pertinent content related to specific issues on the Web. Although this was a strictly academic endeavour, students agreed that this newly acquired capacity for effectively managing massive amounts of information would benefit their professional futures. There are a few aspects about the term "content curator" that are worth highlighting, such as the fact that content curators are people and not robots. Effective content curation cannot be performed solely with the use of an algorithm. In order to obtain high-quality information, it's best to have a domain expert administering the curation in order to ensure finely tuned selectivity of content. This knowledge management process should be implemented continually and administrators should be consistently up-to-date with the domain that they are focusing on. Third, a curator is not simply regurgitating any content that they come across as they must be very discerning, discriminative, and selective in only sharing the "best and most relevant" content. Lastly, a curator focuses on "specific issues". They do not curate on all of the topics available. Instead, they specialize on specific topics and over time they may have the opportunity to become an authority and perhaps even a thought leader on those topics.

The landing page on our platform was linked to a Wordpress CMS platform. Landing pages are an essential element in online marketing. The first element of the landing page is to appear under the shape of a responsive design Web. The responsive design Web consists of creating a website which adapts itself to the browser whatever is the size of the screen. Creating Web sites in this way allows the display to perfectly adapt for Smartphones and Tablets. This includes the techniques of conception which allow auto-adaptable contents according to the interfaces of consultation used by the visitor. So, a web page or an image can resize according to the size of the screen of the used terminal (computer, tablet, smartphone, etc.). The responsive web design allows adapting the contents of the site to the technical environment without having multiple versions of the same content. It is based generally on style sheets and latest techniques and standards of Web programming such as the HTML 5.

The objective of the landing page consists in converting an Internet user towards:

- A sale, the Internet user is then converted into customer;
- A lead, the Internet user is then converted into potential customer or prospect.

The first objective of the landing page being to play on the behaviour of the consumer, the Internet user. It is based on an attractive offer, putting forward the value of the product or service presented with the aim of pushing the Internet user to the action. On a platform such as an e-learning or MOOC, objectives are similar: an aesthetically attractive lecture can help learners to better learn. Thus, the offer has to correspond to the expectations of the visitor, and the use of attractive colours (Pelet, 2010; Pelet & Papadopoulou, 2013) can play in favour to these expectations. Precise typographies tempting the user to read or not the contents of the landing page will be important too (Pyke, 1926), with the objective to create a legible message encouraging the individual to pursue reading (Rieck, 1997). The intrinsic value of the landing page constitutes the main reason for which the Internet user clicks on posts, pages or links posted on social media. Thus, it must not only be understandable, but mostly convincing. Finally, the invitation to call for action can convert the Internet user into a learner or a student interested to pursue the lecture on the MOOC. They constitute the outcomes of the landing page. The first goal of the landing page is then to convince the user to act and the same happens in an e-learning context. Students must be convinced and involved when studying, especially on a MOOC where nobody is there to instruct their actions. The landing page was made using a responsive web design, adapting to all screen sizes. In such cases, learners benefit with access to content on any device.

The Wordpress CMS platform is easy to manage once it is created. It also provides users with lots of widgets enabling curation, use of RSS, Search Engine Optimization tools and so on. The latter can be designed to mimic or resemble the landing page, in order to keep learners in a homogeneous online atmosphere. The landing page and the CMS represent an interesting combination for creating efficient online lectures and MOOCs adapted to ubiquitous learners.

Students were evaluated after the completion of two exercises:

- Creation of a website (Web 2.0 Strategy and Fundamentals of e-business and e-marketing).
- Preparation of a platform aimed at collecting RSS feeds and curating information on the Web (i.e. lectures on New Marketing as well as Communitarian and sensorial marketing).

Students were also required to write a synthesis on the 4 lecture topics, using a Tumblr platform (Table 1).

This part of the course included peer-review and assessment and counted as a part of the student's participation grade. Students were also asked to complete a short online questionnaire in order to get feedback with regards to the methods used in teaching the course.

RESULTS

The act of building a website proved to be very beneficial to students as they engaged in a hands-on approach to learning which is one of the success factors of this pedagogy. The ability students have to write content on the Internet, whether on social media, UGC (User Generated Content) such as TripAdvisor, to give an opinion, mark a service or product, or comment another comment, seems to represent THE facilitator. This enabled students to express their opinion very easily and participate in the whole process of the course more instinctively, without the fear of being judged by peers. We present a synthesis of the most common responses given by the students:

- Students appreciated the facility of accessing information in a ubiquitous form. The websites had a very responsive character and offered an easy-to-read interface and facilitated mobile consultation.
- Students stated that the user interface facilitated the memorization of content, and the finding of information. Due to an ergonomic layout with good colour contrast ratio, user-friendly graphical fonts, good font spacing and width of paragraphs also facilitated reading. These factors also facilitated the sharing of the information and knowledge management, particularly on mobile devices.
- The use of quick loading photography enhanced the quality of the information and facilitated understanding of the course content by reducing cognitive workload and providing graphic representations of information.

The ability of accessing content (e.g. websites, lectures, PowerPoint presentations, etc.) while students where constructing their own websites and RSS curating platforms, offered a form of ubiquitous mobile support. The term RSS is an abbreviation for Really Simple Syndication or Rich Site Summary as it provides a rich summary of a websites new content without the need to manually check the website. The fact that our CMSs were supported by mobile devices was a pertinent factor in the success of this educational initiative. It enabled students to ask questions and get responses easily, without temporal-spatial barriers.

Our post-course survey provided results on student's satisfaction and overall experience using the MOOC interface and its social media components. As shown in Table 2, students overall provided positive feedback to the course. The highest satisfaction was related to ease-of-use and learning compared to other courses (M=4.0). This was followed by students satisfaction with their level of involvement in the course and its preparation for their future professional practice (both M=3.9). Students also felt that this form of teaching was accessible, reasonable and overall felt satisfied (as shown in Table 2). In

Table 1. Screenshots of the MOOC on e-marketing

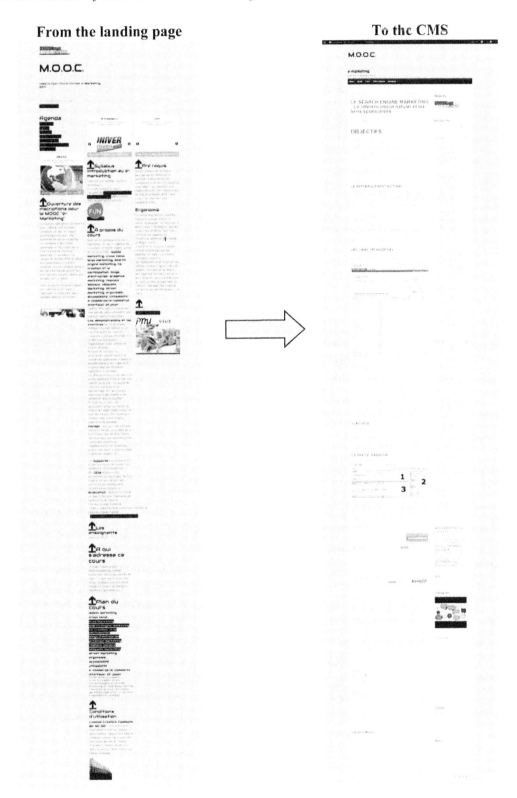

Table 2. Student satisfaction and overall experience using the MOOC interface and its social media components

Student Feedback	N	Mean*	Std. Dev	Min	Max
Ease-of-use and learning compared to other courses	19	4.0	0.9	2	5
Level of involvement in this course (homework, participation...) was enough	19	3.9	1.0	2	5
Course was adequate in relation to professional practice	19	3.9	0.6	3	5
This form of teaching appears accessible	19	3.7	0.7	3	5
Workload was reasonable	19	3.7	0.9	2	5
This form of learning accessible	19	3.7	0.7	3	5
In general, this form of education is satisfactory	19	3.7	0.9	2	5
Documents submitted and teaching materials were satisfactory	19	3.1	1.1	2	5
Prerequisites were sufficient	19	2.9	1.0	2	5
The number of exercises and illustrative examples supporting the course was sufficient	19	2.7	1.1	1	5

*1 = strongly disagree to 5 = strongly agree

addition, 58% of respondents who participated in these mobile e-learning courses agreed or strongly agreed that it was a satisfactory experience. Student productivity was also enhanced due to the flexible nature of the courses. Ease-of-use, flexibility and adequation to professional practices seem necessary when learning on this new form of support (mobile + social media). Areas which students felt the least satisfied was the lack of resources supplied.

DISCUSSION AND CONCLUSION

This chapter presented an exploratory analysis regarding the use of a MOOC and m-learning with strong implementation of social media content creation tools in the context of university business school courses. The analysis allowed us to gain a better understanding of student perceptions on using MOOCs in m-learning situations, as well as their capacity to adapt to new learning environments strongly anchored in collaborative and constructivist learning. As social media usage increases, we find that it is in the best interests of students to integrate m-learning opportunities into traditional higher education. The use of social media to share, exchange information and ideas is increasing, especially in the field of online learning. The study conducted on Twitter by Zaina et al. (2014) states that students act as receivers of information: although few students have published messages containing the relevant words in their posts, all the students admitted having acquired new and interesting information. Messages posted by professors have therefore not motivated students to share the received information, even if this information was relevant. The experience shows that Twitter, as a social media tool, acts more like a receptive resource of knowledge than as a transmitting one. Our study shows that the use of a mobile supported MOOC facilitated mobile knowledge management system created a flexible and effective learning environment. Although the students in this study rarely met with the professor, there was constant community support provided by other students as well as the content provided through the CMS. The digital learning setting provoked the active participation of students in a collaborative working architecture that one could easily refer to as "social learning". Students who were more fluent in the operation of the various development mechanisms provided support to the others as "technological stewards". The term "technology stewards" refers to technology savvy members of the learning community with excellent comprehension of the digital atmosphere. Although the teacher primarily assumes this role, learners who are highly fluent in the use of mobile and Internet ICT also acted as technology stewards. This type of leader oriented behaviour is typical in the digital learning environment; it empowered students with a sense of gratification and motivation while fostering a sense of a united academic micro-community. As discussed earlier, Self-Determination Theory views individuals identifying themselves as the main cause of their behaviour. Individuals also need to feel effective in the activities they want to achieve and in dealing with their environment. This is evidenced where students developed autonomous working habits, as well as community oriented collaborative working skills. They successfully developed their own websites based on the themes provided by the instructor as well as a unique knowledge management tool with the function of curating RSS feeds on topics specified by the instructor. The RSS feeds, also called web feeds, are a type of content delivery vehicle used for syndicating news or other web content. The tools that the students created contributes to their individual lifelong learning processes and granted them new capacities as seen in the theory of expansive learning.

A new relationship between students and professors is developing characterized by collaboration and attributing new value to communication amongst students and with administrators. Social media and mobile Internet technologies reinforce the potential for effective communication between all of the participating parties. Computer supported constructivist learning is a hands-on approach that equips learners with fine-tuned research skills and nurtures educational development in the lifelong learning continuum. The computer mediated setting facilitates the creation of visual representations of information, which reduces cognitive workload required by learners to understand knowledge in a more expedient manner. The implementation of digitized learning is reciprocally beneficial to teachers as evaluation processes become increasingly automated. Course administrators have detailed analytics that provide graphic representations of information which are much easier to understand than traditional grading methods. This further supports Kaplan (2014) where it is the learning environment (design, objectives, activities, subject, personal variables) that determines how participants will regulate their learning. Having access to graphic visualization of student results also contributes to the individualization of learning in the digital environment as students and teachers alike are able to identify strengths and shortcomings much more easily than in a face-to-face educational setting. It's a win-win situation!

The advantages of MOOCs are to learn independently, to make participants work together from geographically distant places with different social levels and workship. The drawbacks are risk of isolation, difficulty using the tools, inability to self-manage his work and its progress (Daniel, 2012; McAuley, Stewart, Siemens, & Cormier, 2010). The literature shows that the dropout rate of online courses is higher by 10-20% than those of group lessons, and the retention rate MOOC (participants who will fully complete) is very low: 0.4% against 3 to 8% (Liu et al., 2014). Research by Gütl et al. (2014) is based on a 4-week MOOC on e-learning, with 1680 registered students, where the authors seek to determine the motivation to register for a MOOC and causes of high dropout rates. The analysis compares the responses of those who completed the MOOC (only 8.5% of registered voters) and those who drop out (91.5%). Results of this experiment regarding the main motives: to live the experience of a MOOC (33.5%) and fully complete it (22.4%). All participants expressed motivated to learn more on the subject of the course; but those who left were not motivated to learn with a new teaching method that is to say, the MOOC format. Major causes of abandonment were the difficulty to study the online course after a day of work (70%), change in job responsibilities (69%), lack of support from family or employer (35%), lack of returns on their assessments by teachers (32%) and lack of interaction with other participants and teachers (28% and 24%). As a result, Gütl et al. (2014) found that the main factor that brings a student to follow a MOOC is his curiosity, but perseverance may be stopped for personal reasons, academic or learning environment. The organization of work (1 to 2 hours, evenings or weekends) was similar for both groups, but the perception of MOOC, emotional attachment and ease of use were more negative for those who abandoned. Despite a high dropout rate, 98% of registered voters found that MOOC are an effective way to learn and are willing to re-register. Participants expressed no preference as to the educational content (video, text, graphics).

Despite the specificities of the host platforms of MOOCs (Round, 2013), there are a marked increase in the number of registered students and course offerings. Due to the free aspect of MOOCs, their business model is not perennial unless registrations, certifications and / or equivalencies become chargeable. In addition to this economical aspect, other issues threaten the sustainability of MOOCs: the individual rating (impossible given the many registered unless to hire or to resort to computerized logs), cheating (unclear to know who actually complete the review or pass the exam), problems of interactions (lack of interaction in face-to-face and of quality between learners and teachers (Zimmerman, 2011), equivalence with credits such as ECTS[2] credits (even when the MOOC is completed with full success, few universities give in) and prerequisites (how to verify them). Billington and Fronmueller (2013) present several possible scenarios about its future in higher education:

- MOOCs cause the disappearance of universities unlikely as these are places of socialization, interaction and family estrangement remain attractive (Smutz, 2013).
- Disappearing of MOOCs unlikely because the high number of registrations shows that they offer the opportunity to enhance their skills at any location.
- The MOOC as a complement or replacement of the course: allow more class time for discussion and problem resolutions, and improve the learning experience.
- Universities become examination centers for MOOC: students enjoy the benefits of working online and meetings with the professors on campus to pass the exam.
- Partnership between universities and MOOCs: similar but examination unnecessary on the campus.
- **Accreditation of MOOCs:** Universities lose their reliability to give ratings and diplomas.
- The MOOC become a tool for development of professional skills.
- Development of credit transfer in case of success of a MOOC: why universities should generate similar course to graduate.

In conclusion, the MOOC seem to persist and could eventually replace group lessons within universities as they meet the requirement of lower tuition fees. The MOOC has the advantage of freeing teachers from their classes, but to be of quality, require a lot of interactivity, as our results highlight the importance of involvement in the course. In addition, homework and participation have to be adequate and engaging. The ideal would be to involve students to reduce costs while increasing the quality of the learning experience. The Future of MOOC appears uncertain, however interactivity appears a key success factor.

FUTURE RESEARCH

MOOCs are still in their infancy and many uncertainties exist regarding their future role in traditional higher learning. Future research will help to shed light on the uncertainties surrounding MOOCs and embrace their potential to be a transformative educational innovation of the 21st century. Results from this exploratory study demonstrates that success can be achieved with the use of MOOCs in combination with social media constructivist tools (i.e. website development and content curation applications) in a mobile supported format. Additional research is to be conducted with the objective of identifying motivating factors behind student commitments and overall success in e-learning and m-learning environments. Future research will also strengthen the external validity of our preliminary results, which indicate a successful outcome with the use of social media constructivist tools for the purpose of knowledge management in a mobile supported MOOC scenario. In addition, further research is required to examine the use of other rewards and incentives according to Coetzee et al. (2014).

REFERENCES

Adamopoulos, P. (2013). *What makes a great MOOC? An interdisciplinary analysis of student retention in online courses.* Paper presented at the 34th International Conference on Information Systems, Milan, Italy.

Bady, A. (2013). The MOOC moment and the end of reform. *Liberal Education, 99*(4), 6–15.

Billington, P. J., & Fronmueller, M. P. (2013). MOOCs and the future higher education. *Journal of Higher Education Theory and Practice, 13*(4), 37–43.

Bosch, T. E. (2009). Using online social networking for teaching and learning: Facebook use at the University of Cape Town. *Communication, 35*(2), 185–200.

Bourcieu, S., & Léon, O. (2013). Les MOOC, alliés ou concurrents des business schools? *LExpansion Management Review, 149*(2), 14–24. doi:10.3917/emr.149.0014

Bowen, W. G. (2012). The "cost disease" in higher education: Is technology the answer. *The Tanner Lectures.* Retrieved 02/02/15, from http://edf.stanford.edu/sites/default/files/Bowen%20lectures%20SU%20102.pdf

Boyatt, R., Joy, M., Rocks, C., & Sinclair, J. (2014). *What (use) is a MOOC?* Paper presented at the The 2nd International Workshop on Learning Technology for Education in Cloud. doi:10.1007/978-94-007-7308-0_15

Cernezel, A., Karakatic, S., & Brumen, B. (2014). *Predicting grades based on students' online course activities.* Paper presented at the Ninth International Knowledge Management in Organizations Conference, Santiago, Chile. doi:10.1007/978-3-319-08618-7_11

Chatti, M. A., Jarke, M., & Frosch-Wilke, D. (2007). The future of e-learning: A shift to knowledge networking and social software. *International Journal of Knowledge and Learning, 3*(4), 404–420. doi:10.1504/IJKL.2007.016702

Clow, D. (2013). *Moocs and the Funnel of Participation*. Paper presented at the The Third International Conference on Learning Analytics and Knowledge, Leuven, Belgium. doi:10.1145/2460296.2460332

Coetzee, D., Fox, A., Hearst, M. A., & Hartmann, B. (2014). *Should your MOOC forum use a reputation system?* Paper presented at the Computer on supported cooperative work & social computing, Baltimore, MD. doi:10.1145/2531602.2531657

Corbeil, J. R., & Valdes-Corbeil, M. E. (2007). Are You Ready for Mobile Learning. *Education Quarterly, 30*(2), 51–58.

Cormier, D., & Siemens, G. (2010). The Open Course: Through the Open Door – Open Courses as Research, Learning and Engagement. *EDUCAUSE Review, 45*(4), 30–32.

Cormode, G., Krishnamurthy, B., & Willinger, W. (2010). A manifesto for modeling and measurement in social media. *First Monday, 15*(9). doi:10.5210/fm.v15i9.3072

Daniel, J. (2012). Making sense of MOOCs: Musings in a maze of myth, paradox and possibility. *Journal of Interactive Media in Education, 3*, 1–20.

De Waard, I. (2011). *Explore a New Learning Frontier: MOOCs*. Learning Solutions Magazine.

Deci, E. L., & Ryan, R. M. (1985). *Intrinsic motivation and self-determination in human behaviour*. New York: Plenum. doi:10.1007/978-1-4899-2271-7

Dunlap, J. C., & Lowenthal, P. R. (2009). Tweeting the night away: Using Twitter to enhance social presence. *Journal of Information Systems Education, 20*(2), 129–136.

Eurostat. (2014). *Internet use statistics - individuals - Statistics Explained*. Retrieved from http://epp.eurostat.ec.europa.eu/statistics_explained/index.php/Internet_use_statistics_-_individuals#

Gaebel, M. (2013). *MOOCs: Massive Open Online Courses*. Retrieved from http://www.eua.be/Libraries/Publication/EUA_Occasional_papers_MOOCs.sflb.ashx

Gütl, C., Hernández Rizzardini, R., Chang, V., & Morales, M. (2014). Attrition in MOOC: Lessons learned from drop-out students. In L. Uden, J. Sinclair, Y.-H. Tao, & D. Liberona (Eds.), *Learning Technology for Education in Cloud. MOOC and Big Data* (Vol. 446, pp. 37–48). Springer International Publishing. doi:10.1007/978-3-319-10671-7_4

Hall, R. H., & Hanna, P. (2003). The impact of web page text-background colour combinations on readability, retention, aesthetics and behavioural intention. *Behaviour & Information Technology, 23*(3), 183–195. doi:10.1080/01449290410001669932

Hartnett, M., St. George, A., & Dron, J. (2011). Examining Motivation in Online Distance Learning Environments: Complex, Multifaceted, and Situation-Dependent. *International Review of Research in Open and Distance Learning, 12*(6), 20–37. doi:10.19173/irrodl.v12i6.1030

Heutte, J., Kaplan, J., Fenouillet, F., Caron, P., Rosselle, M., & Uden, L. (2014). MOOC User Persistence. In L. Uden, J. Sinclair, Y.-H. Tao, & D. Liberona (Eds.), *Learning Technology for Education in Cloud. MOOC and Big Data* (Vol. 446, pp. 13–24). Springer International Publishing.

Ito, M., Horst, H., Bittanti, M., Boyd, D., Herr-Stephenson, B., Lange, P. G., & Robinson, L. (2008). *Living and Learning with New Media. Summary of Findings from the Digital Youth Project. The John D. and Catherine T. MacArthur Foundation Reports on Digital Media and Learning*. The MIT Press.

Kaplan, J. (2014). Co-regulation in technology enhanced learning environments. In L. Uden, J. Sinclair, Y.-H. Tao, & D. Liberona (Eds.), *Learning Technology for Education in Cloud. MOOC and Big Data* (Vol. 446, pp. 72–81). Springer International Publishing.

Karsenti, T. (2013). The MOOC: What the research says. *International Journal of Technologies in Higher Education, 10*(2), 23–37.

Khalil, H., & Ebner, M. (2013). *How satisfied are you with your MOOC? - A Research Study on Interaction in Huge Online Courses.* Paper presented at the World Conference on Educational Media and Technology.

Kim, K.-J., & Frick, T. W. (2011). Changes in Student Motivation During Online Learning. *Journal of Educational Computing Research, 44*(1), 1–24. doi:10.2190/EC.44.1.a

Kiritani, Y., & Shirai, S. (2003). Effects of background colours on user's experience in reading website. *Journal of the Asian Design International Conference, 1*(64).

Knowledge@Wharton (Producer). (2013). MOOCs on the Move: How Coursera Is Disrupting the Traditional Classroom. *Innovation and Entrepreneurship*. Retrieved from http://knowledge.wharton.upenn.edu/article.cfm?articleid=3109

Kolowich, S. (2013). The professors who make the MOOCs. *The Chronicle of Higher Education*. Retrieved from http://chronicle.com/article/The-Professors-Behind-the-MOOC/137905/?cid=at&utm_source=at&utm_medium=en#id=overview

Kop, R. (2011). The challenges to connectivist learning on open online networks: Learning experiences during a massive open online course. *International Review of Research in Open and Distance Learning, 12*(3), 19–38. doi:10.19173/irrodl.v12i3.882

Lane, J., & Kinser, K. (2015). *MOOC's and the McDonaldization of Global Higher Education*. Retrieved 11/12, 2015, from http://chronicle.com/blogs/worldwise/moocs-mass-education-and-the-mcdonaldization-of-higher-education/30536

Lewin, T. (2012, March 4). Instruction for masses knocks down campus walls. *New York Times*.

Lewis, S., Pea, R., & Rosen, J. (2010). Beyond participation to co-creation of meaning: Mobile social media in generative learning communities. *Social Sciences Information. Information Sur les Sciences Sociales, 49*(3), 351–369. doi:10.1177/0539018410370726

Liu, M., Kang, J., Cao, M., Lim, M., Ko, Y., Myers, R., & Weiss, A. S. (2014). Understanding MOOCs as an emerging online tool: Perspectives from the students. *American Journal of Distance Education, 28*(3), 147–159. doi:10.1080/08923647.2014.926145

Luca, J., & Mcloughlin, C. (2004). *Using Online Forums to Support a Community of Learning*. Paper presented at the EdMedia: World Conference on Educational Media and Technology 2004, Lugano, Switzerland. Retrieved from http://www.editlib.org/p/12668

Masters, K. (2011). A Brief Guide to Understanding Moocs. *The Internet Journal of Medical Education, 1*(2).

McAuley, A., Stewart, B., Siemens, G., & Cormier, D. (2010). *The MOOC model for digital practice.* Retrieved from http://www.elearnspace.org/Articles/MOOC_Final.pdf

McLuhan, M. (1994). *Understanding Media: The extensions of man: Critical Edition.* London: The MIT Press.

Motiwalla, L. F. (2007). Mobile learning: A framework and evaluation. *Computers & Education, 49*(3), 581–596. doi:10.1016/j.compedu.2005.10.011

Pahl, C., & Donnellan, D. (2002). *Data minning technology for the evaluation of web-based teaching and learning systems.* Paper presented at the World conference on e-learning in Corporate, Government, Healthcare and Higher Education.

Pelet, J.-É. (2010). Effets de la couleur des sites web marchands sur la mémorisation et sur lintention dachat. *Systèmes dInformation et Management, 15*(1), 97–131. doi:10.3917/sim.101.0097

Pelet, J.-É. (Ed.). (2013). *E-learning 2.0 technologies and web applications in higher education.* Hershey, PA: Information Science Reference.

Pelet, J.-É., Khan, J., Papadopoulou, P., & Bernardin, E. (2014). m-learning: exploring the use of mobile devices and social media. In N. Baporikar (Ed.), Handbook of Research on Higher Education in the MENA Region: Policy and Practice (pp. 261-296). Hershey, PA: IGI Global.

Pelet, J.-É., & Papadopoulou, P. (Eds.). (2013). *User Behavior in Ubiquitous Online Environments.* Hershey, PA: IGI Global.

Pelet, J.-É., & Uden, L. (2014). *Mobile learning platforms to assist individual knowledge management.* Paper presented at the 9th International Conference on Knowledge Management in Organizations, Santiago, Chile. doi:10.1007/978-3-319-08618-7_26

Piech, C., Huang, J., Chen, Z., Do, C., Ng, A., & Koller, D. (2013). *Tuned Models of Peer Assessment in MOOCs.* Retrieved from http://www.stanford.edu/~cpiech/bio/papers/tuningPeerGrading.pdf

Pritchard, D., & Warnakulasooriya, R. (2005). *Data from a Web-based Homework Tutor can predict Student's Final Exam Score.* Paper presented at the World Conference on Educational Media and Technology.

Pyke, R. L. (1926). *Report on the legibility of print* (Vol. 110). London: HM Stationery Office.

Redecker, C., Ala-Mutka, K., & Punie, Y. (2010). *Learning 2.0 - The impact of social media on learning in Europe.* Retrieved from http://ftp.jrc.es/EURdoc/JRC56958.pdf

Resnick, P., Kuwabara, K., Zeckhauser, R., & Friedman, E. (2000). Reputation systems. *Communications of the ACM, 43*(12), 45–48. doi:10.1145/355112.355122

Rieck, D. (1997). Design, legibility and unnatural acts. *Direct Marketing, 60*(6), 23–25.

Rivard, R. (2013). Coursera's contractual Elitism. *Inside Higher Education*. Retrieved from the Internet December 11, 2015 at https://www.insidehighered.com/news/2013/03/22/coursera-commits-admitting-only-elite-universities

Rodriguez, O. (2013). The concept of openness behind c and x-MOOCs. *Open Praxis*, *5*(1), 67–73. doi:10.5944/openpraxis.5.1.42

Round, C. (2013). *The Best MOOC Provider: A Review of Coursera, Udacity and edX. Skilledup.com*. Skilledup.

Ryan, R. M., & Deci, E. L. (2000). Self-determination theory and the facilitation of intrinsic motivation, social development, and well-being. *The American Psychologist*, *55*(1), 68–78. doi:10.1037/0003-066X.55.1.68 PMID:11392867

Ryan, R. M., & Deci, E. L. (2006). Self-regulation and the problem of human autonomy: Does psychology need choice, self-Determination, and will? *Journal of Personality*, *74*(6), 1557–1585. doi:10.1111/j.1467-6494.2006.00420.x PMID:17083658

Selwyn, N. (2012). Social media in higher education. *The Europa World of Learning*. Retrieved from http://sites.jmu.edu/flippEDout/files/2013/04/sample-essay-selwyn.pdf

Setaro, J. L. (2001). *If you build it, will they come? Distance learning through wireless devices*. Retrieved from http://www.unisysworld.com/monthly/2001/07/wireless.shtml

Sinclair, J., Boyatt, R., Foss, J., Rocks, C., & Uden, L. (2014). A tale of two modes: Initial reflections on an innovative MOOC. In L. Uden, J. Sinclair, Y.-H. Tao, & D. Liberona (Eds.), *Learning Technology for Education in Cloud. MOOC and Big Data* (Vol. 446, pp. 49–60). Springer International Publishing. doi:10.1007/978-3-319-10671-7_5

Smutz, W. (2013). MOOCs Are No Education Panacea, but Here's What can Make Them Work. *Forbes*. Retrieved from http://www.forbes.com/sites/forbesleadershipforum/2013/04/08/moocs-are-no-education-panacea-but-heres-what-can-make-them-work/?cid=dlvr.it

Stahl, G., Koschmann, T., & Suthers, D. (2006). Computer-supported collaborative learning: an historical perspective. In R. K. Sawyer (Ed.), *Cambridge Handbook of the Learning Sciences* (pp. 409–426). Cambridge, UK: Cambridge University Press.

Stone, A., Briggs, J., & Smith, C. (2002). *SMS and interactivity-some results from the field, and its implications on effective uses of mobile technologies in education*. Paper presented at the Wireless and Mobile Technologies in Education. doi:10.1109/WMTE.2002.1039238

Traxler, J. (2007). Defining, Discussing and Evaluating Mobile Learning: The moving finger writes and having writ.... *International Review of Research in Open and Distance Learning*, *8*(2). doi:10.19173/irrodl.v8i2.346

Vaufrey, C. (2015). *MOOC "Economie du Web"*. Thot Cursus.

Willging, P. A., & Johnson, S. D. (2004). Factors That Influence Students' Decision to Drop out of Online Courses. *Journal of Asynchronous Learning Networks*, *8*(4), 105–118.

Xu, D., & Jaggars, S. S. (2014). Performance Gaps Between Online and Face-to-Face Courses: Differences Across Types of Students and Academic Subject Areas. *The Journal of Higher Education, 85*(5), 633–659. doi:10.1353/jhe.2014.0028

Yang, D., Sinha, T., Adamson, D., & Rose, C. P. (2013). *Turn on, tune in, drop out: Anticipating student dropouts in massive open online courses.* Paper presented at the NIPS Data-Driven Education Workshop.

Yang, D., Wen, M., & Rose, C. (2014). Peer influence on attrition in massively open online courses. *Educational Data Mining.* Retrieved from http://www.cs.cmu.edu/~mwen/papers/edm2014-39.pdf

Zaina, L. A. M., Ameida, T. A., & Torres, G. M. (2014). Can the Online Social Networks Be Used as a Learning Tool? A Case Study in Twitter. In L. Uden, J. Sinclair, Y.-H. Tao, & D. Liberona (Eds.), *Learning Technology for Education in Cloud. MOOC and Big Data* (Vol. 446, pp. 114–123). Springer International Publishing. doi:10.1007/978-3-319-10671-7_11

Zheng, L., Junfeng, Y., Wei, C., & Ronghuai, H. (2014). Emerging approaches for supporting easy, engaged and effective collaborative learning. *Journal of King Saud University - Computer and Information Sciences, 26,* 11-16.

Zimmerman, L. (2011). Critical Importance of Social Interaction in Online Courses. *ETC Journal.* Retrieved from http://etcjournal.com/2011/01/02/7050/

ENDNOTES

[1] For an extensive review of curation platforms, compared according to their particular functions, please visit: http://socialcompare.com/fr/comparison/curation-platforms-amplify-knowledge-plaza-storify.

[2] European Credits Transfer System.

Chapter 2
College Students' Perception on the Use of Social Network Tool for Education Learning in USA

Mohammed Alfadil
University of Northern Colorado, USA

Hamzah Alhababi
University of Northern Colorado, USA

Ali Buhamad
University of Northern Colorado, USA

ABSTRACT

This study presents a survey research design on college students' perception on the use of social network tools for education in USA. The survey research design is a very valuable tool for assessing opinions and trends and it includes both graduate and undergraduate students, as well as male and female respondents. The data collection was via University of Northern Colorado (UNC) website (list-serves and Facebook groups). Composite 2 of the survey was eliminated because the responses orders of the items were different from composite 1 as the responses in composite 1 went from strongly disagree to strongly agree while in composite 2 the responses went the other direction and we were afraid the participants did not pay attention to that.

INTRODUCTION

College Students' Perception on the Use of Social Network Tool for Education Learning Proliferation of Internet has revolutionized virtually all aspects of human life. One of the primary benefits of the internet is the ability to provide different platforms where information exchange occurs. Social media platforms such as Facebook, YouTube, and Twitter among others, provides supportive learning environments if well utilized. However, the success of social networks as a pedagogical tool primarily depends on the users' perception.

DOI: 10.4018/978-1-5225-2469-4.ch002

With the rapid increase in the number of Social Media users in tools such as Facebook, Twitter, I think there is a need to integrate social media in all the fields of education. It is a convenient aid to stimulate students to brainstorm and connect students to real-world issues. Social media can help both teachers and students communicate, collaborate, and teach. Moreover, it is a good opportunity for students and teachers to be connected and exchange information in real time. Many challenges are solved through social media in education. Furthermore, students can be very motivated to learn through E-learning. For example through Facebook teachers can communicate with other faculty members. Also vocational learning can be done through mobile phones, and other digital devices (Brooks, 2009).

BACKGROUND

Social media can help in many ways to improve higher education. Through social media, students can use digital literacy in every academic environment. Social media can influence learning through computers and video (Kozma, 1994). Usually used for information students need, the Internet helps in many different ways. One of the ways Internet helps through education is by exchanging knowledge with others. Aghaee Naghmeh (2010), found a significant number of students appreciated use of social media platforms such as Facebook because of its flexible nature. The study found many students were strongly supportive of Internet learning platforms because the platforms made learning easier (Aghaee, 2010). Aghaees findings correlate with Beltran and Belle's conclusions, which found most participants strongly agreed and agreed that they had better engagement using social networks (Beltran & Belle, 2013).

Just because technology is brought into classrooms does not always mean that the technology will be used; many teachers are not comfortable with the skills of integration and active learning using technology (Gorder, 2008). Gorder (2008), attempted to determine teacher perceptions of instructional technology integration in the classroom and what factors contributed to teachers actually using the technology with their students. The findings suggest that the teachers who used technology regularly in their own lives were more likely to integrate technology into their classrooms. The study also found that significant differences existed for technology use and integration based on grade level, showing that some teachers may have found technology to be less effective with younger children.

Modern technological innovations have drastically changed the way individuals live and communicate within an increasingly global society. However, technological advancements within our society have not translated to technological advancements within educational settings. Teaching practices and student learning have been minimally influenced by modern technological innovations (Levin & Wadmany, 2006). When technology is implemented within the school setting, technological interventions are often used in a narrow fashion and in a manner that does not fully take advantage of potential benefits related to student engagement and achievement (Levin & Wadmany, 2006). Given the practical uses of technology in everyday life and, incorporating technology into the school setting can increase curricular relevance and prepare students for success in their local communities as young adults.

A different study by Mok, established "91.7% of students going online daily" (Hoe, 2012) with largest number opting to use Facebook. These statistics strongly cement the fact that many students find social networking platforms part of life. Therefore, use of social networks for learning purposes is likely to be met minimal resistance since students perceive it as an aid to learning. Irwin, Ball, Desbrow and Leveritt (2012), however, seem to raise the question on the effectiveness in blending social platforms into school curriculum in order to have meaningful performance impact (Irwin & Leveritt, 2012). Enriquez Angelo

(2014) on the other hand investigated the effectiveness of a virtual network as a learning tool. He concluded a significant percentage of study participants perceived Edmodo as an invaluable pedagogical tool. However, he also noted that difficulties in operating Edmodo could act as an impediment to use of Edmodo as a learning platform (Enriquez, 2014).

According to Ryan Lytle's 2012 Michigan State University study, Twitteracy: Tweeting as a New Literary Practice, courses where students use Twitter and are actively engaging with their colleagues and instructors are observed to have a higher interaction and higher grades. They determined that this happen because students did not learn for learning, but they used something real to engage, answer questions, brainstorm, learn concise writing styles, and contact each other outside of the classroom more than what happens in the classroom (Lytle, 2012). This is an interesting finding. It might suggest that it is not simply Twitter that encourages learning, but using technology to engage students in various ways. This supports a growing trend toward real world and engaging learning.

The students can gain and share personal learning goals through social media. Many studies done through the Institute for Perspective Technological Studies (IPTS) proclaim social media brings out new opportunities in education (Christine Redecker & Punie, 2010). This can benefit institutions to prepare their students for better E-Learning. All students and teachers can also communicate using e-mails to discuss any learning related issues.

A recent study by Buzzetto(2014) on students' perception on YouTube suitability as learning augmenting tool provided interesting scholarly findings, primarily that participants perceived YouTube as a passive rather than active learning tool. This is because YouTube provides recorded information in video form. This is unlike Facebook video or Skype that provides live feeds and direct communication. However, students perceived YouTube as a rich source of educative content especially for educational videos. With Facebook being the most visited social site in the world, it does not come by surprise that it is among the first choices for blended learning (Buzzetto-More, 2014). Indeed, according to Irwin, Ball, Desbrow and Leveritt (2012), most students perceive benefits through enhanced communication, interaction, and flexibility in course content delivery (Irwin & Leveritt, 2012). It can be argued that the reason why Facebook receives a high regard by students is the fact that most students actually spend most of their time on Facebook. Bringing education into the same platform where people meet makes it acceptable for students. However, it is worth considering that use of uncontrolled social sites for learning calls for high levels of discipline. Students' multitasking between Facebook comments and Facebook classroom are likely not to concentrate on the learning process. Whereas most of the literature available suggests a strong perception of social network sites by students, little research has been done to correlate the specific use of social media sites and performance grades. A more comprehensive research problem would be to investigate the relationship between augmenting social media and improved grades.

Overall, these studies highlight the need for social media in the learning process. The evidence presented in this section indicates the social media has a great impact on education. Through E-learning students have an easier chance accessing and succeeding school. Teachers and other school faculty can use social media sites such as Facebook to communicate. Internet can also be used for many different ways to help the process of education. Some people may have the wrong idea about social media, but if it is used the right way it does more good than bad and benefits who ever used it the right way.

The rationale for using undergraduate and graduate level students is because they are the most active group on the use of social media. Whereas the use of social media seems fairly divided across both genders, the research aims to investigate interesting outcomes in regard to gender. The perception on use of social media has a large dependence on the gender of users. The fact that the study will be carried out in a college setting sets the age bracket to 18-50 years. The study also puts into consideration the ethnic diversity across the college. While collecting study sample, care will be taken to consider different ethnic groups. It is expected that perception will show some variation based on gender. Considering the students level have already adapted to college life, this level will be the most appropriate for the study. This study will use survey research to collect information about the use of social media for learning by college students.

The purpose of this investigation is to understand how students informally use social media. The idea for this study is rooted in searching for ways to improve the learning process for students. The following are research questions:

1. Is the mean perception of social media use for learning different between undergraduate students and graduate students who currently use social media?
2. Is the mean perception of social media use for learning different between undergraduate male and female students?
3. Is the mean perception of social media use for learning different between graduate male and female students?

METHODOLOGY

Survey Research

This study is a survey research design, an invaluable tool for assessing opinions and trends. Even on a small scale, such as local government or small businesses, judging opinions with carefully designed surveys can dramatically change strategies.

Methods employed in this study include quantitative processes. Bogdan and Biklen (2006), notes that quantitative researchers sometimes find it useful to measure numerical data. Quantitative data collection methods rely on structured data collection instruments that fit diverse experiences into predetermined response categories. They produce results that are easy to summarize, compare, and generalize. Quantitative data are often included in qualitative writings in the form of descriptive statistics.

Participants

According to G-power software version 3.1, the recommended sample size was 64 in each group, however the total sample size in this current study was 58. The probability sample was used in this analysis. Both undergrad and graduate students had an equal chance to be in the analysis. The study sample was randomly collected with age, ethnic background, and history of use of social media as factors for consideration.

The participants of this research will be college students, including both international and American college students. The participants will be gender non-specific, graduate and undergraduate students currently attending the university. The participants will be solicited via electronic communications (email, Facebook, etc.). Participation in this research is voluntary and participants can withdraw at any time after signing a consent form electronically.

The participants are not required to provide personal information, and any information collected will be anonymous. Further, no obligations to participate in the survey are made. This study is a helpful tool for the researchers to investigate the students perceptions of the use of social networking tools on education learning and expand the study for future application toward technology in education through the participants feedback. Participants comfortably give their feedback to the researchers and they can discuss any concern or input related to the topic throughout the email. The maximum risk for the participants is just answering the survey questions.

Procedures

The use of surveys has been one of the most effective tools for data collection. In this study, structured survey was administered to the study participants. After An institutional review board (IRB) had been approved by the instructor, the consent form was sent along with the survey electronically through Qualtric website. Participants need from five to ten minutes to complete the survey. The data collected in this study was sent through a survey emailed to the participants through University of Northern Colorado (UNC) list-serves and Facebook groups. The survey consists of nine items in a Likert scale answered by the participants. Each item has 5 responses from strongly disagree to strongly agree. The survey consists of four sections; biography, Social media background, Devices used and time spent and experiences and beliefs. After conducting the survey quantitatively, data was processed through a statistical code program such as SPSS. After that, data was analyzed and interpreted accordingly. The email that contains the survey was open from the period of March 23 May 1st, 2015.

Data Analysis

A quantitative research method has been selected for this study to collect the data for better understanding of whom will use more of the social network between grade and undergrad). In this study, data were analyzed using statistical software resulting in descriptive statistics to determine the mean and standard deviation for closed-ended survey questions and a correlation matrix to determine statistically significant relationships. Also, a t-test was used to compare variables that include university level (undergraduate or graduate). A T-test was also used to compare the mean perception different perception between undergraduate and graduate males and females students separately. The statistical software SPSS was used to complete statistical analysis.

After successful data collection, a T-test method of data analysis was used to analyze the raw data from the survey. The t-test is very helpful when a study involves both dependent and independent variables. In this study, t test was used to check the sample mean against population mean. T test is also helpful here to compare two sample means. In this study, t test was used to analyze the variables; (perception) and (experience) of students (undergraduate and graduate) to see the mean difference among the students.

The survey was created to help investigate the students' perception of the use of social networking tools for their educational learning. In the survey, there is one composite of eleven items. Responses of all items are answered in a Likert scale. Responses are recoded as following: Strongly agree=1, Disagree=2, Neither=3, Agree=4, Strongly agree=5. Also, students level is coded as under- graduate=1, graduate=2. Negatively worded answers are recoded to positively and vice versa. After that, data was analyzed to run t test and correlation accordingly.

For the reliability, reliability was run to report Crohnbachs alpha to increase reliability as the researchers developed the survey items.

Future recommendations were constructed based on this study for effective educational environment learning using the technology for current and next generation.

RESULTS

Composite 2 of the survey was eliminated because the response orders of the items oppositely ordered compared to composite 1 and there was concern participants may not have been aware of the difference. There were 58 participants in the final sample for this study (N=58), N=23 for undergraduate, N=35 for graduate, N=39 for males, N=19 for females in this study. For the composite experience, the average score of the university level was 4.2 (SD=0.38) for undergraduate students and 3.4 (SD=1.06) for graduate students.

Group Statistics

Research Question 1: Is the mean perception different between under- graduate students and graduate students who currently use social media?

We ran a t-test to answer that question. The t-test indicated that the mean difference is statistically significant between undergraduate and graduate students in their perception of the use of social networking for educational purposes, F=14.69, p<0.05, df=45,66, and the confidence interval of the difference at 95% =.39 for lower, .1.18 for upper. The effect size based on was large (Cohen's D = 0.97).

Independent Samples Test

Research Question 2: Is the mean perception on the Use of Social Network Tools for Education in USA different between undergraduate male and female?

Research Question 3: Is the mean perception on the Use of Social Network Tools for Education in USA different between graduate male and female?

Table 1. Group statistics

	University Status	N	Mean	Std. Deviation	Std. Error Mean
Mean_exp	Undergraduate	23	4.2253	0.3787	0.0799
	Graduate	35	3.4338	1.0684	0.1806

Table 2. Independent samples test

		Levene's Test for Equality of Variances		t-test for Equality of Means	
		F	Sig.	t	df
Mean_exp	Equal variances assumed	14.618	.000	3.406	56
	Equal variances not assumed			4.016	45.663

A t-test was used to answer these questions concluding that there is a statistically significant difference in student's perception on the Use of Social Network Tools for Education in USA between both undergraduate and graduate males and females students.

Gender=Male: Based on Levens test for equality of variances, F=7.17, p=.035, df=34.31, with 95% confidence interval.

Gender=Female: Based on Levens test for equality of variances, F=9.56, p=.001, df=10.35, with 95% confidence interval

Gender = Male

See Table 3 and Table 4.

Gender=Female

See Table 5 and Table 6.

Table 3. Independent samples test[a]

		Levene's Test for Equality of Variances		t-test for Equality of Means	
		F	Sig.	t	df
Mean_exp	Equal variances assumed	7.166	.011	1.669	37
	Equal variances not assumed			2.195	34.305

Table 4. Independent samples test[a]

		t-test for Equality of Means			
		Sig. (2-tailed)	Mean Difference	Std. Error Difference	95% Confidence Interval of the Difference Lower
Mean_exp	Equal variances assumed	.104	.51049	.30589	-.10931
	Equal variances not assumed	.035	.51049	.23260	.03794

Table 5. Independent samples test[a]

		t-test for Equality of Means			
		Sig. (2-tailed)	Mean Difference	Std. Error Difference	95% Confidence Interval of the Difference Lower
Mean_exp	Equal variances assumed	.104	0.5105	0.3059	-.1091
	Equal variances not assumed	.035	0.5105	0.2320	.0374

Table 6. Independent samples test[a]

		Levene's Test for Equality of Variances		t-test for Equality of Means	
		F	Sig.	t	df
Mean_exp	Equal variances assumed	9.586	.007	4.590	17
	Equal variances not assumed			4.410	10.349

RELIABILITY

From the output, the reliability is high across all the items of the survey, N=58, N of items=11 Cronbach's Alpha=.915.

Reliability Statistics

See Table 7.

DISCUSSION

The purpose of this study was to see the student's perception of the use of social media for their educational learning. We expected students welcome the idea of using social networking interfaces such as Facebook for their educational learning. We found the responses show strong positive in support of the idea of using the social networking for learning. Statistical analysis indicated there is a statistically significance difference between undergraduate and graduate students who currently use social media in their perception. It could be from age rather than from school level. Also, there results showed the significant difference between males and females in terms of perception of use of social networking.

As we expected in the literature review, these findings support the idea that undergraduate students differ from graduate students who support the use of social media for educational learning due to the savvy use of undergraduate students and the time spent on social media in the daily basis. Results showed in the cross tabulation that 17 undergraduate students from the total of 23 strongly agree with the statement they use social media to keep in touch with their families and friend to indicate that there is a chance to incorporate social media for learning. In the experience composite of the survey, both undergraduate and graduate show high usage of social networking in the daily life and some already use it for education purposes.

Table 7. Reliability statistics

Cronbach's Alpha	Cronbach's Alpha Based on Standardized Items	N of Items
.915	.916	11

LIMITATION/RECOMMENDATIONS

This study support previous studies that mentioned earlier in the literature review. It is recommended that social networking interfaces are incorporated for both graduate and undergraduate for both males and females as this study shows that there is a significance difference between both undergraduate and graduate for their perception that support the idea of using social networking for their learning purposes. We believe that the results turned out that way because most of students at college age are interested in having social media in their classes because they already own smartphones, laptops, access to the Internet so easily so it was expected to find significance.

This study fell into some limitation. For example, the suggested sample size based on G Power was 138 where the real study was conducted on only 58 participants. This could be due to the limited time as this is a study conducted during a semester. We suggest more time for this study to increase the sample size and then the results should be different. Additionally, composite 2 of the survey was eliminated after collecting the data so the results are not affected due to a mistake while writing the responses of the item. The second composite was focused to grasp student's beliefs. If the composite was added to the analysis, that could have given different results to support the study. This is a pilot study. The study will be redone to show different results.

REFERENCES

Aghaee, N. M. (2010). *Social media use in academia; campus students perceptions of how using social media supports educational learning* (Unpublished master's thesis). Uppsala University.

Beltran, M., & Belle, S. (2013, December). The use of internet-based social media as a tool in enhancing student's learning experiences in biological sciences. *Higher Learning Research Communications, 3*(4).

Bogdan, R., & Biklen, S. K. (2006). *Qualitative research for education: An introduction to theories and methods* (5th ed.). Pearson.

Brooks, L. (2009, May-June). Social learning by design: The role of social media. *Knowledge Quest, 37*(5), 58–60.

Buzzetto-More, N. A. (2014). An examination of undergraduate student's perceptions and predilections of the use of youtube in the teaching and learning process. *Interdisciplinary Journal of E-Learning and Learning Objects, 10*, 17–32.

Enriquez, M. A. S. (2014). Students' perceptions on the effectiveness of the use of edmodo as a supplementary tool for learning.*The DLSU Research Congress.*

Gorder, L. M. (2008). A study of teacher perceptions of instructional technology integration in the classroom. *Delta Pi Epsilon Journal, 50*(2).

Hoe, J. M. C. (2012). Facebook and learning: Students' perspective on a course. *The Journal of the NUS Teaching Academy, 2*(3), 131–143.

Irwin, L. D. B., Ball, L., Desbrow, B., & Leveritt, M. (2012). Students perceptions of using Facebook as an interactive learning resource at university. *Australasian Journal of Educational Technology, 28*(7), 1221–1232. doi:10.14742/ajet.798

Kozma, R. B. (1994, June). Will media influence learning? Reframing the debate. *Educational Technology Research and Development, 42*(2), 7–19. doi:10.1007/BF02299087

Levin, T., & Wadmany, R. (2006). Listening to students voices on learning with information technologies in a rich technology-based classroom. *Journal of Educational Computing Research, 34*(3), 281–317. doi:10.2190/CT6Q-0WDG-CDDP-U6TJ

Lytle, R. (2012, October). Twitter improves student learning in college classrooms (Study). *U.S. News and World Report and World Report. Education.*

Redecker, C. K. A.-M., & Punie, Y. (2010). Learning 2.0 - the impact of social media on learning in Europe. *European Communities*, 1-13.

APPENDIX

Participants will be asked the following questions in the survey:
Gender: M - F
Age:
Race/Ethnicity:

- American Indian or Native Alaskan
- Asian
- Black or African American
- Caucasian
- Hispanic or Latino
- Native Hawaiian or Other Pacific Islander
- Other/None

University Status

- Undergraduate
- Graduate

Please indicate your level of agreement with each of the following statements (1 being extremely disagree and 5 being extremely agree).
Experience and Engagement:
I have access to social networking interfaces (Internet)
I use a computer to access social networking interfaces.
I use a smart phone or tablet to access social networking interfaces.
I use social networking interfaces daily.
I use social networking interfaces to keep in touch with family and friends.
While using social networking interfaces, I make frequent comments.
I use social networking interfaces to play games.
I use social networking interfaces in other course/classes use/have used social networking interfaces for my own teaching.
(as a teaching and learning tool). I feel confident navigating and interacting with one or more social networking interfaces.
Beliefs:
I think social networking interfaces are effective tools to communicate with people around the world.
I think social networking interfaces are helpful in classes to engage students in learning.
I think students may feel enthusiastic using social networking interfaces for learning.
I think using social networking interfaces can enhance peer relationships.
I think using social networking interfaces can enhance relationship with instructors.
I think using social networking interfaces can enhance learning.

Chapter 3
Heuristic Evaluation on M-Learning Applications:
A Comparative Analysis of Two Heuristic Sets

Christofer Ramos
Universidade do Estado de Santa Catarina, Brazil

Flávio Anthero Nunes Vianna dos Santos
Universidade do Estado de Santa Catarina, Brazil

Monique Vandresen
Universidade do Estado de Santa Catarina, Brazil

ABSTRACT

Heuristic evaluation stands out among the usability evaluation methods regarding its benefits related to time and costs. Nevertheless, generic heuristic sets require improvements when it comes to specific interfaces as seen on m-learning applications that have acquired considerable evidence within the current technologic context. Regarding the lack of studies aimed at interfaces of this sort, the authors propose, through a systematic methodology, the comparative study between a heuristic set specific to the assessment on e-learning interfaces and other, on mobile. The identified usability problems were matched with the aspects of coverage, distribution, redundancy, context and severity, in a way that it was possible to understand the efficiency of each set in covering m-learning issues. Among the findings, e-learning's heuristic set could detect a larger number of usability problems not found by mobile's.

INTRODUCTION

The popularization of mobile devices such as tablets and smartphones, fostered at the beginning of this decade has reflected significant impacts on society and made people's daily lives more dependent on the convenience provided by this type of technology. In fact, technological evolution manifests according to the context of each epoch and is able to transform the way society behaves, a process intrinsic to human

DOI: 10.4018/978-1-5225-2469-4.ch003

progress. In this scenario, the influences established by technology in the educational sphere from the emergence of e-learning in the 1990s, has reconfigured the way people acquire knowledge. The learning mediated by platforms of this type is often seen as appropriate to the expectations of those classified by Prensky (2001) as "digital natives", the generation born from the 1990s, interested in learning processes that are active, interactive, socially shared and based on findings from the error.

Obviously, the convenience offered by mobile devices has made e-learning systems gradually migrate from web platforms to mobile applications, inaugurating the category called mobile learning or simply m-learning.

According to Molenet, mobile learning can be broadly defined as 'the exploitation of ubiquitous handheld technologies, together with wireless and mobile phone networks, to facilitate, support, enhance and extend the reach of teaching and learning. [...] There is no agreed definition of 'mobile learning', partly because the field is experiencing rapid evolution, and partly because of the ambiguity of 'mobile' – does it relate to mobile technologies, or the more general notion of learner mobility? (Hashemi et al., 2011, pp. 2478)

M-learning not only manifests as a trend in this particular era when knowledge acquisition is facilitated by mobile devices, but is also justified in aspects as social inclusion, allowing economically disadvantaged people or the ones isolated by geographic distances to have access to education; economic rise of countries, leveraging quantitative and qualitatively the production of operational and intellectual workforce; and savings for corporations, reducing costs associated with personnel training and professional capacitation. Furthermore, as the ubiquous characteristic of mobile devices provide spontaneity in the sense that users are able to access everything they want whenever they want, m-learning offers more opportunities for educational companies to encourage users to purchase learning products such as courses, full classes, extra content and certificates by fostering the system's m-commerce appeal.

The Implications of Usability to M-Learning Systems

It's known that, lacking appropriate usability design, the interface of a m-learning can, for example, require from its users excessive cognitive load while they perform certain tasks, affecting their learning or increasing drop outs, since usability attributes directly influence user satisfaction. "Users want systems that are easy to learn and to use as well as effective, efficient, safe, and satisfying" (Preece, Rogers & Sharp, 2002, p. 318). In this respect, the usability evaluation's purpose is to investigate problems that might prejudice the dialogue between a system's interface and its user, and to assist the evaluator with concrete and relevant information in order to propose adjustments or improvements able to render the system more friendly. Therefore, because it considers the attributes relevant to the context of use and to the end-user, usability evaluation is advantageous not only for systems in development, avoiding costs with future adjustments, but also for those already released to the public in which design intervention is required.

In fact, the development of interfaces endowed with usability features is an adequate way to provide the users meaningful experiences which, therefore, is converted into profits for companies (Rubin & Chisnell, 2008). On the other hand, as stated by Nielsen and Budiu (2015), when users frequently face problems that impoverish their experience, it tends to unconsciously influence their perception about the m-commerce's credibility and security, which also influences their disposition to engage in a first purchase or to buy again. Thus, it's of extreme importance for m-learning companies to understand the need of performing usability testing throughout the system's development, bearing in mind that it's crucial for profiting and standing out from their market competitors.

Heuristic Evaluation

The determination of a usability evaluation method needs to take into account the specifics of each project, and regarding to this, expert evaluation stands out among the diverse techniques mainly due to its low cost and for not demanding the engagement of neither the users nor the use of a laboratory. Besides that, inspecting the interface can be time saving because it helps anticipating problems that would emerge in user testings only after a number of sessions. One of the most usual methods among experts is the Heuristic Evaluation (HE) that stands for "an informal usability technique […] in which experts, guided by a set of usability principles known as heuristics, evaluate whether user-interface elements, such as dialog boxes, menus, navigation structure, online help, etc., conform to the principles" (Preece, Rogers & Sharp, 2002, p. 409). Nielsen (1994, pp. 155), the most influent author of this technique, states that Heuristic Evaluation is "a sistematic inspection of a user interface design for usability. The goal of heuristic evaluation is to find the usability problems in a user interface design so they can be attendend to as part of an iterative design process".

Therefore, according to Stone et al. (2005, pp. 539), in contrast to the usability methods in which the users are protagonists, performing tasks and verbalizing information about their own point of view, during an HE the evaluators put themselves in the users' shoes to interact with the product, searching for possible interaction issues guided by the heuristics' guideliness or even their own expertise. Also, the specialist can point out which heuristic was violated, describing the issue's location, context, characteristic and severity. The authors also state that, alternatively, in some cases it can be interesting to settle a list of tasks in order to make the evaluator walk through the same paths a real user would explore, yet it can enhance the chances of camouflaging the issues not involved with those specific tasks. Besides, as the evaluator reports issues immediately, there's no need to record the session as its common in a user evaluation, once the benefits of watching the expert taking notes and alternating among screens are minimal. Although a unique evaluator can find significant issues on an interface, Nielsen (1994, pp. 155-156) argues that empiric evidences suggest that five evaluators are able to assure the results an accuracy of 75%.

Somehow, HE holds disadvantages that must be taken into account while establishing the usability evaluation technique. As it doesn't comprehend the real user, mistakes related to erroneous prediction can occur, though it's not advisable to make users perform the analysis due to their lack of expertise in conducting usability evaluations. Similarly, the judgement of an evaluator concerning a given problem, due to personal preferences and bias matters, can diverge from a real user's opinion and, therefore, omit important issues or disclose irrelevant aspects of the interface. Notwithstanding, the quality of the outcome is strictly dependent on the evaluators' expertise, a factor that can cast doubt on the validity of the analysis, supposing the one in charge for the evaluation is not enough familiar with the system's domain and task's specifics (Stone, Jarret &Woodroffe, pp.532).

While heuristic evaluation is characterized as a convenient and often advantageous technique for usability evaluation, the most consolidated model, proposed by Nielsen and Molich (1990), may not be the most appropriate in many situations, especially when it comes to specific domains and emerging technologies such as an m-learning.

Different sets of heuristics for evaluating toys, WAP devices, online communities, wearable computers, and other devices are needed, so evaluators must develop their own by tailoring Nielsen's heuristics and by referring to design guidelines, market research, and requirements documents. Exactly which heuristics are the best and how many are needed are debatable and depend on the product (Preece, Rogers & Sharp, 2002, pp. 409).

Indeed, evaluating the usability of an m-learning application by using generic heuristic sets or any specific to other classes of system such as web applications, minimally contribute towards the experience of users, since they are unable to cover particular aspects related to the contexts of virtual learning and mobility. What concerns to the domain of virtual learning environments, Zaharias and Koutsabasis (2012) reffer to "pedagogical usability" as the particularities related to the complex nature of learning processes, stating that the evaluation of the usability on this kind of systems should be able to "assess whether the pedagogical design of the learning environment is based on learning theories and whether other important factors such as motivation, diversity and growth are taken into consideration" (Zaharias & Koutsabasis, 2012, p. 47). The importance of considering the aspects related to pedagogical usability is highlighted by Salles Júnior et al. (2016). The authors allegate that the students tend to blame themselves whilst facing problems associated with the absence of pedagogical guideliness on an interface, and that's a key issue for loss of energy and enthusiasm, which can culminate in high dropout rates.

Notwithstanding, the mobility domain also holds interaction requirements that render the generic heuristic sets insufficient for the evaluation on m-learning apps. Zhang and Adipat (2005) argue that the advent of mobile devices has brought significant challenges to the process of usability evaluation. According to them, connectivity, small screen size, different display resolutions, limited processing capability and power, and data entry methods (pp. 295-296) are key issues that the usability evaluater can encounter and that can directly influence the interaction quality on systems of this nature. Other authors agree with this premise stating that one can face issues related to small screens, limited input mechanisms, dynamic contexts of use (Kjeldskov, 2002) and that the usability on mobile devices is not restricted only between the interface and the user, also comprising factors associated with the environment and the characteristics of the device (Lee, Shineider, & Schell, 2005).

Heuristic Sets for Specific Domains

Although, as stated, in specific domains it's crucial for a system to count on a particular set of heuristics in order to provide the heuristic evaluation substantial effectiveness and representative results, the emerging domain of m-learning still lacks studies concerned with this end. That's clear observed in Hermawati and Lawson (2016)'s research that reviewed seventy studies related to usability heuristics for specific domains. They identified through the outcomes from the studies analysed that specific sets of heuristics, when subjected to comparative validation against sets of general nature, were able to identify not only a greater number of usability problems, but also a higher frequency of problems classified as severe. Somehow, it's critical that there couldn't be found any study concerned with the development of a specific tool for heuristic evaluation on m-learning applications, here in question. It's believed, somehow, that this gap in bibliography may be assigned to the fact that the hardware technologies for mobile devices have evolved more representatively in the last decade, and that m-learning applications are yet a novelty. This condition must reinforce the reason why the studies have only concentrated on exploring either e-learning applications (Reeves et al., 2002; Ardito et al., 2004; Costabile et al., 2005; Dringus

& Cohen, 2005; Kemp, Thompson, & Johnson, 2008; Alsumait & Al-Osaimi, 2010; Omar, Yusof, & Sabri 2010) or mobile softwares/applications (Bertini, Gabrielli, & Kimani, 2006; Korhonen & Koivisto, 2006; Väänänen-Vainio-Mattila & Wäljas, 2009; Somro, Ahmad, & Sulaiman, 2012; Inostroza et al., 2012a,b, 2013; Neto & Pimentel, 2013; Al-Razgan, Al-Khalifa, & Al-Shahrani, 2014; Joyce & Lilley, 2014; Inostroza & Rusu, 2014).

Whereas e-learning and mobile domains are of substantial importance for m-learning, two other studies stand out. The first one, performed by Zaharias and Koutsabasis (2012) investigated heuristics sets singular to the evaluation of e-learning environments, comparing the instruments developed by Reeves et al. (2002) that considered aspects of instructional design, against the conceptual framework for design and evaluation of instructional environments proposed by Mehlenbacher et al. (2005), hereinafter referred to as Heuristic Set 1 or simply HS1, and presented in Table 1. According to the authors, the heuristic sets studied represent two different strands. The first strand is known as the field that studies Nielsen's heuristics and proposes ways to expand and adapt them for the new technological contexts, somehow it is criticized by the lack of theoretical research. The second strand is concerned with models and theories of learning and can be characterized as more appropriate to the e-learning context, despite empirical validation of their studies is still required. The study's outcome did not show significant differences between the two heuristic sets and defined both as effective for the proposal, as they were able to identify more than 90% of the usability problems present in the analyzed system. Despite the insignificance of differences between sets, HS1 showed to be slightly more efficient in terms of numbers of identified usability problems and, for that reason, it was chosen for the purposes of this study, as it is explained further on.

Similarly, other researchers also analyzed the effectiveness of two distinct heuristic sets but focused on the domain of mobile application. To that intention, Rocha, Andrade and Sampaio (2014) assessed

Table 1. Usability heuristics for e-learning design

	Learner Background and Knowledge
Accessibility	Has the WBI been viewed on different platforms, browsers, and modem speeds? Is the site ADA compliant (e.g., red and yellow colors are problematic for visually-challenged users)? Have ISO-9000 standards been considered?
Customizability and maintainability	Does printing of the screen(s) require special configuration to optimize presentation and, if so, is this indicated on the site? Are individual preferences/sections clearly distinguishable from one another? Is manipulation of the presentation possible and easy to achieve?
Error support and feedback	Is a design solution possible that prevents a problem from occurring in the first place? When users select something does it differentiate itself from other unselected items? Do menu instructions, prompts, and error messages appear in the same place on each screen?
Navigability and user movement	Does the site clearly separate navigation from content? How many levels down can users traverse and, if more than three, is it clear that returning to their initial state is possible with a single selection? Can users see where they are in the overall site at all times? Do the locations of navigational elements remain consistent? Is the need to scroll minimized across screens and frames within screens?
User control, error tolerance, and flexibility	Are users allowed to undo or redo previous actions? Can users cancel an operation in progress without receiving an error message? Are multiple windows employed and, if so, can they be manipulated easily?
	Social Dynamics
Mutual goals and outcomes	Are learners rewarded for using the communication tools? Are communication tools provided that allow synchronous and asynchronous interaction? Do communication tools allow information revision, organization, and management? Are interactions organized around instructional objectives and task deliverables?

continued on following page

Table 1. Continued

Communication protocols	Are learners rewarded for using the communication tools? Are communication tools provided that allow synchronous and asynchronous interaction? Do communication tools allow information revision, organization, and management? Are interactions organized around instructional objectives and task deliverables?
Instructional Content	
Completeness	Are levels clear and explicit about the "end" or parameters of the site? Are there different "levels" of use and, if so, are they clearly distinguishable?
Examples and case studies	Are examples, demonstrations, or case studies of user experiences available to facilitate learning? Are examples divided into meaningful sections, e.g., overview, demonstration, explanation, and so on?
Readability and quality of writing	Is the text in active voice and concisely written (> 4 < 15 words/sentence)? Are terms consistently plural, verb + object or noun + verb, etc., avoiding unnecessarily redundant words? Do field labels reside on the right of the fields they are closely related to? Does white space highlight a modular text design that separates information chunks from each other? Are bold and color texts used sparingly to identify important text (limiting use of all capitals and italics to improve readability)? Can users understand the content of the information presented easily?
Relationship with real world tasks	Is terminology and labeling meaningful, concrete, and familiar to the target audience? Do related and interdependent functions and materials appear on the same screen? Is sequencing used naturally, if sequences of common events or narratives are expected? Does the site allow users to easily complete their transactions or selections?
Interaction Display	
Aesthetic appeal	Does the screen design appear minimalist (uncluttered, readable, memorable)? Are graphics or colors employed aesthetically? Are distractions minimized (e.g., movement, blinking, scrolling, animation, etc.)?
Consistency and layout	Does every screen display begin with a title/subject heading that describes contents? Is there a consistent icon design and graphic display across screens? Is layout, font choice, terminology use, color, and positioning of items the same throughout the site (< 4 of any of the above is usually recommended)?
Typographic tools and structuring	Does text employ meaningful discourse cues, modularization, chunking? Is information structured by meaningful labeling, bulleted lists, or iconic markers? Are legible fonts and colors employed? Is the principle of left-to-right placement linked to most-important to least-important information?
Visibility of features and self-description	Are objects, actions, and options visible? Do users have to remember information from one part of a dialogue to another? Are prompts, cues, and messages placed where users will be looking on the screen? Do text areas have "breathing space" around them? Is white space used to create symmetry and to lead the eye in the appropriate direction?
Instructor Activities	
Authority and authenticity	Does the site establish a serious tone or presence? Are users reminded of the security and privacy of the site? Are humor or anthropomophic expressions used minimally? Is direction given for further assistance if necessary?
Intimacy and presence	Is an overall tone established that is present, active, and engaging? Does the site act as a learning environment for users, not simply as a warehouse of unrelated links?
Environment and Tools	
Help and support documentation	Does the site support task-oriented help, tutorials, and reference documentation? Is help easy to locate and access on the site? Is the help table of contents or menu organized functionally, according to user tasks?
Metaphors and maps	Does the site use an easily recognizable metaphor that helps users identify tools in relation to each other, their state in the system, and options available to them?
Organization and information relevance	Is a site map available? Is the overall organization of the site clear from the majority of screens? Are primary options emphasized in favor of secondary ones?
Reliability and functionality	Do all the menus, icons, links, and opening windows work predictably across platforms? Have important interactive features and multimedia elements been tested across platforms and browsers?

Note: Heuristic Set developed by Mehlenbacher et al. (2005).

the usability of an application designed for field exploration in geology which, among other features, relates collected information with geographical coordinates. In this study, the heuristic evaluations were performed by using the sets proposed by Neto and Pimentel (2013), which hereinafter is referred to as either Heuristic Set 2 or HS2, and by Inostroza et al. (2012a). According to the authors, the development of the first set also originated from the perceived inadequacy in generic sets for evaluation in specific domains and comprehended eleven heuristics, which are presented in table 2, and were based on a previous analysis that identified common mobile usability problems. The later, on the other hand, was led from the principle that traditional methods of interface evaluation didn't consider the specific nature of touchscreen devices and, thus, the researchers gathered information concerned with the challenges of evaluating systems of this sort to generate eleven heuristics. The results of the study showed that

Table 2. Heuristics for evaluating the usability of mobile device interfaces

Use of screen space	The interface should be designed so that the items are neither too distant, nor too stuck. Margin spaces may not be large in small screens to improve information visibility. The more related the components are, the closer they must appear on the screen. Interfaces must not be overwhelmed with a large number of items.
Consistency and standards	The application must maintain the components in the same place and look throughout the interaction, to facilitate learning and to stimulate the user's short-term memory. Similar functionalities must be performed by similar interactions. The metaphor of each component or feature must be unique throughout the application, to avoid misunderstanding.
Visibility and ease access of all information	All information must be visible and legible, both in portrait and in landscape. This also applies to media, which must be fully exhibited, unless the user opts to hide them. The elements on the screen must be adequately aligned and contrasted.
Adequacy of the component to its functionality	The user should know exactly which information to input in a component, without any ambiguities or doubts. Metaphors of features must be understood without difficulty.
Adequacy of the message to the functionality and the user	The application must speak the user's language in a natural and non-invasive manner, so that the user does not feel under pressure. Instructions for performing the functionalities must be clear and objective.
Error prevention and rapid recovery to the last stable state	The system must be able to anticipate a situation that leads to an error by the user based on some activity already performed by the user. When an error occurs, the application should quickly warn the user and return to the last stable state of the application. In cases in which a return to the last stable state is difficult, the system must transfer the control to the user, so that he decides what to do or where to go.
Ease of input	The way the user provides the data can be based on assistive technologies, but the application should always display the input data with readability, so that the user has full control of the situation. The user should be able to provide the required data in a practical way.
Ease of access to all functionalities	The main features of the application must be easily found by the user, preferably in a single interaction. Most-frequently-used functionalities may be performed by using shortcuts or alternative interactions. No functionality should be hard to find in the application interface. All input components should be easily assimilated
Immediate and observable feedback	Feedback must be easily identified and understood, so that the user is aware of the system status. Local refreshments on the screen must be preferred over global ones, because those ones maintain the status of the interaction. The interface must give the user the choice to hide messages that appear repeatedly. Long tasks must provide the user a way to do other tasks concurrently to the task being processed. The feedback must have good tone and be positive and may not be redundant or obvious.
Help and documentation	The application must have a help option where common problems and ways to solve them are specified. The issues considered in this option should be easy to find.
Reduction of the user's memory load	The user must not have to remember information from one screen to another to complete a task. The information of the interface must be clear and sufficient for the user to complete the current task.

Note: Heuristic set developed by *Neto & Pimentel (2013)*.

both sets missed some usability problems that have been detected by the confronted one and, therefore, needed refinement in order to be able to contemplate the unidentified problems. Nevertheless, under the analyzed conditions, HS2 was able to detect 64% more problems than the confronted set.

Purpose of the Study

Faced to what has been debated, it's believed that the development of a heuristic set specific for the evaluation on m-learning applications could be a significant step towards enhancing the user experience in this new domain. Somehow, as mentioned, a corpus of researchers has focused on developing and studying heuristic sets for both e-learning and mobile systems, that together comprise the new recent domain of m-learning. Therefore, the authors are confident that a comparative analysis of these different kinds of heuristic sets is a substantial first step, once it can show whether one of them is more suitable for assessing m-learning interfaces.

In order to achieve the objective of the current study, which stands for "compare the effectiveness of heuristic sets developed for usability evaluation on e-learning and mobile domains" the authors adopt the two heuristic sets that showed to be more efficient in the last two aforesaid studies. Therefore, it was used HS1 that is meant for evaluations on e-learning environments (see table 1) and, HS2 that was developed for assessing mobile applications (see Table 2). Thus, this research was based on the hypothesis that, in heuristic evaluations of a m-learning system, one of the heuristic sets analyzed is able to show a better level of efficiency in identifying usability problems.

METHODS

Participants

The subjects of the experiment comprehended 4 usability evaluators with at least two years of expertise. For heuristic evaluation, the literature doesn't state an appropriate minimum time of expertise, somehow, Nielsen (1992), divide specialists in three different levels: novice evaluators, the ones holding knowledge related to computational systems; regular specialists, with knowledge regarding usability; and double specialiast, with expertise related to both usability and the specific interface domain.

Heuristic evaluation was originally developed as a usability engineering method for evaluators who had some knowledge of usability principles but were not necessarily usability experts as such. Subsequent research has shown the method to be effective also when the evaluators are usability experts. Unfortunately, usability experts are sometimes hard and expensive to come by, especially if they also need to have expertise in a particular kind of application. (Nielsen, 1992, pp. 373)

Even though the evaluators involved in this study were familiar with the m-learning application chosen for the analysis, as each of them had already been engaged in at least one course before the experiment took place, they weren't specialists in the domain. Thus, in an attempt to reduce bias, the evaluators were subjected to a training session that lasted three hours and covered the specifics of the application. That was also aimed at controlling the knowledge of the participants about the interface.

Materials and Procedures

The m-learning application chosen to be evaluated was MOOC[1] EdX, due to its popularity and for being endowed with complex features such as social dynamics, peer reviews, and asynchronous features. Besides, it was used Samsung Galaxy S5, a smartphone which holds the following characteristics: 5.2" touchscreen display, version 5.1 of Android operational system, and a 1.6 Ghz processor. Heuristic evaluation sessions were performed by 4 evaluators equally and randomly divided into two groups: one group used HS1, and the other group used HS2. They were asked to identify as many numbers of Usability Problems (hereinafter is referred to as UP) as possible and to associate them with the heuristics in use. In order to report the UPs in a fast, ease and intuitive way, it was used the structured document proposed by Cockton & Woolrych (2001), following Zaharias and Koutsabasis' (2012) same methodology.

Each evaluator performed the evaluation alone and later, the results were consolidated and interpreted by the authors of this study. In order to make the evaluators familiar with the system, before the evaluation they explored it freely, taking notes they deemed important. This step provided the evaluators a global overview of the basic functions available to end-users. Moreover, it was believed that a substantial number of problems could also emerge at this stage, showing important aspects for the official analysis. Subsequently, during the actual evaluation session, the evaluators identified and documented the UPs using the structured document (Table 3). On the structured document the evaluators needed to fill out the following information for each identified problem: 1) in column "Usability number", number the identified problem in order of appearance; 2) in column "Heuristic code", provide the code of heuristic, from the sheet regarding the heuristic set in use, which made possible detect that problem; 3) in column "Problem description", provide a detailed description of the problem; 4) in column "Problem context", mark whether the problem was related to the context of system structure (SS), course content (CC), or global (G) meaning both contexts; 4) in column "Problem severity", mark whether the severity degree of the identified problem was considered minor (Min.), moderate (Mod.) or severe (Sev.); 5) in column "Redundancy", indicate, if applicable, which other heuristic(s) from the set were also able to cover that same problem.

A minimum time for the evaluation wasn't set, once the time would vary according to the heuristic set in use or level of specialist expertise. Also, task scenarios and a unique course for exploration were set in order to control the outcomes of the analyses. The chosen of tasks considered paths holding a significant number of UPs. The tasks are detailed below the same way they were presented to the evaluators.

1. Search for a course in the field of Design, select the one called "Product Design: The Delft Design Approach" and apply for it;
2. Take a complete module of a class;

Table 3. Structured document for the heuristic evaluation sessions

Usability Number	Heuristic Code	Problem Description	Problem Context			Problem Severity			Redundancy
			SS	CC	G	Min.	Mod.	Sev.	

Note: The structured document for the heuristic evaluation sessions in the current study is adapted from *Zaharias & Koutsabasis' (2012) methodology*. SS = system structure, CC = Course Content, G = global, Min. = Minor, Mod. = Moderate, Sev. = Severe.

3. Post a comment or question in the course's discussion session;
4. Log off the system and then log in again. After, take back the course from where you stopped;
5. Perform an activity, essay or evaluation in the course.

RESULTS

Diagnoses of the System's Usability

The time taken by evaluators to perform the heuristic evaluation sessions ranged from 114 to 151 minutes, somehow, the results didn't make possible associate standards of time with heuristic sets. As it can be seen on Table 4, evaluator 2 holding HS2 performed the fastest evaluation while evaluator 1 holding HS1 took the longest time to finish the session. Even though it seems that HS2 allowed their evaluators to perform the lowest times, it is also possible to observe similar performance times (133 and 137 minutes) related to both heuristic sets.

Overall, the evaluators concluded that the mobile application EdX holds good technical and pedagogical structure, and cares to employ significant usability aspects for user experience, however, the evaluations made possible perceive a substantial number of problems that can often influence the level of student satisfaction. Among problems, there were inconsistent buttons or commands that don't refer to their functions; unnecessary leading to the web version; system errors, paths and commands that require excessive user cognitive load; long paths to perform frequent tasks; excessive time to load sessions; unavailability of sessions important to the learning process (ie.: evaluation of knowledge and discussion list), forcing the user to visit the web version; lack of feedback about progress of the course and system/user errors; problems related to readability in video lessons; among others.

Table 5 shows the frequency of UPs identified during the evaluations from different perspectives. As one can observe, by grouping similar UPs among experts, HS1 could detect 25 problems and HS2, 17. Thus, if the results from the two sets are considered together, it's possible to observe that 11 of UPs

Table 4. Time (minutes) taken by evaluators to complete the heuristic evaluations

	HS1	HS2
Evaluator 1	151	133
Evaluator 2	137	114

Table 5. Frequency of usability problems identified in the heuristic evaluation sessions

Total UPs detected by Heuristic Set 1	25
Total UPs detected by Heuristic Set 2	17
Detected UPs common to both sets	11
UPs detected only by Heuristic Set 1	14
UPs detected only by Heuristic Set 2	6
Total UPs detected by both sets	**31**

are similar between them and therefore, it leads one to believe that the EdX mobile application holds, in fact, 31 UPs within the context analyzed. Consequently, this fact implies that, HS1 was more efficient in terms of numbers of identified UPs as, alone, it was able to cover 14 UPs omitted from the evaluations performed with HS2 and, similarly, the later could detect only 6 UPs missing in the evaluations performed with HS1.

Relation of Usability Problems with Coverage, Distribution, and Redundancy Aspects

Figure 1 presents the results from the assessment performed by evaluators who used HS1, and Figure 2 those identified by evaluators endowed with HS2. Based on the approach from Zaharias and Koutsabasis (2012) for exposure of results, this session is focused on discussing the relationships between heuristic sets and their characteristics of coverage, distribution and redundancy; and then, the relationships between heuristic sets and aspects of context and severity.

What regards to coverage of heuristics, within the 31 usability problems identified in the system, 80% was detected by HS1 and 54% by HS2, suggesting that the first set holds greater ability to cover usability problems in m-learning applications, and provides significant guidelines to guide usability evaluations and development of platforms of this sort. However, this set was innefficient in covering one of the identified problems, as reported by evaluator 2 (Figure 1), although this condition is not critical, taking into account the total number of UPs.

What concerns to the distribution of heuristics, it is observed that some of them did not indicate any usability problem in both sets, while others alone identified a large portion of them; and that the distribution is similar between sets and among evaluators. Specifically, for HS1, the most important heuristics were "navigability and user movement" and "completeness", representing respectively 29% and 25% of UPs identified by evaluator 1, and 25 and 30% by evaluator 2. For HS2, the most important heuristics were "reduction of the user's memory load" and "immediate and observable feedback", representing respectively 25% and 19% of UPs related by evaluator 1, and 30% and 18% by evaluator 2. This condition may have occurred mainly because the evaluations were limited to specific tasks, camouflaging, therefore possible usability problems that could have been identified by the heuristics that showed few or no coverage. Moreover, it's clear that the heuristics with higher levels of coverage comprise more general usability principles. Nevertheless, the fact that some heuristics have not indicated any usability problem can also mean that the analyzed interface is efficient in that regard, and should not necessarily be used as parameters for inferring that the heuristic set which presented less coverage is inefficient, since this study did not consider the similarity of heuristics between sets and was essentially meant to detect usability problems.

Finally, redundancy has to do with the usability issues that could not be classified by only 1 heuristics, because the problem was too complex or because the heuristic wasn't enough to cover it. Therefore, with regard to redundancy, it was found that HS2 exhibited better results, with 2 redundant heuristics from the evaluation performed by evaluator 1, and 3 by evaluator 2, while evaluators holding HS1 considered redundant 4 and 2 heuristics respectively. The occurrence of redundancies varies according to the system and, in this study, was unable to generate significant interference, somehow, it's suggested that redundant heuristics be better observed and, if necessary, adjusted so as to reduce this characteristic.

Heuristic Evaluation on M-Learning Applications

Figure 1. Frequency distribution of evaluations with Heuristic Set 1 matched to usability problems, context and severity

Heuristic Set 1	Evaluator 1									Evaluator 2							
	UPs	%	SS	CC	G	Min.	Mod.	Sev.		UPs	%	SS	CC	G	Min.	Mod.	Sev.
1 Learner Background and knowledge	9	37.5	5		4	2	4	3		8	40	5	1	2	3	2	3
1.1 Accessibility	1	4.16	1					1		1	5	1			1	1	
1.2 Customizability and maintainability	1	4.16			1		1			2	10	2			1		1
1.3 Error support and feedback	3	12.58	2		1	1	1	1		2	10	1	1		1		1
1.4 Navigability and user movement	4	16.6	2		2	1	2	1		3	15	1		2			
1.5 User control, error tolerance, and flexibility																	
2 Social dynamics	2	8.32			2			2									
2.1 Mutual goals and outcomes	1	4.16			1			1									
2.2 Communication protocols	1	4.16			1			1									
3 Instructional content	9	37.51	5	2	2	4	2	3		6	30	4		2	3	1	2
3.1 Completeness	6	25	4	1	1	2	2	2		6	30	4		2	3	1	2
3.2 Examples and case studies																	
3.3 Readability and quality of writing	1	4.16	1			1											
3.4 Relationship with real-world tasks	2	8.35		1	1	1		1									
4 Interaction display	2	8.35	2			1	1			4	20	1	1	2	1	1	2
4.1 Aesthetic appeal																	
4.2 Consistency and layout										2	10		1	2	1	1	
4.3 Typographic cues and structuring										1	5						1
4.4 Visibility of features and self-description	2	8.35	2			1	1			1	5	1					1
5 Instructor activities																	
5.1 Authority and authenticity																	
5.2 Intimacy and presence																	
6 Environment and tools	2	8.32	1		1	1	1			1	5	1			1		
6.1 Help and support documentation																	
6.2 Metaphors and maps																	
6.3 Organization and information relevance	1	4.16			1		1			1	5	1			1		
6.4 Reliability and functionality	1	4.16	1			1											
UPs not comprised by any heuristic (N/A)										1	5		1			1	
TOTAL	24	100	13	2	9	8	8	8		20	100	11	3	6	8	5	7
Redundancy	4									2							

Figure 2. Frequency distribution of evaluations with Heuristic Set 2 matched to usability problems, context and severity

Heuristic Set 2	Evaluator 1									Evaluator 2							
	UPs	%	Context			Severity			UPs	%	Context			Severity		
			SS	CC	G	Min.	Mod.	Sev.			SS	CC	G	Min.	Mod.	Sev.
01 Use of screen space									1	5.89		1		1		
02 Consistency and standards	1	6.25		1					2	11.76		1	1	1	1	
03 Visibility and easy access to all information	2	12.5	2						2	11.76	1		1	2		
04 Adequacy of the component to its functionality																
05 Adequacy of the message to the functionality and to the user	1	6.25	1			1			1	5.89			1		1	
06 Error prevention and rapid recovery to the last stable state									1	5.89			1			1
07 Ease of input																
08 Ease of access to all functionalities	2	12.5	1	1				2								
09 Immediate and observable feedback	3	18.75	1		2	1	1	1	3	17.84	1	1	1	1	2	
10 Help and documentation																
11 Reduction of the user's memory load	4	25	2	1	1	1	1	2	5	29.41	3	1	1	2	2	1
UPs not comprised by any heuristic (N/A)																
TOTAL	13	100	7	3	3	3	4	6	15	100	5	4	6	7	6	2
Redundancy	2		-	-		-	-		3		-	-		-	-	

Relation of Usability Problems With Context and Severity

As it's also possible to be observed, during evaluations the experts provided information related to the context in which usability problems were found, as well as the severity of problems. Thus, the contexts were categorized in System Structure (SS), Course Content (CC) and Global (G); and the severity ratings contemplated the dimensions Minor (Min.), Moderate (Mod.), and Severe (Sev.). This information was important to identify whether each of the heuristics sets would be more efficient in aspects related to the purposes for which they have been developed, since HS1 was meant for e-learning systems, and HS2, for mobile ones. In addition, they would be able to provide more specific knowledge through the usability diagnosis of the evaluated system.

According to the results, the evaluators guided by HS1, the set aimed at evaluating e-learning interfaces, identified respectively 5 and 3 problems (21% and 15% of the total problems detected by each) specifically related to course content, a condition that leads one to believe that this set, was unable to detect usability problems mainly related to e-learning context. Likewise, observing this same context, the evaluators endowed with HS2, the one aimed at assessing mobile systems, were able to identify respectively 3 and 4 usability problems (23% and 27% of the total problems detected by each), a significant number for a set that wasn't built exclusively for e-learning purposes, and that holds fewer number of heuristics. In addition, HS2 was also efficient in detecting a substantial portion of problems related to system structure context, for which it was developed. It is important to note, however, that it does not nullify the possibility of these results varying if the type of mobile application, the tasks and the evaluators are changed.

Regarding the severity of identified usability problems, the results were similar between the evaluators holding HS1, who defined degrees of severity equally distributed. Conversely, among the evaluators holding HS2, the difference was substantial: evaluator 1 rated more problems as severe, whereas evaluator 2 categorized more problems as minor and moderate. Because sharp convergences are observed between experts and between heuristic sets, the results are not able to provide conclusive information concerning the degree of severity of the problems identified. The manifestation of this kind of result can be derived from the subjective nature of such information.

DISCUSSION

As the interest in learning mediated by m-learning arises, especially in current technological context where such platforms represent significant responses to the expectations of a generation increasingly virtually connected and conditioned to the use of mobile devices, it becomes imperative to ensure that these applications' interface be efficient enough to provide satisfaction and hence, quality of learning. Nevertheless, it's known that both users and educational companies can benefit from a m-learning developed based on usability principles, once user experience directly influences profitability. In this scenario, heuristic evaluation is an advantageous usability evaluation technique in relation to time and cost, even though, generic sets widely used by experts are not able to provide relevant outcomes in specific domains such as m-learning applications. Nevertheless, in the literature, there cannot be found a developed and validated heuristic set specific for heuristic evaluation on this order of interfaces and, therefore, this study proposed a comparative analysis of two instruments that, to some extent, are engaged in this sense: the heuristics set for evaluation on e-learning domain developed by Mehlenbacher et al. (2005), which has been referred to as HS1, and Neto and Pimentel (2013)'s heuristic set for evaluation on mobile applications, here entitled HS2.

It's been concluded that the two sets identified significant usability problems in a m-learning application and are useful for evaluating this type of interface, however, within the limits of this study, HS1 could detect a larger number of usability problems not found by HS2, suggesting a better efficiency. It was also noted that the distribution pattern of heuristics in identifying problems is similar between sets and evaluators and, it is manifested as follows: some heuristics attract a significant number of problems while others attract few or none. Moreover, the analysis of redundancy showed no significant results that would justify the re-adaptation of any of the sets.

In the results concerning the usability problem context, HS1 that was developed to analyze e-learning systems could not detect a significant number of problems closely related to pedagogical aspects, as one would expect. In contrast, even though HS2 identified a smaller number of usability problems, it did well in covering not only problems related to system structure – the context it was developed for – but also those comprising pedagogical aspects. Moreover, analysis comprising severity of the problems presented marked divergence between sets and between evaluators, a condition enhanced by the subjective nature of this aspect and which manifestation did not allow clear interpretation of results.

Finally, in studies engaged with comparing different instruments, it would be expected that, in a certain level, the outcomes leaded to recommendations for the instruments' improvement. Somehow, the instruments involved in the current study were not primarily developed for assessing m-learning systems and, therefore, it's expected that, only to some extent, each of them be able to cover the conditions here controlled. Consequently, it is not the intention of this study to state that the instruments are susceptible of refinements based on the results, knowing it would be risky. Conversely, it's believed that the comparison provided relevant knowledge about usability evaluation on m-learning applications and, specifically, showed that the available instruments for mobile and e-learning domains (that together constitute the m-learning domain) are not sufficient alone to cover this sort of system. Finally, the results conditions strengthened the author's belief that there is a pressing need to develop a specific heuristic set able to comprise m-learning systems, regarding its relevance to current society.

Limitations and Future Studies

It's not out of consideration that, as discussed, the results presented in this study, as well as any interpretation that would be derived from data extractions in comparative analysis, were often related to the limits used to control the variables, to the subjective nature of this kind of evaluation, to the experience of evaluators, and to the small sample of subjects involved. For future studies it can be suggested that a larger number of evaluators be applied and which expertise is focused on both usability and domain's aspects, as these characteristics can enhance the representativeness of the results.

Also, as previously discussed, the current work indicates the need of a study engaged with proposing a specific set for the assessment of m-learning systems, that could also consider, through the process, the heuristics from the heuristic sets here evaluated, once they showed to be efficient in some aspects of the m-learning application analyzed.

REFERENCES

Al-Razgan, M. S., Al-Khalifa, H. S., & Al Shahrani, M. D. (2014). Heuristics for evaluating the usability of mobile launchers for elderly people. *LNCS, 8517*, 415-424.

Alsumait, A., & Al-Osaimi, A. (2010). Usability heuristics evaluation for child E-learning applications. *J. Softw., 5*(6), 654–661. doi:10.4304/jsw.5.6.654-661

Ardito, C., De Marsico, M., Lanzilotti, R., Levialdi, S., Roselli, T., Rossano, V., & Tersigni, M. (2004, May). Usability of e-learning tools. In *Proceedings of the working conference on Advanced visual interfaces*. Gallipoli, Italy: ACM. doi:10.1145/989863.989873

Bertini, E., Gabrielli, S., & Kimani, S. (2006). Appropriating and assessing heuristics for mobile computing. In *Proceedings of the Working Conference on Advanced Visual Interfaces*. Venezia, Italy: ACM. doi:10.1145/1133265.1133291

Cockton, G., & Woolrych, A. (2001). Understanding inspection methods: lessons from an assessment of heuristic evaluation. In People and Computers XV. Interaction without Frontiers, 171-191. doi:10.1007/978-1-4471-0353-0_11

Costabile, M. F., De Marsico, M., Lanzilotti, R., Plantamura, V. L., & Roselli, R. (2005). On the usability evaluation of E-learning application. In *Proceedings of the 38th Hawaii International Conference on System Sciences*. IEEE. doi:10.1109/HICSS.2005.468

Dringus, L. P., & Cohen, M. S. (2005). An adaptable usability heuristic checklist for online courses. In *Proceedings of 35th Annual Conference Frontiers in Education*. IEEE. doi:10.1109/FIE.2005.1611918

Hashemi, M., Azizinezhad, M., Najafi, V., & Nesari, A. J. (2011). What is Mobile Learning? Challenges and Capabilities. *Procedia: Social and Behavioral Sciences, 30*, 2477–2481. doi:10.1016/j.sbspro.2011.10.483

Hermawati, S., & Lawson, G. (2016). Establishing usability heuristics for heuristics evaluation in a specific domain: Is there a consensus? *Applied Ergonomics, 56*, 34-51. doi:10.1016/j.apergo.2015.11.016

Inostroza, R., & Rusu, C. (2014). Mapping usability heuristics and design principles for touchscreen-based mobile devices. In *Proceedings of the 7th Euro American Conference on Telematics and Information Systems*. Valparaiso, Chile: ACM. doi:10.1145/2590651.2590677

Inostroza, R., Rusu, C., Roncagliolo, S., Jimenez, C., & Rusu, V. (2012a). Usability heuristics for touchscreen-based mobile devices. In *Proceedings of the 9th International Conference on Information Technology*. IEEE.

Inostroza, R., Rusu, C., Roncagliolo, S., Jimenez, C., & Rusu, V. (2012b). Usability heuristics validation through empirical evidences: a touchscreen-based mobile devices proposal. In *Proceedings of the 31th International Conference of the Chilean Computer Science Society*. IEEE. doi:10.1109/SCCC.2012.15

Inostroza, R., Rusu, C., Roncagliolo, S., Jimenez, C., & Rusu, V. (2013). Usability heuristics for touchscreen-based mobile devices: update. In *Proceedings of the 1st Chilean Conference of Computer-Human Interaction*. Temuco, Chile: ACM. doi:10.1145/2535597.2535602

Joyce, G., & Lilley, M. (2014). Towards the development of usability heuristics for native smartphone mobile applications. LNCS, 8517, 465-474. doi:10.1007/978-3-319-07668-3_45

Kemp, E. A., Thompson, A. J., & Johnson, R. (2008). Interface evaluation for invisibility and ubiquity e an example from E-learning. In *Proceedings of International Conference on Human-Computer Interaction*. Wellington, New Zealand: ACM.

Kjeldskov, J. (2002). "Just-in-Place" information for mobile device interfaces. In *International Conference on Mobile Human-Computer Interaction*. Pisa, Italy: Springer Berlin Heidelberg.

Korhonen, H., & Koivisto, E. M. I. (2006). Playability heuristics for mobile games. In *Proceedings of Mobile Human-Computer Interactions*. Helsinki, Finland: ACM. doi:10.1145/1152215.1152218

Lee, V., Shineider, H., & Schell, R. (2005). *Aplicações móveis: arquitetura, projeto e desenvolvimento*. São Paulo: Pearson.

Mehlenbacher, B., Bennett, L., Bird, T., Ivey, M., Lucas, J., Morton, J., & Whitman, L. (2005). Usable e-learning: A conceptual model for evaluation and design. *Proceedings of HCI International 2005: 11th International Conference on Human-Computer Interaction, 4*.

Neto, O. M., & Pimentel, M. D. G. (2013). Heuristics for the assessment of interfaces of mobile devices. In *Proceedings of the 19th Brazilian symposium on Multimedia and the web*. Salvador, Brazil: ACM. doi:10.1145/2526188.2526237

Nielsen, J. (1992). Finding usability problems through heuristic evaluation. In *Proceedings of the SIGCHI conference on Human factors in computing systems*. Monterey, CA: ACM.

Nielsen, J. (1994). Heuristic evaluation. In Usability Inspection Methods (pp. 155-163). New York: John Wiley & Sons.

Nielsen, J., & Budiu, E. (2015). *Usabilidade Móvel*. Rio de Janeiro: Elsevier.

Nielsen, J., & Molich, R. (1990). Heuristic evaluation of user interfaces. In *Proceedings of ACM CHI'90 Conference*. Seattle, WA: ACM. doi:10.1145/97243.97281

Omar, H. M., Yusof, Y. H. M., & Sabri, N. M. (2010). Development and potential analysis of heuristic evaluation for courseware. In *Proceedings of the 2nd International Congress on Engineering Education*. IEEE.

Preece, J., Rogers, Y., & Sharp, H. (2002). *Interaction Design: beyond human-computer interaction*. New York: John Wiley & Sons Inc.

Prensky, M. (2001). Digital natives, digital immigrants part 1. *On the horizon, 9*(5), 1–6. doi:10.1108/10748120110424816

Reeves, T. C., Benson, L., Elliott, D., Grant, M., Holschuh, D., Kim, B., & Loh, S. et al. (2002). Usability and Instructional Design Heuristics for E-Learning Evaluation. In *Proceedings of World Conference on Educational Multimedia, Hypermedia and Telecommunications*. Norfolk, VA: AACE.

Rocha, L. C., Andrade, R. M., & Sampaio, A. L. (2014). Heurísticas para avaliar a usabilidade de aplicações móveis: estudo de caso para aulas de campo em Geologia. *Proceedings of XIX Conferência Internacional sobre Informática na Educação.*

Rubin, J., & Chisnell, D. (2008). *Handbook of usability testing: how to plan, design and conduct effective tests.* John Wiley & Sons.

Salles Junior, F. M., Pinho, A. L., Santa Rosa, G. J., & Ramos, M. A. (2016). Pedagogical usability: a theoretical essay for e-learning. *Holos, 32*(1), 3-15. doi:10.15628/holos.2016.2593

Somro, S., Ahmad, W. F. W., & Sulaiman, S. (2012). A preliminary study on heuristics for mobile games. In *Proceedings of International Conference on Computer & Information Science.* IEEE. doi:10.1109/ICCISci.2012.6297177

Stone, D., Jarret, C., Woodroffe, M., & Minocha, S. (2005). *User interface design and evaluation.* San Francisco: Elsevier.

Väänänen-Vainio-Mattila, K., & Wäljas, M. (2009). Developing an expert evaluation method for user experience of cross-platform web services. In *Proceedings of MindTrek.* Tampere, Finland: ACM. doi:10.1145/1621841.1621871

Zaharias, P., & Koutsabasis, P. (2012). Heuristic evaluation of e-learning courses: A comparative analysis of two e-learning heuristic sets. *Campus-Wide Information Systems, 29*(1), 45–60. doi:10.1108/10650741211192046

Zhang, D., & Adipat, B. (2005). Challenges, methodologies, and issues in the usability testing of mobile application. *International Journal of Human-Computer Interaction, 18*(3), 293–308. doi:10.1207/s15327590ijhc1803_3

ENDNOTES

[1] A particular type of Virtual Learning Environment comprising courses offered by the most celebrated universities around the world which is mainly characterized by active engagement of a massive number of students. It's best known for stimulating the breakdown of the traditional learning standards.

Chapter 4
Mobile News Apps in India:
Relocating News in the Mobile Platform

Saayan Chattopadhyay
University of Calcutta, India

ABSTRACT

This chapter on the consumption of mobile news in developing country draws on the limited but growing scholarship on journalism and mobile media. India, becomes an emblematic instance, as India's mobile phone subscriber base peaked to more than 1 billion users in late 2015, making India the second largest mobile phone user base. In recent years, mobile media in India have also penetrated individuals' news consumption and sharing behaviors. These emerging practices can be posited in relation to the ubiquitous presence of mobile devices and a steadily expanding digital ecosystem. Utilizing both quantitative and qualitative methods, this chapter seeks to explore how mobile apps position themselves into wider news media assemblages in a developing country like India and what are the factors that influence mobile apps usage for news consumption in India? Hence, broadly, the article aims to explore, how these emerging practices are transforming not only the dominant ways of distributing the news but also the very nature of the relationship between news media and its audience.

INTRODUCTION

The multipurpose mobile phone represents an evolving but inadequately regulated medium for distributing news and information in developing countries like India. Although, there are a number of studies that explore the use of mobile phones for news consumption in other Asian countries, there are hardly any studies that explored the use of mobile apps for news consumption in the Indian context (Chan, 2015; Li, 2013; Wei, R., et al., 2013). In regard to everyday news consumption, news consumers are alternating between several online platforms, and mobile media have gained significant importance as a rich site for studies (PEW, 2010; Chan-Olmsted et al., 2013). Bearing in mind, their extensive ubiquity, mobile news media allow audiences to access news content almost anywhere, anytime, and in various formats (Dimmick et al., 2011). In the process, the news content becomes an adaptable commodity that is no longer associated with a particular media format (Hartmann, 2013). Parallel to multiplatform commercial tactics,

DOI: 10.4018/978-1-5225-2469-4.ch004

news items are constantly modified so that it may reach hitherto non-distributed audiences (Westlund, 2013). While the news audience is witnessing an array of options, they are also expected to continuously navigate within a growing digital media ecosystem. Presently, even in developing countries, emerging practices of accessing news is no longer an inconspicuous preference between legacy news media and new digital media. Rather, the individuals consciously create a blend of different news access points into multifarious arrangements of media use (Yuan, 2011). Thus the conflicting relation between legacy media and new media is increasingly getting blurred, both in terms of the use of technology (choice between newspaper and mobile news apps) and preference of content (established news media house vs. alternative/independent news media), as these emerging practices and usage patterns become more commonplace. News audiences in different countries are thus making their way through the assortment of news messages across print, broadcasting, web, and mobile media platforms (Schroder, 2014). This study seeks to invest in the understanding of the emerging practices and patterns of consuming mobile news in developing nation-states.

In recent years, mobile media in India have also penetrated individuals' news consumption and sharing behaviors. These emerging practices can be posited in relation to the ubiquitous presence of mobile devices and a steadily expanding digital ecosystem that have produced a set of conditions in which users rapidly acquire a set of behaviors connected to the medium, effectually incorporating mobile apps within the digital media ecosystem of news media consumption. Within this context, this chapter aims to explore mobile news media consumption practices in India in response to two essential questions: firstly, how mobile apps position themselves into wider news media ecosystem in a developing country? and second: what are the factors that influence mobile apps usage for news consumption in India? The following section briefly outlines the patterns of mobile news consumption with reference to existing literature; the second section describes the Indian scenario in relation to rapidly expanding mobile phone culture and a thriving app ecosystem in India; the third section elucidates the method employed for this study and the following sections comprise of detailed discussion of the main findings; limitations and future scope of research is discussed in the concluding section.

MOBILE NEWS AND THE EXPANDING NEWS MEDIA ECOSYSTEM

Even though the personal computer or laptop continues to be the primary digital device for accessing news, but the most important recent development is the increasing access through multiple devices (Schrøder & Christian, 2010). Hence, the question (RQ1) remains, how mobile apps locate themselves into the broader spectrum of news media ecosystem in a developing country? Besides the general understanding of mobile news consumption it is crucial to question (RQ2) what are the factors that impact the usage of mobile news apps for news consumption? This is particularly pertinent considering ubiquitous attributes of the mobile phone, contesting conventional news audience research.

The practice of reading news through mobile phones have greatly influenced the ways audience usually access news in traditional media (Peters, 2012). Presently, in different national contexts, a number of news media are competing with each other for audience's attention. This condition may remind of niche theory, which argues every medium need to offer specific prospects of gratification so that it may sustain its niche position (Dimmick et al, 2011). In other words, if a particular medium has to survive in this highly competitive market it has to provide distinctive benefits. Thus, for every audience member, media have to strive to cohabit and in the process an ecosystem of media access points evolve (Feaster, 2009).

Moreover, studies particularly focusing on niche news media in developed Asian countries showed that news media may eat into one another's share, which is also applicable to mobile media (Sarrina Li, 2001). Similarly, scholars argue that although mobile media is expected to provide a potentially complementary medium for distribution of news but there is always the threat of "displacing effects" (Nel & Westlund, 2012, p. 743). However, studies on mobile news media also points to an evident synergy.

According to the Reuters Institute for the Study of Journalism, growing use of smartphones among news audience does not necessarily translate into decreasing engagement with other media, rather it expands the scope of choices (Newman & A.L. Levy, 2013). Hence, what is useful is to find out patterns of multiple media use, rather than a particular media selection (Van Damme et al., 2015). In this densely populated media ecosystem the objective should be to study the amalgamation of diverse practices of news media use, rather than looking to the options discretely, primarily because the news audiences are "inherently cross-media" (Thorson et al., 2015, p.164).

Traditionally, news invested in the shaping of household routines. News is generally accessed in a comparatively settled spatial and social arrangement. While, newspapers were habitually read in the morning, predominantly during breakfast, television news were watched in the evening within the comfort of the living room. The "disentanglement of object and context" is caused by the improved degrees of autonomy, and the occurrence of media saturation (Picone et al., 2014). As studies show, the everyday habit of media consumption obtains its meaning from the intricate triadic interaction of media texts, their affording media technologies, and the socio-spatial context in within which they are surrounded. Such conceptions can be useful to understand multi-platform news consumption and for interpreting different news media ecosystem through the study of common practices (Taneja et al., 2012). Importantly, it also underscore the fact that at any such act of media consumption interacts with the dynamic notions of context, comprising of its spatial, technological, and socio-cultural features (Schrøder, 2014).

Against this background, to reiterate, the main objective of this chapter is twofold. Firstly, it aims to gain insight in how mobile apps, in the context of a developing country like India, fit into broader news media ecosystem. Secondly, the study intends to understand what the factors that affect mobile news app access, within these consumption practices.

MOBILE CULTURE IN INDIA

However, at the onset it is necessary to discuss briefly the specificity of India. For India, the mobile phone swiftly turned out to be a necessity for billions of people, the single largest category of consumer goods in the country (Doron & Jeffery, 2013). In 2011, census revealed nearly half of India's 1.2 billion people (46.9%) do not have toilet at home, but 53.2% own a mobile phone (BBC, 2012). However, in India, because of enduring prejudices and structures of authority, the mobile has proved even more disruptive than elsewhere (Doron & Jeffery, 2013).

Getting a telephone connection, even in early 1990s was a troublesome task (Kumar, & Amos, 2006). However, the scenario improved quickly and, according to Telecom Regulatory Authority of India (TRAI), the total number of telephones in the country stands at 1035.18 million and the total numbers of mobile phone subscribers have reached 1009.46 million as of May 2015 (TRAI, 2016). In the mid-1990s mobile phone was first introduced in the Indian market. Although since early 2000 multimedia phones were already existing in India but with the launch of Android's open source platform in 2010, the popularity of smartphones began to rise in India. In 2016, India became the second largest smartphone market in terms of active unique smartphone users, crossing 220 million users, outshining the US market (TH, 2016). Although more than eighty-one percent people, in India, own a mobile phone but a mere seventeen percent people use smart phones (Poushter, 2016). Smartphone penetration in India is presently low, nonetheless smartphones are already the most popular computing platform in India. However, India is considered as one of the fastest growing markets for smartphones, and in the Indian metropolitan cities, more than one in five people now carry a smartphone (Singh & Pant, 2014). Smartphones have steadily increased the mobile internet subscribers as well, and according to report by IAMAI and KPMG, India will reach 236 million mobile Internet users by 2016, and 314 million by 2017 (IE, 2015).

In relation to mobile apps, it is important to note, the app ecosystem in India has been able to flourish mainly due to three interconnected factors: firstly, the prices of mobile devices are constantly plummeting, as presently in India smartphones can be bought for less than $15; second, mobile internet data plans are offering greater bandwidth under a highly competitive price war; and third, there is an growing demand for mobile ready regional content. This has resulted in the growth of a flourishing developer culture sustained by promising app entrepreneurs in India, making India as one of the top countries across app stores. India has also emerged as the second largest user base of Android users in the world (Chattopadhyay, 2015).

It is worth noting that smartphone owners in India, according to report, are least likely to frequently access all categories of apps, but the most popular tend to be more entertainment driven, like social networking and games. According to survey, thirty-nine percent Indian app users use mobile apps for gaming, while twenty-nine percent use apps for social networking; the same percentage also use apps for watching online videos and only thirteen percent access news using mobile apps in India (Nielsen, 2013). Thus, India can be a useful site to explore in what way emerging practices in a developing nation are transforming not only the prevailing ways of delivering the news but also the very nature of the relation between news media and its audience.

METHOD

For the purpose of this study, news consumption means any interaction with news items, including conscious news accessing and unintentional connection; for example, popup notification. In the context of this research, news is defined in the widest possible meaning, including hard news, soft news and service news. This also comprises of editorial news, micro-news or short news updates. Utilizing both quantitative and qualitative methods, this chapter seeks to explore the emerging practices of using mobile news apps in response to wider news media assemblages in a developing country like India.

To explore a range of issues related to mobile news apps and consumption of mobile news in a developing country a qualitative research method is used. Semi-structured interviews with 20 adults living in the Kolkata and its suburban area were conducted. The sample involved a range of ages from 18 to 70 (M= 30.7, SD= 11.3). As an exploratory study, participants were selected to generate an equal distribution among the parameters. Interview questions were designed to explore everyday practices of mobile news consumption. All the interviews were firstly recorded and then transcribed for the study. To gain insight in the mobile news practices, questions on the use of mobile news apps, access habits, and news media preferences and other news apps related practices were asked. The purpose of the interview was to understand, through which mobile app, in what circumstances, and with what motivation the audience consume news. The data was broadly coded and organized manually into thematic frames to check for evolving patterns in a critical manner. The coding of empirical data was done to reveal connections between the mobile news apps, news consumption practices and socio-economic and techno-cultural factors.

Further, to empirically substantiate the research questions, a survey was conducted among the participants (N= 380) who are owners of a smartphone, phablet or tablet and lived in different regions within Kolkata district, eastern India. The sample used in this study was a non-probability convenience sample, where participants were recruited on the basis of convenience in terms of availability, reach and accessibility. Of the 826 individuals who responded to the initial questionnaire, 713 were suitable according to the criteria and were asked to participate in the main research. Of those who were requested, 416 (58.3%) completed the main research procedure. A number of cases were eliminated due to problems of clearly understanding the responses or concern about duplication. A section of the sample (8.6%) refused to state specific information requested in the questionnaire, hence, the final sample used for this study included 380 respondents. The self-administered questionnaires were distributed during a twenty four-week period from February 2015 to July 2015. Respondents were assured of anonymity and participation was entirely voluntary. The questionnaire was distributed both online and on paper, and it comprised of specific questions on the time and location specific mobile apps use, social aspects of mobile apps, monthly expenses of mobile data, preferences for news consumption etc. Among the participants who completed the survey, 74% were male. The participants had the median age of 36 (SD= 11.35) and were 83.4% urban residents.

FINDINGS

All of the respondents were mobile phone users, and among them, 83.9% owned only smart phone or phablet, 11.3% used only tablet and 54.3% used both. 51.3% of the participants said they accessed news on these platforms several times a day, while 23.4% of the participants revealed that they edited the list of news topics or sources they followed on theses platforms every few months. Also 18.7% of the participants said they clicked on links to related news content on these devices several times a day and 33.4% claimed to do it about once a day. In terms of general news content, topics on breaking news (sometimes termed as 'Spotlight') (83.2%), national (80.4%) and entertainment (61.2%) news constituted the main types of news for the participants. Facebook was the most popular mobile social media platform used for getting micro-news, followed by Twitter and messaging app WhatsApp.

In response to the first research question: how mobile apps position themselves into wider news media ecosystem in a developing country, the study revealed that the majority of audience (84%) prefer news items on their mobile phone from established news brands. The Times of India (TOI) news app (76.3%), NDTV (64%), AajTak (53.2%) remained the most popular news apps from established media houses, while Google News & Weather (83%) DailyHunt (76%), NewsDog (73.4%), and NewsRepublicIndia (71%) were the most used news aggregator apps.

In the Indian context, news audience have a unique relation with the news aggregator apps. While only 11% of individuals actually installed news aggregator apps such as Flipboard, DailyHunt, NewsDog or NewsPoint but 63.2% people use news aggregators, primarily Google news & weather app or Google Play Newsstand, or any other apps as they come as bloatware, which means they are pre-installed app that cannot be uninstalled. Hence, the use of news aggregator apps in India seems to be driven by the standard software that hardware makers install onto their devices along with the operating system.

However, when it comes to choosing news source within the news aggregator app, there is an overwhelming preference towards established news media brands. This corroborates the argument that because of the news diversity and fragmentation, trust in news media is rested on the most established, most visible news brands (Starkey, 2013). Moreover, a number of participants expressed their difficulty in actually editing and configuring the news aggregator apps according to their preference and hence selecting particular news apps that require less configurations.

Two specific trends which are in line with developed countries are the use of service based news and preference for short-news or micro-news. Service-based news (mainly weather information in the Indian context) is substantially consumed more (96.4%) through mobile applications. A number of news audience claim that it is much easier to check the weather forecast through mobile news app. Moreover, the automatic notification option often let the user know about the weather irrespective of his or her interest. Similarly, there is a significant inclination towards micro-news or short-news (89.1%) in relation to accessing news through news media apps. One participant explained the reason for getting micro-news: "This news app presents the news within one or two lines. I can definitely get more if I tap on the link but I prefer to read the gist. It saves time and keeps me interested."

When it comes to searching for a particular news, browser based searching is more commonplace (73.2%) than app based search among Indian news consumers. The easy accessibility of pre-installed Google search widget in most of the standard android phones in India has motivated news audiences to use Google search for a particular news topic rather than searching through the news app's search option.

The practice of news sharing is predominately dominated by patterns of social media use, however in the context of the present study it is evident that there is lack of interest in sharing news content through mobile news apps, as only 7.4% admitted to have shared news content at least once in six months. However, news accessed through social media apps like Facebook, Twitter and popular messaging app like WhatsApp are shared more frequently. Although, this trend indeed merit a separate analysis but a possible explanation is that the process of sharing in social media app is much easier (often through a single click), while the news app's share button only links it to the installed social media app or it opens a new browser window with the link. Hence, it is perhaps not surprising that 47% participants admitted that they have never shared a single news item either through specific news apps or news aggregator apps.

Language becomes a crucial issue when it comes to mobile news apps, especially in a multilingual country like India, which is also the country with the largest English-as-second language population (TOI, 2010). In India, English language content accounts for 56% of the content on the Internet, while Indian languages account for less than 0.1%. According to estimates, currently, about 100-120 million, accounting a 10% of India's population, is comfortable consuming content in English (Mallya, 2015). This has recently opened doors for new market opportunities for mobile news app developers. News aggregators such as InShorts, DailyHunt and vernacular news media brands like ABP, Aajtak, Navbharat Times, Dainik Bhaskar are focusing on the vernacular Indian language sector. In this study, only 9.4% participants claimed to use a news app in vernacular language, although 57.2% participants said they would prefer to use a vernacular news app but have not downloaded it, primarily because of lack of information and limited options. Although android has drastically improved its Indian language support in 2014 and major keyboard improvements are also being made regularly so that app developers may tap into the market with a strong vernacular strategy (Ruddra, 2015).

In response to the second research question— what are the factors that influence mobile apps usage for news consumption— in line with audience-centred research on media consumption, the factors impacting news app usage can be divided into five interrelated trajectories: Spatiality, Temporality,

Individuality, and Affordability and Capability. These factors are derived from the data by manually coding and organizing the interview responses into thematic frames to check for evolving patterns in a critical manner. Each factor has a possible influence on the selection and usage pattern of mobile news apps or news type. Although the choice for news source and news media brand remain largely traditional, mobility relocates the usage of these apps into an entirely variable context.

Spatiality

The participants claim that their news consumption practices through mobile apps is largely contingent on the spatial conditions of the news consumption, as it influences the need and accessibility of the news. While 67.3% use mobile apps to access news in the morning, primarily on the way to work, 58% use mobile to access news in the evening, mainly on their way back home. However, 24.7% use mobile to access news in the afternoon or post-lunch period and 32.1% use mobile to access news at night at home. Interestingly, relatively high use of mobile apps during the morning is driven by service based news as it is usually delivered linked with news apps on android devices. Mobile is largely used for games and entertainment during the afternoon and evening period. The spatial conditions also influence screen size choice, as a participant commented: "At home I usually use the laptop or tablet to read news as I don't have a large screen phone. However, at office I use my phone, as it is more convenient". While commuting, the mobile phone is the only available device and hence mobile apps are used to access news. On public transportation, mobile app is used for news access more than browser based news access since it is easier to navigate the app interface in crowded train or bus. As another participant explained: "While I am traveling on bus, during office hours, it is much convenient to go through the news just by sliding my fingers, instead of typing words and sentences". Thus while digital media and particularly mobile phone offer the freedom to wander and is often referred to as a detachment from physical space and time, but the spatial conditions evidently effect the use of mobile apps and devices.

Temporality

The second factor, temporality, is characterized by an interest in engaging with news as it unfolds. A number news audience (46.3%), who participated in this study stated that they feel the need to continuously stay updated by checking the news app: "Whenever there is a breaking news, I feel the urge to constantly check my Google news app for latest updates. Even at home, I check the news app for updates and let others know." Particularly, in relation to sports news, disaster news and breaking news the news app plays a crucial role in keeping the news audiences updated. However, local political and crime news is also 'followed' through mobile apps. This temporal sensitivity can be a mark of serious news audiences but the narrowing and the urgency of the time frame need to be emphasized here. The importance accorded to these temporal cycles, points to the newer patterns of the consumption of news as such, and the anxious browsing of the news updates. According to one participant, "Why would I install a news app, unless I want to check the news at regular intervals? It's not newspaper. I don't have to wait 24 hours for it to change". Hence, temporality is not merely a trait but becomes the defining factor of news consumption as it offers provisional autonomy of a temporal form. Thus it can be said that the functions of mobile news apps are irreducibly temporal.

Individuality

Based on the interviews with news audience, results show that individuality has a significant impact on the consumption of news. Generally mobile news is considered to be consumed personally, even in public places and with others around. One of the participants said, "My app is personalized for me, so I check the news personally even when my family members are present with me". She hereby points out the relationship between personalization and individuality in the context of mobile news consumption. While entertainment apps are sometimes consumed collectively, the news apps are hardly ever (0.3%) used with others. Although laptops and tablets are often shared with family members and friends at home and workspace but the main variable here is the screen size and choice of app. As 21% participants said that they only share their news app with others when it contains a video. Thus individuality does influence what kind of app is being consumed, (for example, news app with videos and large images) and also it influences the selection of devices. Another emerging aspect is the use of multiple screens and viewer's participation; as some of the news apps offer a way of sending comments visible on live tickers on screen of television news programs, thus providing the audience a news experience that legacy news media cannot offer. Although, most of the time social media apps are used more to send these comments but a limited number of news audiences are enthusiastic about news apps that let them send comments to live news and panel discussion shows on television. As they claims, "being able to send my personal comments through my mobile phone while watching a newscast on television gives me a sense of empowerment".

Affordability

Analyzing the economic aspects of using mobile news apps in a developing country, it becomes clear that affordability becomes an important, influential factor. Unlike device costs, mobile Internet data costs in India have been swiftly rising over the years due to the change in dynamics of Indian telecom market after the 2G spectrum scam. "This is a hundred and fifty percent price hike over a period of six years" (Damani, 2016). However, news audience tend to utilize the free Wi-Fi services available at office and public spaces. Participants, admitted to turn off their mobile data and turn on Wi-Fi in their devices whenever they entered such a Wi-Fi zone. This shift is very recent as free Wi-Fi hotspots in eastern India is still a new phenomenon, however, free wireless Internet at the workplace is relatively common. However, it is usually limited to information technology sector. 73.4% participants claimed that when they are consuming mobile data they prefer short news updates or micro-news but while using free Wi-Fi they also access news videos and image based news apps. At home, those who have broadband Internet connections also typically use router for Wi-Fi connection and prefer to use data heavy news apps on that connection. "If I am using my mobile data then I only check the news headline. On a WiFi connection I prefer to read articles, and watch videos. Sometimes I add an interesting news content to 'watch-later' tab so that I can watch it whenever I find a WiFi service". Thus affordability plays a critical role in deciding which news app would be used and what kind of news consumption would take place.

Capability

The fifth factor influencing mobile news app consumption is the technological capability or aptitude of the news audience in a developing country. While gender, educational level and age plays a crucial role in the degree of capability of using digital media devices, it becomes a pivotal factor in a postcolonial, developing country like India. Drawing from the fieldwork more broadly, a significant number of participants (33.1%) expressed their inability to navigate through most of the news apps. It is important to remember, how technologies often fabricate gendered hierarchies, through their design, accessibility, and patterns of use. While news apps are installed independently by many, the refresh rate, news topic selection and data management options remain either untouched or they are edited with consultation from younger, male individuals. Participants in the age group of above 35 have repeatedly articulated that they have to seek help, not only in using mobile news app, but in different aspects of a smart phone use. While this study took place in a predominantly urban region the capability factor is bound to be more instrumental among less educated and non-English speaking population in rural India.

DISCUSSION

As stated in the introduction, the main purpose of this study is twofold. First, this study sought to investigate how mobile apps position themselves into wider news media ecosystem in a developing country. In relation to the first research question it is revealed that firstly, the majority of news audience favor news items on their mobile phone from established news brands; second, the use of news aggregator apps in India seems to be propelled by bloatware; third, similar to developed countries, service based news and short-news or micro-news is preferred among news app consumers; fourth, among Indian news audience, browser based news search is more commonplace than app based search; fifth, there is a marked lack of interest in sharing news content through mobile news apps, and finally, significant number of news audience are interested in using a vernacular news app but have not used it, primarily because of lack of information and limited options. These findings point out how the usage of these mobile news apps are linked to one another, in addition to the legacy news media in the wider field of using and managing news media in India. The findings emphasized the fact that these emerging practices are transforming not only the dominant ways of distributing the news but also the very nature of the relationship between news media and its audience.

Second, the study attempted to explore the factors that influence mobile apps usage for news consumption in India. In other words, the objective of this study was to understand how the appropriation of these factors in everyday use of mobile news apps relates to emerging practices of news media consumption. Based on the fieldwork and survey results, five different factors are identified that affect the mobile news app consumption in a developing country. Temporality, is characterized by an interest in engaging with news as it unfolds. It is not simply a trait rather a defining factor of news consumption as it offers provisional autonomy. Spatiality, is characterized by dependence on the spatial conditions of the news consumption, as it influences the need and accessibility of the news. Individuality, is marked by the practice of consuming mobile news independently, even with friends and relatives around; this also points to the relationship between personalization and individuality in the context of mobile news consumption. Affordability, as a critical factor in a developing country, shapes the decision about which news app to be used and what kind of news consumption may take place. Finally, capability underlines the technological aptitude of the news audience in a developing country. Each factor has a formative impact on the selection and usage pattern of mobile news apps or news type. Even though, the preference of news media brands remains mostly traditional, these factors shape the engagement with these apps within an entirely adaptable context. These findings points to the increasingly fragmented use patterns of news apps in different contexts and by different users. Different influential factors associated with the mobility of mobile media also contribute in the fragmented consumption patterns in developing country.

CONCLUSION

The mobile phone has turned out to be a popular platform to deliver news and information even in societies, where the technology is still in its early stages of adoption. Overall the findings suggest that mobile news app usage in India is steadily growing irrespective of the fact that there is low penetration of smart phones and increasing mobile data cost. In addition, there is lack of general awareness regarding different mobile apps. Nevertheless, the usage of mobile news may rise among this generation as their familiarity with accessing news via the mobile apps develops and they acquire a mobile phone-oriented news-access routine. The study tried to examine the emerging practices of news app use in a setting that endures economic, technological and infrastructural constraints.

However, it is necessary to mention some of the limitations of this study. The survey sample is not a nationally representative sample, as the representativeness is constrained by the fairly small number of participants, who were residing in the same geographic region (eastern India) at the time that the surveys were conducted. It is probable that these specific attributes of the sample affected the responses and hence a random sample drawn from the national population strata would have produced different results. Hence, this study should be considered as exploratory in nature. Because this study only examined mobile news use in Kolkata, eastern India, it would be difficult to generalize the findings in the context of other developing countries. Hence, cross-cultural studies involving other developing nations can be helpful to underline important factors that influence the use of mobile news apps. Moreover, the study only focused on smart phone users in predominantly urban region, whereas the number of basic feature phone users are substantial. Hence, future study may also include basic phones and look at rural news audience to conduct a comparative study, which would provide an even clearer picture of mobile news usage in developing nations. Another limitation of this study is that the data is not generated through user independent logging software, instead the study is based on information reported by participants. The validity of such data may be low and may not correctly signify news app usage. Finally, the technological capability aspect discussed in this study remains a less developed section. Future studies may also look into factors related to age, educational level and gender as they impact mobile news app use in less developed countries.

Broadly, the present study shows, while the preference for news media brands remains mostly traditional, several critical factors shape the use of these news apps within a completely flexible context, resulting in increasingly fragmented use patterns of news apps by different users. The findings underscores that these evolving practices are not only altering the dominant ways of news dissemination but also the dynamics of the interactions between news media, technology and its audience.

REFERENCES

BBC. (2012, March 14). *India census: Half of homes have phones but no toilets*. Retrieved from http://www.bbc.com/news/world-asia-india-17362837

Chan, M. (2015). Examining the Influences of news use patterns, motivations, and age cohort on mobile news use: The case of Hong Kong. *Mobile Media & Communication, 3*(2), 179–195. doi:10.1177/2050157914550663

Chan-Olmsted, S., Rim, H., & Zerba, A. (2013). Mobile news adoption among young adults: Examining the roles of perceptions, news consumption, and media usage. *Journalism & Mass Communication Quarterly, 90*(1), 126–147. doi:10.1177/1077699012468742

Chattopadhyay, S. (2015). Mobile news culture: news apps, journalistic practices and the 2014 Indian General Election. In E. Thorsen & C. Sreedharan (Eds.), *India Election 2014: First Reflections* (pp. 143–157). The Centre for the Study of Journalism, Culture and Community: Bournemouth University.

Damani, V. (2016, May 15). *What's Holding Back Mobile Data Uptake in India, and How Will It Change?* Retrieved from http://gadgets.ndtv.com/telecom/features/whats-holding-back-mobile-data-uptake-in-india-and-how-will-it-change-818783

Dimmick, J., Feaster, J. C., & Hoplamazian, G. J. (2011). News in the interstices: The niches of mobile media in space and time. *New Media & Society, 13*(1), 23–39. doi:10.1177/1461444810363452

Dimmick, J., Powers, A., Mwangi, S., & Stoycheff, E. (2011). The fragmenting mass media marketplace. In W. Lowrey & P. J. Gade (Eds.), *Changing the news. The forces shaping journalism in uncertain times* (pp. 177–192). London, UK: Routledge.

Doron, A., & Jeffery, R. (2013). *The Great Indian Phone Book: How the Cheap Cell Phone Changes Business, Politics, and Daily Life.* Cambridge, MA: Harvard University Press. doi:10.4159/harvard.9780674074248

Feaster, J. C. (2009). The repertoire niches of interpersonal media: Competition and coexistence at the level of the individual. *New Media & Society, 11*(6), 965–984. doi:10.1177/1461444809336549

Hartmann, M. (2013). From domestication to mediated mobilism. *Mobile Media & Communication, 1*(1), 42–49. doi:10.1177/2050157912464487

Kumar, K. J., & Amos, T. (2006). Telecommunications and development: The cellular mobile revolution in India and China. *Journal of Creative Communications, 1*(3), 297–309. doi:10.1177/097325860600100306

Li, X. (2013). Innovativeness, personal initiative, news affinity and news utility as predictors of the use of mobile phones as news devices. *Chinese Journal of Communication, 6*(3), 350–373. doi:10.1080/17544750.2013.789429

Mallya, H. (2015, Nov 26). *Is regional language content the next frontier to reach India's 1.2 billion people?* Retrieved from http://yourstory.com/2015/11/news-aggregators-vernacular/

Nel, F., & Westlund, O. (2012). The 4 Cs of mobile news. Channels, conversation, content and commerce. *Journalism Practice, 6*(5–6), 744–753. doi:10.1080/17512786.2012.667278

Newman, N., & Levy, A. L. D. (Eds.). (2013). Reuters Institute digital news report 2013: Tracking the future of news. Oxford, UK: Reuters Institute for the Study of Journalism.

Nielsen. (2013). *The Mobile Consumer: A Global Snapshot.* Retrieved from http://www.nielsen.com

Peters, C. (2012). Journalism to go. *Journalism Studies*, *13*(5-6), 695–705. doi:10.1080/1461670X.2012.662405

Pew Research Centre. (2010, March 15). *The state of the news media 2010: Audience behavior*. Retrieved from http://stateofthemedia.org/2010/online-summary-essay/audience-behavior/

Picone, I., Courtois, C., & Paulussen, S. (2014). When news is everywhere. Understanding participation, cross-mediality and mobility in journalism from a radical user perspective. *Journalism Practice*, *9*(1), 35–49. doi:10.1080/17512786.2014.928464

Poushter, J. (2016, Feb 22). Smartphone Ownership and Internet Usage Continues to Climb in Emerging Economies. *Pew Research Center*. Retrieved from http://www.pewglobal.org/2016/02/22/smartphone-ownership-and-internet-usage-continues-to-climb-in-emerging-economies/

Ruddra, A. (2015, Feb 12). *Why Do We Make Apps in English in India?* Retrieved from http://anshumaniruddra.com/2015/02/11/make-apps-english-india/

Sarrina Li, S. C. (2001). New media and market competition: A niche analysis of television news, electronic news, and newspaper news in Taiwan. *Journal of Broadcasting & Electronic Media*, *45*(2), 259–276. doi:10.1207/s15506878jobem4502_4

Schrøder, K. C. (2014). News media old and new. *Journalism Studies*, *16*(1), 60–78. doi:10.1080/1461670X.2014.890332

Schrøder, K. C., & Christian, K. (2010). Towards a typology of cross-media news consumption: A qualitative-quantitative synthesis. *Northern Lights*, *8*(1), 115–137. doi:10.1386/nl.8.115_1

Singh, P., & Pant, S. (2014, Jan 24). *Unstoppable! Smartphone surge in India continues*. Retrieved from http://www.nielsen.com

Starkey, G. (2013). Trust in the diverging, convergent multi-platform media environment. *Communication Management Quarterly*, *26*(26), 73–98. doi:10.5937/comman1326073S

Taneja, H., Webster, J. G., Malthouse, E. C., & Ksiazek, T. B. (2012). Media consumption across platforms: Identifying user-defined repertoires. *New Media & Society*, *14*(6), 951–968. doi:10.1177/1461444811436146

The Hindu. (2016, Feb. 3). *With 220mn users, India is now world's second-biggest smartphone market*. Retrieved from http://www.thehindu.com/news/cities/mumbai/business/article8186543.ece

IE The Indian Express. (2015, July 21). *IAMAI says India will have 500 Million Internet users by2017*. Retrieved from http://indianexpress.com/article/technology/tech-news-technology/iamai-says-india-to-have-236-million-mobile-internet-users-by-2016/

The Times of India. (2010, March 14). *Indiaspeak: English is our 2nd language*. Retrieved from http://timesofindia.indiatimes.com/india/Indiaspeak-English-is-our-2nd-language/articleshow/5680962.cms

Thorson, E., Shoenberger, H., Karaliova, T., Kim, E., & Fidler, R. (2015). News use of mobile media: A contingency model. *Mobile Media & Communication*, *3*(2), 160–178. doi:10.1177/2050157914557692

TRAI. (2016, July 31). *Highlights of Telecom Subscription Data*. Retrieved from http://trai.gov.in/WriteReadData/PressRealease/Document/PR-TSD-Nov-15.pdf

Van Damme, K., Courtois, C., Verbrugge, K., & Marez, L. D. (2015). Whats APPening to news? A mixed-method audience-centered study on mobile news consumption. *Mobile Media & Communication*, *3*(2), 196–213. doi:10.1177/2050157914557691

Wei, R., Lo, V., Xu, X., Chen, K., & Zhang, G. (2013). Predicting mobile news use among college students: The role of press freedom in four Asian cities. *New Media & Society*, *16*(4), 637–654. doi:10.1177/1461444813487963

Westlund, O. (2013). Mobile news: A review and model of journalism in an age of mobile media. *Digital Journalism*, *1*(1), 6–26. doi:10.1080/21670811.2012.740273

Yuan, E. (2011). News consumption across multiple media platforms: A repertoire approach. *Information Communication and Society*, *14*(7), 998–1016. doi:10.1080/1369118X.2010.549235

Chapter 5
Are We Ready to App?
A Study on mHealth Apps, Its Future, and Trends in Malaysia Context

Sharidatul Akma Abu Seman
Universiti Teknologi MARA, Malaysia

Ramayah T.
Universiti Sains Malaysia, Malaysia

ABSTRACT

In Malaysia, adoption of the mobile application for smartphones and tablet computers are growing in number and are actively applied in healthcare. However, limited studies were found looking at mHealth app that is focusing on Malaysia context. This study aims to examine the current mHealth app that is available in Malaysia. This study also seeks to rank the pricing of top paid apps from two major platforms, Apple iOS, and Android PlayStore. In mid-2016, the authors overviewed the Medical app and Health and Fitness category from two dominant platforms; Apple iOS and Android Play Store. The only app that was related to human healthcare, described in Bahasa Malaysia or English, was examined. Most app that is designed specifically for Malaysia is informational apps, which provide information on healthcare and medical information. The study also reveals that most consumers in Malaysia are ready and are willing to pay for mHealth app. Majority of app price is between RM10.01 to RM25.

INTRODUCTION

The evolution of smartphone production has led to an enormous increase in the demand for mobile applications. The simplest tools for individuals who seek a modern lifestyle, mobile devices have become ubiquitous. Clear evidence shows that out of 5.3 billion worldwide mobile phone subscriptions, 3.8 billion (73 percent) are from developing countries (Marshall, Lewis, & Whittaker, 2013). The recent statistic also revealed that the numbers is projected to grow from 2.08 billion in 2016, to be 2.69 billion of users by the year of 2019 (Statista, 2014).

DOI: 10.4018/978-1-5225-2469-4.ch005

A mobile application is defined as a "software program for a computer or phone operating systems" (Chun, Chung, & Shin, 2013). In the healthcare environment, a mobile app also called an mHealth app offers assistance to individuals by providing health info and personal self-administration tools (Handel, 2011). It also acts as an aids to the professionals who engage with patients, expanding the reach of their practices and possibly bringing down health care costs (Boyce, 2014). An mHealth app makes self-monitoring of ailments and illnesses possible (Marshall et al., 2013; Sidney et al., 2012; Cole-Lewis & Kershaw, 2010) in an interactive way (Aitken, 2013). Current statistics (Abt, Bader, & Bonetti, 2012) suggest that there are currently 40,000 available medical apps and that the figure is gradually increasing.

As this ubiquitous technology takes root and grows, little known about how consumers are currently using mobile technologies in healthcare (Atienza & Patrick, 2011), and far too little consideration has been paid to Asian countries, particularly Malaysia. Malaysia is one of the prominent countries in using the mobile technology, yet there has there has been little discussion on the mHealth app. The overall smartphone adoption market in Malaysia is unclear, and the statistically evidences are hardly available (Osman, Zawawi Talib, Sanusi, Yen, & Alwi, 2011). The research to date on mHealth app has tended to focus on a literature review (Hussain et al., 2015), the usage perception by medical students or expertise (Kwee Choy Koh, MMed et al., 2014, Ming et al., 2016), a conceptual framework (Fadzilah & Arshad, 2015), and general smartphone usage. No previous studies have dealt with pricing issues in mHealth app in Malaysia.

This chapter will provide a review of the mHealth app with an emphasis on the Malaysian context. The central purpose of this chapter is to identify the minimum and maximum prices in Malaysia Ringgit, on the most downloaded mHealth apps, as most literature is based on the US dollar. All of apps that were evaluated were selected from Apple iTunes Store and Android Play Store marketplace. Designated platforms are chosen as they were the global leaders in app marketing and distribution (BinDihm, Freeman, & Trevena, 2012). The term "mHealth application" and "mHealth app" will be interchangeably used in this study.

There are two primary aims of this study:

- To identify the mHealth apps that are designed specifically for the Malaysian context and to categorize them based on functionality.
- To analyze the top paid mHealth app in Malaysia from two major platforms, Android Play Store and Apple iOS platform, and to rank pricing in the Malaysian Ringgit (RM).

LITERATURE REVIEWS

Mobile Technologies and Healthcare

Mobile technology is described as "wireless devices and sensors (including mobile phones) that are intended to be worn, carried, or accessed by the person during normal daily activities" (Kumar et al., 2013). First digital mobile phones appeared in the early 1990s, and since then mobile technologies have continued to advance with various features and better networks (Fiordelli, Diviani, & Schulz, 2013).

With the growth in adoption of mobile technologies, the number of health-related applications has also grown. The concept of mHealth refers to the use of "mobile computing, medical sensor, and communication technology for healthcare" (Liu, Zhu, Holroyd, & Seng, 2011). Closely connected to the fields of "mServices" and "eHealth," mHealth is an evolving sub-field and the term is frequently used informally, making it difficult to measure its effectiveness (Mechael, 2009). "mHealth" is a term frequently used in literature, but to date there is no consensus on its definition. Akter, D'Ambra, Ray, & Hani (2016) defined mHealth as "the application of mobile communications such as mobile phones and PDAs – to deliver right time health services to customers (or patients)." A further definition is given by Bull & Ezeanochie (2015), who describe mHealth as "health promotion and disease management programs delivered via text message, social media, and applications (or 'apps') downloaded to phones with programs targeting individuals to change behaviors, and efforts to facilitate provider care delivery."

A survey by the World Health Organization (WHO) found that 83 percent of member reported at least one mHealth initiative in their country, while most of them reported the four or more (Marshall et al., 2013). mHealth is frequently noted as an effective tool for its benefits in managing community healthcare. A key aspect of mHealth is to help communities to make decisions about personal healthcare via real-time monitoring, since it is easily assessable in remote areas where medical facilities are not available (Gleason, 2015). The technology allows patients or practitioners to create, store, retrieve, and transmit data that will potentially automate and expedite the healthcare delivery processes. In addition, the technology helps to reduce costs, spearheading better engagement with patients and more convenience and generating appeal through its unique services (Akter, Ray, & D'Ambra, 2013; Akter & Ray, 2010; Vital Wave Consulting, 2009). For the community, mHealth makes it possible to manage non-critical care, lessening in the hospitalization statistic, enlightening patients quality of life, and saving costs (Norris, Stockdale, & Sharma, 2009).

mHealth Application (mHealth App)

Formerly, mHealth technology provided services in inaccessible areas, but with the evolution of 3G and 4G network technologies, the idea of mHealth has been modernized to be more flexible and thus more widely applicable. mHealth is now a part of a mobile application in smartphone devices, providing personal healthcare and helping patients to achieve a healthy lifestyle. mHealth applications have proven their potential for reducing healthcare expenditures (Kumar et al., 2013) without neglecting transparency and accountability of information (Marshall et al., 2013; Schweitzer & Synowiec, 2012; Mechael, 2009). The application itself creates an opportunity for individuals to handle their healthcare in more manageable and assessable ways while also reducing cost (Shahriar Akter et al., 2012).

Among app categories, two core classes are related to healthcare: Health & Fitness and Medical. Fitness and nutrition were the most common categories of mHealth app used, with most respondents using them at least once per day (Krebs & Duncan, 2015). As of June 2016, Android users were able to choose among 2.2-million apps, while Apple's App Store remained the second- largest app store, with 2-million available apps (Statista, 2016). Apple's platform offers the highest number of health-related apps of any platform (Bidmon, Terlutter, & Röttl, 2014), with more than 160,000 mHealth apps: the Medical category contained 44% of all mHealth apps, while Health & Fitness contained 56% (Research2Guidance, 2015). In 2016, it is expected that mHealth apps will become a $20-billion industry (Fadzilah & Arshad, 2015). It is also predicted that, in 2017, more than 1.7-billion people will be utilizing the healthcare application with projection market revenue of a total of USD 26 billion (Jahns, 2013). In the mobile app market, any developer can label their app as medical; neither Apple nor Google provide criteria to define the term (Moodley, Mangino, & Goff, 2013). Apple charges a $99 developer fee per year, whereas Google charges a one-time $25 developer fee (Viswanathan 2014). In the Google Play Store, the review process is automated, and the app is verified with automated tests for malware, making the app within a few hours available on the store (McIlroy, Ali, & Hassan, 2015). On the other hand, in the Apple Store, the review process involves a manual review of the app, and the process can take several days or even weeks. It is not uncommon for apps to be rejected (McIlroy et al., 2015).

Like other categories of a mobile app in the Apps Store, the monetization of downloaded mobile apps come from a free app, freemium, in-app purchase of premium apps. In general, as for June 2016, the average price of each app and games was one (1) U.S. dollar. A Recent report by AppBrain (2016), on Android platform, the average price for medical paid apps is the most expensive with $10.54. Health & Fitness category is among the lowest price with the mean of $3.57. For others, some are in the low price (e.g personalization $1.89, puzzle $2.08,whether $2.32, action $2.32) while most are on the average (e.g educational $3.12, music $2.96,travel & local $3.67 and social $4.90).

Malaysia, Healthcare, and Technologies

Malaysia has a multicultural population with approximately 30.4-million residents in 2015 with Bumiputeras (i.e. the Malays, Sabah and Sarawak, and the natives of the Landmass) representing 67.4%, Chinese 24.6%, Indians 7.3% and others 0.7% (Mohd-Tahir, Paraidathathu, & Li, 2015). As an upper-middle-income country, Malaysian society has been transformed by rapid economic growth in the latter half of the 20th century (Jaafar et al., 2013). Concerning technological usage, 56% of individuals use computers, 57% of people use the Internet, and 94.2% of individuals have a mobile phone, as of 2013. Among all internet users, 60.3% used at least once a day, 71.2% used at home, 55.3% used via mobile phone, and 37.8% used at the workplace. 59.4% of households have access to a computer, 58.6% have access to the Internet, and 97% have access to a mobile phone. Among ethnic groups in Malaysia, 61% Malays topped the mobile phone usage (Malaysian Communications And Multimedia Commission, 2015)

In general, healthcare in Malaysia is mainly a responsibility of the government's Ministry of Health. As Malaysia tries to achieve high-income nation status, demand for healthcare among the population will continue to rise. The expenditure of the Malaysia healthcare system was nearly RM43 million in the year of 2012 (Ministry of Health Malaysia, 2015). Moreover, in Malaysia Budget 2015, the Malaysian government announced that up to RM 5K had been given to individual taxpayers for medical expenses incurred for the treatment of acute diseases (PWC, 2014).

The Malaysia Government is passionately implementing new technology in the healthcare context. In an example, the Telemedicine Act 1997 is introduced to promote the development of information communication technology (ICT) in the health sector. Jaafar et al., (2013), highlighted the ICT blueprint which consists of three broad initiatives: Multimedia Super Corridor Flagship, Facility-based initiatives, and Program/function-based initiatives. Among them, initiatives related to healthcare are:

- Telehealth,
- PMS (patient management system).
- Facility-based initiatives include rolling out the hospital information system to all hospitals and other national facilities.
- Program/function-based initiatives:
    - Nationwide Health Management Information System,
    - Statewide Teleprimary Care and
    - Public/Client Access (Health online).

Are We Ready to App?

On the mobile application, Malaysia Government has launched the 1Gov Appstore. Under the management of Malaysia Administrative Modernisation and Management Planning Unit (MAMPU), it's one of the innovative initiatives to improve Government Service delivery via mobile devices. Using the "One Stop Centre" concept, Malaysians are free to download all applications provided by the government of Malaysia in the Gallery of Malaysian Government Mobile Application (GAMMA) (MAMPU, 2015). All apps can able to download from App Store, Google Play Store, and Windows Phone Store. Among those that have been developed in healthcare are:

While the mHealth app may have played a vital role in bringing about a new paradigm in healthcare management, evidence of its effectiveness remains unclear. To date, studies investigating mHealth app usage and adoption in Malaysia have produced equivocal results in the most literature. A review by Statista (2016) revealed that among the top 10 of most popular mobile application from both platforms in Malaysia, none of them were health related. Whether Malaysians are ready and prepared to utilize mHealth application in the management of their personal health care or remain with the face-to-face systems, remain unanswerable.

METHODS

Stage 1: Identify mHealth App in Malaysia Context

Data Sources: The Selection Criteria

As mentioned above, among app categories, two core categories are related to healthcare: Health & Fitness and Medical. The search was conducted in June 2016 on Apple IOS and Android Play Store,

Table 1. Mobile applications on healthcare offered by Malaysia government

App Name	App Description	Offered By
MyHealth	Provide information on health facilities, health risk & assessment, list of registered practitioners, tips and FAQ	MAMPU
myMahtas	Provide Information on: • Medical & cosmetic procedures • Regenerative medicines • Pharmaceutical products & more	MAMPU
MyNutriApps II: MyNutriDiari	A tool to manage food intake, to users to self-manage bodies and health and to develop a new healthy mindset and lifestyle especially the new Generation Ys.	Nutrition Division, Ministry of Health, Malaysia
My Blue Book	Help the Medical students and the Medical Professions in achieving a better health care and act as a quick references tool.	Malaysia & Pharmaceutical Services Division MOH
MyFoodSafe	Provide information regarding: • companies products approved with the HaCCP & GMP certification. • Approved list of bottled water & mineral water manufacturers.	MAMPU
iDengue	A platform that is build to provide the public with the latest information on dengue in the country displays daily and cumulative Dengue cases, location for all epidemic and hotspot area.	MAMPU

which are the most prominent platforms in smartphone operating systems. A search query in both of the app categories was conducted with the keyword "Malaysia health" AND "Malaysia healthcare." From the results obtained, we later performed a descriptive analysis of the applications. Each app comes with a name, publisher, price, in-app price information, and privacy policy availability, all of which are captured and recorded.

Inclusion and Exclusion Criteria

The inclusion criteria are base on the scrutiny of the titles and the substances of applications in both app categories. In ensuring that only apps that are related to managing healthcare or that work as healthcare resources, we have set few criteria for app selection. As the goal of this study is to look at the accessibility of mHealth applications in Malaysia, we do not constrain any type or functionality of the chosen mHealth app. However, we unequivocally reject applications that are not related for health-care purposes, for example, shopping, business (except those for insurance, healthcare providers), cooking (non-health related), and restaurants. Only applications in English and Bahasa Malaysia were chosen.

All applications are classified into categories based on roles and functionality. All types of categories were adapted from various resources in the literature (Benferdia & Zakaria, n.d.; Research2Guidance, 2015; Silva, Rodrigues, de la Torre Díez, López-Coronado, & Saleem, 2015). Four new categories were added: healthcare providers, insurance company, health game, and magazine. Although most literature excluded the categories with regards to the mHealth app, as Malaysia is yet at the entrance level, its belief this category may be a beginning step to inspire the users to use the application.

Stage 2: Classifying Top 100 Apps in Malaysia Ringgit (RM)

The second purpose of our study was to rank the top 100 paid apps as charged in the Malaysia Ringgit. To achieve that, we conducted a search of the top hundred apps for each platform in both categories (Medical and Health & Fitness). Applying the same idea of inclusionary criteria like mentioned above, we later analyzed the descriptive information using IBM SPSS 20.

To rank the pricing of mHealth apps, we have grouped the app prices into five categories a) Less than RM5.00 b) RM 5.01 to RM 10 c) RM 10.01 to RM 25 d) RM 25.01 to RM 50 e) above RM 50.00. For clarity, the conversion is made at the state of US$1 = RM4.12. In classifying the apps, we have adapted various categories of functionality as described in previous literature.

FINDINGS

Principle Findings

Stage 1: mHealth App in Malaysia

After carefully examining all the app criteria, we have extracted information from thirty-five apps. The finding revealed that more than 90% of the evaluated apps provided a privacy policy that clarified that the developers are aware of the security of the data that is stored during the downloading process. The majority of the apps in the Malaysian context were designed for public users. Out of thirty-five apps, only five apps were specifically intended for medical professionals or medical students.

On the other hand, regarding functionality, more than half served to inform and provide information directly to the public. One and only one app was a health game, and one application was for searching for private doctors and setting up an arrangement with specialists at home.

Stage 2: App Ranking in Ringgit Malaysia

In ranking the apps, among 400 evaluated applications, twenty-eight applications were omitted, leaving 372 applications. Some applications were excluded for a few reasons: for example, the application is not in English or Bahasa Malaysia, the application is not designed for humans, or the application is not related to healthcare. In the Medical category, four apps were excluded from iOS App Store, while fourteen were also omitted from Android Google Play. Meanwhile, in the Health & Fitness category only three apps were excluded from the iOS App Store and six apps were excluded from the Android Google Play Store.

In general, the overall mean price of the most frequently downloaded paid apps for both platform and categories was RM 17.29 (SD RM 21.44) (Refer to Table 3). Among the top 100 favored application, Medical apps seem to be more expensive relative to Health & Fitness apps. The rationale is that Medical apps contain more complex calculations and compound terms that require more expertise validation and advanced coding. Among the top 100 Medical apps reviewed, the highest price is RM247.16, which is found on the Android platform. In the Health & Fitness category, both platforms provide the same maximum pricing value for the top 100 paid apps, at RM 41.16.

Referring to Table 4, it can be concluded that the most common price paid by the consumers is in the range of RM10.01 to RM25. In the Medical category, among the top 100 apps, only a few apps cost more than RM50, with six applications from the iOS Store and seven from the Google Play Store. Conversely, none of the apps in Health & Fitness went above RM50.

DISCUSSION

This study aims to assess the availability of mHealth apps that are provided and designated specifically for a Malaysian audience. The first question in this study sought to identify mHealth app and categorize them based on functionality. From the search, a total of thirty-five applications were found and further analyzed. More than 50% of the evaluated applications have limited functionality; many are just used for delivering information. A majority of apps that are offered are available freely, with a few offering in-app purchases. Most in-app purchase comes from apps that are intended to be read like a magazine or a book. The diversity of the applications is mostly the same on both platforms, as most apps are available on both platforms. Compared to other countries, Malaysia is still at the entry level of mHealth adoption. More evidence needs to be captured to confirm practices and usage of mHealth apps among the citizens. Although the Malaysia government did encourage the usage of the technology, there is much room for improvement. Rather than solely serving as an information delivery platform, the mHealth app should be optimized to its full potential. The burden should also rest with the private sector. Collaboration among app developers and mobile network providers may eventually lead to a high-quality, unique, and beneficial application. It is understandable that with the strict budget, minimal facilities, and basic skills

Table 2. Number of mHealth apps in Malaysia context

Category	Application Name	Price	In-App Purchase	Platforms	
				iOS	Android
Government/Healthcare Providers & Insurance	PLUX Healthcare	Free	X	√	√
	CompuMed	Free	X	√	√
	KPJ Healthcare	Free	X	√	√
	AIA Vitality Malaysia	Free	X	√	√
	Pantai Hospital Ayer Keroh	Free	X	√	√
	myHealth	Free	X	√	√
	myMahtas	Free	X	√	√
Doctor's Finder & Appointment	BookDoc - Find & Book a Doctor	Free	X	√	√
Women's Health & Pregnancy	Jom Mama eHealth	Free	X	√	√
	My Pink Health	Free	X	√	√
	Being Mom by Huggies® Malaysia	Free	X	√	√
Baby & Kids Management	MYVaksinBaby	Free	X	√	√
Disease Specific	COMBIS	Free	X	√	√
	Denguide: Malaysia Dengue CPG	Free	X	√	√
	iDengue	Free	X	√	√
Fitness	My Aone Swimmer - Swimming App in Malaysia	Free	X	√	√
	Malaysian Weight Loss Plan App	Free	X	√	√
Diet & Nutritions	MyNutriApps II: MyNutriDiari	Free	X	√	√
	MyFoodSafe	Free	X	√	√
Magazine	Men's Health Malaysia	Free	√	√	√
	Health Today Malaysia	Free	√	√	√
	Health Today Magazine Malaysia	Free	X	√	√
	Urban Health	Free	√	√	√
	Women's Health Malaysia	Free	√	√	√
	Health Holidays in Malaysia	Free	√	√	√
	Shape Malaysia	Free	√	√	√
	Running Malaysia	Free	√	√	√
	Swim Bike Run Malaysia	Free	√	√	√
Books / Medical References	CPG Malaysia	Free	X	√	√
	DocDx	Free	√	√	√
Medical news, journals and health education	MIMS Malaysia	Free	X	√	√
	Malaysian Medical Gazette	Free	X	√	√
Medicine Information	MediQuest	Free	X	√	√
	My Blue Book	Free	X	√	√
Game	Phosphorus Mission for Malaysia	Free	X	√	√

Are We Ready to App?

Table 3. Overview of top 100 iOS Apple Store and Android Play Store apps in medical and health & fitness categories (N= number of apps)

Category	N	Min Price (RM)	Max Price (RM)	Mean (RM)	Std. Deviation (RM)
Medical: iOS	95	4.08	102.96	17.74	19.13
Medical: Android	86	4.08	247.16	27.25	36.77
Health&Fitness: iOS	97	4.08	41.16	11.47	6.29
Health&Fitness: Android	94	4.08	41.16	13.71	6.88
Total	372			17.29	21.44

Table 4. Price ranking for top 100 iOS Apple Store and Android Play Store apps in Medical and Health & Fitness categories (N= number of apps).

Price Ranking	Medical		Health & Fitness	
	iOS (N)	Android (N)	iOS (N)	Android (N)
<RM5	30	7	15	6
RM 5.01-RM10	13	13	33	30
RM10.01-RM25	34	42	45	53
RM 25.01-RM50	12	17	4	5
>RM50	6	7	0	0
Total	95	86	97	94

available here, the process is not a one-month job. However, with the experiences and knowledge of Malaysian culture, unique apps will attract more consumers to use the application, and sooner or later the app, developers should generate a profit.

The second objective in this chapter was to analyze and rank the top paid mHealth app in the Malaysian Ringgit (RM). The most obvious finding to emerge from the analysis is that the price ranking for Medical app category is higher than that in the Health & Fitness category. Another important finding in all categories is that the majority of apps in both the iOS Store and the Android Play Store fall between the prices of RM10.00 to RM 25.00. Besides, the results of this study indicate that the availability of mHealth apps in Malaysia continues to be concentrated in the areas of healthcare providers, magazines, women's health & pregnancy, and fitness. Surprisingly, most of the apps were provided freely to the users, and only a few offer in-app purchases. A possible explanation for this might be that is Malaysia is still at the entry level of adopting the mHealth app. Compared to other categories of mobile applications such as mobile commerce, the mHealth app still has a long journey ahead.

Directions of mHealth Apps in Malaysia

Clearly, Malaysia has a very dynamic and bright future in the mHealth app technology. Even though there are several apps available on the market, the apps that specifically focused on the Malaysian context are still far behind. More work is required to affirm how this ubiquitous technology is most likely to have an impact on actual health care. More confirmation is needed with regards to how Malaysians utilize the mHealth application when dealing with their healthcare and how the apps may eventually change their behavior. Regardless of that, a careful look ought to be given to the readiness of Malaysian to pay for the application. As said by Boudreaux et al., (2014) before designing and testing a new mHealth app, researchers should assess the apps that already exist in their target domain.

The results will provide a key opportunity to advance the understanding of the Malaysian mHealth app market price, as each country differs in their market readiness for mHealth app adoption. It will offer some important insights for mHealth app developers in their go-to-market approach. Understanding this issue is important to understand the important metrics, including the percentage of the population that uses health apps, reasons for adoption, and reasons for continued use (Krebs & Duncan, 2015). Although the mHealth app industry is still in its infancy, its future looks bright (Powell, Landman, & Bates, 2014). Even with the increasing number of apps available, consumers are not buying or pursuing m-health app as a regular product (Dwivedi, Shareef, Simintiras, Lal, & Weerakkody, 2015), compare to other categories of application such as social networking and email processing, which are the most dominant apps.

CONCLUSION

This chapter provides insight into mHealth apps availability in Malaysia. Although there are available studies on it, none delve into the context of pricing. While most research evaluates the top ranking in each platform from other currency, we concentrate on how Malaysian is partaking in this advancement. As far as the author is concerned, this is the first study looking at mHealth app environments that are uniquely designed for Malaysia. Although the findings are limited to Malaysia context, it will be a good benchmark to look up the trends in mHealth app development in Malaysia. Such information is critical for both scholars and professionals either in Malaysia or other regions to understand how an upper-middle-income country, are persisting in using the technologies to increase healthcare adoption.

From the academic viewpoint, the review of the smartphone market in Malaysia will establish the current usage trends and pattern of mHealth app market store, helping future researchers to focus on ways to improve the usage of this app category among Malaysian citizens. On the other hand, for specialists in the industry, application developers, and all business partners would understand the real needs of Malaysia. Following the trend and pattern will not only produce some insight into what Malaysian citizen are looking for but also help the developers to be more productive in customizing the apps in the future.

LIMITATION AND FUTURE RESEARCH

Like any other research, this study carries a few limitations. It does not involve real data collection among the app users. All findings are based on the observation of the download pattern and trends in the market store. Only two platforms were used, and the different outcome may be revealed if all platforms were studied together. Therefore, to obtain a more accurate and focus answer, future research needs to be conducted to identify all critical factors from real consumer data examining on mHealth app adoption among Malaysia citizens. The findings above reveal that most mHealth apps in Malaysia are magazine-based or government-operated.

Further research should also be undertaken to investigate the financial elements that are connected to the mHealth app from the Malaysian client's point of view. As all apps are a conversion from US Dollar ($), the pricing is a bit expensive in Malaysia, meaning that the pricing is somewhat costly in Malaysia. While numerous researchers from various countries have likewise viewed upon the budgetary issues of mHealth apps (Hebly, 2012; Wu, Kang, & Yang, 2015) more evidence should be captured on Malaysian standpoint.

There are still many unanswered questions and curiosity on mHealth app privacy and security concerns among Malaysians. More than half of Malaysian is said to be more cautious when it comes to paying online by credit card (Purcell et al., 2010). Similar issue has been frequently raised by researchers from many other countries (Bussone, 2016; Faudree & Ford, 2013; Frenkel, 2013; Li, Ren, Cheng, Xiang, & Liu, 2014; Nilashi, Ibrahim, Reza Mirabi, Ebrahimi, & Zare, 2015; Reinfelder, Benenson, & Gassmann, 2014). Even if the application is disseminated, security is not guaranteed (Hussain et al., 2015) as the information collected may be reused for marketing or product development purposes. In Malaysia, despite the fact that there is a Personal Data Protection Act 2010 (PDPAA) and other laws to protect consumers, concerns on the privacy and security of mHealth apps remain inexplicable.

REFERENCES

Abt, E., Bader, J. D., & Bonetti, D. (2012). A practitioners guide to developing critical appraisal skills: Translating research into clinical practice. *The Journal of the American Dental Association*, *143*(4), 386–390. doi:10.14219/jada.archive.2012.0181 PMID:22467699

Aitken, M., & Gauntlett, C. (2013). *Patient apps for improved healthcare: from novelty to mainstream*. Parsippany, NJ: IMS Institute for Healthcare Informatics.

Akter, S., Dambra, J., Ray, P., & Hani, U. (2013). Modelling the impact of mHealth service quality on satisfaction, continuance and quality of life. *Behaviour & Information Technology*, *32*(12), 1225–1241. doi:10.1080/0144929X.2012.745606

Akter, S., & Ray, P. (2010). mHealth-an ultimate platform to serve the unserved. *Yearbook of Medical Informatics*, *2010*, 94–100. http://doi.org/me10010094 PMID:20938579

Akter, S., Ray, P., & DAmbra, J. (2013). Continuance of mHealth services at the bottom of the pyramid: The roles of service quality and trust. *Electronic Markets*, *23*(1), 29–47. doi:10.1007/s12525-012-0091-5

AppBrain. (2016). *Top categories*. Retrieved September 18, 2016, from http://www.appbrain.com/stats/android-market-app-categories

Atienza, A. A., & Patrick, K. (2011). Mobile health: The killer app for cyberinfrastructure and consumer health. *American Journal of Preventive Medicine*, *40*(5), S151–S153. doi:10.1016/j.amepre.2011.01.008 PMID:21521588

Benferdia, Y., & Zakaria, H. (n.d.). A Systematic Literature Review of Content-Based Mobile Health. *Journal of Information Systems Research and Innovation*. Retrieved from http://seminar.utmspace.edu.my/jisri/

Bidmon, S., Terlutter, R., & Röttl, J. (2014). What explains usage of mobile physician-rating apps? Results from a web-based questionnaire. *Journal of Medical Internet Research, 16*(6), e148. doi:10.2196/jmir.3122 PMID:24918859

BinDhim, N. F., Freeman, B., & Trevena, L. (2014). Pro-smoking apps for smartphones: The latest vehicle for the tobacco industry? *Tobacco Control, 23*(1), e4–e4. doi:10.1136/tobaccocontrol-2012-050598 PMID:23091161

Boudreaux, E. D., Waring, M. E., Hayes, R. B., Sadasivam, R. S., Mullen, S., & Pagoto, S. (2014). Evaluating and selecting mobile health apps: Strategies for healthcare providers and healthcare organizations. *Translational Behavioral Medicine, 4*(4), 363–371. doi:10.1007/s13142-014-0293-9 PMID:25584085

Boyce, B. (2014). Nutrition apps: Opportunities to guide patients and grow your career. *Journal of the Academy of Nutrition and Dietetics, 114*(1), 13–15. doi:10.1016/j.jand.2013.10.016 PMID:24342602

Bull, S., & Ezeanochie, N. (2015). From Foucault to Freire Through Facebook: Toward an Integrated Theory of mHealth. *Health Education & Behavior, 43*(4), 399–411. doi:10.1177/1090198115605310 PMID:26384499

Bussone, A. (2016). *Disclose-It-Yourself: Security and Privacy for People Living with HIV*. Academic Press.

Chun, S. G., Chung, D., & Shin, Y. B. (2013). Are Students Satisfied with the Use of Smartphone Apps? *Issues in Information Systems, 14*(2), 23–33.

Cole-Lewis, H., & Kershaw, T. (2010). Text messaging as a tool for behavior change in disease prevention and management. *Epidemiologic Reviews, 32*(1), 56–69. doi:10.1093/epirev/mxq004 PMID:20354039

Fadzilah, F. M., & Arshad, N. I. (2015). Evaluating the Impact of Non-Clinical M-Health Application: Towards Development of a Framework Akademia Baru. *Journal of Advanced Research Design, 14*(1), 28–38.

Faudree, B., & Ford, M. (2013). Security and Privacy in Mobile Health. *CIO Journal*. Retrieved May 20, 2016, from http://deloitte.wsj.com/cio/2013/08/06/security-and-privacy-in-mobile-health/

Fayaz-Bakhsh, A., & Geravandi, S. (2015). Medical Students Perceptions Regarding the Impact of Mobile Medical Applications on their Clinical Practice. *Journal of Mobile Technology in Medicine, 4*(2), 51–52. doi:10.7309/jmtm.4.2.8

Fiordelli, M., Diviani, N., & Schulz, P. J. (2013). Mapping mhealth research: A decade of evolution. *Journal of Medical Internet Research, 15*(5), e95. doi:10.2196/jmir.2430 PMID:23697600

Frenkel, K. A. (2013). Mobile Apps Need Better Security. *CIO Insight*, 1.

Gleason, A. W. (2015). mHealth — Opportunities for Transforming Global Health Care and Barriers to Adoption. *Journal of Electronic Resources in Medical Libraries, 12*(2), 114–125. doi:10.1080/15424065.2015.1035565

Handel, M. J. (2011). mHealth (mobile health)-using Apps for health and wellness. *Explore (New York, N.Y.), 7*(4), 256–261. doi:10.1016/j.explore.2011.04.011 PMID:21724160

Hebly, P. (2012). *Willingness to pay for mobile apps*. Erasmus Universiteit Rotterdam. Retrieved from http://thesis.eur.nl/pub/13913/Heblij-P.G.-375422ph-.docx

Hussain, M., Al-Haiqi, A., Zaidan, A., Zaidan, B., Kiah, M., Anuar, N. B., & Abdulnabi, M. (2015). The landscape of research on smartphone medical apps: Coherent taxonomy, motivations, open challenges and recommendations. *Computer Methods and Programs in Biomedicine*, *122*(3), 393–408. doi:10.1016/j.cmpb.2015.08.015 PMID:26412009

Jaafar. (2012). Malaysian Health System Review. *Health Systems in Transition*, *3*(1), 1–103. Retrieved from http://www.wpro.who.int/asia_pacific_observatory/hits/series/Malaysia_Health_Systems_Review2013.pdf

Jahns, R.-G. (2013). *The market for mHealth app services will reach $26 billion by 2017*. Retrieved from http://research2guidance.com/the-market-for-mhealth-app-services-will-reach-26-billion-by-2017/

Koh, , Wan, J. K., Selvanathan, S., Vivekananda, C., Lee, G. Y., & Ng, C. T. (2014). Medical students perceptions regarding the impact of mobile medical applications on their clinical practice. *Journal of Mobile Technology in Medicine*, *3*(1), 46–53. doi:10.7309/jmtm.3.1.7

Krebs, P., & Duncan, D. T. (2015). Health App Use Among US Mobile Phone Owners: A National Survey. *JMIR mHealth and uHealth*, *3*(4), 1–12. doi:10.2196/mhealth.4924

Kumar, , Nilsen, W. J., Abernethy, A., Atienza, A., Patrick, K., Pavel, M., & Swendeman, D. et al. (2013). Mobile health technology evaluation: The mHealth evidence workshop. *American Journal of Preventive Medicine*, *45*(2), 228–236. doi:10.1016/j.amepre.2013.03.017 PMID:23867031

Li, X., Ren, S., Cheng, W., Xiang, L., & Liu, X. (2014). Smartphone: Security and Privacy Protection. *Pervasive Computing and the Networked World*, 289–302. doi:10.1007/978-3-319-09265-2_30

Liu, C., Zhu, Q., Holroyd, K. A., & Seng, E. K. (2011). Status and trends of mobile-health applications for iOS devices: A developers perspective. *Journal of Systems and Software*, *84*(11), 2022–2033. doi:10.1016/j.jss.2011.06.049

Malaysian Communications And Multimedia Commission. (2015). Hand phone users survey 2014. *Malaysian Communication and Multimedia Commission*, 48.

MAMPU. (2015). *GAMMA : Galeri Mudah Alih Kerajaan*. Retrieved from http://www.moh.gov.my/mgpwa2015/Mesyuarat Teknikal ICT 2015/GAMMA.pdf

MAMPU. (n.d.). *Gallery of Malaysian Government Mobile Application (GAMMA)*. Retrieved October 1, 2016, from http://gamma.malaysia.gov.my/#/home

Marshall, C., Lewis, D., & Whittaker, M. (2013). *mHealth technologies in developing countries: a feasibility assessment and a proposed framework*. Working paper series (Vol. 25). Retrieved from http://www.uq.edu.au/hishub/docs/WP25/WP25 mHealth_web.pdf

McIlroy, S., Ali, N., & Hassan, A. E. (2015). Fresh apps: An empirical study of frequently-updated mobile apps in the Google play store. *Empirical Software Engineering*, *21*(3), 1346–1370. doi:10.1007/s10664-015-9388-2

Mechael, P. N. (2009). The Case for mHealth in Developing Countries. *Innovations: Technology, Governance, Globalization, 4*(1), 103–118. doi:10.1162/itgg.2009.4.1.103

Ming, . (2016). *Use of Medical Mobile Applications Among Hospital Pharmacists in Malaysia*. Therapeutic Innovation & Regulatory Science. doi:10.1177/2168479015624732

Ministry of Health Malaysia. (2015). *Malaysia National Health Accounts Health Expenditure Report 1997-2013*. Retrieved from http://www.moh.gov.my/index.php/file_manager/dl_item/554756755a584a 6961585268626939515a57356c636d4a7064474675494656305957316

MOH. (2016). *My Blue Book*. Retrieved from https://play.google.com/store/apps/details?id=air.mypharmacisthouse.mybluebook

Mohd-Tahir, N. A., Paraidathathu, T., & Li, S. C. (2015). Quality use of medicine in a developing economy: Measures to overcome challenges in the Malaysian healthcare system. *SAGE Open Medicine, 3*. doi:10.1177/2050312115596864 PMID:26770795

Nilashi, , Ibrahim, O., Reza Mirabi, V., Ebrahimi, L., & Zare, M. (2015). The role of Security, Design and Content factors on customer trust in mobile commerce. *Journal of Retailing and Consumer Services, 26*, 57–69. doi:10.1016/j.jretconser.2015.05.002

Norris, A. C., Stockdale, R. S., & Sharma, S. (2009). A strategic approach to m-health. *Health Informatics Journal, 15*(3), 244–253. doi:10.1177/1460458209337445 PMID:19713398

Nutrition Division, Ministry of Health, Malaysia. (n.d.). Retrieved from http://nutrition.moh.gov.my/

Osman. (2011). An exploratory study on the trend of smartphone usage in a developing country. Communications in Computer and Information Science, 387–396. http://doi.org/ doi:10.1007/978-3-642-22603-8_35

Powell, A. C., Landman, A. B., & Bates, D. W. (2014). In search of a few good apps. *Journal of the American Medical Association, 311*(18), 1851–1852. doi:10.1001/jama.2014.2564 PMID:24664278

Purcell, K., Project, P. I., Entner, R., President, S. V., Practice, T., Company, T. N., & Henderson, N. (2010). *The Rise of Apps Culture. Group*. Retrieved from http://www.pewinternet.org/~/media//Files/Reports/2010/PIP_Nielsen Apps Report.pdf

PWC. (2014). *2014/2015 Malaysian Tax and Business Booklet*. Retrieved from https://www.pwc.com/my/en/assets/publications/2015-malaysian-tax-business-booklet.pdf

Reinfelder, L., Benenson, Z., & Gassmann, F. (2014). Differences between Android and iPhone users in their security and privacy awareness. Lecture Notes in Computer Science, 8647, 156–167. doi:10.1007/978-3-319-09770-1_14

Research2Guidance. (2015). *mHealth App Development Economic 2015*. Retrieved from http://research-2guidance.com/r2g/r2g-mHealth-App-Developer-Economics-2015.pdf

Research2Guidance The App Market Specialist. (2015). *mHealth App Developer Economics 2015*. Retrieved from www.mHealthEconomics.com

Schweitzer, J., & Synowiec, C. (2012). The Economics of eHealth and mHealth. *Journal of Health Communication*, *17*(October), 73–81. doi:10.1080/10810730.2011.649158 PMID:22548602

Sidney, K., Antony, J., Rodrigues, R., Arumugam, K., Krishnamurthy, S., Dsouza, G., & Shet, A. et al. (2012). Supporting patient adherence to antiretrovirals using mobile phone reminders: Patient responses from South India. *AIDS Care*, *24*(5), 612–617. doi:10.1080/09540121.2011.630357 PMID:22150088

Silva, B. M. C., Rodrigues, J. J. P. C., de la Torre Díez, I., López-Coronado, M., & Saleem, K. (2015). Mobile-health: A review of current state in 2015. *Journal of Biomedical Informatics*, *56*, 265–272. doi:10.1016/j.jbi.2015.06.003 PMID:26071682

Statista. (2014). *Number of smartphone users worldwide from 2014 to 2019 (in millions)*. Retrieved July 10, 2016, from http://www.statista.com/statistics/330695/number-of-smartphone-users-worldwide/

Statista. (2016). *Number of apps available in leading app stores as of June 2016*. Retrieved June 1, 2016, from http://www.statista.com/statistics/276623/number-of-apps-available-in-leading-app-stores/

Vital Wave Consulting. (2009). mHealth for Development: The Opportunity of Mobile Technology for Healthcare in the Developing World. *Technology*, *46*. http://doi.org/10.1145/602421.602423

Wu, L., Kang, M., & Yang, S. B. (2015). What makes users buy paid smartphone applications? Examining app, personal, and social influences. *Journal of Internet Banking and Commerce*, *20*(1), 2–22. doi:10.1007/978-3-531-92534-9_12

Chapter 6
Models of Privacy and Security Issues on Mobile Applications

Lili Nemec Zlatolas
University of Maribor, Slovenia

Marjan Heričko
University of Maribor, Slovenia

Tatjana Welzer
University of Maribor, Slovenia

Marko Hölbl
University of Maribor, Slovenia

ABSTRACT

The development of smart phones and other smart devices has led to the development of mobile applications, which are in use frequently by the users. It is also anticipated that the number of mobile applications will grow rapidly in the next years. This topic has, therefore, been researched highly in the past years. Mobile applications gather user data and that is why privacy and security in mobile applications is a very important research topic. In this chapter we give an overview of the current research on privacy and security issues of mobile applications.

INTRODUCTION

The aim of this Chapter is to present the current research topics in the field of privacy and security of mobile applications. Mobile technologies enable people to use mobile applications on their smart phones and smart devices constantly. The main operating systems used in the majority of mobile devices are Android with 70% market share and iOS with 23% market share ("Mobile/Tablet Operating System Market Share," 2016). Both most popular operating systems offer an app market for users with various mobile applications. Android users can download their mobile applications on Google Play Store and iOS users on Apple's App Store (Degirmenci, Guhr, & Breitner, 2013). The app markets are centralised systems that offer applications to users, which means that the users cannot download the applications from other websites like they can on desktop computers, unless they change their security settings on their mobile devices. The mobile device users have downloaded over 100 billion mobile applications from app markets in 2015 (Statista, 2016). Mobile applications are usually put on the app markets by the third-party developers and with different purposes of the applications. When an application is uploaded to the Google Play Store it is not checked and usually the application is then available for download

DOI: 10.4018/978-1-5225-2469-4.ch006

within a few hours. On the other hand, in Apple's Apps Store applications are checked and approved before they are put on the market ("iOS app approvals," 2016). Even though Apple checks applications on its market, researchers have found that Apple App Store enables developers of applications to download all the user's photos and calendars, meaning that being approved by the Apple App Store does not necessarily mean that the application respects the user's privacy (Weintraub, 2012). A German study compared users of Android and iPhones and discovered that Android users are more concerned that the applications could charge them with hidden costs and they more often mention security and privacy issues as important in comparison to iPhone users (Reinfelder, Benenson, & Gassmann, 2014). These results could either mean that the Android users do not trust their application market, because it is a more open system of applications upload, or that the users trust Apple more, because of Apple's general reputation of being more secure. Gilbert, Chun, Cox, and Jung (2011) proposed an automated security validation of mobile applications in application markets, but their proposal has not been implemented in practice in any app market. Therefore, applications on the app market could present a security problem and can be vulnerable. Implementing a better system of security and privacy checking in app markets would mean more security and privacy for users' data.

When users install a new mobile application from the app market, they are asked to read and confirm the terms of agreement of the application. Usually this means that users are requested to give their context information to applications or other third parties. Y. Liu (2014) argues that this way of getting users' consent is not the most adequate, because users usually just press continue without even reading which permissions they are giving the application. Users actually should have some control over what personal data they provide to third-party applications but, due to design and other restrictions, users just accept the terms of the application. A proposed solution by Y. Liu (2014) is the use of privacy by design concept with clearer and more user-friendly controls for privacy settings. Another group of researchers have conducted a study among 168 users using the Nokia N95 and, based on the research results, they presented a new business model for mobile platforms which would make mobile applications more privacy-friendly (Z. Liu, Bonazzi, & Pigneur, 2016). Another study showed that smartphone users are not much concerned about security when they install third-party applications to their smartphones (Mylonas, Kastania, & Gritzalis, 2013). The collaborators in the study trusted the application repositories and they disregarded security while adding applications to their smartphones.

Mobile applications are in use in many areas in people's lives as well as in corporate environments. The flow of data in mobile applications is enormous and there are numerous possibilities of the analysis and use of collected data by third-parties. It is important that data anonymity, confidentiality, integrity and authentication is provided for users of mobile applications. It is also important that protection of personal data is assured for the users of mobile applications. Research shows that users are often concerned about the privacy and security of their data, but some users would still not take any actions for protection (D'Ambrosio et al., 2016). On the other hand, Katell, Mishra, and Scaff (2016) have found that prompting users about permissions on data security and privacy might encourage users to take care of their privacy. Other possibilities to protect data also include biometric security measures when mobile applications are used and, in cases where personal authentications are required (Guerra-Casanova, Sánchez-Ávila, de Santos Sierra, & del Pozo, 2011). Such methods could provide better security for the applications that do not require personal authentication and collect sensitive information about the user.

In this Chapter a literature review is conducted to get a clear state-of-the-art on research in the field of privacy and security of mobile applications and what models were presented in considering this topic.

BACKGROUND

In his research, S. Kang (2014) found that people use mobile applications for different reasons. The factors which are mostly influencing the intention of mobile application use are entertainment, social utility, communication, social influence, performance expectancy and effort expectancy. Other research has also shown that people have an intention to use mobile applications in continuance also when they perceive usefulness, because of satisfaction and when they get used to it as a habit (Harrison, Flood, & Duce, 2013; Hsiao, Chang, & Tang, 2016). Since mobile applications are collecting a lot of user data, the application developers should be the ones addressing the issues of privacy and security and making adjustable settings for applications added by the users. Mobile devices often send user location, audio, mobile camera footage etc. to application developers, sometimes without the users' consent or users' acknowledgement (Shilton, 2009). That is why it is important to address research in the topics of privacy and security of user data when using mobile applications.

Researchers are constantly exposing privacy and security threats of mobile applications. Studies have shown that the majority of mobile application users tend to ignore the permissions that an application is requesting when using mobile applications (Felt et al., 2012; J. Kang, Kim, Cheong, & Huh, 2015; J. Khan, Abbas, & Al-Muhtadi, 2015). Users also find it difficult to understand the terms and words used to describe permissions (Kelley et al., 2012). Jain and Shanbhag (2012) found that insecure mobile application can cause security and privacy issues with users and organisations. It was also found that people disclose information in mobile applications without considering the risks, if they trust the application and that the mobile applications security awareness of users is not as high as expected (Morosan & DeFranco, 2015; Sari, Candiwan, & Trianasari, 2014).

Different categories of mobile applications are usually being researched in most research. An overview of different categories, concerning mobile applications are described next.

Categories of Mobile Applications

Different categories of mobile applications are defined in research. Mahatanankoon, Wen, and Lim (2005) analysed multiple mobile applications and divided mobile applications into five classifications: Content delivery, transaction-based, location-based, emergency-assisted and entertainment. Content delivery applications are intended for searching for information on the Internet. Transaction-based applications support, for example, transferring money between customers and businesses. Location-based applications record user location and use this data for advertisement or similar purposes. Emergency-assisted applications include, for example, health care applications. Entertainment applications are for enjoyment in free time. These categories were defined before the applications markets were open and the applications were usually integrated in the mobile devices or otherwise pre-installed on mobile devices.

New classifications of mobile applications have been set recently, because the application markets are more open and new mobile applications with different purpose are developed and put to use frequently. In their research, Kennedy-Eden and Gretzel (2012) divided tourism applications into navigation, social, mobile marketing, security or emergency, transactional, information and entertainment, covering a lot of areas also for applications in general. Al-khamayseh, Lawrence, and Zmijewska (2006) have researched mobile government applications developed by the government. The applications evolved from the concept of electronic government and might try to replace electronic government on desktop computers. Medical and health care applications are also an example of applications and can be used

Models of Privacy and Security Issues on Mobile Applications

for patient care and monitoring, communication and education and as health applications for laypersons (Meulendijk, Meulendijks, Jansen, Numans, & Spruit, 2014; Ozdalga, Ozdalga, & Ahuja, 2012). Medical and health care applications can be applied in professional and personal settings, both for patients, as well as professional medical staff.

Emmanouilidis, Koutsiamanis, and Tasidou (2013) have studied the context in mobile guides and have divided the different context in these mobile guides for users to user, system, social, service and environment categories. These categories can be applied to mobile applications. Another classification was done by Schilit, Adams, and Want (1994) who defined context aware applications as applications that examine and react to the user's changing context. In the mobile application environment this means that location, light, sound and other behavioural patterns of the users are being used to produce a better content, tailored to a specific user (Pattan & Madamanchi, 2010). Similar types of such applications are referred to as sensing applications, which mostly capture the location and time information in different formats (D. Christin, Reinhardt, Kanhere, & Hollick, 2011).

A number of different classifications for different mobile applications categories exist and researchers usually choose one type of application to research. We will present a review of the literature published on privacy and security in different categories of mobile applications in the next Section.

Privacy and Security in Different Categories of Mobile Applications

Mobile applications can be location-based, which means that the location of the user is reported to the application on the user's mobile device. This feature can have either positive or negative impact. On one side it enables users to see who are their nearby friends and, on the other side, it sends location data to third-party application developers or other third-parties (Ajami, Qirim, & Ramadan, 2012). An example of mobile applications that incorporates location detection and recording is Foursquare. In this application users can post a picture or status update with a location tag that can be turned on automatically without the user being aware of it.

In a study at Carnegie Mellon University an application was developed with the possibility of sharing the location of students with their friends (Cornwall et al., 2007). Sadeh et al. (2009) used an application to understand how people perceive privacy policies in such a mobile social networking application. In the study they found that people loosened their privacy settings for location sharing over time, especially when they realised that sharing their location with their friends and others had more positive aspects then negative.

Research shows that users can encrypt their location on mobile devices and only share it with selected persons by exchanging the encryption keys. In this way, users can verify if the location was generated by a friend or not (Melinger, Bonna, Sharon, & SantRamet, 2004; Puttaswamy & Zhao, 2010). With such options, the user location can stay private from all non-friends and location data should be stored on the user's device, and not in the mobile application. Location-based applications are often used by advertisers who deliver more focused context appropriate for the end user by using their location (Xu & Teo, 2005).

Location-based mobile applications are also a part of sensing or context aware mobile applications. Sensing and context aware mobile applications are applications which use mobile sensors and include a wide range of applications such as car tracking, noise pollution, diets etc. In such applications it is important that the development of the mobile application takes security and privacy into account. Research shows that privacy benefits, privacy assurance and privacy risks have a great effect on the user and how they disclose personal information in location-based services on mobile applications (Xu, Rosson, &

Carroll, 2008). Another study proposed a more flexible privacy architecture with the combination of a distributed and centralised approach to achieve higher privacy value of the user (D. Christin et al., 2011). A study by Pattan and Madamanchi (2010) tested the participants of the study on their perception of privacy and security in the application Friend View. They used an improved prototype of this application to achieve better user protection. The participants were keener on the improved prototype, which led to the conclusion that the mobile applications should focus more on privacy and security aspects and, consequently, users would be using these applications with more pleasure.

Cloud-based mobile applications like Dropbox and Google Drive are also among the most used applications for mobile devices ("List of most downloaded Android applications," 2016). Users sometimes use them to exchange important documents with their collaborators and, therefore, data processed by these applications can be very sensitive. That is why it is important that cloud-based mobile applications provide a high level of security and privacy (A. N. Khan, Mat Kiah, Khan, & Madani, 2013; Kumar & Lu, 2010). In a study conducted by Anastasopoulou, Tryfonas, and Kokolakis (2013), the authors argue that most users do not want a payable option for protecting their privacy and security on cloud-based mobile applications and that users believe that their information is not valuable enough to be protected.

Some types of mobile applications are used for entertainment purposes, such as playing different games and interacting on social networking sites. A research by Jeong, Kim, Yum, and Hwang (2016) has shown that both games and social networking sites are addictive and that users who use their mobile devices for these entertainment purposes are often addicted to smartphone use. However, not only users are vulnerable by being addicted, application developers also have their faults. A free mobile application with social networking site features Path, which is available from Google Play store and Apple Apps market, was installed by over 23 million people (Weber, 2014). It was later discovered that this application did not request the permission to access a user's address book, but did it anyway (Baldwin, 2012). Also, Twitter confessed to using the entire address book of users to find their friends (Whittaker, 2012). Mobile applications with entertainment purposes can seem harmful, but because of their success and a great number of users, these applications can be very unprotective of users' privacy and security.

Medical applications are used widely by patients and medical staff. Important issues of privacy in this field are often addresses, because patients' health records are very sensitive. Such data can be exposed by sharing patients' information with mobile applications. In a study about the requirements of medical mobile applications, researchers have tested potential users of such applications to see which are the most important functional requirements for the development of such mobile applications (Meulendijk et al., 2014). The requirements which they evaluated were accessibility, certifiability, portability, privacy, safety, security, stability, trustability and usability. The results showed that privacy was considered the least important attribute among the requirements in this study on medical applications functionalities. The security of the mobile applications was considered important for younger participants in the study. Research also shows that new technologies should not rely on the old standards and laws regarding privacy and security and that policy-makers should review the current standards and laws and adjust them to mobile health applications (Martínez-Pérez, de la Torre-Díez, & López-Coronado, 2014). There are different medical applications, but most record very sensitive data and, therefore, it is of the utmost importance to consider guidelines for protecting everyone's privacy and security.

It can be concluded that users of mobile applications care about privacy and security of their data and that this is an important topic for research. In the next Section, a presentation of the selected research methods will be done.

RESEARCH METHODS

The method of literature review was selected to get a thorough insight of the research already published in the field of privacy and security issues of mobile applications. The selected publications were limited to the topic of presented statistical models regarding mobile applications or similar domains that could be applied to mobile applications. For a focused literature review, we have set the research questions which are presented below.

Research Questions

RQ1: Which constructs besides privacy or security are used in models for understanding mobile applications?

All models involving privacy and security related to mobile applications or similar will be defined in the search process. An analysis will be conducted of constructs or factors in models involving privacy or security of mobile applications.

RQ2: Which other constructs are found in relation to privacy and security constructs in the researched models on mobile applications?

All the constructs which appear in the selected models which include privacy and security constructs will be analysed. We will identify the constructs in the models and test the significance of the constructs in the model. We will also analyse the paths to privacy or security constructs in the models.

RQ3: What types of measurements are used in research regarding models, involving privacy and security of mobile applications?

Research papers measure privacy and security of mobile applications differently. It is going to be investigated if different factors are measured and constructed in models. Normally, these factors determine how users perceive privacy and security in mobile applications.

RQ4: Which mobile applications categories are researched in the field of privacy and security models?

Most scientific papers focus their studies on one type of mobile applications, which is mostly connected with its purpose. This can be either entertainment, business, communication etc. (Delphine Christin, Roßkopf, & Hollick, 2013; Lupton & Jutel, 2015; Morosan & DeFranco, 2015; Puttaswamy & Zhao, 2010). The publications included in the literature review will be selected based on categories of mobile applications.

The literature review search is limited to the most important references, dating between 2008 and 2016, because the first applications markets were opened in 2008 and set a state of applications as they are today. Prior to 2008, applications were pre-installed on mobile devices or were installed by users manually.

Search Process

The most important publications, explaining privacy and security issues of mobile applications, were collected through a thorough literature review. The search was done in scientific publications databases, which are presented in Table 1.

The search strings used for the searches in the scientific publications databases included combination of the keywords "Mobile applications", "Mobile apps", "Security" and "Privacy".

Data Collection

The total number of search results in each scientific publications database is presented in Table 2. In the first phase we searched each search string separately to see how many results appear in each scientific publications database. The term "mobile apps" is sometimes used instead of "mobile applications" in research, but "mobile applications" is the dominant term as seen in Table 2. The search terms "privacy" and "security" were also searched in scientific publications databases as standalones. One interesting fact is that, in all the scientific publications databases except in Google Scholar, there were more results for the term "security" than "privacy". As Google Scholar is a large database which includes searching through various scientific publications databases, this shows that there are more privacy-related publications in the research than security related publications.

In the next phase of search, two of the search strings were combined when searching to get a better view of the more specific results. In the search, either "mobile applications" or "mobile apps" was used combined with "security" or "privacy" to get the best results possible. The search with two search strings already showed far less results in the scientific publications databases. In this phase a more thorough search for appropriate publications was conducted on the topic of mobile applications, security and privacy. If the title and abstract were related to the topic, the publication was selected for further investigation.

In the last search phase, a combination of "privacy" and "security" with "mobile applications" or "mobile apps" was searched for in scientific publications databases. As seen in the search results, there are not that many publications in this field, because it is also a new field in the research area. From the search results, the most appropriate publications were selected for further investigation.

Table 1. Scientific publications databases used in the literature review

Scientific Publications Databases	Url
ScienceDirect (SD)	http://www.sciencedirect.com
Springerlink (Springer)	http://link.springer.com
IEEExplore (IEEE)	http://ieeexplore.ieee.org
ACM Digital Library (ACM)	http://dl.acm.org
Scopus	http://www.scopus.com
Emerald Insight (Emerald)	http://www.emeraldinsight.com
Web of Science (WoS)	https://apps.webofknowledge.com
Google Scholar	https://scholar.google.com

In the last row in Table 2, there are numbers of selected publications per each scientific publications database. As seen, Google Scholar has the most search results, while other scientific publications data bases show significantly fewer results.

After the search was conducted the abstracts in the selected papers were read. The paper was selected for further investigation if there were enough indications that the content of the paper related to mobile applications and security or privacy. If the content of a paper included the presentation of a statistical model with factors, affecting privacy or security in mobile applications, it was selected for a more thorough reading and was put in the final list of publications.

Out of all the papers, 22 publications were included for a detailed investigation. There were some duplicate papers among these 22 publications and, after the exclusion of the duplicates, 16 publications were used for the analysis. The final selection of papers was dated between 2009 and 2016, although most of them are very recent (between 2015 and 2016). The number of publications by source, publication type and year is presented in Table 3. As seen from this Table, two publications were selected from the International Conference on Information Systems (ICIS) and International Journal of Information Management. Other publications selected represent one selected publication per source.

RESULTS

To evaluate the selected papers in the literature review the defined research questions were used for the analysis.

RQ1: Which constructs besides privacy or security are used in models for understanding mobile applications?

Table 2. Total number of search results with each scientific publications database

Key-words	Science Direct	Sprin-ger	IEEE	ACM	Scopus	Emerald	WoS	Google Scholar
1	8.630	13.377	4.320	2.041	37.645	534	9.441	262.000
2	1.763	2.207	481	574	3.271	191	731	38.000
3	79.514	119.218	25.402	17.871	199.420	10.161	95.933	4.610.000
4	381.348	524.734	147.475	26.942	997.088	47.397	630.127	2.910.000
1 + 3	1.317	3.080	214	190	3.633	141	219	58.000
2 + 3	512	671	30	75	648	64	38	22.600
1 + 4	2.421	5.284	496	241	5.954	196	435	52.900
2 + 4	680	969	71	74	807	60	37	14.600
1 + 3 + 4	975	2.292	137	138	2.704	94	96	38.900
2 + 3 + 4	359	499	18	52	507	36	12	9.580
No. of selected publications	5	1	1	0	2	0	3	10

1 – Mobile applications, 2 – Mobile apps, 3 – Privacy, 4 – Security

Table 3. Number of publications by source, publication type and year

Source	Publication Type	2009	2010	2012	2013	2014	2015	2016
Computers in Human Behavior	Journal article	0	0	0	0	0	0	1
International Journal of Hospitality Management	Journal article	0	0	0	0	0	1	0
International Journal of Human-Computer Studies	Journal article	0	0	0	1	0	0	0
International Journal of Information Management	Journal article	0	0	0	0	0	0	2
International Journal of Mobile Communications	Journal article	0	1	0	0	0	0	0
International Journal of Mobile Learning and Organisation	Journal article	0	0	0	0	0	1	0
MEDINFO 2015: eHealth-enabled Health	Journal article	0	0	0	0	0	1	0
Journal of Cyber Security and Mobility	Journal article	0	0	0	0	0	0	1
New Contributions in Information Systems and Technologies: Volume 1	Book section	0	0	0	0	0	1	0
iConference (iSociety: Research, Education, and Engagement)	Conference proceedings	1	0	0	0	0	0	0
Hawaii International Conference on System Sciences	Conference proceedings	0	0	0	0	1	0	0
International Conference on Information Systems (ICIS)	Conference proceedings	0	0	1	1	0	0	0
ACM Conference on Computer-Supported Cooperative Work & Social Computing	Conference proceedings	0	0	0	0	0	0	1
International Conference on Intelligent User Interfaces	Conference proceedings	0	0	0	0	0	0	1
Publications by Year		**1**	**1**	**1**	**2**	**1**	**4**	**6**

In the search processes 16 publications were identified with presented statistical models regarding privacy and security in mobile applications. Based on the selected publications, an analysis of all the constructs in the models was done. The constructs were divided into groups of independent, mediating, dependent and moderating constructs as in Table 4. Independent variables produce an effect or result on dependent variables (Neuman, 2011). Mediator variables are set logically or temporarily between independent and dependent variables and moderating variables affect the direction or strength of relationship between independent and dependent variable or criterion variable (Baron & Kenny, 1986).

In Table 4, the privacy and security constructs in each paper are presented in bold. As seen in the Table, privacy and security constructs either affect other constructs or are defined by other constructs.

The column before the last one in Table 4 presents the number of citations per publication as reported by Google Scholar on September 2016. This is to better illustrate the impact of individual publications within the research cluster. As can be seen from Table 4, publications released before 2013, have much more citations than more recent publications.

A more thorough analysis of Table 4 was done in Table 5 regarding the used constructs for privacy and security. As seen in Table 5, privacy constructs appear more often in publications than security constructs. There were 20 privacy constructs identified in the selected publications and only 7 security constructs. The most common privacy and security constructs present independent variables in the models.

Table 4. Constructs used in the models based on the role in the models

Independent Constructs	Mediator Constructs	Dependent Constructs	Moderating Constructs	Publication
Global information privacy concern	**Privacy risk literacy**, Coping behaviuor literacy	Intention to install the application	Applications' characteristics	De Santo and Gaspoz (2015)
Access to personal information (Personal identity, Location, Device content, System and network settings)	**Mobile users information privacy concerns**	Secondary use of personal information, **Perceived surveillance**, Perceived intrusion		Degirmenci et al. (2013)
Perceived benefits, **Perceived privacy risk**	Trusting beliefs, **Privacy concern**	Actual disclosure (Amount of profile data disclosed, Accuracy of profile data submitted, Number of people referred to visit site, Total clues found **Level of privacy settings**, The randomly assigned direction of change in benefits, The total number of times the participant logged into a website, Number of updates to profile data)		Keith, Babb, and Lowry (2014)
Privacy risk awareness, Perceived benefits	**Perceived privacy risk, Privacy concern**	Registration info provided, **Privacy settings**	Honest info, Employment	Keith, Thompson, Hale, Lowry, and Greer (2013)
Risk, Benefit, Positive emotion, Negative emotion, Trust in organisation	Value of disclosure, Trust in system	Willingness to disclose		Morosan and DeFranco (2015)
Privacy concern, Perceived enjoyment, Perceived ease of use, Perceived usefulness, Perceived risk	Satisfaction	Continuance		Ofori, Larbi-Siaw, Fianu, Gladjah, and Boateng (2016)
Trust	**Perceived risk**, Perceived usefulness, Perceived ease of use	Behavioural intention. mHealth use		Schnall, Higgins, Brown, Carballo-Dieguez, and Bakken (2015)
Personalised service, Self-presentation, **Perceived severity**, Perceived control	Perceived benefits, Perceived risks	Information to disclose via mobile applications	Use	Wang, Duong, and Chen (2016)
Prior privacy experience	**Mobile users information privacy concern**	Secondary use of personal information, **Perceived surveillance**, Perceived intrusion, Behavioural intention		Xu, Gupta, Rosson, and Carroll (2012)
Privacy concern, Effort expectancy	Performance expectancy	Intention to use	Innovativeness	Xu, Gupta, and Pan (2009)
Social nudge, Frequency nudge	Creepiness, Perceived control	**Privacy concern**, Disclosure comfort		Zhang and Xu (2016)
Privacy concern	Quality	Satisfaction, Acceptance		Zhao, Ye, and Henderson (2016)
Direct Network Externalities (Number of IoT services, Perceived critical mass), Indirect network externalities (Perceived compatibility, Perceived complementarity)	Perceived benefits, **Concern for information privacy**, Attitude	Continued intention to use, Collection, **Unauthorised secondary use**, Improper access, Errors		Hsu and Lin (2016)

continued on following page

Table 4. Continued

Independent Constructs	Mediator Constructs	Dependent Constructs	Moderating Constructs	Publication
Perceived security, Perceived reputation, Application characteristics, Familiarity, Desensitization, Consumer disposition to trust, Consumer disposition to risk	Trust, Perceived risk	Intention to install	Perceived benefit	Harris, Brookshire, and Chin (2016)
Behavioural control, Personal innovativeness, Enjoyment, Usefulness, Ease of use, Social, External, Image, **Security**, Intention to adopt				Hill and Troshani (2010)
Availability, Accessibility, **Awareness**	**Trust**	Satisfaction		Sheila, Faizal, and Shahrin (2015)

Table 5. Privacy and security constructs used in the models

	Independent Constructs	No.	Mediator Constructs	No.	Dependent Constructs	No.	No.
Privacy constructs	Global information privacy concern, Perceived privacy risk, Privacy risk awareness, Risk, Privacy concern (three times), Perceived severity, Prior privacy experience	9	Privacy risk literacy, Mobile users information privacy concerns (twice), Privacy concern (twice), Perceived privacy risk (twice), Concern for information privacy	8	Level of privacy settings (twice), Privacy concern	3	20
Security constructs	Perceived security, Security, Awareness	3	Trust	1	Perceived surveillance (twice), Unauthorised secondary use	3	7
Number of constructs		12		9		6	27

121 constructs were defined in all models from the selected publications. These constructs were divided into groups according to the importance and number of times they appear in different models. As seen in Table 6, after combining the constructs with the same name or very similar meaning, altogether 62 different constructs were identified from 16 publications that discussed models related to privacy and security in mobile applications.

RQ2: Which other constructs are found in relation to privacy and security constructs in the researched models on mobile applications?

In this research question, an analysis will be done of the selected publications which were focused on mobile applications will be done. Privacy and security constructs have a direct effect on many different constructs. These constructs set the basics in the models, which are being researched. In the paper by

Models of Privacy and Security Issues on Mobile Applications

Table 6. Constructs used in the models by the number of appearance and groups of constructs

No.	Construct	Number of Appearances in the Publications	Group of Constructs
1.	Privacy concern	10	Privacy
2.	Privacy risk	5	
3.	Privacy settings	2	
4.	Privacy risk awareness	1	
5.	Perceived severity	1	
6.	Prior privacy experience	1	
7.	Perceived security	2	Security
8.	Perceived surveillance	2	
9.	Unauthorised secondary use	1	
10.	Awareness	1	
11.	Trust	1	
12.	Perceived benefit	7	Others, occurring more than once
13.	Trust	6	
14.	Continued intention to use	5	
15.	Ease of use	5	
16.	Satisfaction	5	
17.	Information to disclose via mobile applications	5	
18.	Perceived risk	4	
19.	Intention to install	3	
20.	Usefulness	3	
21.	Social nudge	3	
22.	Behavioural control	2	
23.	Perceived intrusion	2	
24.	Secondary use of personal information	2	
25.	Perceived control	2	
26.	Applications characteristics	2	
27.	Self-presentation	2	
28.	Personal innovativeness	2	
29.	System and network settings	1	Others, occurring once
30.	Perceived compatibility	1	
31.	Perceived critical mass	1	
32.	Number of IoT services	1	
33.	Quality	1	
34.	Acceptance	1	
35.	Frequency nudge	1	
36.	Creepiness	1	
37.	Performance expectancy	1	

continued on following page

Table 6. Continued

No.	Construct	Number of Appearances in the Publications	Group of Constructs
38.	Effort expectancy	1	
39.	Personalised service	1	
40.	Negative emotion	1	
41.	Location	1	
42.	Device content	1	
43.	Positive emotion	1	
44.	Registration info provided	1	
45.	Honest info	1	
46.	Employment	1	
47.	Accuracy of profile data submitted	1	
48.	Total clues found	1	
49.	The total number of times the participant logged into a website	1	
50.	Number of updates to profile data	1	
51.	Desensitization	1	
52.	Familiarity	1	
53.	Perceived reputation	1	
54.	Attitude	1	
55.	Errors	1	
56.	Improper access	1	
57.	Collection	1	
58.	Perceived complementarity	1	
59.	Availability	1	
60.	Accessibility	1	
61.	External	1	
62.	Image	1	

De Santo and Gaspoz (2015), global information privacy concerns have a significantly positive effect on privacy risk literacy and a negative effect on the intention to install the mobile application. Privacy risk literacy has a significantly negative effect on the intention to install the mobile applications. These two variables show that the more users are concerned and educated about privacy, the less possibilities there are that they will install a vulnerable mobile application.

In another publication by Degirmenci et al. (2013), mobile users' information privacy concerns are being significantly positively affected by the personal identity of the user, their location and mobile device content. The users' privacy concern has a significantly positive effect on secondary use of personal information, perceived surveillance and perceived intrusion. These results show that the access to users' personal information has an important effect on users' privacy concerns, as well as on the perceivement of security breaches into the users' data.

Models of Privacy and Security Issues on Mobile Applications

Harris et al. (2016) incorporated perceived security as a dependent variable in their model. It has a significantly positive effect on users' trust and significantly negative on perceived risk. The higher the user perceives security, the less risk the user perceives and the more they trust the mobile application.

Morosan and DeFranco (2015) analysed when do users disclose their information to hotel applications? The privacy construct of risk has a significantly negative effect on the value of disclosure. A high value of this construct was also defined by users' benefit, positive emotions and negative emotions. The model shows that the higher the users' value of disclosure is, the more he is willing to disclose data to hotel applications, and the higher he perceives the risks for his privacy, the lower the value of disclosure is.

Social media applications are used widely and developed recently. Ofori et al. (2016) analysed factors influencing the continuance of use of mobile social media. The privacy concern of the user in this model is an independent variable, which has a significantly negative effect on the user's satisfaction while using mobile social media and a non-significant effect on continuance of mobile social media application. On the other hand, satisfaction has a significantly positive effect on continuance of use. So to simplify, the higher the privacy concern is, the lower the satisfaction of the user is when using social mobile applications and, on the other hand, the higher the satisfaction of the user, the better the possibility is of the user continuing using mobile social media applications.

Another study was done on mobile applications and tested the effect of privacy nudges on privacy concerns of the user (Zhang & Xu, 2016). This study has shown a significantly negative effect of perceived control on privacy concerns and a significantly positive effect of creepiness on privacy concerns. The privacy concerns of the user have a significantly negative effect on disclosure comfort. The results of this study show that the higher the user's privacy concern is, the less the user is comfortable with disclosing information. When the user feels that they have control over their application they feel they have less privacy concerns, and the more the users feel uncomfortable by their friends, the higher their concern about their privacy.

The study shows that privacy and security constructs have an important part in discovering various mobile applications use.

RQ3: What types of measurements are used in research regarding models, involving privacy and security of mobile applications?

From the selected papers, an analysis of the measures used in the papers was done (Table 7). The most used measures are PLS – Partial Least Squares regression and SEM – Structural Equation Model-

Table 7. Publications by types of measures

Type of Measures	No. of Publications	Sources
PLS	7	De Santo and Gaspoz (2015), Harris et al. (2016), Keith et al. (2014), Keith et al. (2013), Ofori et al. (2016), Wang et al. (2016), Xu et al. (2009)
SEM	6	Degirmenci et al. (2013), Hsu and Lin (2016), Morosan and DeFranco (2015), Sheila et al. (2015), Zhang and Xu (2016), Zhao et al. (2016)
Model analysis	1	Xu et al. (2012)
Regression model	1	Hill and Troshani (2010)
Focus groups	1	Schnall et al. (2015)

ling. PLS is a form of SEM. 3 out of 16 papers are not using these measures. One paper is using model analysis, another presented a regression model and one tested the model with a focus group.

RQ4: Which mobile applications categories are researched in the field of privacy and security models?

From the selected papers an analysis was done of the application categories in the papers. Researchers usually analyse a specific field, such as hotel applications and security of information in mobile applications. The selected publications were divided into multiple categories considering the application domain while including privacy or security in mobile applications (Table 8). Some categories are not related directly to mobile applications, but are from a field that is connected closely to mobile applications. These papers present a good analysis of privacy and security issues that can then be transferred to the field of mobile applications. The additional fields are mobile devices in general, mobile personalisation services, internet of things services and location sharing services. These were used in 8 out of 16 papers. 8 papers are focused specifically on mobile applications; out of these 4 are in the field of mobile applications in general and one publication is in the field of health applications. One research is considering the field of social applications such as Facebook and WhatsApp, one mobile hotel applications, and the last one consumers who use mobile applications.

Limitations

As is common in research, our literature review has also limitations. A very small amount of papers was found to be appropriate for the thorough analysis of the publications' content regarding privacy and security in mobile applications. Mobile application and security and privacy is quite a new research field, as the first mobile applications were available for the broader public when Apple presented its App store in 2008 (Bowcock & Pope, 2008). Additionally, some publications might have been omitted due to limited resources.

Table 8. Publications by categories

Categories	Subcategories	Sources
Mobile applications	General	Degirmenci et al. (2013), De Santo and Gaspoz (2015), Wang et al. (2016), Zhang and Xu (2016)
	Health applications	Schnall et al. (2015)
	Social applications	Ofori et al. (2016)
	Hotel applications	Morosan and DeFranco (2015)
	Consumers application	Harris et al. (2016)
Mobile devices (not specific to mobile applications)		Keith et al. (2013); Xu et al. (2012), Keith et al. (2014), Sheila et al. (2015), Xu et al. (2009)
Mobile personalisation services		Hill and Troshani (2010)
Internet of Things services		Hsu and Lin (2016)
Location sharing services		Zhao et al. (2016)

DISCUSSION

The results acquired with this systematic literature review show the current state-of-the-art in research in the field of mobile applications. A specific focus was given on users' privacy and security aspects when using mobile applications. In the literature review, a special focus was given to publications that included presented statistical models of the effects of privacy and security on users of mobile applications. In the literature review it was found that a lot of papers are focused on the field of privacy and security in mobile applications. Our focus in the search were models incorporating privacy and security of mobile applications. However, not a lot of studies presented such models. Only 16 publications, containing models with privacy and security were found during a wide search on popular scientific publications databases. Mobile applications have been wide-spread only since 2008 and this is still quite a new research topic. However, based on the results of the literature review, researchers do not consider this a very important topic. On the other hand, users are most often well aware that their data's privacy and security should be assured. In some research users even report being worried about their privacy, but do not take any actions in protecting their privacy (Nemec Zlatolas, Welzer, Heričko, & Hölbl, 2015). Studies have also shown that it is important that users have control over their privacy settings in mobile applications (Delphine Christin, Engelmann, & Hollick, 2014).

From the 16 selected papers included in the literature review, an analysis was done of the confirmed statistical models in those papers. Altogether 121 constructs were identified in these models and, by combining similar or same constructs, 62 different constructs were identified finally. Out of the identified constructs, 6 were privacy constructs and 5 were security constructs. Privacy constructs appeared in the models 20 times and security constructs 7 times. The most often occurring construct was privacy concerns, which was presented in 10 of the selected papers. The most important aspects of privacy being researched in selected publications are privacy concerns of the users, perceived privacy risks and awareness of privacy risks, the possibility of adjusting privacy settings, perceived severity of privacy problems and prior privacy experience. The most important aspects of security are perceived security, perceived surveillance, unauthorised secondary use, security awareness and trust in security. It was found that most of the privacy and security constructs are independent or mediator variables in models, meaning, that they have an effect on other dependent variables. This also shows that privacy and security constructs are very important in defining the users' perception of mobile applications. In the selected papers, mostly Structural Equations Modelling and Partial Least Squares were used for the analysis of the models.

The literature review also revealed that researchers usually focus on one type of mobile applications. Some researchers have focused on mobile applications in general, and some on health applications, social applications, hotel reservation applications and consumers' applications. No clear differences have been identified between different types of mobile applications, but health applications carry very sensitive information about the user and this has to be considered when developing new applications.

Based on the results, further research could include the presentation of new models, concerning privacy and security while using mobile applications.

CONCLUSION

An overview of privacy and security in mobile applications was presented in this Chapter. This topic has been researched in recent years and an adequate number of papers were available to conduct such a review. This research focused on the analysis of publications related to models of privacy and security in regard to mobile applications.

The contribution of this study is a clearer knowledge on what privacy and security comprehensions of the users have an effect on the mobile applications field. The research could help mobile applications developers identify users' issues with mobile applications and get a clearer view of what the users are missing in the developed applications. The recommendations for setting new privacy and security rules in application markets could be extracted from this Chapter and be also in use by application developers. The research could also help users of mobile applications understanding of how their perception of privacy and security influences their actions in general. For example, users often report that privacy and security are important to them but, on the other hand, they publish a lot of personal information on mobile applications and share them with third-parties.

For the next research, we propose a wide literature review study with more detailed analysis of the search results.

REFERENCES

Ajami, R., Qirim, N. A., & Ramadan, N. (2012). Privacy Issues in Mobile Social Networks. *Procedia Computer Science*, *10*, 672–679. doi:10.1016/j.procs.2012.06.086

Al-khamayseh, S., Lawrence, E., & Zmijewska, A. (2006). *Towards Understanding Success Factors in Interactive Mobile Government* Paper presented at the Consortium International.

Anastasopoulou, K., Tryfonas, T., & Kokolakis, S. (2013). Strategic Interaction Analysis of Privacy-Sensitive End-Users of Cloud-Based Mobile Apps. In L. Marinos & I. Askoxylakis (Eds.), *Human Aspects of Information Security, Privacy, and Trust: First International Conference, HAS 2013, Held as Part of HCI International 2013, Las Vegas, NV, USA, July 21-26, 2013. Proceedings* (pp. 209-216). Berlin: Springer Berlin Heidelberg. doi:10.1007/978-3-642-39345-7_22

Baldwin, R. (2012). *Path Uploads Your Entire Address Book to Its Servers Without Your Explicit Permission*. Retrieved 18 June, 2016, from http://gizmodo.com/5883118/path-uploads-your-entire-address-book-to-find-your-friends

Baron, R. M., & Kenny, D. A. (1986). The moderator-mediator variable distinction in social psychological research: Conceptual, strategic, and statistical considerations. *Journal of Personality and Social Psychology*, *51*(6), 1173–1182. doi:10.1037/0022-3514.51.6.1173 PMID:3806354

Bowcock, J., & Pope, S. (2008). iPhone App Store Downloads Top 10 Million in First Weekend. *Apple Press Info*. Retrieved 10 June, 2016, from http://www.apple.com/pr/library/2008/07/14iPhone-App-Store-Downloads-Top-10-Million-in-First-Weekend.html

Christin, D., Engelmann, F., & Hollick, M. (2014). Usable Privacy for Mobile Sensing Applications. In D. Naccache & D. Sauveron (Eds.), *Information Security Theory and Practice. Securing the Internet of Things: 8th IFIP WG 11.2 International Workshop, WISTP 2014, Heraklion, Crete, Greece, June 30 – July 2, 2014. Proceedings* (pp. 92-107). Berlin: Springer Berlin Heidelberg. doi:10.1007/978-3-662-43826-8_7

Christin, D., Reinhardt, A., Kanhere, S. S., & Hollick, M. (2011). A survey on privacy in mobile participatory sensing applications. *Journal of Systems and Software*, *84*(11), 1928–1946. doi:10.1016/j.jss.2011.06.073

Christin, D., Roßkopf, C., & Hollick, M. (2013). uSafe: A privacy-aware and participative mobile application for citizen safety in urban environments. *Pervasive and Mobile Computing*, *9*(5), 695–707. doi:10.1016/j.pmcj.2012.08.005

Cornwall, J., Fette, I., Hsieh, G., Prabaker, M., Rao, J., Tang, K., . . . Hong, J. (2007). *User-controllable security and privacy for pervasive computing.* Paper presented at the 8th IEEE workshop on mobile computing systems and applications (HotMobile 2007). doi:10.1109/HotMobile.2007.9

D'Ambrosio, S., De Pasquale, S., Iannone, G., Malandrino, D., Negro, A., Patimo, G., . . . Spinelli, R. (2016). Energy consumption and privacy in mobile Web browsing: Individual issues and connected solutions. *Sustainable Computing: Informatics and Systems*. doi: 10.1016/j.suscom.2016.02.003

De Santo, A., & Gaspoz, C. (2015). Influence of Users' Privacy Risks Literacy on the Intention to Install a Mobile Application. In A. Rocha, M. A. Correia, S. Costanzo, & P. L. Reis (Eds.), *New Contributions in Information Systems and Technologies* (Vol. 1, pp. 329–341). Cham: Springer International Publishing. doi:10.1007/978-3-319-16486-1_33

Degirmenci, K., Guhr, N., & Breitner, M. H. (2013). *Mobile Applications and Access to Personal Information: A Discussion of Users' Privacy Concerns.* Paper presented at the International Conference on Information Systems, Milan, Italy.

Emmanouilidis, C., Koutsiamanis, R.-A., & Tasidou, A. (2013). Mobile guides: Taxonomy of architectures, context awareness, technologies and applications. *Journal of Network and Computer Applications*, *36*(1), 103–125. doi:10.1016/j.jnca.2012.04.007

Felt, A. P., Ha, E., Egelman, S., Haney, A., Chin, E., & Wagner, D. (2012). *Android permissions: user attention, comprehension, and behavior.* Paper presented at the Eighth Symposium on Usable Privacy and Security, Washington, DC. doi:10.1145/2335356.2335360

Gilbert, P., Chun, B.-G., Cox, L. P., & Jung, J. (2011). *Vision: automated security validation of mobile apps at app markets.* Paper presented at the second international workshop on Mobile cloud computing and services, Bethesda, MD. doi:10.1145/1999732.1999740

Guerra-Casanova, J., Sánchez- Ávila, C., de Santos Sierra, A., & del Pozo, G. B. (2011). Score optimization and template updating in a biometric technique for authentication in mobiles based on gestures. *Journal of Systems and Software*, *84*(11), 2013–2021. doi:10.1016/j.jss.2011.05.059

Harris, M. A., Brookshire, R., & Chin, A. G. (2016). Identifying factors influencing consumers intent to install mobile applications. *International Journal of Information Management*, *36*(3), 441–450. doi:10.1016/j.ijinfomgt.2016.02.004

Harrison, R., Flood, D., & Duce, D. (2013). Usability of mobile applications: Literature review and rationale for a new usability model. *Journal of Interaction Science*, *1*(1), 1–16. doi:10.1186/2194-0827-1-1

Hill, S. R., & Troshani, I. (2010). Factors influencing the adoption of personalisation mobile services: Empirical evidence from young Australians. *International Journal of Mobile Communications, 8*(2), 150–168. doi:10.1504/IJMC.2010.031445

Hsiao, C.-H., Chang, J.-J., & Tang, K.-Y. (2016). Exploring the influential factors in continuance usage of mobile social Apps: Satisfaction, habit, and customer value perspectives. *Telematics and Informatics, 33*(2), 342–355. doi:10.1016/j.tele.2015.08.014

Hsu, C.-L., & Lin, J. C.-C. (2016). An empirical examination of consumer adoption of Internet of Things services: Network externalities and concern for information privacy perspectives. *Computers in Human Behavior, 62*, 516–527. doi:10.1016/j.chb.2016.04.023

iOS app approvals. (2016). Retrieved 20 June, 2016, from https://en.wikipedia.org/wiki/IOS_app_approvals

Jain, A. K., & Shanbhag, D. (2012). Addressing Security and Privacy Risks in Mobile Applications. *IT Professional, 14*(5), 28–33. doi:10.1109/MITP.2012.72

Jeong, S.-H., Kim, H., Yum, J.-Y., & Hwang, Y. (2016). What type of content are smartphone users addicted to?: SNS vs. games. *Computers in Human Behavior, 54*, 10–17. doi:10.1016/j.chb.2015.07.035

Kang, J., Kim, H., Cheong, Y. G., & Huh, J. H. (2015). Visualizing Privacy Risks of Mobile Applications through a Privacy Meter. In J. Lopez & Y. Wu (Eds.), *Information Security Practice and Experience: 11th International Conference, ISPEC 2015, Beijing, China, May 5-8, 2015, Proceedings* (pp. 548-558). Cham: Springer International Publishing. doi:10.1007/978-3-319-17533-1_37

Kang, S. (2014). Factors influencing intention of mobile application use. *International Journal of Mobile Communications, 12*(4), 360–379. doi:10.1504/IJMC.2014.063653

Katell, M. A., Mishra, S. R., & Scaff, L. (2016). *A Fair Exchange: Exploring How Online Privacy is Valued.* Paper presented at the 2016 49th Hawaii International Conference on System Sciences (HICSS).

Keith, M. J., Babb, J. S., & Lowry, P. B. (2014). *A Longitudinal Study of Information Privacy on Mobile Devices.* Paper presented at the 2014 47th Hawaii International Conference on System Sciences.

Keith, M. J., Thompson, S. C., Hale, J., Lowry, P. B., & Greer, C. (2013). Information disclosure on mobile devices: Re-examining privacy calculus with actual user behavior. *International Journal of Human-Computer Studies, 71*(12), 1163–1173. doi:10.1016/j.ijhcs.2013.08.016

Kelley, P. G., Consolvo, S., Cranor, L. F., Jung, J., Sadeh, N., & Wetherall, D. (2012). *A conundrum of permissions: installing applications on an android smartphone.* Paper presented at the 16th international conference on Financial Cryptography and Data Security, Bonaire. doi:10.1007/978-3-642-34638-5_6

Kennedy-Eden, H., & Gretzel, U. (2012). A Taxonomy of Mobile Applications in Tourism. *e-Review of Tourism Research, 10*(2).

Khan, A. N., Mat Kiah, M. L., Khan, S. U., & Madani, S. A. (2013). Towards secure mobile cloud computing: A survey. *Future Generation Computer Systems, 29*(5), 1278–1299. doi:10.1016/j.future.2012.08.003

Khan, J., Abbas, H., & Al-Muhtadi, J. (2015). Survey on Mobile Users Data Privacy Threats and Defense Mechanisms. *Procedia Computer Science, 56*, 376–383. doi:10.1016/j.procs.2015.07.223

Kumar, K., & Lu, Y. H. (2010). Cloud Computing for Mobile Users: Can Offloading Computation Save Energy? *Computer*, *43*(4), 51–56. doi:10.1109/MC.2010.98

List of most downloaded Android applications. (2016). Retrieved 2016, 20 June, from https://en.wikipedia.org/wiki/List_of_most_downloaded_Android_applications

Liu, Y. (2014). User control of personal information concerning mobile-app: Notice and consent? *Computer Law & Security Report*, *30*(5), 521–529. doi:10.1016/j.clsr.2014.07.008

Liu, Z., Bonazzi, R., & Pigneur, Y. (2016). Privacy-based adaptive context-aware authentication system for personal mobile devices. *Journal of Mobile Multimedia*, *12*(1&2), 159–180.

Lupton, D., & Jutel, A. (2015). Its like having a physician in your pocket! A critical analysis of self-diagnosis smartphone apps. *Social Science & Medicine*, *133*, 128–135. doi:10.1016/j.socscimed.2015.04.004 PMID:25864149

Mahatanankoon, P., Wen, H. J., & Lim, B. (2005). Consumer-based m-commerce: Exploring consumer perception of mobile applications. *Computer Standards & Interfaces*, *27*(4), 347–357. doi:10.1016/j.csi.2004.10.003

Martínez-Pérez, B., de la Torre-Díez, I., & López-Coronado, M. (2014). Privacy and Security in Mobile Health Apps: A Review and Recommendations. *Journal of Medical Systems*, *39*(1), 1–8. doi:10.1007/s10916-014-0181-3 PMID:25486895

Melinger, D., Bonna, K., Sharon, M., & SantRamet, M. (2004). *Socialight: A Mobile Social Networking System*. Paper presented at the 6th International Conference on Ubiquitous Computing, Nottingham, UK.

Meulendijk, M. C., Meulendijks, E. A., Jansen, P. A. F., Numans, M. E., & Spruit, M. R. (2014). *What concerns users of medical apps? Exploring non-functional requirements of medical mobile applications*. Paper presented at the ECIS 2014 22nd European Conference on Information Systems. http://aisel.aisnet.org/cgi/viewcontent.cgi?article=1004&context=ecis2014

Mobile/Tablet Operating System Market Share. (2016). Retrieved 2016, 8 June, from https://www.netmarketshare.com/operating-system-market-share.aspx?qprid=8&qpcustomd=1

Morosan, C., & DeFranco, A. (2015). Disclosing personal information via hotel apps: A privacy calculus perspective. *International Journal of Hospitality Management*, *47*, 120–130. doi:10.1016/j.ijhm.2015.03.008

Mylonas, A., Kastania, A., & Gritzalis, D. (2013). Delegate the smartphone user? Security awareness in smartphone platforms. *Computers & Security*, *34*, 47–66. doi:10.1016/j.cose.2012.11.004

Nemec Zlatolas, L., Welzer, T., Heričko, M., & Hölbl, M. (2015). Privacy antecedents for SNS self-disclosure: The case of Facebook. *Computers in Human Behavior*, *45*(0), 158–167. doi:10.1016/j.chb.2014.12.012

Neuman, W. L. (2011). Social Research Methods: Qualitative and Quantitative Approaches (D. Musslewhite Ed.; 7th ed.). Pearson.

Ofori, K. S., Larbi-Siaw, O., Fianu, E., Gladjah, R. E., & Boateng, E. O. Y. (2016). Factors Influencing the Continuance Use of Mobile Social Media: The effect of Privacy Concerns. *Journal of Cyber Security and Mobility, 4*(2), 105–124. doi:10.13052/jcsm2245-1439.426

Ozdalga, E., Ozdalga, A., & Ahuja, N. (2012). The Smartphone in Medicine: A Review of Current and Potential Use Among Physicians and Students. *Journal of Medical Internet Research, 14*(5), e128. doi:10.2196/jmir.1994 PMID:23017375

Pattan, N., & Madamanchi, D. (2010). Study of Usability of Security and Privacy in Context Aware Mobile Applications. In T. Phan, R. Montanari & P. Zerfos (Eds.), *Mobile Computing, Applications, and Services: First International ICST Conference, MobiCASE 2009, San Diego, CA, USA, October 26-29, 2009, Revised Selected Papers* (pp. 326-330). Berlin: Springer Berlin Heidelberg. doi:10.1007/978-3-642-12607-9_21

Puttaswamy, K. P. N., & Zhao, B. Y. (2010). *Preserving privacy in location-based mobile social applications*. Paper presented at the Eleventh Workshop on Mobile Computing Systems and Applications, Annapolis, MD. doi:10.1145/1734583.1734585

Reinfelder, L., Benenson, Z., & Gassmann, F. (2014). Differences between Android and iPhone Users in Their Security and Privacy Awareness. In C. Eckert, S. K. Katsikas & G. Pernul (Eds.), *Trust, Privacy, and Security in Digital Business: 11th International Conference, TrustBus 2014, Munich, Germany, September 2-3, 2014. Proceedings* (pp. 156-167). Cham: Springer International Publishing. doi:10.1007/978-3-319-09770-1_14

Sadeh, N., Hong, J., Cranor, L., Fette, I., Kelley, P., Prabaker, M., & Rao, J. (2009). Understanding and capturing peoples privacy policies in a mobile social networking application. *Personal and Ubiquitous Computing, 13*(6), 401–412. doi:10.1007/s00779-008-0214-3

Sari, P. K., Candiwan, & Trianasari, N. (2014). *Information security awareness measurement with confirmatory factor analysis*. Paper presented at the Technology Management and Emerging Technologies (ISTMET), 2014 International Symposium on.

Schilit, B., Adams, N., & Want, R. (1994). *Context-aware computing applications*. Paper presented at the Mobile Computing Systems and Applications.

Schnall, R., Higgins, T., Brown, W., Carballo-Dieguez, A., & Bakken, S. (2015). Trust, Perceived Risk, Perceived Ease of Use and Perceived Usefulness as Factors Related to mHealth Technology Use. *MEDINFO 2015: eHealth-enabled Health, 216*. doi: 10.3233/978-1-61499-564-7-467

Sheila, M., Faizal, M. A., & Shahrin, S. (2015). Dimension of mobile security model: Mobile user security threats and awareness. *International Journal of Mobile Learning and Organisation, 9*(1), 66. doi:10.1504/IJMLO.2015.069718

Shilton, K. (2009). Four billion little brothers? Privacy, mobile phones, and ubiquitous data collection. *Communications of the ACM, 52*(11), 48–53. doi:10.1145/1592761.1592778

Statista. (2016). *Number of mobile app downloads worldwide from 2009 to 2017 (in millions)*. Retrieved 26 February, 2016, from http://www.statista.com/statistics/266488/forecast-of-mobile-app-downloads/

Wang, T., Duong, T. D., & Chen, C. C. (2016). Intention to disclose personal information via mobile applications: A privacy calculus perspective. *International Journal of Information Management, 36*(4), 531–542. doi:10.1016/j.ijinfomgt.2016.03.003

Weber, H. (2014). *Path has just 5 million daily active users globally*. Retrieved 17 June, 2016, from http://venturebeat.com/2014/09/10/path-has-just-5-million-daily-active-users-globally/

Weintraub, S. (2012). *Apple's iOS problem: Contacts uploading is just the tip of the iceberg. Apps can upload all your photos, calendars or record conversations*. Retrieved 15 June, 2016, from http://9to5mac.com/2012/02/15/apples-ios-problem-contacts-uploading-is-just-the-tip-of-the-iceberg-apps-can-upload-all-your-photos-calendars-or-record-conversations/

Whittaker, Z. (2012). *Twitter uploads contact list data without consent; retains for 18 months*. Retrieved 19 June, 2016, from http://www.zdnet.com/article/twitter-uploads-contact-list-data-without-consent-retains-for-18-months/

Xu, H., Gupta, S., & Pan, S. (2009). *Balancing User Privacy Concerns in the Adoption of Location-Based Services: An Empirical Analysis across Pull-Based and Push-Based Applications*. Paper presented at the iConference (iSociety: Research, Education, and Engagement), University of North Carolina, Chapel Hill, NC. http://hdl.handle.net/2142/15224

Xu, H., Gupta, S., Rosson, M. B., & Carroll, J. M. (2012). *Measuring Mobile Users' Concerns for Information Privacy*. Paper presented at the International Conference on Information Systems.

Xu, H., Rosson, M. B., & Carroll, J. M. (2008). *Mobile User's Privacy Decision Making: Integrating Economic Exchange and Social Justice Perspectives*. Paper presented at the AMCIS.

Xu, H., & Teo, H.-H. (2005). Privacy Considerations in Location-Based Advertising. In C. Sørensen, Y. Yoo, K. Lyytinen & J. I. DeGross (Eds.), *Designing Ubiquitous Information Environments: Socio-Technical Issues and Challenges: IFIP TC8 WG 8.2 International Working Conference, August 1–3, 2005, Cleveland, Ohio, U.S.A.* (pp. 71-90). Boston, MA: Springer US. doi:10.1007/0-387-28918-6_8

Zhang, B., & Xu, H. (2016). *Privacy Nudges for Mobile Applications: Effects on the Creepiness Emotion and Privacy Attitudes*. Paper presented at the 19th ACM Conference on Computer-Supported Cooperative Work & Social Computing, San Francisco, CA. doi:10.1145/2818048.2820073

Zhao, Y., Ye, J., & Henderson, T. (2016). *The Effect of Privacy Concerns on Privacy Recommenders*. Paper presented at the 21st International Conference on Intelligent User Interfaces, Sonoma, CA. doi:10.1145/2856767.2856771

Chapter 7
Virtualization in Mobile Cloud Computing (VMCC) Environments

Raghvendra Kumar
LNCT College, India

Prasant Kumar Pattnaik
KIIT University, India

Priyanka Pandey
Lakshmi Narain College of Technology, India

ABSTRACT

Unfortunately, most of the widely used protocols for remote desktop access on mobile devices have been designed for scenarios involving personal computers. Furthermore, their energy consumption at the mobile device has not been fully characterized. In this chapter, we specially address energy consumption of mobile cloud networking realized through remote desktop technologies. In order to produce repeatable experiments with comparable results, we design a methodology to automate experiments with a mobile device. Furthermore, we develop an application that allows recording touch events and replaying them for a certain number of times. Moreover, we analyze the performance of widely used remote desktop protocols through extensive experiments involving different classes of mobile devices and realistic usage scenarios. We also relate the energy consumption to the different components involved and to the protocol features. Finally, we provide some considerations on aspects related to usability and user experience.

INTRODUCTION

Mobile devices discussed by authors Rajkumar Buyyaa, Chee Shin Yeoa, Srikumar Venugopala, James Broberga, and Ivona Brandicc (2009), ranging from smart phones to tablets, have recently become so pervasive that they are increasingly replacing personal computers in everyday activities related to both entertainment and work. However, due to their limited resources, mobile devices cannot or the same

DOI: 10.4018/978-1-5225-2469-4.ch007

performance of personal computers and workstations. To this regard, one of the prominent approaches to overcome such limitations consists in adding computational and storage resources to the cloud. With adding, the mobile device runs only a thin layer of software which interfaces with application-specific services in the cloud. For instance, software for picture organization and categorization can exploit powerful and accurate face recognition algorithms discussed by author Huaglory Tianfield (2011) running on the cloud without the need of any computation at the mobile device. However, such an approach requires that source data are available to the remote service. This may require transferring data from the mobile device to the cloud, which incurs in both communication and energy consumption overheads, discussed by authors P. Sefton (1980). A different option is given by remote desktop access. In this case, the mobile device uses thin client software which connects to a remote desktop server providing an operating system and its applications. The thin client shows the desktop user interface and handles the related interactions. Specially, input events captured by the client are transferred to the server and the display of the mobile device is updated according to the received response, in order to match the content of the desktop screen. To a certain extent, remote desktop access can be seen as an extreme case of mobile cloud computing, wherein the mobile device only acts as remote display and input device, while all the rest is demanded to the remote system. When the remote server is virtualized, this access scheme corresponds to a special case of mobile cloud networking. While research targeted to mobile cloud computing has considered resource utilization as the primary design objective, most of the commonly used solutions for remote desktop access were originally designed for personal computers. As a consequence, the reference scenario was represented by systems which have enough resources, are static and access the Internet through a wired connection. Even though there are some solutions specially designed for mobile devices, they are usually not publicly available, or they cannot be easily integrated in the existing infrastructure. As a consequence, the vast majority of remote desktop protocols available for mobile devices are still those designed for personal computers. In this chapter, we aim at characterizing the energy consumption of mobile cloud computing realized through remote desktop technologies. Our goal is to analyze the performance of widely used remote desktop protocols through experiments involving different classes of mobile devices and realistic usage scenarios. Consequently, we have to ensure the reliability and consistency of all experiments discussed by author D. Rowe (2011). Finally, we also seek to relate the energy consumption to the different components involved and to the protocol features discussed by author D. E. Y. SAR NA (2011).

Cloud is a parallel and distributed computing system consisting of a collection of inter-connected and virtualized computers discussed by authors M. Cafaro and G. Aloisio (2010) that are dynamically provisioned and presented as one or more unified computing resources based on service-level agreements (SLA) established through negotiation between the service provider and consumers discussed by authors Souvik Pal and P.K.Pattnaik (2012).

CONCEPT OF MOBILE CLOUD COMPUTING

The mobile cloud computing shows Figure 1 was outlined because the convenience of cloud computing services during a mobile scheme discussed by authors Judith Hurwitz, Robin Bloor, Marcia Kaufman, Fern Halper (2010), which contains several components together with shopper, enterprise, femtocells, transcending, end-to-end security, home gateways, and mobile broadband-enabled services. In fact, the researchers take into account the mobile cloud computing because the combination of mobile computing, mobile web and cloud computing discussed by authors Souvik Pal, Suneeta Mohanty, SisirKunar Jena, Prasant Kumar Pattnaik, (2012).

Mobile computing technology is delineating as a variety of human-computer interaction by that a laptop is anticipated to be transported throughout traditional usage in Wikipedia . Mobile computing relies on a group of 2 major concepts: hardware and package. The ideas of hardware are often thought-about as mobile devices, like good phone and laptop computer, or their mobile parts. Package of mobile computing is that the varied mobile applications within the devices, like the mobile browser, anti-virus package.

Mobile web technology is that the combination of mobile communication and web, the essence of that is to let shoppers get real time network resources and network service, together with the infrastructure of mobile networks, protocols and knowledge delivery in their use.

Cloud computing could be a large-scale economic and business computing paradigm with virtualization as its core technology. The cloud automatic data processing system is that the development of multiprocessing, distributed and grid computing on the web, that provides varied QOS secured services like hardware, infrastructure, platform, package and storage to totally different web applications and users.

As shown is that the Figure three, mobile cloud computing are often merely divided into mobile computing, mobile web, cloud computing. Those mobile devices connect with the hotspot or base station by the mobile web. The "cloud" plays a job within the computing and major processing. The mobile devices may also accomplish mobile cloud computing by employing a cross platform mid-ware. Mobile users send service requests to the cloud by the mobile web, so the management element of cloud allocates resources to the request to ascertain association.

Features of Mobile Cloud Computing

1. **Effectiveness of Task Processing:** Rapidly deploys and increase employment by speedy providing physical machines or virtual machines. This advantage clearly indicates that users will see the results of tasks directly by mobile devices though the interface of input and output isn't ok discussed by authors A. Kivity et al. KVM 2007.
2. **Convenience of Sharing Data:** A great deal of knowledge is holding on within the cloud finish of servers, sanction active sharing knowledge handily. If the information measure is wide enough, it'll work as fluently as domestically, that is straightforward to comprehend for mobile devices.
3. **Elimination of Rationality:** Mobile cloud computing eliminates the limitation of rationality, sanction active individuals to get what they need at anytime and anyplace from the mobile web. At identical time, MCC will monitor real-timely resources usage and rebalance the allocation of resources once required.
4. **Mobile Devices Independence of MCC System:** The entire computation area unit carried on within the cloud-far finish servers, therefore, mobile cloud computing doesn't have demand for mobile devices, even imbecilic cell phones may also notice mobile cloud computing.
5. **Robust Accommodative Ability:** Mobile cloud computing manages a spread of various workloads, together with the batch of back-end operations and user-oriented interactive applications discussed by authors B. Rochwerger et al 2009. It support for redundancy, self-healing and extremely ascendable programming model, in order that employment are often pass though a spread of inevitable hardware/software failure.

Figure 1. Architecture of mobile cloud computing

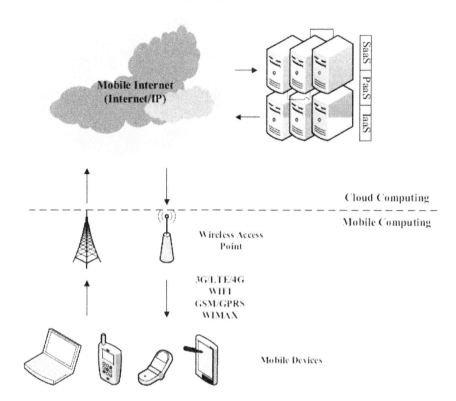

SECURITY THREATS PRESENT IN MOBILE CLOUD COMPUTING (MCC)

There are a unit three vital things to be considered the safety of mobile cloud computing discussed by authors Souvik pal, Suneeta Mohanty, Dr. P.K. Pattnaik, and Dr. G.B. Mund (2012). Firstly, the safety of mobile cloud computing is nearly specifically just like the security of mobile web. The safety tools that area unit used these days to guard the inner network cloud, conjointly accustomed defend knowledge within the mobile cloud. Secondly, for remaining financially competitive, a number of these security technologies ought to be touched to the mobile cloud. Thirdly, if a top quality mobile cloud service supplier is chosen discussed by authors Pooja Malgaonkar, Richa Koul, Priyanka Thorat, Mamta Zawar (2011), the safeties within the mobile cloud are pretty much as good as or higher than this security in most cases. We tend to offer the attainable security threats in step with the system structure of the mobile cloud computing discussed by authors V. Sarathy, P. Narayan, RaoMikkilineni (2010).

EXPERIMENTATION IN VMCC

From the challenges as discussed above, the major issue in implementing mobile cloud computing is to divide the application on the mobile device in such a way so that it will require least resources on mobile device and will apply minimum load on network to get the processing done on the cloud and display results to the user. It also faces the problem of security of the data from the network users and over the cloud.

This chapter is offering to provide security of the mobile data for both the security over cloud and encryption / decryption of the data over the network. The algorithm for this work is as follows:

Algorithm 1

```
Step 1: A User interface shall be created over cloud for adding the mobile id
online, which is added after password authentication by the mobile user.
Step 2: Mobile application shall be created using android application devel-
opment tool, which shall connect to the cloud database using SaaS application
developed over the cloud.
Step 3: As the mobile app tries to communicate with the cloud application, it
will require authenticating internally with the mobile id which was added by
the user through web application.
Step 4: After authentication, mobile will communicate with the cloud applica-
tion using encrypted data.
Step 5: For encryption of the mobile data poly alphabetic substitution cipher
will be used which is faster and secure as require n26 permutations by the
brute force attack.
Step 6: The work shall have following implementation modules:
Mobile App Module
Authentication
Encryption using Poly Alphabetic Substitution Cipher
Decryption using Poly Alphabetic Substitution Cipher
SaaS Layer Application Module
Web Interface for Mobile Users
Encryption using Poly Alphabetic Substitution Cipher
Decryption using Poly Alphabetic Substitution Cipher
Communication module with Mobile Application
Step 7: Poly Alphabetic Substitution Cipher requires O(n) complexity to en-
crypt or decrypt the data.
```

The data or contacts earlier were stored in the mobile, then the mobile's internal memory is been used. In this we are saving mobile data in cloud and can be retrieving quickly when required. The data is been stored in clouds database.

Algorithm 2: Secured Communication Model Using Mobile Cloud Computing

```
1.      Firstly we have to register our information with cloud, by the means
of user name and password.
2.      After registration we have to sign in our account. If unable to sign
in repeat step 1.
3.      After login to your account we have to select that contact which is
```

Virtualization in Mobile Cloud Computing (VMCC) Environments

```
to be stored in to the cloud.
4.        After selecting that contact, the data is been encrypted and sent to
the cloud.
5.        Now that contact is been saved to the cloud and is fetched only when
user name and password is entered correctly.
6.        Now you have data saved in the cloud and the memory is not been oc-
cupied by the data in mobile.
7.        In order to fetch data we have to just type the name and data will
be in your phone.
```

This is a home page of the application shows Figure 2 "Secured Communication Model using Mobile Cloud Computing"

This is a login page of the application shows Figure 3. If user is already registered user just have to login.

This is the registration page, if user is not registered; the user has to register first to the cloud application. After registering, user can start saving contacts to the cloud. In saving contacts to cloud the data or contact will be encrypted and will be send to cloud. User just has to add name and contact number and that contact will be encrypted and will be save in the cloud. This is how contact looks when it will be save in the cloud. When we want to retrieve contact we just have to type name or contact number and data will be retrieving from the cloud. This is how the user interface shows figure 4 will look when it will be run on mobile.

For searching contacts in mobile application shows Figure 5.

Number of contacts saved in a mobile device.

Figure 2. Snapshot 1 of cloud server

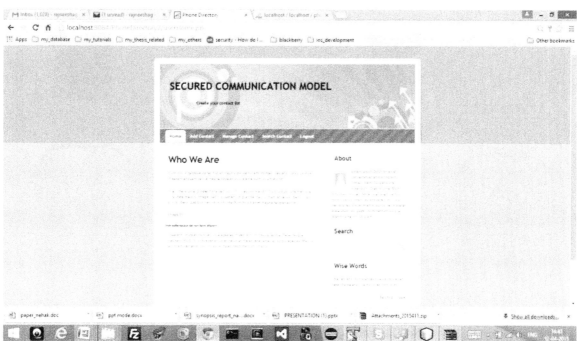

Figure 3. Snapshot 2 of cloud server

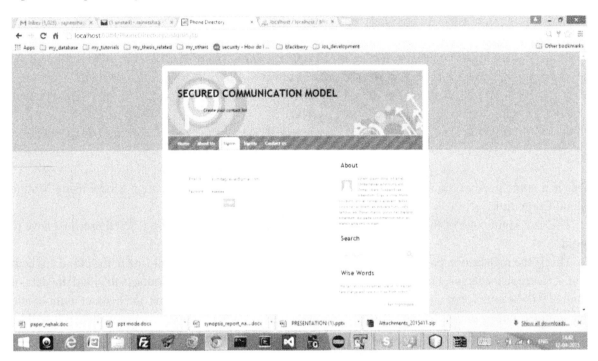

Figure 4. Snapshot 1 of mobile application

Figure 5. Snapshot 2 of mobile application

CONCLUSION

This chapter introduced mobile cloud computing as the latest emerging technology. Besides there are several critical drawbacks in devices, which include battery life time, security issues etc. This chapter presents an approach for improving the security and authenticity of mobile cloud computing in an android environment. In future, next step will be to improve the battery lifetime of devices, which is using mobile cloud computing.

REFERENCES

Buyyaa, R., Yeoa, C. S., Venugopala, S., Broberga, J., & Brandicc, I. (2009, June). Cloud computing and emerging IT platforms: Vision, hype, and reality for delivering computing as the 5th utility. *Future Generation Computer Systems*, *25*(6), 599–616. doi:10.1016/j.future.2008.12.001

Cafaro, M., & Aloisio, G. (2010). *Grids, Clouds and Virtualization*. Springer-Verlag New York, Inc.

Hurwitz, J. (2010). *Cloud computing for dummies*. Wiley Publications.

Kivity, A. (2007). KVM: the linux virtual machine monitor. *Proceedings of the Linux Symposium*, *1*, 225–230.

Malgaonkar, Koul, Thorat, & Zawar. (2011). Mapping of Virtual Machines in Private Cloud. *International Journal of Computer Trends and Technology*, *2*(2), 54-57.

Pal, S., & Mohanty, S. (n.d.). *An Approach to Cross-Cloud Live Migration of Virtual Machines in Cloud Computing Environment*. Unpublished.

Pal & Pattnaik. (2012). Efficient architectural Framework of Cloud Computing. *International Journal of Cloud Computing and Services Science*, *1*(2), 66-73.

Pal, Mohanty, Pattnaik, & Mund. (2012). A Virtualization Model for Cloud Computing. *Proceedings of International Conference on Advances in Computer Science*, 10-16.

Rochwerger, B., Breitgand, D., Levy, E., Galis, A., Nagin, K., Llorente, I. M., & Galan, F. et al. (2009). The RESERVOIR model and architecture for open federated cloud computing. *IBM Journal of Research and Development*, *53*(4), 1–11. doi:10.1147/JRD.2009.5429058

Rowe. (2011). The Impact of Cloud on Mid-size Businesses. In *Implementing and Developing Cloud Computing Applications*. Taylor and Francis Group, LLC.

Sarathy, Narayan, & Mikkilineni. (2010). Next generation cloud computing architecture -enabling real-time dynamism for shared distributed physical infrastructure. *19th IEEE International Workshops on Enabling Technologies: Infrastructures for Collaborative Enterprises (WETICE'10)*, 48-53.

Sefton, P. (1980). *Privacy and data control in the era of cloud computing. OECD Guidelines on the Protection of Privacy and Transborder Flows of Personal Data.* Organisation for Economic Cooperation and Development.

Tianfield, H. (2011). Cloud Computing Architectures. *Proceedings of Systems, Man, and Cybernetics (SMC), 2011 IEEE International Conference*, 1394 – 1399.

Chapter 8
The SMAC Opportunity Contracts:
Generating Value From Responsive Agile Risk-Oriented Techniques

Mohammad Ali Shalan
Jordan Engineers Association, Jordan

Nebal Abdulrazzak Anaim
SJ Group, Saudi Arabia

ABSTRACT

The concept of Social, Mobile, Analytics and Cloud (SMAC) is increasingly asserted as the phenomena with the potential to change technology and business relationships. In this SMAC era plenty of Middle Circle Contractors (MCCs) are being introduced as principal suppliers, integrators or outsourced contractors. This is reducing the Client Enterprise (CE) controls over their technology assets. Because it is not mature yet, SMAC nature is very disruptive and agile, thus rapidly changing the landscape of contracting, and ultimately turning the long-held promise of utility based computing into a reality. Such changes necessitate contracting transformation with innovated approaches to get targeted benefits, reduce risks and enhance operational controls. The main objective of this chapter is to provide guidelines to generate SMAC Opportunity Contracts (OCs) that are responsive and agile to provide the maximum business value, enhance risk governance and re-invent the roles and obligations in an ever-changing environment.

INTRODUCTION

The SMAC (Social, Mobile, Analytics and Cloud) phenomenon has emerged as a growing trend of scalable, flexible and powerful paradigm shift that is affecting how technology is delivering value to businesses. It has emerged as a journey to achieve the Client Enterprise's (CE) technology expectations through plenty of Middle Circle Contractors (MCCs) including its staff in the internal Technology Organization (TO), SMAC Service Providers (SSPs), integrators and subcontractors.

DOI: 10.4018/978-1-5225-2469-4.ch008

All parties need to act jointly to enhance service mission and support functions on a shared basis in an agile, correlated and involved manner. This will enable designing and selecting solutions that provide more business value that is purchased based on consumption in a shared model. Consequently, with significant changes and challenges, SMAC is completely revolutionizing the contracting mindset, and promoting new collaborated, responsive and agile approaches.

Technology is moving today from "complicated" bulk acting units into "complex" system behaviors comprising multiple independent units. Complex systems is like bird flocks and fish schools that are separate units having a leader to orchestrate their joint behavior. These separated and integrated units are mostly acquired from different SSPs, generating several risks, security concerns, contracting, compliance and many other issues.

Additionally, systems are evolving continuously with plenty of components being removed or integrated in an agile, rapid and correlated form. Traditional contracting can no longer fit the SMAC era. Legal practitioners need to introduce more agile contracting mechanisms to fit the new era and the partnership requirements.

One more contracting challenge is introduced by moving the CEs' key applications and corporate information to be processed in multiple environments that are publicly owned. Add to this the technology systems and SMAC services that can be adopted outside the internal TO. Both may form a shadow environment causing plenty of loose activities and associations.

This chapter aims to portray a picture -for Opportunity Contracts (OCs)- that supports agile responsive contracting to smartly harmonize such an exponentially growing heterogeneous environment. The target is to realize value from SMAC platforms which requires plenty of iterations to become mature enough. Solid performance management tools are still needed for further development.

The chapter insight is putting value, agility, risk and ethics in the heart, while providing highly valuable experience to those looking for guidance to write contracts for SMAC services comprising MCC outsourcing. It aims to increase navigation clarity through the chaos, complex features and multi-layer contracting. It will also provide answers to legal practitioners who are trying to generate a structured and systematic contracting approach. This chapter characterizes many SMAC features that should be addressed. It also discusses the components of associated agreements and the shift in change management behaviors.

The main objective of this chapter is to establish a jump start to support the move from a "traditional contract" which is bounded often to a rigid fixed price based on specific requirements, to an "OC" that has a different approach. This will help understand and capture the operational and delivery processes into the contract, thus mirroring real-world successful projections and effects. Additionally, this chapter aims to help CEs to manage some of the risks surrounding SMAC contracting in a smooth and effective manner.

BACKGROUND

The opportunities being created are seemingly unlimited and driving a digital surge of exponential impact. SMAC is ranked high with a significant capability to shift the integration between business and technology. It is expected to influence all business factors toward maximizing the benefits and limiting the risks in this utility based era (Cornelius, 2013). However, the technology-business coupling still lacks a universal way to measure the alignment (HBR, 2011; De Haes & Grembergen, 2009). Literature

The SMAC Opportunity Contracts

is trying to clarify the key benefits and provide measures to recognize when the CE intended goals are met (Laakkonen, 2014).

The four SMAC dimensions (Social, Mobile, Analytics and Cloud) are heavily discussed in literature, but evolving separately or bi-combined, with plenty of research dedicated to each topic individually (Ackermann, 2012). For instance, "The mobile cloud computing paradigm is set to drive technology over the next decade and integrate the resources availability through the 3As (Anywhere, Anything, Anytime)" (Mastorakis, Mavromoustakis, & Pallis, 2015, p. xvii). However, while cloud computing is considered as a steward in the SMAC era, its standards are still evolving.

"Digital transformation occurs when the physical and the digital worlds join forces" (Shelton, 2013, p. 9). Authentic act and trust are considered as important factors. Literature provides guidelines about outsourcing decisions, vendor selection and other issues that need to be addressed when opting for services (Hoda, Noble, & Marshall, 2009). An evolutionary and holistic view of technology outsourcing draws an extensive literature and theory (Tiwana, & Bush, 2007). Any agreement between CEs and SSPs should be cascaded to MCCs to enforce common acceptable rules across all parties (Metheny, 2013). Outsourcing and contracting call out for various technical and legal issues (Siepmann, 2014; Marks, 2015).

"The motivation for using agile methods is first and foremost to make the development process as efficient as possible" (Ganes & Nævdal, 2008, p. 13). Agile started with a manifesto which describes the main principles to prioritize certain values over others (Beck et al., 2001). Agile, somehow, is promoting the move into a utility based model of computing where an application can start small and grow enormously overnight. "This democratization of computing means that any application has the potential to scale, and that even the smallest seed planted in the cloud may be a giant" (Sosinsky, 2011, p. xxv). The agile concept affects the contracting behaviors at every level to encourage innovation and sets a side time for the SSPs and CEs to collaborate regularly (Laakkonen, 2014).

To establish a structured approach for SMAC contracting with a reduced business, legal practitioners are trying to understand features that affects SMAC contracts (Ferrier, 2015; Jayaswal et al., 2014). Such features include the flexible scope, agile delivery, smart documentation, acceptance criteria and payment scheduling (Opelt et al., 2013; Hoda et al., 2009). Change management, value proposition and maturity considerations are also discussed but with a lot of noise around (Subramaniam & Hunt, 2006; Franklin, 2008). This chapter is trying to tailor these concepts and features to cope with the SMAC era.

SMAC era is shifting and uplifting the business. Thus, it is gaining more space to reshape the business as a combined phenomenon. More studies are discussing SMAC related topics. The main differentiator of this chapter is to jointly address most of the aspects and to orchestrate the wide spectrum of different elements. This effort is intended to help practitioners and researchers understand necessary correlations and plan enhancements toward an OCs for SMAC services with supplementary agreements to generate the necessary business value.

CONTRACTING FOR THE SMAC

In traditional contracting, time and material (T&M) contracts provide the least amount of risk for service providers, but it is difficult to sell to CEs who fear an open ended agreement and the latent costs. On the other hand, fixed price (FP) contracts are not acceptable by the suppliers who are required to absorb all the scope risks, consequently leading to extremely expensive projects if all required contingency is made. "Clients also lose out in fixed price because they commit themselves to a scope too early in the project and then suffer the cost and difficulty in making changes once they realize their needs are different" (Eckfeldt et al., 2005, p. 161). OCs aim to produce the right mix in view of the SMAC agility, prospects and risks.

SMAC acumens necessitate the separation of requirements, results and design paths to facilitate a mechanism in analogy to the "no cure, no pay" concept. This is important because:

- Most of the SMAC services are acquired for direct buy to sell activities in which the final purchase cannot be committed unless the sell activity is complete.
- Outsourcing and MCCs are heavily and creatively utilized.
- Most SMAC projects refract from the original path or product that was initially contracted.

Considering the lean principles, OCs should enable learning, discovery, innovation and re-evaluation capabilities. OCs can derive powerful prioritization and optimization to achieve the key business objectives while considering the metrics that matter. Outcomes can be measured over time in a quantified manner to reduce cost and increase value to various stakeholders. This section defines few contracts' alphabet and looks at differentiators related to the SMAC era.

The Purpose of Contracts

Per Ganes & Nævdal (2008), "Contracts are used in order to reduce perceived risk between the buyer and the supplier" (p. 21). A contract is a written or spoken agreement, with specific terms between two or more persons or entities in which there is a promise to do something in return for a valuable benefit known as consideration. This definition compromises three elements which are:

- Offer and acceptance in the meeting of minds;
- Promise to perform with commitment; and
- Consideration for payment.

A contract is a legally binding agreement that act as protective blankets against legal action. They also help formulate the plan to complete a project, and determine how the client and supplier will interact with one another. Contracts usually define the delivery date, associated fees and costing, requirements, responsibilities, stipulations and penalties for missing requirements. There are concerns that must be agreed upon between the parties before service deployment starts. Two schools of thought identify the purpose of contracts (Charette, 2005, pp. 42-49):

- Claims to protect each party from opportunistic behavior of the other party.
- Set up appropriate incentives for companies to work together in a synergistic manner.

Buyer-supplier associations evolve over time, as business relationships that are indeterminate in terms of changes materialize over time (Pierantonelli, Perna, & Gregori, 2015, pp. 144-152). This happens in a non-linear process that is moving towards an ideal state. Four stages can be identified in a business relationship comprising: pre-relationship, exploratory, development, and stable (H°akansson et al., 2011, p. 45). Learning, investment, adaption, distance, trust and commitment are concepts behind these four stages that explain the development of the business relationship.

The SMAC Opportunity Contracts

Some times when a client is under time pressure, a letter of agreement (LOA) can be used to start the work, while a parallel process continues to reach a master services agreement (MSA) that covers the remaining contractual issues (Eckfeldt et al., 2005, p.163). A contracting strategy is to use an LOA to cover project related details while the MSA is to cover all non project specific information regarding the terms of the relationship. Contracts typically include defining confidential information and non-disclosure terms, ownership of work product, indemnification, any non-compete agreements, and several specific legal sections related to the administration of contracts.

Shifting the Paradigm

Some people assume the nature of SMAC projects is alike to construction projects which is not the case. Construction projects are relatively predictable, while SMAC projects are highly uncertain and variable. In traditional projects, legal practitioners spend plenty of time studying problems arising from silo mentality assuming great problems if the project is stopped at any arbitrary point in time. SMAC projects should begin with an agile contract, which explains the same limitations and liabilities as in a traditional contract, but is written in a way that requires time for team collaboration and feedback.

The manifesto for agile software development clearly states that agile methods value working software over comprehensive documentation (Lander, 2014). This means there is less planning and estimation in agile projects, enforcing a contractual challenge. As compared to traditional projects, where detailed specifications are used to price the contract (Awad, 2005). The structural and legal aspects of SMAC contracts vary a little from traditional styles. However, "the key difference is the approach to, and understanding of, how the operational process and delivery, and how this is captured in or intersects with contracts" (Larman, & Vodde, 2010, p. 500).

SMAC projects are not always predictable, and they don't have the entire plan mapped out from beginning to end. There is a starting point, research and development midway checkpoints, followed by a solution process. For such projects where the entire target isn't visible from the outset of the project, bad contracts can lead to poor project planning, risk exposure, liability and weak team structure.

One exhaustive area in traditional contracting lies in the proliferation of complex, multi-document contract structures which are often poorly updated and oddly worded. Thus, customers need to wade through the many pieces of paper, that usually lack consistency leading to frustration. These multi-layered contract structures are unwieldy and often, when quizzed, even the providers' representatives cannot navigate their way around them. On the other side, SMAC contracts are using a much-relaxed approach which results in less expensive, less time consuming, but higher value deliverables with a flexible structure that can adapt to change and match real-life expectations.

Typical contracts can hinder productivity, thus incorporating agility in SMAC contract management offers several benefits. It allows contract managers to write contracts that are focused on quality, rather than punishing vendors for not delivering the terms of the contract. It also allows for adjustments and enables change to benefit the business so that the final product is what the client desires and benefits from.

Agility encourage innovation and sets a side time for SSPs, MCCs and CEs to collaborate regularly to make sure that all are on the right path toward project success (Laakkonen, 2014). Visibility is increased with plenty of checkpoints along the way in the form of demos, deployments and tests, so that team members and stakeholders are aware of how the project is unfolding. OCs are focused on maintaining high-quality team performance rather than simply defining the delivery date.

Agility Motivations

The word "agile" by itself means that something is flexible and responsive. An "agile method" implies its ability to survive in an atmosphere of constant change and emerge with success. Whether doing agile or being agile, agile is an umbrella term for several different models with the key feature that they are iterative. Agile approach applies a timebox in place of the traditional phase-gate to managing progress. Consequently, CEs will be involved in frequent evaluations of deliveries rather than successive intermediate work products that lead up to a single delivery.

Per the agile manifesto (Beck et al., 2001), agile methodology focuses on four main principles, while prioritizing a value over one another:

- Valuing individuals and interactions over processes and tools;
- Working software over comprehensive documentation;
- Customer collaboration over contract negotiation; and
- Responding to change over following a pre-defined plan.

While the world of SMAC is changing frequently, agile provides value to cope up with such changes in an effective manner. Agile contracting models are needed to balance the risks between both parties and give both the incentive to work together towards the success of the joint project. An agile contract has a clear goal to discover and execute a viable product or business. It puts the entire business in the Deming cycle of Plan-Do-Check-Act and forces to confront business assumptions and risks while writing a contract. This gradual effect provides an ideal opportunity to respond to changes and feedback to avoid a big bang delivery. "An Agile approach, rather than the traditional waterfall approach, would give it a better chance to deliver the new services to challenging timescales and targets" (UK_NAO, 2014, p. 20).

Risk Motivations

"Risk management practice has presented asthmatic behavior when it comes to keeping the pace with enterprise strategy and objectives" (Shalan, 2010, p. 115). Per Larman & Vodde (2010), there are three general areas to be concerned with when drafting a contract:

- Risk and exposure,
- Flexibility to allow for change, and
- Clarity regarding obligations, deliverables, and expectations

SMAC will require a contractual approach that embraces agile methods to reduce risks and advance relative interests. This will help CEs avoid the very problems that a lawyer is worried about in traditional contracts. Consequently, it is important to keep a widely open eye on the area of requirement gathering and testing. An agile approach expect that requirements will be articulated in an iterative and evolutionary manner.

Understanding technology concepts is important to determine what efforts are reasonable to address risks. Thus, CEs must do their due diligence or consult with a technology expert who can make trustworthy contractual recommendations. "Technology in general, and cloud computing in particular are promoting the shift to risk based approaches" (Shalan, 2017, p.54). This has been brewing for some time and is crucial because SMAC has no clear casted forms or checks that can cover its entire spectrum. Consequently, business heads must start thinking in terms of acceptable risk levels versus precast requirements.

To be reasonable, the CEs efforts must be commensurate with the risks presented. As CEs cannot be protected against every conceivable risk when using SMAC services, numerous factors should be considered in real life practices. For example, CEs should ask themselves how bad it would be if unauthorized parties gained access to information on a case by case basis. The experience and reputation of the SSP is duly important, where CEs should choose providers with proven records.

Value Motivations

The software industry is always providing new techniques for business. Value contracting is turning a single contract into successive contracts saving the time required to award successful individual contract. Even the administrative cost is outweighed by the potential cost and technical benefits derived from the real value delivery. Also, it can maintain the contractor continuity if a predominant concern of risks occurs among multiple phases. This means that if any phase encounters a certain risk, it will be addressed in the next relevant phase.

A value contract reduces the change requests and the different types of incentives. It motivates greater cost control and better performance, nevertheless, incentives can still be applied. A value based contracting enables money to be better spent on requirements that were not recognized at the beginning rather than wasting time and money in developing requirements that are not ultimately needed.

From a legal point of view, value based contracting can protect a client from threats that are not known at the contracting period. "All of the industry experts agreed that trust between the supplier and customer is very important" (Ganes & Nævdal, 2008, p.78). The value contract allows for inevitable change and focuses negotiations on the actual delivery that accommodate inevitable changes during the lifetime of the contract.

Outsourcing Motivations

Understanding the roles and responsibilities of all stakeholders involved in deploying a SMAC solution is critical to successful implementation. The National Institution of Science and Technology (NIST) reference architecture describes five major actors with their roles and responsibilities in the cloud computing taxonomy: the cloud consumer, cloud provider, cloud broker; cloud auditor; and cloud carrier (Hogan & Sokol, 2013).

Outsourcing arrangements are increasingly encompassing SMAC computing elements. CEs should be prepared to invest more time and effort evaluating an ever-expanding roster of providers and solutions. While using subcontractors, there are five rules of the road worth keeping in mind that is to know your traveling partners well, protect your passengers, keep everyone on course, chart the itinerary together and plan for arrival (Trant & Ravi, 2014).

Outsourcing is a common method whereby a third party performs a function on behalf of the CE, often when it doesn't have the resources (either time or expertise) to undertake by itself. In SMAC era there exist two dimensions for outsourcing, one is related to a subcontractor hired by the CE to integrate internal technologies with a SMAC service, while the other is related to a subcontractor who is hired directly by an SSP. The CE may procure services directly from an SSP and separately with an MCC. In all scenarios, any agreement should be cascaded to enforce common acceptable use of standards across all parties effectively.

Agency theory is concerned with the coordination and motivation issues that are inherent in a relationship between a principal and an agent. A basic assumption of agency theory is that opportunism is an inherent characteristic of such a relationship and an important risk factor in an outsourcing contract. The most common issue is by far communication, since it is very easy to misunderstand or interpret something in the wrong way because of culture, language or other difficulties. (Tiwana, & Bush, 2007).

SMAC Related Agreements

Combined with SMAC contracts there should be some provisions which are important to both CEs and SSPs to define the acceptable behavior of personnel from both parties. Such agreements should be fully contemplated by both prior to procuring SMAC services. All such agreements need to be incorporated, either by full text or by reference, to avoid the usually costly and time-consuming process of negotiating these details after the enactment of a SMAC contract.

SSPs enforce common acceptable use of standards across all users to effectively maintain how CEs use the SSP environment in a bounded behavior as stated by Terms of Service Agreements (TOSAs). The TOSA usually includes provisions that detail how end-users may use the services, responsibilities of the SSP, and how the SSP will deal with customer data. On the other hand, the Non-Disclosure Agreements (NDAs) are required to enforce acceptable SSP personnel behavior when dealing with CEs' data (Metheny, 2013).

Service Level Agreements (SLAs) represent an important component under the umbrella of the overall SMAC contracts. SLAs define acceptable service levels to be provided by the SSP in measurable terms. Software Escrow Agreements (SEAs) are used to ensure continued use of critical business systems by depositing the software source code with a neutral third party escrow agent (Olson & Peters, 2011). They mitigate the risks that are inherent when an organization relies on third party technology vendors. The software source code is released to the licensee if the licensor vanishes out or fails to maintain and update the software as specified in the license agreement.

BUILDING THE OPPORTUNITY CONTRACTS

SMAC has launched a new utility based era that correlates and integrates various services and components to create unique values for CEs in an effective manner. However, adopting SMAC services does not change the CE legal duties. It is important to understand the division of rights and responsibilities to make an informed judgement about the legal risks. Preparing an opportunity contracts requires a knowledge of the SMAC nature and its unique characteristics which is partially discussed in this section.

Thinking Journeys

SMAC is an agile journey, not an end feature. Journeys typically ignite plenty of improvement areas that require improved collaboration and alignment during the contracting process. Such journeys require agile contracts capable of introducing high level scopes that are detailed enough to depict the service as required by CEs per this contracting. Such high-level scope that is linked to the business value is better than getting lost in many details, which may be deprecated over time (Opelt et al., 2013).

The journey mindset necessitates an agile contracting to solve the alleged contradiction between fixed and progressive scenarios. This mindset helps to get out of the rigid contracting trap. Yet, contracting a trusted SSP, choosing the appropriate services and deployment models are critical initial steps in the SMAC journey. Nonetheless, as more time elapses the service itself will be more mature, helping CEs to continuously acquire different services or jump between multiple providers. Effective contracting should always support such a dynamic behavior.

The Joint Venture Attitude

Enterprises inevitably look out for their own interests, trying to take advantage of opportunities without regard toward the consequences to others (Poppendieck & Poppendieck, 2013). Some of them even seek immediate advantage with little regard to ultimate consequences. Contracts are needed to limit opportunistic behaviors. SMAC presents a paradigm shift that motivates more technology involvement with substantive issues that need to be addressed in the business and contracting models.

This new paradigm requires CEs to re-think the way they acquire technology services. Such paradigm can provide mission and support functions on a shared basis between CEs, TOs and SSPs who are becoming more agile, correlated and involved. This helps in designing and selecting solutions that provide more business value while allowing adaptive payment based on consumption in a shared model. Assuming all parties would act in good faith, a long relationship attitude is required to limit opportunism and use the contract to set up incentives that align the best interests of each party.

From Complicated to Complex

In the systems theory, "complicated" refers to something with many parts, whereas "complex" refers to unpredictable, emergent behavior (Wells, 2013). A complex system is like bird flocks and fish schools that are separate units having a leader to orchestrate the flock joint behavior. However, this leader has nothing to do with every individual. On the other hand, a complicated system has many parts, acting altogether to produce an effect. Every part has a functionality that cannot be replaced by any other part. Therefore, complicated systems are harder to understand and every part has no advantage when working separately.

Complex systems are not constructed; they are grown from interdependent components. They are not fully predictable as they are made up of many components with a lot of coupling and feedback. "In a complex system, the same starting conditions can produce different outcomes, depending on the interactions of the elements in the system" (Sargut & McGrath, 2011, p. 69). SMAC nature has promoted the usage of complex systems where CEs utilize large numbers of independent components that can be contracted, operated and managed independently, to create value from technology and enable business agility. During its lifetime, a large and complex software must evolve to remain useful. SMAC contracting mindset is heavily affected by this phenomenon.

Business Requirements

SMAC components are exponentially growing and becoming more versatile. Nevertheless, the CE may expect to find enhancements or alternatives every day before the project is complete. In addition, business people may have new requirements that affect the specifications of each component. SMAC contracting should be flexible enough to use business cases or user stories to understand the required CE features. Throughout the deployment, requirements must be verifiable by the CE at certain check points to confirm that they are met and that the acceptance is ready to be signed based on the acceptance criteria defined beforehand.

SMAC contracting promotes the user experience enhancements. "The description of the feature must be fully understood by the customer and by the development team. The description uses a vocabulary that is familiar to the customer, no technical mumbo-jumbo! "(Cauwenberghe, 2016, p. 3). The CE should include the business value of every feature. Business people should spare enough time to discuss, review and improve the specifications of certain component. Else, planning cannot move forward, and bad impression about that component importance will be generated.

Managing Integration

SMAC contracting needs to consider the integration between multiple providers, services or components, with each having its own characteristics. Such behaviors generate an ever-changing uncertainty or risk profiles. Also, it is reflected in multi-phase frameworks for involved parties that are drifting or shifting over the passage of time. Good contracts produce increased alignment of motivations since all parties have "skin in the game" and they may improve fundamental fairness and relationship building.

The concept of a win-win approach is located at the heart of SMAC contracting philosophy. It creates the trust and relationships that will foster further business and avoid the blame-game of shifting pain to the other party. Risk should be borne by the party best able to manage it, thus uncertainty in the domain needs to be managed by CE, while uncertainty in the technology will be under the SSP management (Poppendieck & Poppendieck, 2013). Various agreements should be considered to support solid integration, associated risks and effectively manage SMAC components. These agreements should include agile roles and responsibilities to provide careful delineation between the accountabilities and relationships among all parties.

Measurements, Forensics, and Auditing

OCs help stakeholders focus on what they are good at, to make explicit value delivery to business through an iterative, incremental delivery. To enhance the ultimate outcome, OCs should include means to measure progress in terms of business value not delivered features. This will provide context to design performance requirements and business constraints. CEs should require SSPs to allow forensic investigations for both criminal and non-criminal purposes. These investigations need to be conducted without affecting data integrity and without interference from the SSPs.

SMAC standards are not yet mature, but they include conceptual models, reference architectures, and criteria to facilitate communication and contracting. International or voluntary standards can be utilized except if the legislations are inconsistent or impractical (Metheny, 2013). Internal and external auditors may require to preserve audit logs. Therefore, CEs must work with SSPs to ensure audit logs of an SSP

The SMAC Opportunity Contracts

environment are preserved with the appropriate standards. Any changes to the SSP environment should follow a pre-agreed terms and conditions.

FEATURES OF SMAC OPPORTUNITY CONTRACTS

Technically there exist plenty of terms and conditions that need to be associated and clarified in the SMAC contracts such as jurisdiction and ownership. This section is more concerned with unique features that are directly related to SMAC contracting.

Starting at the Provider Terms

Contracts for SMAC services are generally implemented on the provider's terms, thus the degree of negotiability pales in comparison with the contracting model in traditional services (Opelt et al., 2013). The providers' terms are what they are. There is a general recognition that CEs must start at SSP terms, especially those well established large providers who dictate their rules. Thus, if a CE expects customization with genuine negotiation of service terms, then there should be special considerations related to the acquired solutions or components.

The contracting areas perceiving most points of negotiation tend to be commercially oriented issues such as price, scope, service levels and liability caps. These are areas where CEs often show their naivety by asking for changes that directly contradict the commoditized nature of the service offering. Little changes are based on the in-variability of the technical solution, even when an issue is plainly commercial and not technical.

Flexible Scope

Agile contracts do not define an exact project scope; however, the degree of scope specificity and change vary from low to high based on the pricing scheme (Larman, & Vodde, 2010). Near one end of the spectrum are target-cost contracts, in which the overall project scope and details are identified at the start as best as possible along with change mechanisms. At the other end are progressive contracts, in which no necessary scope is defined beyond one iteration.

OCs suggests that the contract framers, including the legal staff, must be involved at early stage, with active participation in project visioning workshops and other project-engaged activities. Participation will help the legal professionals to creatively write from their own understanding to achieve a comprehensible contract that avoids demonstrating legalistic, locally optimized silo mentality. Such phenomenon leads to rigid scope, limited vision and mysterious language traditional contracts.

Multi-Layered Structure

SSPs encourage CEs to contract for few iterations to begin with, instead of signing a comprehensive contract up front (Opelt et al., 2013). This allows the CE to use a service or product on a trial basis to build confidence with the SSP and avail a sort of risk coverage. Once the CE has tried few iterations, then he is offered the option to buy more iterations or features as needed. Such behavior will help the SSPs to communicate a message that there are no risks to the CEs.

Additionally, the client will have the option of terminating the project within a one sprint notice. Consequently, the CE can lose a maximum of one sprint only. The CE does not have to make up the entire vision. It can include changes in sprints, making incremental benefits (Hoda et al., 2009). Sometimes providers insert a termination clause in the contract such that clients have the option to quit on predefined iteration notices.

Incremental Deliverables

SMAC contracts promote incremental delivery of services and features, either through development or acquisition in a pre-defined delivery cycle and milestones. They include how every iteration delivery is done, programmed, tested, and so on, and how it will be deployed, within a time-boxing of fixed scope. Traditional project contracts often include a detailed, prescriptive list of what should be delivered. Conversely, OCs need to avoid such specificity and rigidity, but to focus on artifacts that are valuable.

To change mindsets, SMAC contracts should focus on cooperation to manage integration and reinforce learning and responding toward creative discovery work. This is better than the illusory command-control behavior over planning as in traditional contracting which generates a sense that a fully defined system can be predictably ordered and delivered, just like pizza and its ingredients.

Responsive Change Management

The issue of change is largely inherently addressed within the overall philosophy of the agile approach to allow re-prioritization of phases and adaptive iteration without a harsh change management process. This area requires significant care in OC to prevent subverting agility and to make the change easy and frequent. SSPs often show the CEs the features that are seldom used, giving them more control of the product. This responsiveness change the mindset of CEs and encourage them to look beyond the constraints of contracts toward the bigger picture.

The concept of a change for free clause in contracts, allows CEs to change feature priorities for free if the total contract work remains the same. It also enables CEs to add new features if low priority items of equal work are removed from the contract (Hoda et al., 2009). However, this does not mean that all kinds of change management are dispensed within contractual form. Pertinent variations may result due to changes in relationships between parties, such as when a party is being acquired by another entity. A fundamental transformation in corporate direction may also occur.

Creative Contract Termination

The concept of termination is linked with change control, in that a SMAC service should be amenable to changing course. This may be to the point of stopping further effort at the end of any iteration. In contrast to conventional project thinking, legal professionals need to understand that early termination should be viewed as a positive, desirable event in an agile project. That is because such activity may not be a failure, it can mean that success was achieved quickly.

Arguably the ideal termination model in an agile contract is to allow the CE to stop, without penalty, at the end of any iteration. Termination can be one of the most difficult areas to negotiate in any contract. The key qualifying differences in an agile approach is either the CE has a working system each

The SMAC Opportunity Contracts

iteration, or both parties will have clear and up-to-date views on the state of the deliverable. These are crucial points for legal professionals to grasp.

Acceptance Criteria

As the SMAC paradigm allows shifting between the service providers, behaviors of CEs can be liberated from sticking to one vendor. Such an attitude attenuates the acceptance criteria conditions which can be incremental and adaptively agreed upon for each service. This attitude relaxes service acceptance terms and eases negotiations during the contracting phase, avoiding massive comprehensive exercises. The framework for acceptance must be contractually clear. The contract language should help considerably in soothing the project progress and encourage collaboration.

Acceptance builds upon itself such that the final acceptance is the culmination of several acceptances (for sprints or services) that have occurred earlier, with enough clarity regarding acceptance and correction. Another element of acceptance in SMAC era is to include candidate users when defining acceptance testing in the initial contracting exercise. Legal professionals concerned with a successful project should ask, are the right people involved in acceptance and are they collaborating with the supplier? (Larman & Vodde, 2010).

Liability and Warranty

Negotiation of liability clauses is perhaps the most difficult area in any contract, and the SMAC era does not change that. However, it reshapes liability to have a lesser impact on operations by discovering any negative consequence sooner, and addresses the knock-on effects through the liability paradigm. OCs would proactively reduce the damage exposure of several elements such as cost and goodwill, and minimize consequences. It might be cheaper to fix because the system would have various independent components with fewer entanglements.

Similarly, the concerns related to warranty are attenuated in SMAC era. As the deliverables of every sprint will be under warranty, the risk profile associated with the final product' warranty is considerably less. The confidence and acceptance will be in the deliverables themselves. This is due to accumulated acceptance and multiple components. Warranty should be tied to each incremental working package, though there should be still an overall warranty to the final product.

Smart Documentation

In the SMAC era the technical documentation to support system maintenance is less valuable to the CE as that support will be conducted by the same provider who created the system which is evolving. Accordingly, it could be wasteful to require over documentation even if the provider accepts to deliver it. Conversely, there may be a demonstrated need for such documentation if the customer takes over the maintenance work. In this case the contract should consider creating a joint agile documentation for the final system. This represent smart documentation.

This is contradictory to traditional methods which place much emphasis on documenting the whole process, from the requirement specification through design, implementation and testing, all the way to Go Live. Specific documentation related to user training and support could be of more importance as early deliverables that are going to be continuously updated all over the project lifecycle.

Payment Scheduling

SMAC contracts should discuss billing and payments including bonus and penalty clauses. Although the most popular practice is to pay a periodic charge in addition to setup fees if required, more complex payment schemes can be tied to complex portfolios. Hence, some shared pain or gain accountabilities may be applied. Pay-per-use models should be designed to align the interests of the CE and the SSP. There are multiple means and business cases that contracts should care about to achieve fair payment schedules, just to mention a few:

- In average business scenarios, a CE can be regularly invoiced based on frequency of service usage or number of transactions in addition to occasional maintenance or update investments. SSPs can benefit from average business scenarios by shifting the loads. For example, the low utilization of school systems during vacations can be compensated with the high utilization of tourism systems in the same period.
- In odd business scenarios, i.e. exceptional seasons, CEs and SSPs alike should be protected against extreme conditions highly deviating away from average service usage (below-average or above average). For example, during seasonal and natural disasters. A specific example can be: A bidding real estate company sells a huge number of apartments twice a year only, all at once in a few days with very high system usage, while the reminder of the year the system is only used for planning with very low system usage.
- In distinctive business scenarios, CEs and SSPs alike should share the costs of developing custom enhancements requested by a specific CE then sold by the SSP to other CEs based on the future revenue. For example, based on unique requirements, an SSP develops an innovative mobile service for a hospital, with a strategic vision to become paperless, to make its operations conveniently available at the fingertips of doctors and patients. If this SSP sells this service to other hospitals, then profits can be shared with the original hospital as a return on the capital employed (RoCE) on the developed custom system.

Additional Contract Elements

There exist few points which belong to every contract such as the objectives of the engagement and cooperation. The outline of the project structure including processes, key roles and applied variances should be included also in the contract. Key personnel who are responsible at the operational and escalation levels should be named including what is required out of them.

Contracts in the SMAC era can be negotiated for plenty of issues. For instance, some CEs need to have control and visibility over SSP subcontractors. Other CEs might negotiate the SSP ability or limitation to modify the services provided based on validated requirements. In such scenarios, the contract negotiation ought to focus on the commercial implications of such changes rather than the basic right itself (Opelt et al., 2013). SSPs should acknowledge such a negotiation right of CEs.

Privacy and data security commitments by SSPs toward CEs are crucial to determine who is responsible and how obligations should be allocated between them. One scenario may be that an SSP cares about the security of its network, while its CEs care about the security of their data. Another scenario may define the rights to suspend services under circumstances such as non-payment or violation of an acceptable use policy. Contracts should also layout provisions of limitation of liability and termination assistance. Such points are negotiable in a manner that ensures business continuity, allows the CE to extend SSP services for a convenient period when required and to facilitate migration to the replacement solution.

Data location and transferability may be a disputed issue. For instance, most of the large SSPs react to commercial pressures from Europe-based CEs by offering services from ring-fenced European data centers only. Furthermore, CEs should set performance and measurement metrics jointly with SSPs to professionally execute SMAC services and avoid the "take it or leave it" approach. Finally, some conditions related to security, privacy, e-discovery or e-records management may be added to SMAC contracts where necessary.

FUTURE RESEARCH DIRECTIONS

Nowadays, deep technology penetrations are creating fast shifts in business paradigms. For instance, SMAC can be overlooked as a complex system that combine multiple SSPs, MCCs, subcontractors and integrators to provide a terrain of technology services. Such system should be studied further to harmonize and correlate its ingredients. Role changes are viable today as well. For example, the CIO is being transformed from Chief Information Officer to Chief Innovation Officer who should be engaged with business people from early discussion to add value to business and align with the CE strategies. Likewise, legal practitioners need to be involved at earlier stages of service acquisition to understand the evolving requirements and practices before drafting a contract.

SMAC opens a wide research terrain and vast opportunities presenting both theories and their applications in real-life. Traditional contracts are no longer sufficient; or even worse, they can hinder business development. Agile behaviors including OCs should be further studied as the new norm for SMAC services that is aligned with new business-technology paradigm. OCs are very important to handle a lot of challenges and provide self-standing services that can be sustained during business transformation. A wide discussion is required to re-phrase the SMAC OCs terms characteristics and methods.

There are interdependent relationships between internal and external stakeholders. Within a CE there are relationships which ought to be aligned strategically, tactically and operationally between business people and the TO. These relationships are extended and expanded to external parties, such as SSPs, MCCs among other providers. This should be researched and rephrased to enhance involvement and collaboration among these teams. This is increasingly affecting the legal practice behaviors and tactics to get more involved and to perform necessary amendments to the OCs.

Jointly, all the above topics requires plenty of research which is considered part of the SMAC arena that have plenty of dependencies and correlations. The development of effective and acceptable OCs methods for SMAC era requires the participation of a very wide range of stakeholders. To be successful, the research should benefit from various contracting behaviors including the agile, traditional and off-shore contracting while avoiding pit falls and rigid structures. This chapter generates thorough conclusions, and number of research paths that researchers can follow to advance the knowledge in SMAC arena and the OCs.

CONCLUSION

Business-technology relationship is transforming the future. The 3As (Anywhere, Anything, Anytime) represent a convenient practice that users are looking for. SMAC is leading a major reform that is affecting all business engagement models. It has unique features including joint venture attitudes and complex systems' adoption. Practice has shown that plenty of SMAC features should be well-understood prior to SMAC contracting processes. This includes, but not limited to, flexible scope, responsive change management, iterative processing, multi-layered structure and agile implementation.

Traditional contracts are no longer capable of providing the desired value in the SMAC era. SMAC contracting is unique; it should utilize best practices to enable operational flexibility, satisfy integration and isolation requirements, manage various providers and orchestrate services. This is widely promoting the concept of OCs where agility and smart features are utilized to correlate with stakeholders' continuously evolving requirements. However, without direct involvement of legal practitioners into the planning and design workshops, there will be no point of reference for prioritizing and drafting a working SMAC contract.

SMAC transactions have triggered a major dependency on MCCs promoting the "man-in-the-middle" phenomenon, which create new dimensions into the contracting mechanisms and processes. CEs, SSPs and MCCs need to act faithfully to their best interest and to be involved across the spectrum. These activities should also be cascaded to engage integrators and subcontractors in addition to the main SSPs. Where necessary, a code of conduct should be composed to outline responsibilities, practices and data handling.

The agile value based approach offers a broad view of associated risks and business expectations. Stakeholders including the legal practitioners need to be involved earlier to widen their scope and evaluate the emerging challenges, risks, values and practices. As part of their due diligence the CE's contracting team should assess involved providers for sufficiency, maturity, reputation and ethics. This contracting team should include members from relevant departments such as strategy, procurement, technology, legal, quality and other concerned divisions. Also, they should manage the associated contracts and service agreements such as non-disclosure agreements (NDAs), service level agreements (SLAs) and software escrow agreements (SEAs).

The bottom line is that SMAC contracting must achieve what seems to be impossible to support the ever-changing agile business requirements while protecting the rights and benefits of CEs, SSPs and MCCs. This chapter attempts to present such an effort and aspires to ignite fruitful discussions that will eventually lead to an increased number of research studies in the field of SMAC contracting. Eventually this is expected to pave the way for a smooth and fruitful trusted relationship between all stakeholders.

REFERENCES

Ackermann, T. (2012). *IT Security Risk Management: Perceived IT Security Risks in the Context of Cloud Computing*. Berlin, Germany: Springer-Gabler.

Awad, M. A. (2005). *A Comparison between Agile and Traditional Software Development Methodologies* (Unpublished master dissertation). The University of Western Australia, Perth, Australia.

Beck, K. (2001). *Manifesto for Agile Software Development*. Retrieved February 5, 2016, from http://agilemanifesto.org/

Cauwenberghe, P. V. (2016). *Agile Fixed Price Projects part 1: "The Price Is Right"*. Retrieved February 5, 2016, from http://www.nayima.be/html/fixedpriceprojects.pdf

Charette, R. (2005). Why software fails. *IEEE Spectrum, 42*(9), 42–49. doi:10.1109/MSPEC.2005.1502528

Cornelius, D. (2013). *SMAC and transforming innovation*. Paper presented at the meeting of the 2013 PMI Global Congress, New Orleans, LA.

De Haes, S., & Grembergen, W. (2009). *Enterprise governance of information technology: Achieving strategic alignment and value*. New York: Springer. doi:10.1007/978-0-387-84882-2

Eckfeldt, B., Madden, R., Horowitz, J., & Grotta, E. (2005). Selling Agile: Target-Cost Contract. In *Proceedings of 2012 Agile Conference* (pp. 160-166). Denver, CO: IEEE Computer Society.

Ferrier, M. (Ed.). (2015). *Leading in the SMAC age*. Bangalore, India: Wipro.

Franklin, T. (2008). Adventures in Agile Contracting: Evolving from Time and Materials to Fixed Price, Fixed Scope Contracts. In *Proceedings of IEEE Agile 2008 Conference* (pp. 269-273). Los Alamitos, CA: IEEE Computer Society doi:10.1109/Agile.2008.88

Ganes, A., & Nævdal, S. (2008). *Software Contracting and Agile Development in the Norwegian ICT Industry: A Qualitative Survey* (Unpublished Master dissertation). Norwegian University of Science and Technology (NTNU), Trondheim, Norway.

Hakansson, H., Ford, D., Gadde, L., & Snehota, I. (2011). *Managing Business Relationships*. Hoboken, NJ: John Wiley & Sons.

Hoda, R., Noble, J., & Marshall, S. (2009). Negotiating Contracts for Agile Projects: A Practical Perspective. In *Proceedings of 10th International Conference on Agile Processes in Software Engineering and Extreme Programming* (pp. 186-191). Berlin: Springer. doi:10.1007/978-3-642-01853-4_25

Hogan, M., & Sokol, A. (Eds.). (2013). NIST Cloud Computing Standards Roadmap, NIST Special Publication, 500-291. Gaithersburg, MD: National Institute of Standards and Technology (NIST).

Jayaswal, K., Kallakurchi, K., Houde, D., & Shah, D. (2014). *Cloud Computing Black Book*. New Delhi, India: Dreamtech Press.

Laakkonen, K. (2014). *Contracts in Agile Software Development* (Unpublished Master Dissertation). Aalto University, Helsenki, Finland.

Lander, K. (2014). *Contracted to deliver outcomes*. Retrieved February 5, 2016, from http://www.energizedwork.com/weblog/2014/09/agile-beach-2014-contracted-deliver-outcomes

Larman, C., & Vodde, B. (2010). *Practices for Scaling Lean & Agile Development*. Boston, MA: Addison-Wesley & Pearson Education Inc.

Marks, N. (2015). *The myth of IT risk*. Retrieved September 09, 2015, from https://normanmarks.wordpress.com/2015/08/28/the-myth-of-it-risk/

Mastorakis, G., Mavromoustakis, C., & Pallis, E. (2015). *Resource Management of Mobile Cloud Computing Networks and Environments*. Hershey, PA: IGI Global. doi:10.4018/978-1-4666-8225-2

Metheny, M. (2013). *Federal Cloud Computing: The Definitive Guide for Cloud Service Providers.* Waltham, MA: Elsevier.

National Audit Office. (2014). *A snapshot of the use of Agile delivery in central government.* London: Author.

Olson, D. & Peters, S. (2011). Managing Software Intellectual Assets in Cloud Computing, Part 1. *Journal of Licensing Executives Society International, H*(3), 160-165.

Opelt, A., Gloger, B., Pfarl, W., & Mittermayr, R. (2013). *Agile Contracts: Creating and Managing Successful Projects with Scrum.* Hoboken, NJ: John Wiley & Sons. doi:10.1002/9781118640067

Pierantonelli, M., Perna, A., & Gregori, G. L. (2015). Interaction between Firms in New Product Development. *International Conference on Marketing and Business Development Journal, 1*(1), 144-152.

Poppendieck, M., & Poppendieck, T. (2013). *The Lean Mindset: Ask the Right Questions.* Westford, MA: Addison-Wesley.

Review, H. B. (2011). *Harvard Business Review on Aligning Technology with Strategy.* Boston, MA: Harvard Business School Publishing.

Sargut, G., & McGrath, R. (2011). Learning to Live with Complexity. *Harvard Business Review, 89*(9), 68–76. PMID:21939129

Shalan, M. A. (2010). Managing IT Risks in Virtual Enterprise Networks: A Proposed Governance Framework. In S. Panios (Ed.), *Managing Risk in Virtual Enterprise Networks: Implementing Supply Chain Principles* (pp. 115–136). Hershey, PA: IGI Global. doi:10.4018/978-1-61520-607-0.ch006

Shalan, M. A. (2017). Ethics and Risk Governance for the Middle Circle in Mobile Cloud Computing: Outsourcing, Contracting and Service Providers Involvement. In K. Munir (Ed.), *Security Management in Mobile Cloud Computing* (pp. 43–72). Hershey, PA: IGI Global. doi:10.4018/978-1-5225-0602-7.ch003

Shelton, T. (2013). *Business Models for the Social Mobile Cloud: Transform Your Business Using Social Media, Mobile Internet, and Cloud Computing.* Indianapolis, IN: John Wiley & Sons. doi:10.1002/9781118555910

Siepmann, F. (2014). *Managing Risk and Security in Outsourcing IT Services: Onshore, Offshore and the cloud.* Boca Raton, FL: Taylor and Francis Group.

Sosinsky, B. (2011). *Cloud Computing Bible.* Hoboken, NJ: Wiley Publishing, Inc.

Subramaniam, V., & Hunt, A. (2006). *Practices of an Agile Developer.* Frisco, TX: Pragmatic Bookshelf.

Tiwana, A., & Bush, A. A. (2007). A Comparison of Transaction Cost, Agency, and Knowledge-Based Predictors of IT Outsourcing Decisions: A U.S.–Japan Cross-Cultural Field Study. *Journal of Management Information Systems, 24*(1), 259–300. doi:10.2753/MIS0742-1222240108

Trant, M., & Ravi, R. (2014). *Cloud bound: Advice from organizations in outsourcing relationships.* Retrieved February 5, 2016, from http://www.ibm.com/smarterplanet/us/en/centerforappliedinsights/article/cloudbound.html

Wells, J. (2013). *Complexity and Sustainability*. New York, NY: Routledge.

ADDITIONAL READING

Abram, T. (2009). The hidden values of it risk management. *Information Systems Audit and Control Association (ISACA) Journal*, 2009(2), 40-45.

Ben Halpert, B. (2011). *Auditing Cloud Computing: A Security and Privacy Guide*. Hoboken, NJ, USA: John Wiley & Sons. doi:10.1002/9781118269091

Dallas, G. (Ed.). (2004). Governance and risk. New York, USA: McGraw-Hill companies Inc.

Dutta, A., Peng, G. C., & Choudhary, A. (2013). Risks in enterprise cloud computing: The perspective of IT experts. *Journal of Computer Information Systems*, 53(4), 39–48. doi:10.1080/08874417.2013.11645649

Grembergen, W. (Ed.). (2003). *Strategies for information technology governance*. Hershy, Pennsylvania, USA: Idea Group Publishing.

Kark, K., & Vanderslice, P. (2015). CIO as Chief Integration Officer. *Deloitte Journal of Tech Trends 2015*, 6 (1), 04-19.

Kunreuther, H. (2006). Risk and reaction: Dealing with interdependencies. *Harvard International Review: Global Catastrophe*, 28(3), 17–23.

Moran, A. (2015). *Managing Agile: Strategy, Implementation, Organization and People*. Berlin, Germany: Springer. doi:10.1007/978-3-319-16262-1

Tarkoma, S. (2012). *Publish / Subscribe Systems: Design and Principles*. Hoboken, NJ, USA: John Wiley & Sons. doi:10.1002/9781118354261

Westerman, G., & Hunter, R. (2007). *IT risk: Turning business threats into competitive advantage*. Boston, Massachusetts, USA: Harvard Business School Publishing.

Zhao, F. (Ed.). (2006). *Maximize business profits through e-partnerships*. Hershey, PA: IRM Press. doi:10.4018/978-1-59140-788-1

KEY TERMS AND DEFINITIONS

Agile Methodology: A development method and a leadership philosophy that encourages teamwork, self-organization and accountability to develop a dynamic service that can respond to change and continuously deliver business value.

Client Enterprise (CE): An organization that uses the professional, networking or computing services provided by SMAC Service Providers (SSPs) per a signed contract against some agreed financial charges.

Middle Circle Contractor (MCC): An external or internal person, group or organization that is appointed to perform work or to provide goods/services at a certain price or within a certain time. The MCC appears as a middle person who usually disappears after the specified task is complete.

Opportunity Contracts: A suggested contracting structure that enables successful business engagement, through the maximization of opportunities and reduction of risk based on agile and smart features that are being utilized to correlate with stakeholders' continuously evolving requirements. This contracting methodology focuses on sharing pains/gains between involved stakeholders, setting all possible mechanisms to expand and convey quality and value to stakeholders.

Outsourcing: A common method whereby a third party performs a function on behalf of the Client Enterprise (CE), often when additional resources (either time, expertise, human resources, service, etc.) are needed.

Risk Management: The act of handling the risk exposure through mitigation, acceptance, sharing and avoidance. It includes the ability to handle information and technology risks based on stakeholders' risk parameters.

SMAC Service Provider (SSP): An entity that supplies SMAC services usually based on their existing platforms from a remote facility connected via the internet. They apply certain rules for these services.

SMAC: An acronym generated from the first letters of Social, Mobile, Analytics and Cloud words. These four technologies are currently driving business innovation, with multiple service categories and scenarios.

Technology Organization (TO): A team either inside the client enterprise (CE) or outside it, that is in-charge-of establishing, monitoring and maintaining technology systems and services. In addition, the "TO" supports strategic planning to ensure that all technology initiatives support business goals. It was referred to as the information technology (IT) department which does not reflect the entire spectrum of technology usage nowadays.

Chapter 9

Irritating Factors While Navigating on Websites and Facebook and Its Reactions Using Different Devices

Sana El Mouldi
IAE Université Bordeaux IV, France & ISG Tunisia, Tunisia

Norchene Ben Dahmane Mouelhi
Université de Carthage, Tunisia

ABSTRACT

The research presented in this chapter identifies sources of the irritation felt by internet users while browsing websites and Facebook. A qualitative approach was taken, including 40 individual interviews, enabled the authors to determine the irritating factors and user reactions when using different devices such as smartphones, computers and tablets to navigate websites and Facebook. The implications of this research will help marketers and web developers to reduce internet user irritation and better understand their behavior to better meet their expectations.

INTRODUCTION

The internet is considered as the biggest invention of 21th century. Indeed, the internet has permitted individuals to evolve from a simple information receptor status to an active searcher status. According to Belk (2013, p.477), the internet is "a cornucopia of information, entertainment, images, films, and music mostly all free for accessing, downloading, and sharing with others". Nowadays, "information and communication technology (ICT) has a large impact on the society in which we live and on the development and interactions of individuals, communities, corporations" (Vošner & al., 2016, p.230). Today, there are 3.7 billion internet users – roughly half of the world's 7.4 billion population.[1] Besides, more than 60 percent of internet users are drawn to social networks every month. Sakas & al. (2015)

DOI: 10.4018/978-1-5225-2469-4.ch009

stipulate that customers are co-creators of a company's marketing approaches and communication strategies through social networks.

Consumers have access to all kinds of information on the internet especially with the development of mobile devices, which reduces the use of other media. The growth in the prominence of digital, social media and mobile marketing has conducted to several technological innovations such as the increasing penetration of home internet and affordable high-speed broadband connections, the development of social media platforms such as Facebook, and widespread consumer adoption to "smart" mobile devices (Lamberton & al., 2016). Through "the advent of smartphones and social media, accessibility of information is higher than it ever has been before" (Agnihotri, 2016, p.173). The use of mobile devices is increasing exponentially threw different devices such as tablets and smartphones. Global Net Index study in 2015 revealed that the most used devices are laptops, smartphones and tablets but there are other new devices that are taking important places in consumer's internet use habits such as smart TVs, Smart Watches and smart wristband. Since the digitalization of our lifestyle, internet became a world a world in itself and like the real world it has its own irritating factors. Indeed, the internet users is frequently facing irritating factors as he's surfing on websites and mobile applications, and it leads generally to a negative emotion.

This chapter aims to identify the sources of irritation experienced by internet users while navigating on websites and social media, depending on the device used (i.e., smartphone, tablet, or laptop). Moreover, this study will reveal the reactions to the identified irritation sources.

BACKGROUND

Evolution of Internet User's Behavior

In the era of web 1.0, the internet user was passive, had access to static web pages, and could only do research with simple words, until the advent of the web 2.0 which provided the opportunity to become active with access to various tools such as blogs, wikis, and social networks. Individuals can research with tags and participate to the diffusion and creation of the information. According to Byrne & al. (2016, p. 456), "internet facilitates work, social connections, and education". Social networks, which are an essential tool of Web 2.0, allowed users to take power. The social networking phenomenon appeared in 1997 with the launch of the website "Six Degrees.com" (Boyd and Elisson, 2007). During the 90's, many social networks emerged, such as Asian Avenue, Blackplanet, and MoveOn (Edosomwan & al., 2011). Since the 2000s, social networks have increased in number, taking an important place in consumers' lives. Today, social networks have become an essential communication platform for many companies. Boyd and Ellison (2009) define the social networks as web services that allow individuals to create a public or semi-public profile, articulate a list of users with whom they are in contact, and view and scroll through the list of their contacts and those of other users. Lenhart and Madden (2007) see social networks as an effective and powerful channel through which consumers create a personal profile, build a personal network and display interpersonal comments publicly. Koh & al. (2007) introduce social networks as virtual community websites where people who are separated by time and space, can share interests, build relationships, exchange information and conduct transactions. Mayol (2011, p. 35) defines social networks as "tools that allow a connection of users with their friends, relationships to create a private relationship and / or professional network." These tools allow exchange of content (audio, video, photo ...), exchange of applications, monitoring of activities and especially the ability to create and integrate groups based on common interests, common cultures, joint opinion or common lifestyles. Wellhoff (2012) has also identified social networks as "a virtual space where people of the same affinity can meet and interact. Social networks allow the exchange between members, by e-mail or instant messaging and sharing their personal information". According to Vošner and al. (2016), "online social networks can be used to connect with people regardless of time or place". According to these definitions, it is possible to note that social networks bring people together regardless of where they are. Felix (2016, p. 2) stipulate that social networks help "stimulating sales, increasing brand awareness, improving brand image, generating traffic to online platforms, reducing marketing costs, and creating user interactivity on platforms by stimulating users to post or share content". Caseway (2016, p. 759) affirms "companies viewed social media as an opportunity to cultivate customer engagement and to strengthen their relationships with the other businesses who were their customers". Dahnil (2014, p. 120) confirms that social media is a unique marketing communication tool as it "allow the production of information and being collaborate among users and leverage mobile and web-based technologies to create interactive medium where users and groups member sharing, co-creating, discussing, and modifying known as user-generated content". An experiment conducted by Stanley Milgram in 1969 demonstrated that to connect two people it takes an average of six intermediaries. This experiment was called the rule of "six degrees" or the hypothesis of the "small world". Forsé (2012) reproduced this experience on Facebook, he found that the average distance between two users is 4.7 intermediaries. Today Facebook is the first social network in the world with over one billion, seven million users[2].

The advent of technology also impacted the consumer behavior while surfing the internet. Indeed, consumers are using numerous devices to access internet such as computers, mobiles, and tablets. With the evolution of the web and infrastructure, individual can use internet wherever he is through mobile networks and using mobile devices. For example, in France over 50 percent of internet users are connected via mobile devices (Figure 1).

Figure 1.

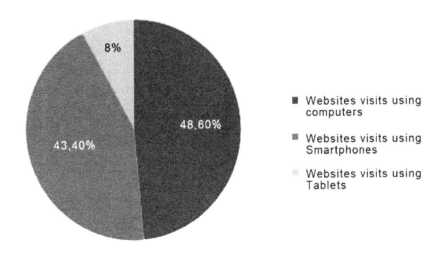

Table 1. Web1.0 VS 2.0

WEB 1.0	WEB 2.0
Passive internet user	Active internet user
Individual web pages	Blogs, Wikis and social networks on internet
Diffusion and information	Participation in the diffusion of information
Research by key words	Apparition of tags which facilitate the research
Broadcast news releases by email	Using RSS Feeds and feed aggregators
Bookmarks via the browser	Social bookmarks (exp: de.li.cio.us)
Content created by the service	Content created by the users
Read only and copyright	Add, edit, delete: some rights protected, open source
Some API's	Tool box (Jquerry or CMS), Open source API (Ajax)
Page views	Cost by clic
DoubleClick	Google Adsense
Ofoto	Flicker

Source: Fayon D., Web 2.0 et au-delà, Economica 2010

Store Atmosphere and Website Atmosphere

It is possible to induce the atmospheric elements of a physical store the virtual store, although Eroglu and al. (2001) emphasize the lack of some atmospheric variables on the internet such as smell, touch, temperature (Volle, 2000), to better understand the impact of the atmosphere on consumer behavior, researchers are interested in the variables that make up that environment by providing different classifications (Turley & Milliman, 2000). As for the physical store atmosphere, several researchers (Eroglu and al., 2001; Childers and al., 2001; Chang and al., 2002) have developed taxonomies for the atmosphere's elements on the internet. Table 2 presents an overview of various classifications related to environmental factors on the internet.

Table 2. Classifications of environmental factors on internet

Author (Year)	Categories	Observations
Eroglu and al. (2001)	2 categories: • High task relevant cues • Low task relevant cues	Eroglu and al. (2001) are the first to suggest a typology of environmental factors on internet.
Childers and al. (2001)	2 categories: • Utilitarian elements • Hedonic elements	This classification was based on internet user motivation
Chang and al. (2002)	2 categories: • Functionnal elements • Symbolic elements	Chang and al. (2002) based their study on the typology of Childers and al. (2001). The functional elements correspond to utilitarian elements and symbolic elements to hedonic elements.

Importance of Irritation Sources in the Study of the Consumer Behavior

Irritation is a derived emotion from anger but which is more moderated (Shaver and al., 1987). Pelet and al (2016) demonstrated that emotions have a positive influence on purchase intent and website recommendation. D'Astous (2000, p. 150) defines irritation as "Elements of the shopping environment that create negative feelings among customers". Touzani & al. (2007) suggest that negative feelings such as discontent or excitement are studied by simple opposition (pleasure/displeasure). The study of the negative influence related to environmental factors on consumer behavior account for a reversal of perspective, according to Helme-Guizon (2002). Many researchers were interested in studying the influence of environment on internet user's behavior, but their research only focused on its positive effects (Jacob, 2005). It is important to explore the dark side of the experience, because negative information plays a larger role in judgment formation than does positive information due to a superior cognitive treatment (Mizerski, 1982).

There are relatively few studies on the influence of the environment on consumer behavior. The works of d'Astous (2000) and Touzani & al. (2007) on sources of irritation felt by consumers in points of sale are notable, as are those of Helme-Guizon (2002) concerning the sources of irritation felt by the user while browsing on merchant sites, and finally, Ben Dahmane Mouelhi & al. (2009), with their research on sources of irritation in hotels.

METHODOLOGY

Given the exploratory nature of this research, a qualitative study seems most appropriate. Thus, semi-structured individual interviews, lasting between 30 and 45 minutes, were conducted among 40 people. These interviews were completed by the critical incident method involves asking the interviewee to talk about a situation already experienced. The 40 interviews were recorded and transcribed. The interviews were conducted using an interview guide previously established. The sample is characterized by the diversity in terms of their professional situations, family, and ages (between 15 and 35 years old). The choice of this age group is justified by the Facebook usage statistics, provided by Facebook stats 2016. Indeed, according to recent research, the 18 to 29 are the most important segment of Facebook users, with 87 percent of users registered in this age bracket[3]. According to Vošner and al. (2016, p. 236), "social networks such as Facebook, Twitter, LinkedIn, etc." are mostly used by youth.

CONTENT ANALYSIS

Thematic analysis was conducted on the transcribed data. Each interview was cut into recording units that were subject to a calculation to know their frequency of occurrence. Then, these recording units were grouped into units of meaning to finally be classified into thematic units to facilitate the understanding and interpretation of results (Bardin, 1996).

Analysis of Sources of Irritation Felt by Users on Websites

Thematic analysis performed for the sources of irritation experienced by users during their navigation on websites helped detecting six major themes that have been classified according to the typology of Childers and al. (2001), hedonic and utilitarian elements (Table 2).

1. **Utilitarian Elements:** In this research, utilitarian elements concerns advertising, the information content and website functionalities.
    a. **Advertising:** Through interviews, advertising proves to be a very irritating element for users. In fact, different forms of advertising were mentioned by respondents. They have cited banner ads, pop-ups, commercials and "spam". Note that pop-ups and advertising were also mentioned by Helme-Guizon (2002).
        i. **Advertising Banners:** For some respondents, the mere presence of banner ads on a website is a source of irritation. Indeed, they found that companies are desperate to make themselves known and to manipulate consumers. One respondent explained that "if I need something I would do a search and I would find it myself". Another clarified: "why showing me ads 10 times while I have no interest?" The number of banner ads on the same website is also an "inconvenient" element for users, they found that "there is more advertising than information. It should be noted that 13 respondents are irritated by banners with sound and one respondent explained that "these are banners like the others, but if you pass it with the cursor they make a weird sound." Faced with the presence of those banners on websites, respondents adopt different reactions. Seven of them try to avoid the banner, for example, one said: "I did not even look at them in order to avoid irritation", while others have a much stronger reaction "it annoys me so much that when I need that stuff I would not go to this brand." The presence of banners is not a source of irritation for some respondents, but when dealing with too large banners 14 respondents say they are very angry. For example, the one respondent explained: "I can only see that giant banner in the middle and nothing else".
        ii. **Pop-Ups:** Through interviews, pop-up were shown to be the most irritating form of advertising for respondents. In fact, 24 respondents do not support the presence of pop-ups on websites and found this form of advertising "very irritating". One respondent noted that "ads that open by themselves and appear in the middle of the page are very annoying". Respondents also insist on the lack of solution to remove these pop-ups which increases irritation. As well, the respondent who stated: "[...] and the worst part, there is no a red button so that I can close those windows." It seems that pop-ups are a very strong source of irritation for respondents as they were cited by more than half of respondents. This result was also confirmed by Helme-Guizon (2002) which found that pop-ups are a very irritating element. Windows that appears on their own and follow the user throughout its navigation were also cited by respondents as the pages of advertisements that open alone. Facing these various forms of pop-up, respondents have almost the same reaction. In fact, three respondents gave answers similar to that of one particular who noted: "I'm too pissed I leave the website immediately", while another respondent said: "I'm already pretty stressed by my work I will not also endure those

advertisements that move". In conclusion, when faced with this kind of advertisements users adopt escape behavior to avoid irritation.

    iii. **Video Advertising:** When discussing advertising spots on websites, respondents mention two types of irritating situations. The first is the presence of advertising spot without the possibility for the user to stop or mute the ad. This was mentioned by four respondents and it causes two types of reaction either the person leaves the site or she deactivates the mute button on his computer. The second situation is forcing the user to watch an advertising spot so he can access the site or watch a video. This was mentioned by 5 respondents and their reaction is unanimous "I leave the website. By what right it makes me look a spot? That's blackmail."

    iv. **Advertising Mail:** According to the interviews, "spam emails" are a source of irritation for respondents. Indeed, nine respondents are bothered by emails sent by companies while they have not asked for anything. For example, the respondent 14 states that "companies are bombarding me with emails while it does not interest me".

b. **The Informational Content:** The informational content is a very important variable because during navigation on internet, users are primarily looking for information whatever their motivation (utilitarian or hedonic). Indeed, Helme-Guizon (2002) have shown that the informational content was the most irritating factor for the user. When respondents evoke the information found on the websites, they insist on five negatives aspects:

    i. The lack of information on websites was mentioned by 16 respondents who do not find the information sought. This lack of information damages the reputation of the website given that respondents say they are disappointed and leave the website.

    ii. The reliability of the information provided by some websites: This issue was raised by 4 respondents who find that certain websites are not reliable and they take information from social networks without checking if it is true or false.

    iii. Websites that are not updated respondents are irritated by browsing a website and not finding any change compared to the course of events. For example, the respondent 22 states "I am surprised that some sites of certain brands of clothes are still in the winter collection when we are in the summer".

    iv. Sites suggested by Google are not in accordance to the research initiated: Three of the respondents found that the Google results are the websites with the words they used, but when they click on these websites are that they don't find what they wanted.

    v. Paying for information on respondent noted: "I am shocked, we are students and in order to access to a course I have to pay 10 euros, this is intolerable." Facing this irritating condition two respondents felt the same "no way to pay I'm sure I can find the same thing for free on another website."

c. **Navigability:** Through interviews and responses of informants, navigability concerns mainly the time of transfer of web pages. According to the study of Helme-Guizon (2002), the transfer time of web pages is not very irritating for users.

In this research, the transfer time seems to be quite irritating since it was mentioned by 18 respondents. It is a very irritating factor for respondents. Indeed, 18 of the respondents struggled with websites that are very slow to open and one explained that "To read an article or watch a video you have to wait an hour." Browsing the internet, users expect to find what they want quickly and this situation is not very appreciated by them. Evoking the transfer time, respondents also talked about videos that are buffering all the time in the streaming websites but in this situation the respondents have patience and wait until the video works. Respondent 37 says "I have no choice I have to wait, it is only on this website that I can see the lastest episodes from the USA."

2. **Hedonic Elements:** Hedonic elements correspond to website design, sound factors and functionality of the website.
    a. **Website Design:** The website design seems to be quite important for users who find nice to navigate on and pleasant websites with a minimum of decoration. During interviews, two factors about web design were mentioned by respondents. These are colors of the website and its architecture.
        i. **Colors:** Through interviews, respondents identified two situations where color provoked in them irritation. The first situation "*too flashy colors that hurt the eyes*" and this was considered very irritating to 4 respondents. The second situation concerns very dull colors that make the text illegible and it was perceived as irritating from seven respondents.
        ii. **Website Architecture:** When respondents talk about the architecture of the website, they insist on four negatives factors. The first aspect is related to congested websites; 10 respondents felt that the "congested" or "loaded" websites are very irritating, the respondent 23 states that "very cluttered websites where there is so much information, images and advertisements make me even forget the purpose of my visit..."
        The second negative aspect concerns the sections of the websites that are not very "clear" or are "put no matter how" or "poorly organized". This aspect has been considered very irritating to the respondents and the issue was raised by 16 of them. Indeed, while browsing a website, users want to find the right information in the right place. One respondent said "I'm sorry, but I don't have time to waste trying to guess where the information sought is".
        The third aspect concerns too "sophisticated" websites. Indeed, four respondents say they are upset by websites very "complicated" to handle. Thus, one respondent explained that "they believe it is innovation, whereas for me it's a true Chinese puzzle." The last negative aspect is the architecture of the website that is perceived by seven respondents as "ugly" or not very "original". The lack of originality was addressed especially to websites that two respondents claimed are "too basic and lacking in originality compared to websites that are much more original and organized."
    b. **Audible Factors:** Respondents expressed primarily on the presence of music on websites. In recalling music on websites, respondents highlighted three irritating or annoying situations. The first situation concerns the presence of music on websites, 7 respondents find it irritating. For example, a respondent explained that "if I need to listen to music, I would put on my computer". The second situation concerns the presence of music with the absence of the "mute" button. Confronted with this situation, the seven respondents say they are very upset and the respondent 29 says "I'm going crazy it stresses me, I look for the button, but there is none, I bangs for it to stop and ultimately I close the website". The third situation is related to the presence of rhythmic music.

c. **Functionality:** The first type of irritation about Web sites with a single language and respondent 2 says "I'm not fluent in English and websites where I can't translate the text into French annoys me". The second type of irritation is related to the "copy / paste" on the website. Indeed, respondents often find interesting information they would like to copy but the site does not allow it.

Analysis of Irritation Sources Felt by Users on Websites Via Smartphone or Tablet

During this research, the interviewees were asked if they encountered sources of irritation during their navigation on websites using a smartphone or tablet. Once the analysis performed, it was possible to note that respondents encountered the same sources of irritation on the computer but there are also others that we will be classified in the category functionality. The first kind of irritation is related to web sites that are not adaptable with smartphones or tablets (non-responsive). One respondent said "websites that are not adapted to phones annoys me, I have to stretch or shrink, I cannot manipulate it". A different type of irritation was taken by our respondents, these websites that do not open and force them to download the application for access. One of our respondent states "it is not normal I do not want to download the application, why forcing me to do it I just need the site".

By connecting via their smartphone or tablet, users are using certain applications. They explained that the applications accounted also sometimes some sources of irritation such as updates, bugs or too heavy to downloaded applications.

Analysis of Irritating Sources Felt by Facebook Users Connected by Computers

During the content analysis on the sources of irritation on Facebook, 4 thematic units were identified. These four themes were then grouped into two broad categories namely "social factors" and "factors related to the use of the site".

1. **Social Factors:** Respondents insist on interaction within groups and pages firstly and relationships with their friends and people they do not know secondly.
    a. **Groups and Pages:** Groups and pages on Facebook are very important as they allow subscribers to talk about topics that interest them and meet new people who have the same interests. However, through interviews, it turned out that there is a lot of irritation caused by certain groups or pages.
        i. **Group's Manipulation:** Through the interviews, respondents mentioned several irritating situations they experienced in Facebook groups. The first situation is related to fake groups. Indeed, 11 respondents were very irritated by fake groups who claim to be something but share and publish other things. Confronted with this situation, respondents left the group and even report it. One respondent stated "I feel betrayed, I left the group and I am disappointed by the dishonesty of the group in question". The second situation concerns irritating videos or photos modified with Photoshop software, respondents find that some groups are having fun mounting photos and videos just to sow discord in the country or harm the reputation of people. A respondent said that "some groups have fun

photoshoping images and it is obvious that this is not real but people believe and continue to criticize." The third irritating situation corresponds to groups that do not stop making publications. Being a fan of certain business pages, "Facebook fans" expect to find news of these companies and not to be bombarded with the same information throughout the day. For example, one respondent stated "I'm a fan of a brand and it turns out that this brand does not stop sharing every two minutes the same promotions it offers for one week, in the end it becomes heavy". The fourth situation concerns adding a person to a group without his permission. Confronted with this situation, respondents say they are very upset because they have the feeling to endure an obstacle to their freedom. For example, one respondent said that "sometimes they add me to a group without my permission, but by what authority? They can suggest me before adding me like that, "besides it happens with groups that are against my principles and my ideological positions". The latest situation concerns the videos of some groups that oblige respondents put "like" to other pages so they can view videos. Confronted with this situation, respondents give up watching the video because they feel manipulated by groups "Why should I like the pages that I do not like?"

b. **Violence in Groups:** Speaking of violence in groups, respondents mention two types of irritation. The first kind of irritation concerns "personal attacks on respectable people". Respondents find that on Facebook it is very easy to damage the reputation of people and spread out rumors about their families while they are "respectable" and "trustworthy". The second type of irritation is related to vulgarity and verbal violence in groups, seven respondents thought that people are frustrated because that they take advantage of not being able to see them and drop on Facebook "free insult".

c. **Friends or People:** Facebook is a social network that allows each subscriber to keep in touch with friends, family or even to meet new people. Through interviews, it seems evident that friends and people on Facebook can cause irritation to the user.

   i. **Friend's Behavior:** In recalling the behavior of their friend, respondents insist on seven irritant factors. The first factor irritating is related to friends who "reveal their lives on Facebook". Indeed, respondents find that Facebook is a way to stay in touch with with friends and exchange ideas and not to "share the lastest thing I've done" and one respondent added that "some people have no modesty even when they go to shower they publish it but this is madness!!!" The second negative aspect relates to the friends who share without verifying the reliability of information. Respondents describe these people as being "too naive" or "unconscious". The third aspect concerns the friends who spend their time on Facebook instead of working. In fact, two respondents say they are very irritated by people who spend their time on Facebook. For example, one respondent specified "[...] but it is amazing every time I log they are there if I did not know I would say they are unemployed while they have a work and family". The fourth aspect is the "mentality" of friends since 20 percent of respondents felt that with Facebook they discovered some faces of the personality of their friend they did not know before and this was a true "disappointment". Respondents describe the behaviors of their friends by using the terms "macho", "extremist", "uneducated", and "rude". The fifth aspect concerns the "curiosity of people", three respondents felt that Facebook has made people too curious. This can be seen in one response that noted: "they want to know everything as soon as I put a picture everybody wants to know when, where and with whom". The last negative point mentioned by respondents concerns the spelling level of their friends; some respondents say they are disappointed with the language level of some people.

 d. **Invitations and Options:** Among the sources of irritation on Facebook, it is possible to mention invitations and options that Facebook offers. First of all, invitations to events and games are a real irritation for nine respondents who have yet to find a solution to avoid it "[...] there is not even an option to block it this is very troublesome". Then the people who send friend requests to strangers are also a source of irritation for 6 respondents who do not understand "why someone who does not know me want to become my friend". Finally, one respondent said that he was shocked by the option "delete a friend" that Facebook offers because for him it is an incitement to hatred "[...] but what is this option? Someone bothers me, I delete her, so if I reason by analogy in real life I conclude that if someone bothers me to kill her?".

2. **Factors Related to the Use of Facebook:** Through interviews, respondents identified two sources of irritation which are related to the use of Facebook; informational content and updates.

 a. **Informational Content:** On Facebook, informational content is presented in different forms. Indeed, there are articles, videos, photos and discussions between subscribers. Respondents insist on two sources of irritation, the first type of irritation concerns the rumors and misinformation. Indeed, 23 respondents felt that Facebook is a source of information "very dangerous" and "not reliable". Thus, one respondent spoke of misinformation and another noted that "seeing the same information in three versions in less than 5 minutes, we do not know which information should be taken". The second type of irritation is related to the excessive presence of political information. Respondents feel that this has become "boring" and "irritating". In fact, for 7 respondents Facebook is basically a means of entertainment and reunion between friends who, over time, turned into "TV news".

 b. **Version and Security:** Facebook versions correspond to the updates made by the leaders of this social network. Indeed, at every update subscriber is facing a new version and through interviews, it appears that the updates irritate some respondents. Safety on Facebook can be controlled by the subscriber through the privacy settings.

  i. **Updates:** Through interviews, Respondents highlighted the irritation caused by the updates made by Facebook. The first one concerns the frequent changes of versions and this was mentioned by 25 respondents who felt that Facebook is changing too often versions and one respondent said that "every two weeks there is a new version, but it is not possible I barely managed to adapt to the previous that Mr. Zuckerberg puts us another". The second is related to the complexity of the new versions. Indeed, another respondent felt that new versions that Facebook offers are "very complicated" and that "I should have a degree in computer science in order to understand them". The third irritating factor concerns the fact of being obliged to adhere to the new version. Indeed, as for "adding to a group without permission" and "Becoming a fan of a group to watch a video", internet users do not like to be forced to do anything because they feel "controlled" and "manipulated". Confronted with this situation, respondents fit although they prefer to have the choice. The latest irritating factor is concerning options of the new version because with the latest "as soon as a person I know doing something I get a notification and vice versa". Respondents find that this option is very irritating and that it "hurts their privacy" because they do not necessarily want others to see what they do.

c. **Confidentiality:** Throughout the interviews, respondents talked of 4 sources of irritation caused by confidentiality. Eighteen respondents are irritated by the change in privacy settings every time Facebook changes version and one respondent stated: "[...] if Facebook modify version and I do not open my account during this time will my data be accessible to everyone given that each change of version of Facebook puts privacy parameters to zero." The second irritating factor on confidentiality on Facebook concerns piracy. Indeed, 12 respondents said they had been hacked at least once and in this situation, respondents blame Facebook privacy settings they describe as "not very reliable". The third situation involves the fact that Facebook has the right to retain all subscriber data. One respondent explained that "from the beginning I am aware that by signing up to Facebook I allow it to have all my data but it annoys me anyway". The latest irritating factor is related to finding pictures "private" on Google images and one respondent explained that "recently I realized that most of my Facebook photos that are normally accessible only my friends are displayed on Google !!! I'm just shocked."

Analysis of Irritation Sources Felt by Facebook Users Via Smartphone / Tablet

During this research, interviewees were asked if they encounter irritating factors during their navigation on Facebook using a smartphone or tablet. Once the analysis performed, it was possible to note that respondents encountered not only the sources of irritation felt using a computer but also other sources of irritation that we will rank in the "factors related to the use of the Facebook". The first irritating factor on Facebook application is that it stops occasionally or blocks on "tablet or smartphone". The respondent 37 says "I prefer to use Facebook on my pc at least there is no bug as with the tablet". The second source of irritating factor is the fact to make a "copy / paste". According to respondents, on some mobile devices it is difficult or sometimes impossible to copy information on Facebook using a smartphone or tablet. Among the encountered irritating factor on the Facebook application, 12 of the respondents mentioned that on tablet or smartphone, the loupe that exists in groups to make a search does not exist. It must be connected by computer to use this loupe. Finally, some interviewees expressed their irritation due to the fact that Messenger (private messages on Facebook) is a separate application which must be downloaded to access to message.

1. Reactions of Internet Users Face of Irritation on Websites

Confronting irritation encountered during their navigation on internet using different devices (computer, tablet, smartphone), respondents adopt different behaviors. Through the analysis of the content, three kinds of reactions occurred.

There are reactions depending on the importance of the website. If the website or application are really important to them, respondents remain despite the irritation. For example, a respondent stated: "I do not have a choice, if I know that I can't not find the information somewhere else, I have to stay". However, if the site is not very important and they know they can find the information somewhere else respondents leave the website directly. These two reactions were also cited in Helme-Guizon (2002). Some respondents reported reacting to irritation depending on their degree. Indeed, if the irritation is not very strong 10 respondents declared continuing navigation without any problems. However, if the irritation is too strong respondents adopt avoidance behavior by leaving the website. It should be noted that by leaving the website, some respondents declared they will never come back in the future (this

Irritating Factors While Navigating on Websites and Facebook

Table 3. Summary of irritants factors on websites and Facebook using different devices (Computer, tablet, Smartphone)

	Irritating Factors on Websites
Irritating utilitarian elements	**Advertising:** Concerns various types of ads, there are banner ads, pop-ups, ad spots and spams.
	Informational Content: Refers to the information and its reliability.
	Navigability: Corresponds to the heaviness of the website and mobile applications and the slowness of download.
Irritating functional elements	**Design:** Regards colors and architecture of the website.
	Audible Factors: Corresponds to the music found on the websites.
	Functionalities: Concern the "featured" options that can suggest a website to be distinguished from others, They also concerns the websites that are not adapted to different devices.
	Irritating Factors on Facebook
Irritating social factors	**Groups/Pages:** Concerns manipulating groups and violence within these groups.
	People/Friends: Refers to the irritating behavior of some people and unwanted invitations.
Irritating factors related to the use	**Informational Content:** Corresponds to distorted news and rumors.
	Updates and Confidentiality: Includes updates made by Facebook for versions and privacy settings that, in principle, ensure account security.
	Irritation of the Facebook Application: Concerns the absence of the loupe in tablet and smartphone, download the messenger apart, no copy / paste, the heaviness of the application.

reaction is the most probable it was mentioned by 25 respondents) others declared they will come back to check if there have been changes. Through content analysis, other reactions occurred. Among these reactions, ask for help to a friend. In fact, two respondents seek help to a close person "I'm so upset that I asked for help because in my state I know I would not be able to continue navigating although it is very important for me". The second type of behavior is very rare and it was mentioned by one respondent.

2. Reactions of Internet Users Face of Irritation on Facebook

When analyzing the reactions of internet users facing irritation on Facebook (website or application), it turned out that they have different reactions.

Dealing with manipulating groups and violence that takes place there, respondents adopt two types of reactions. The first reaction is very common, it is to leave the group by deleting it and respondent 5 says that "I am disappointed and feel betrayed so I delete the group". The second reaction is stronger than the first, it is to leave the group and report it, and the respondent 14 states that "there are Facebook groups so dangerous that I do not hesitate to report them". Confronting irritating behaviors of their friends and unwanted invitations, respondents adopt five types of reactions. The first reaction is to delete the person who has an unpleasant behavior. The second reaction is related to hiding the latest news of the person. The third reaction is the fact of blocking unwanted invitations and the fourth one is related to disabling Facebook, and the last reaction is to block out invitations to games.

Confronting frequent changes of versions and privacy settings that are not very reliable according to some respondents, two types of behavior emerged through interviews, the first reaction concerns security settings it is materialized by constant checking these parameters. Respondent 26 explains that "I become paranoid I did not stop checking the security settings for fear a sick person having access to my information". The second reaction is related to the changes in versions, in this situation the majority of respondents say they have no particular reaction, they adapt. Respondent 8 says, "I have no choice even if it pisses me off I try to familiarize myself with it I cannot do otherwise".

Confronting multiple rumors and information, respondents have become very distrustful of the information they find on Facebook, for this reason, they adopt two different types of behaviors. Some respondents said they check all the information before sharing it and others prefer not to look at the information on Facebook, For example, the respondent 9 states "Facebook is not a reliable source of information if I need information I will read a newspaper". Finally, regarding the irritation felt by using the Facebook application, respondents say they have no particular reaction because they do not really have a choice and there is not a substitute solution.

MANAGERIAL RECOMMENDATIONS

Through interviews, irritation on websites and Facebook seems to be very present. In this section, some recommendations that will help designers of websites and community managers will be suggested.

Sources of Irritation on Websites Using a Computer, Smartphone, or Tablet

Respondents evoked advertising, information, navigability, the website design, music and functionalities as sources of irritation on a website. People are exposed to advertising every day on the radio, on TV, in the streets, in the malls, in their mailboxes and on the internet. Suddenly, they feel stifled and manipulated by companies. So it would be better for websites managers to avoid large banners, to set the option to mute commercials, to not oblige the user to look at advert to access the websites. Information is very important for users, therefore websites managers need to update their website, check the reliability of the information they provide on the website, be sure to reference their site on engines research and provide information to users for free or for a reasonable price. internet users are impatient and want everything immediately so it would be better for websites not charge too much their site with text, images and videos so their website becomes less slow. The website design is important to respondents. For this reason,

Table 4. Summary of results for reactions to irritation on the Facebook website and application

Reactions	Themes
Reactions to irritation on websites	Depending on the importance of the website
	Depending on the degree of irritation
	Other types of reactions
Reactions to irritation on Facebook	Reactions related to irritation in groups
	Reactions related to the irritation caused by friends
	Reactions to Information
	Reactions to confidentiality and updates
	Reactions to the Facebook application

websites designers should not neglect it and it would be appropriate to use warm colors but not very flashy, do not put the information in "bulk", well organized topics put all information in the adequate category and adopt innovative architecture but simple and easy to understand. Music is also a source of irritation for some respondents. We advise web designers to not use rhythmic music and provide a mute button in case the user does not want to listen. Finally, it would be better for websites designers to make their website available in several languages. Thus, it could be visited by several targets (different languages and/or countries) and enable the "copy/paste" if they accept that people have their information. Companies should also make a mobile version of their site that can adapt to all types of devices.

Sources of Irritation on Facebook Using a Computer, Tablet, or Smartphone

By joining a group or becoming a fan of a page, respondents expect to find information concerning the group or _ page in a good mood in an atmosphere of mutual exchange. For these reasons, it would be interesting for directors of groups and pages Indicate the purpose of the group from the beginning, avoid publishing the same thing all the time, check the comments judged "violent", "displaced" from some people, avoid forcing people to join a group or to watch a video and check the reliability of some photos or videos before sharing. According to the interviewees, the Facebook application on tablet or smartphone causes some irritation as the absence of loupe, the application is stopped or the "messenger" which is a program in itself. It would be preferable to find some solutions to avoid any inconvenience to the user.

CONCLUSION

By examining the different results, we conclude that the irritation is quite present on websites and on Facebook using different devices. Today we talk about the customer as a "king" (or "queen") because they are spoiled and know they can have what they want, when they want – especially on the internet. Besides, individuals are surfing the internet, using various devices such as smartphones, computers, tablets, and other smart objects like Smart-TVs and game consoles. So, at the slightest irritating factor they know they can find happiness elsewhere. As explained above, irritating factors can lead to leave the website or app and never use it again, that's why identifying them is crucial. It is somewhat nonsensical to spend money on advertising to attract individuals to use an app or to visit a website, and they permanently leave it because they will find those irritating factors. This chapter highlighted the different irritating factors and their impact on internet users throughout a qualitative research, aiming to enhance the user experience.

The managerial interest from this study is helping managers of websites and apps and community managers to reduce sources of irritation and to better understand the behavior of internet users to try to better meet their needs. The results found can help professionals to better understand the sources of irritation that provoke in the visitor reactions that can harm their site. The theoretical value of this study is to address a problem that, until now to our knowledge, has not been treated before. On the other hand, exploring irritating factors on mobile devices will help marketers to design efficient mobile advertising campaigns.

However, it should be noted that this research holds a certain number of limitations, such as the lack of literature on sources of irritation. Indeed, apart from the work of Helme Guizon (2002) on the sources of irritation on internet there are no other references. The second limitation is related to the age range of interviewees, only youth (15-35 years old).

FUTURE RESEARCH DIRECTIONS

The limitations of this study allow suggestion of some future directions. Indeed, it would be appropriate to verify whether the sources of irritation are the same to over 35 years. Furthermore, a comparison between Irritating factors felt by the 15-35 years old and the over 35 years old might be useful. It will be also interesting to verify and compare the irritating factors felt on other social networks and medias such as Youtube and Twitter using different devices. Finally, developing a measurement scale that would quantify the negative emotions experienced by consumers on the internet would give both practitioners ant researchers a solid tool for their future Market-studies and researches.

REFERENCES

Agnihotri, R., Dingus, R., Hu, M. Y., & Krush, M. T. (2016). Social media: Influencing customer satisfaction in B2B sales. *Industrial Marketing Management*, *53*, 172–180. doi:10.1016/j.indmarman.2015.09.003

Balagué, C., & Fayon, D. (2010). *Facebook, Twitter and the others: integrate social networks into a business strategy*. Pearson Editions.

Bardin, L. (1996). Content analysis. Paris: SAGE Publications Ltd.

Belk, W. (2013). Extended Self in a Digital World. *The Journal of Consumer Research*, *40*(3), 477–500. doi:10.1086/671052

Ben Dahmane Mouelhi, N., Hassen, S., & Souissi, N. (2009). An exploratory approach to sources of irritation felt by the French customer in hotels in Tunisia. *Revue Marocaine de Recherche en Management et Marketing*, (2-3), 137 – 156.

boyd, d., & Ellison, N. B. (2007). Social Network Sites: Definition, History, and Scholarship. *Journal of Computer-Mediated Communication*, *13*(2), 210–230.

Byrne, Z. S., Dvorak, K. J., Peters, J. M., Ray, I., Howe, A., & Sanchez, D. (2016). From the users perspective: Perceptions of risk relative to benefit associated with using the internet. *Computers in Human Behavior*, *59*, 456–468. doi:10.1016/j.chb.2016.02.024

Cawsey, T., & Rowley, J. (2016). Social media brand building strategies in B2B companies. *Marketing Intelligence & Planning*, *34*(Iss: 6), 754–776. doi:10.1108/MIP-04-2015-0079

Chang, J. E., Simpson, T. W., Rangaswamy, A., & Tekchadaney, J. R. (2002). *A good website can convey the wrong brand image! A preliminary report*. Working paper. E-Business Research Center (EBRC), University of Pennsylvania.

Childers, T. L., Carr, C. L., Peck, J., & Carson, S. (2001). Hedonic and Utilitarian Motivations for Online Retail Shopping Behavior. *Journal of Retailing*, *77*(4), 511–535. doi:10.1016/S0022-4359(01)00056-2

Dahnil, M. I., Marzuki, K. M., Langgat, J., & Fabeil, N. F. (2014). Factors influencing SMEs adoption of social media marketing. *Procedia: Social and Behavioral Sciences*, *148*, 119–126. doi:10.1016/j.sbspro.2014.07.025

DAstous, A. (2000). Irritating aspects of the shopping environment. *Journal of Business Research*, *49*(2), 149–156. doi:10.1016/S0148-2963(99)00002-8

Edosomwan, S., Prakasan, S. K., Kouame, D., Watson, J., & Seymour, T. (2011). The history of social media and its impact on business. *Journal of Applied Management and Entrepreneurship*, *16*(3), 79–91.

Eroglu, S. A., Machleit, K. A., & Davis, L. M. (2001). Atmospheric Qualities of Online Retailing: A Conceptual Model and Implications. *Journal of Business Research*, *54*(2), 177–184. doi:10.1016/S0148-2963(99)00087-9

Felix, R., Rauschnabel, P. A., & Hinsch, C. (2016). Elements of Strategic Social Media Marketing: A Holistic Framework. *Journal of Business Research*.

Forsé, M. (2012). Today's Social Networks. *OFCE Revue*, *7*, 155–169.

Helme-Guizon, A. (2002). *Sources and consequences of the irritation felt during navigation on a retail website. An exploratory study*. 18th AFM International Symposium, Lille, France.

Jacob, C. (2005). *The influence of the music of a business website on consumer responses* (PhD Thesis). Rennes1 University.

Koh, J., Kim, Y. G., Butler, B., & Bock, G. W. (2007). Encouraging participation in virtual communities. *Communications of the ACM*, *50*(2), 68–73. doi:10.1145/1216016.1216023

Lamberton, C., & Stephen, A. T. (2016). A thematic exploration of digital, social media, and mobile marketing researches evolution from 2000 to 2015 and an agenda for future research. *Journal of Marketing*, *27*(6), 146–172. doi:10.1509/jm.15.0415

Lenhart, A., & Madden, M. (2007). Social networking websites and teens: An overview. Pew Research Center/Internet.

Mayol, S. (2011). *Marketing 3.0*. Dunod Editions.

Milgram, S. (1969). Interdisciplinary thinking and the small world problem. *Interdisciplinary Relationships in the Social Sciences*, 103-20.

Miserski, R. (1982). An attribution explanation of the disproportionate influence of unfavorable information. *The Journal of Consumer Research*, *9*(3), 301–310. doi:10.1086/208925

Pelet, J. E., Taieb, B., & Ben Dahmane Mouelhi, N. (2016). From m-commerce website's design to behavioral intentions. *Information and Management Association Symposium*.

Sakas, D. P., Dimitrios, N. K., & Kavoura, A. (2015). The Development of Facebooks Competitive Advantage for Brand Awareness. *Procedia Economics and Finance*, *24*, 589–597. doi:10.1016/S2212-5671(15)00642-5

Shaver, P., Schwarz, J., Kirson, D., & OConnor, C. (1987). Emotion Knowledge: Further exploration of a prototype approach. *Journal of Personality and Social Psychology*, *52*(2), 1061–1086. doi:10.1037/0022-3514.52.6.1061 PMID:3598857

Touzani, M., Khedri, M., & Ben Dahmane Mouelhi, N. (2007), An exploratory approach to sources of irritation felt during a shopping activity: case of food predominantly malls. *6th Marketing Trends International Symposium*.

Turley, L. W., & Milliman, R. E. (2000). Atmospheric effects on shopping behavior: A review of the experimental evidence. *Journal of Business Research*, *49*(2), 193–211. doi:10.1016/S0148-2963(99)00010-7

Volle, P. (2000). From marketing of points of sale to the merchant websites: Specificities, opportunities and research questions. *Revue Française du Marketing*, *177 /178*(2/3), 83–101.

Vošner, H. B., Bobek, S., Kokol, P., & Krečič, M. J. (2016). Attitudes of active older internet users towards online social networking. *Computers in Human Behavior*, *55*, 230–241. doi:10.1016/j.chb.2015.09.014

ENDNOTES

[1] http://www.blogdumoderateur.com/50-chiffres-medias-sociaux-2016/
[2] http://www.blogdumoderateur.com/50-chiffres-medias-sociaux-2016/
[3] http://www.leptidigital.fr/reseaux-sociaux/profil-demographique-utilisateurs-reseaux-sociaux-4924/

Chapter 10
Mobile Augmented Reality:
Evolving Human–Computer Interaction

Miguel A. Sánchez-Acevedo
Universidad de la Cañada, Mexico

Beatriz A. Sabino-Moxo
Universidad de la Cañada, Mexico

José A. Márquez-Domínguez
Universidad de la Cañada, Mexico

ABSTRACT

Users who have access to a mobile device have increased in recent years. Therefore, it is possible to use a mobile device as a tool which helps to users in their daily life activities, not only for communication. On the other hand, augmented reality is a growing technology which allows the interaction with real and virtual information at the same time. Mixing mobile devices and augmented reality open the possibility to develop useful applications that users can carry with them all the time. This chapter describes recent advances in the application of mobile augmented reality in automotive industry, commerce, education, entertainment, and medicine; also identifies the different devices used to generate augmented reality, highlights factors to be taken into account for developing mobile augmented applications, introduces challenges to be addressed, and discusses future trends.

INTRODUCTION

Mobile devices with high computing capabilities, communication modules, high-resolution cameras, and several integrated sensors, have changed the way of communication between individuals. Applications developed for those devices allow users to access Internet for searching information, making purchases, performing banking transactions, sharing photos, or just chat. This growing market of devices and applications provides to augmented reality an open field to generate a better user experience, at the same time that new challenges on mobile computing emerge.

DOI: 10.4018/978-1-5225-2469-4.ch010

The interaction with virtual information over the real world extends the user perception and increases the skills to perform complex tasks. These tasks include: manufacturing a product while specific instructions to assemble it correctly are displayed, playing games composed by real objects or persons, learning by visualizing extra information of objects that are being observed through the camera of a mobile device, performing a surgery while the surgeon is aware of other anatomical relationships in the body of patient, sampling garments before physically try them on, perceiving traffic conditions while driving without lose the attention on the road, among others. To get an increase on knowledge, entertainment, and communication of users, all these forms of interaction need to be directed toward the development of well-designed and useful applications; therefore, emphasis has to be put in the content production, by verifying that the content is helpful and not distracting in order to avoid accidents.

With the aim to get the best of this growing technology, it is important to identify the factors which guarantees the effectiveness of proposed solutions; those factors are dependent on the activity to be performed and the devices available for each activity. The development process should consider adaptation to different kinds of mobile devices to increase portability, moderate the interaction to increase usability, and protect personal information to ensure security. A compilation of recent advances in the inclusion of augmented reality in areas like automotive industry, commerce, education, video games, and medicine, is summarized here, the factors to consider for the development of mobile augmented reality applications are described, and finally challenges to be addressed are discussed.

BACKGROUND

Augmented Reality (AR) is an extension of reality by means of the addition of virtual information that complements the reality perceived by the user. This extension can be composed of 3D virtual objects, textual information, sounds, odors, holograms, and any other artifact with which the user can be in contact. Mobile devices, being carried by a vast majority of users, make possible to talk about mobile augmented reality. Mobile augmented reality can be defined as the perceived augmented reality via devices that users always carry with them. Technology used in augmented reality has been evolved during many years. A brief historical review is presented below to give a wide conception of the evolution of the human-computer interaction through augmented reality, and before to close this section, current technology, that makes possible the mobile augmented reality, is described.

1929 was the year when Edwin Link introduced "The Link Flight Trainer" (Committee, 2000); the first flight trainer used with the intention to train pilots at the same time of avoiding accidents and reducing costs. Through this mechanical airplane, the pilot could acquire skills to manipulate all the instruments without risks. Three years later, in 1932, Sir Charles Wheatstone invented the Stereoscope (Bowers, 2001); with this device, it was possible to observe the first 3D scene generated from two slightly different pictures. The possibility to interact with a 3D world was a motivation for researchers to develop new ways of human-computer interaction. Morton Heilig, in 1962 (U.S. Patent No. 3050870, 1962) creates the "Sensorama"; a machine that incorporates stereoscopic images, sounds, movements of the viewer, and odors to produce a feeling of immersion in the virtual scene. Following the idea of generate a virtual world within which the user can interact, Ivan Sutherland in 1965 thought about "The Ultimate Display" (Sutherland, 1965) as a room where a computer could create things that users would perceive, feel, and use.

Technological advances reached until then, allowed the perception of a virtual world which was limited by the fixed location of machines and users. With the purpose of moving the virtual scene according to the user movements, Ivan Sutherland developed a head mounted display (Sutherland, 1968), which gave the possibility to observe the 3D scene from several points of view controlled by the user movements. After that, virtual scenes could be superimposed over the real world, and new ways of interacting with virtual objects were needed. DeFanti and Sandin (1977) developed a glove for tracking finger movements; as consequence, virtual objects can be manipulated with the use of those gloves. But researchers, looking for a more natural interaction, were trying to eliminate intrusive devices at the same time of maintaining the possibility to interact with virtual information; all these without losing contact with the real world.

A proposal for using voice and gestures to control the movements of graphics in a screen was presented in (Bolt, 1980). This approach was one of the first works trying to provide a more natural communication with a virtual environment. Simple commands like move, create, delete accompanied by signaling gestures were used to interact with the graphic environment. In 1999, John Stephen introduced the I/O Bulb (Underkoffler, 1999). A bulb with an integrated camera and controlled light emission. This device allows the tracking of objects through changes in the emitted light. Working with I/O Bulbs makes possible to interact with real objects and display virtual information through a projector; the real information was gathered by the bulb and then virtually displayed.

By opening the possibilities of interaction with computers also complexity is increased; however, there exist activities which always are performed under pre-established constraints. Ullmer, Ishii and Jacob (2005) proposed a token + constraint system where physical objects are restricted to move into physical containers representing the constraints of the system. In 2008, the concept of "Tangible Augmented Reality" was introduced (Billinghurst, Kato, & Poupyrev, 2008). Tangible AR allows to control a virtual object through a physical object; so, every virtual object to be controlled is registered with a physical object. In this approach, physical objects provide a user friendly interaction control. In the same year, Henderson and Feiner (2008) proposed the idea of opportunistic controls. Since not all the activities that a person has to perform are performed in free environments, nonintrusive interaction mechanisms have to be adapted to the physical constraints; opportunistic controls take advantage of specific regions of the environment to map virtual controls related to the activity that is being performed. The new interactive ways of communication also reached to video game industry; in 2008, Nintendo developed the Wii Remote for allowing users to control game characters through the movement of the control device (U.S. Patent No. D559254, 2008), and six years later, Microsoft created a device for tracking body movements (U.S. Patent No. 8744121, 2014). Nowadays, both devices have been used for a wide variety of virtual and augmented reality applications.

Among technologies that make possible the development of mobile augmented reality applications there are cellphones, smart glasses, AR cards, wearable projectors, e-textiles, and smart connected products. Recent advances in miniaturization of electronic devices have been made possible to integrate cameras, sensors, and displays in tiny smart connected products increasing the portability of those devices; additionally, with e-textile technology, devices can be carried not only on body but also as part of clothing. Mobile augmented reality has an open field for applications development which can take advantage of all those devices.

DIFFERENT AREAS WHERE MOBILE AUGMENTED REALITY IS BEING USED

Diverse areas are feasible for introducing mobile augmented reality due to the benefits that can be obtained from it. Industries, education, and health are major areas where researchers have proposed approaches to include this technology. Augmented reality can be used for performing tasks which require extra information to improve performance; based on the nature of the task, there exists specific hardware that can be used to implement augmented reality. The development of software, which includes augmented reality, needs to address specific factors like content, privacy, security, usability, etc.; those factors vary according to the area of application.

Although all factors can be applied to all sectors, there exist factors which are essentials; for this reason, factors which require to be considered are highlighted in each area. To give a broad illustration of this, consider the factor of data protection in commerce and education. Since commerce is performed among different probably unknown people; the personal information has to be carefully manipulated and always protected. On the other hand, education is an environment where applications are developed for learning specific topics and the information is always available for all students. That means that data protection is not as relevant as in commerce, but it can be considered to provide robustness to applications. Content, usability and user experience are factors that always have to be considered for ensuring that applications can be effectively used.

A summary of devices and factors required to implement mobile augmented reality for improving specific tasks is shown in Table 1.

MOBILE AUGMENTED REALITY INTO THE AUTOMOTIVE INDUSTRY

Automotive industry provides a wide field for the incorporation of mobile augmented reality. It can be applied to vehicle design and conception where engineers can observe how different elements could be integrated into the vehicle without manufacturing them and generating a reduction of costs. Sales, where there is not possibility to have all the models exhibited, can be improved by means of augmented reality applications where customers can observe differences among components of several models and get a better perspective of what they want by using a generic model, which could illustrate the components over augmented reality glasses. In production activities, the new workers need to be trained before to be involved in the production line to guarantee quality achievement; through mobile augmented reality, the training cost can be reduced by providing assembly instructions superimposed on the working area. Mechanics is another field where mobile augmented reality brings the possibility to users of identify problems on the car and solve them easily by receiving instructions in a cell phone or a tablet. Finally, driving assistance gets special attention due to the interest in automatic solutions; augmented reality also could help drivers to identify risk situations or get extra information under unfavorable weather conditions and reduce the possibilities to get involved in accidents.

Driving a vehicle requires the attention of the driver on the road and the surrounding environment in order to perceive caution signals. Rain and fog reduce the visibility of driver and increase the possibility to suffer an accident whether no caution is taken. With the aim to help drivers, Abdi, Abdallah, and Meddeb (2015) proposed an augmented reality head-up-display (AR-HUD) to animate traffic signals and present objects which are not visible under unfavorable weather conditions. The information is displayed in the windshield of the vehicle with the intention to maintain the attention of the driver on the road.

Table 1. Available devices and required factors for implementing mobile AR in specific activities

Areas	Activities	Devices	Factors
Automotive industry	• Design. • Driving. • Manufacturing. • Mechanics. • Sales.	• Augmented glasses. • Electronic windshield. • Head-up-display. • Smartphone. • Tablet. • Wearable devices.	• Content. • Data protection. • Privacy. • Security. • Usability. • User experience.
Commerce	• Exhibition. • Shopping.	• AR-cards. • Augmented glasses. • Augmented mirror. • Display. • E-textiles. • Smartphone. • Wearable devices.	• Content • Data protection. • Privacy. • Usability. • User experience.
Education	• Learning. • Teaching.	• AR-cards. • Augmented glasses. • Displays. • Kinect. • Smartphone. • Tablet. • Tracking system. • Wearable devices.	• Content. • Scalability. • Usability. • User experience.
Entertainment	Play video games.	• AR-cards. • Augmented glasses. • Haptic devices • Kinect	• Content. • Scalability. • Usability. • User experience.
Medicine	• Rehabilitation. • Surgery. • Training.	• Haptic devices • Kinect	• Availability. • Content. • Security. • Usability. • User experience.

Traffic signals are captured with an exterior camera, then they are processed with the Viola-Jones (2004) face detector framework. Overload information of signals to the driver can cause the reduction ability to perceive other important elements, to avoid this situation the most important signals are extracted from the scene through a SURF detector (Bay, Ess, Tuytelaars, & van Gool, 2008).

A group of people who is beneficiary of mobile augmented reality are elder drivers. With age, spatial cognition abilities are lost; so, the possibility to perceive dangerous situations are reduced. Most of elder people lose their independence to move, they need the help of other persons to be moved from one place to another. Kim and Dey (2009) developed a navigation system for helping elder drivers. The information generated by the system is displayed on the windshield of the vehicle to reduce the cognitive distance and maintain the divided attention in an optimal level. Divided attention is defined as the ability to attend simultaneously multiple tasks (Wickens, Hollands, Simon, & Parasuraman, 2013). A driving simulator was implemented in order to evaluate the system. In comparison with a GPS-based system, older drivers performed less navigation errors.

The way information is displayed is an important aspect to consider when augmented reality applications are developed for assisting drivers. Inattentive driving behaviors can cause accidents; so, it is important to diminish distraction at the moment to display extra information. Considering those aspects George, Thouvenin, Frémont, and Cherfaoui (2012) presented an interactive driver assistance system.

The system generates information about obstacles on the road for providing to drivers the possibility of avoid them and anticipate driving difficulties. An obstacle detection module is coupled with a system that is in charge of monitoring the driver behavior. This module is composed of a camera mounted on the windshield of the vehicle, and a commercial software for notifying the presence of vehicles, pedestrians, and lane. The position of the driver is monitored by a commercial system in order to identify the current point of view of the user and display the information in the same direction over the windshield to avoid distractions. Signals displayed are classified according to dangerous levels; this approach allows to show only the relevant information to the driver. With the proposed solution, drivers can get information of dangers while keeping the eyes on the road.

Factors

To develop augmented reality applications that help to increase productivity in the tasks carried out into the automotive industry, it is necessary to determine what factors need to be considered in the development process.

- **Content:** The extra information that will be displayed to users must be carefully analyzed in order to provide only relevant information that helps to improve the activity for which it is was designed and not generate distracting content or excessive information that cannot be properly analyzed by the user.
- **Data Protection:** When designs, manuals, personal information, and location information, are digitally treated and transmitted over a wireless medium, it is necessary to ensure data for protecting sensitive information and preventing hacker attacks; so, secure mechanisms and protocols of communication among mobile and wearable devices have to be established.
- **Privacy:** Social applications, which can be used to extend mobile augmented reality, have been developed for keeping drivers informed about traffic conditions and alternate routes. Considering this fact, the identity of persons who share information and their exact location has to be maintained secure in order to avoid the loss of personal privacy.
- **Security:** Being distracted while driving is dangerous; also when assembling, inattention can lead to injuries. Augmented reality requires partial attention of the user; so, bad designed applications can cause injuries or accidents. The way in that information is displayed and the quality of it, are important to ensure the security of users.
- **Usability:** Quality of content, suitable presentation, and nonintrusive devices, make of mobile augmented reality applications supportive and ease of use tools for drivers, workers, designers, and every person involved in the automotive industry. All these aspects need to be considered to increase the usability of the applications.
- **User Experience:** Workers always prefer to do the activities in the way they know, to introduce a new paradigm generates resistance; however, when the new solution provides more benefits and is easy to implement, the change is well accepted.

Considering the factors mentioned before ensures quality on solutions and guarantees the increase of performance on activities which make use of augmented reality. Currently, many efforts to introduce augmented reality into the automotive industry have been carried out, but many challenges remain to continue improving every area of this sector.

Challenges

There have been many efforts to develop useful augmented applications for users within the automotive industry, and they have shown the benefits to include augmented reality for helping users in their daily activities; but, due to the security required in this area, it is essential to establish guidelines which developers must follow for reducing the probability of risk. Also, a framework should be defined to manage the interaction among several kinds of devices and open the possibility to integrate new products. All the information received from sensors needs to be analyzed and classified automatically in order to show the most relevant data to users by considering their behaviors and attention points, and also by predicting future scenarios into the development of the activity. Tracking systems and the kind of sensors required to obtain accurate measures are essential part to show the adequate information, in the precise location, and at the right moment, for that reason an improvement on those systems is required.

AUGMENTED REALITY APPLICATIONS FOR COMMERCE

The way in the users perform purchases has changed with the widespread of mobile devices and the connectivity to Internet. Users can look for products they want on line, order the products, and receive the products at home, with only few clicks. For this reason, marketing techniques have to stimulate users to buy products online and create an engagement relationship between the clients and the store in order to conserve them. Augmented reality is an adequate tool for generating more attractive presentation of products, users can obtain a better perception of the product as real as possible.

One of the main markets over Internet is clothing; several clothing stores offer their products to be acquired online; however, a usually problem which the customer faces is about quality, size, and design of clothes, which sometimes differ between the pictures displayed on the computer and the characteristics of the real product. Wang, Chiang, and Wang (2015) evaluated a traditional online shopping website against the same website but with augmented reality included. The augmented reality was generated through a commercial software embedded in the website for displaying 3D models of headwear. They identified three subjective measurements that users perceived in proofs: attractiveness, stimulation, and novelty. Based on their experiments, it was observed that augmented reality makes users to have more interest in purchase, due to the sensory stimulations caused by the application. When a customer visits a clothing store, he/she usually has to try on many garments to find the right one; also, when concurrency of clients is present, the customer has to wait for proving the clothes. For solving this problem, Kjaerside, Kortbek, Hedegaard, and Gronbaek (2005) proposed an augmented dressing room. The room is equipped with 3 cameras for tracking markers, which are collocated in specific areas of the user's body. The 3D model of clothes is designed with a toolkit for designing augmented reality applications (MacIntyre, Gandy, Dow, & Bolter, 2004). Future directions of this work intend to provide the same benefits offered by the proposed dressing room, to users at home, by tracking markers through a web cam.

The election of a product by a customer several times is influenced by marks and offers; therefore, new products have to attract the attention of customers to be elected. When several products are available for election, choosing the right one sometimes is a difficult task for clients. For aiding customers in the election of products and for obtaining more information about a specific product, Zhu, Owen, Li, and Lee (2004) proposed the PromoPad, an e-commerce system for advertising and shopping assistance. A Tablet PC is used to capture the product; subsequently, the image is augmented with information related to the product and advertising, based on preferences stored in a database which contains a history of the customer purchases. On the same line, Gowda and A.S (2016) proposed an application for allowing users to know about products and offers in a store before to get into it. The application recognizes the store by steering the camera of the cellphone in a frontal direction to the store, to allow the recognition of the picture and display the related information over it. Those results are beneficial for marketing, and they show how mobile augmented reality can be used to attract users to stores. Users, without intention to buy something, can observe by means of the mobile phone the new products and offers available in the store while they are walking in front of it.

Factors

When talking about commerce, the way in that products are presented receives special attention to motivate users to acquire products. Ease of use is essential to engage customers, and should be considered for the development of augmented applications oriented to aid users in the purchase process.

- **Content:** Products and their presentation are the essential part of commerce; so, in order to attract users to acquire those products it is necessary to make them more attractive. Well-designed 3d models are the base for augmented reality applications applied to commerce. Advertisement applications need to retain the preferences of users in order to show offers and products related for increasing the purchase probabilities.
- **Data Protection:** Since preferences and other information related to users are stored by the applications, it is necessary to establish policies for the management of sensible information and provide the necessary controls to ensure the security of data.
- **Privacy:** When geolocation is required to provide information to users and when customer preferences are shared with other customers, the controls for ensuring the privacy of customers have to be established.
- **Usability:** Attract the attention of user is the first step into the process of induce the purchase; once the user is involved its involvement with the product is required to maintain the client. An application which provides a friendly interaction has the potential to concrete sales.
- **User Experience:** When customers feel comfortable with a service, they are inclined to return. A goal in commerce is not only sell products but also retain customers. By offering user experience, this goal can be reached.

Challenges

Buying products online is a risky activity considering that not always the product will be similar to the version presented through pictures; also sizes and textures can generate a disgust to the customer at the moment to receive the product. Augmented reality is a good tool for improving presentation of products; however, challenges that need to be addressed are related to the accuracy in the 3D modeling of garments and products; good tracking in order to allow that customers can wear virtually the garments and obtain a realistic simulation of the fitting based on the body of the customer. The perception of textures is another area under development which allows to feel the product and increase the comfortableness of the user at the moment to buy products online.

EDUCATION ENHANCEMENT THROUGH MOBILE AUGMENTED REALITY

Nowadays, technology is present in the classroom but not always used as a medium to provide knowledge, many times it is seen as a distraction for students. Prensky (2008) suggested that technology should be used to allow students learn by themselves instead of being a complementary tool in dictated lectures. Children since 5 to 6 years are in contact with technology through several devices making it part of their lives; it is important to take advantage of those resources to increase the learning skills of students.

Cognitive Load Theory (Sweller, 1988) suggests that more interactivity imposes heavy cognitive load leading to a lower learning capability due to the amount of cognitive processing that needs to be performed. So, the interactivity added to a learning process has to be well-structured to increase the knowledge acquisition and avoid the addition of elements that could decrease it. On the other hand, Embodied Cognition states that not only the brain is in charge of learning activity, but also the body and its interactions with the environment; those interactions help to increase the learning capability (Shapiro, 2011). According to (Sweller, van Merrienboer, & Paas, 1998), cognitive load can be divided in three types, intrinsic cognitive load, which refers to the nature of the material presented; extraneous cognitive load, which refers to the way that material is presented; and germane cognitive load, which refers to the process of generating schemas from the topic under study.

Skulmowski, Pradel, Kühnert, Brunnett, and Daniel (2016) proposed an extension to both theories where extraneous load is further divided in sensorimotor load and processing load. Sensorimotor load refers to movements and perception, while processing load refers to the activity of integrating the multi-sensor information. By the design of an interactive user interface for learning the components of a human heart, they conclude that learning can be improved by increasing sensorimotor load and reducing processing load. That conclusion makes of augmented reality a great tool for designing learning content in a way that sensorimotor load can be increased.

Among current efforts to include augmented educative material in learning, Kaufmann and Meyer (2008) developed an augmented reality application for classical mechanics teaching, their proposal followed the constructivist theory to present the material. They conclude that the application allows the performance of several kinds of experiments in a more comprehensive manner; however, an evaluation about knowledge acquired in comparison to classical education techniques was not conducted yet. Merino, Pino, Meyer, Garrido, and Gallardo (2015) developed an application to observe reactions in organic chemistry by following the constructivist learning cycle. The cycle considers activities of exploration, introducing new variables, systematization, and application. The learning performance has not been evaluated for identifying the strengths of augmented reality as a technological tool for teaching. Learning of human anatomy requires that students remember names of the compounding parts and relations of human body; this activity involves a high cognitive processing; in order to reduce the difficulty inherent to this process, Konchada et al. (2011) proposed a pelvic trainer to learn about this structure through a mannequin and a virtual interface. The student can manipulate physically the structure while visualizes the name of parts and relations through the interface. They are working in more structures to generate a more understandable human anatomy teaching method.

To provide a quantitative analysis of beneficial provided by mobile technology in education, Sung, Chang, and Liu (2016) performed a meta-analysis of works which have incorporated technology in teaching. The analysis is performed based on activity theory, which considers objects, tools, rules, context of the activity, and interaction in order to analyze the human practices. Their results shown that 69.95% of learners get better achievement through the use of mobile devices. Those results encourage the use of technology in learning activities, but it is necessary to apply education theories to increase the subject content learning. Also, frameworks have to be developed for facilitating the material preparation by teachers.

Factors

Many augmented reality applications have been developed in the education sector; several of them have followed an education theory, which increases the knowledge acquired by students. Now, it is important to define which factors should be considered by developers or researchers to generate frameworks or applications that teachers can adapt to their classes.

- **Content:** Based on the cognitive load theory, the saturation of information or high interactivity leads to the reduction on the capacity of understanding the presented material. Classifying and establishing a methodology, which guides the presentation of contents through an augmented reality application, is crucial to improve the learning activity.
- **Scalability:** A group of students is often constituted of at least 10 members; the participation of every student is required for allowing the understanding of content by everyone. Applications should support the integration of new elements and must allow the interaction of every device such that all the students can perceive the changes performed.
- **Usability:** Students habituate to lose attention easily when the content presented is boring; therefore, it is necessary to create engaging applications that keep the attention of students on the learning activity. Ease of use and well-designed applications are attractive to students, also the way in that content is presented and the allowed interactions are key pieces to capture attention of users.
- **User Experience:** Learning activities commonly are performed when instructor gives the indication. Once completed, the activities are not considered again. With user experience, students can be inducted to research more about a topic by generating the desire to discover more about it.

Education is a sector where mobile augmented reality applications can introduce great benefits. The low use of technology in this area can be increased by the introduction of augmented reality as a teaching tool. According to results of the works described before, knowledge can be increased through virtual and real interactions. Moreover, following education theories to generate content make possible to create better structured applications.

Challenges

When a new technology emerges, the creation of new applications is reserved to individuals with skills to manage those devices; but, to allow mobile augmented reality to be part of education it is necessary to create frameworks that any teacher could use to prepare their classes. Teachers need the access to 3D models that represent the material they want to teach; need to define the interactions which users can perform through the application, and also tools to evaluate the performance of the application on the learning activity and suggestions to improve them. Furthermore, to communicate among devices with a good bandwidth to share resources like 3D models, it is necessary a network infrastructure in the

Mobile Augmented Reality

classroom; however, there exist places where schools doesn't have access to an infrastructure network; in these cases, an ad hoc network should be created among devices, and protocols which ensure the data transmission in a speedy manner while maintain the consumption of battery in a good level are required.

MOBILE AUGMENTED REALITY INTO THE VIDEO GAMES INDUSTRY

Large companies like Sony, Nintendo, and Ubisoft, among others, have started to take advantage of the potential of interactions which are offered by the mobile augmented reality. By merging graphics with elements of the real world, amazing games are developed and the interaction with other gamers is also enhanced. Video game industry has the advantage of using all kind of devices available for generating mobile augmented reality applications, AR cards, haptic devices, mobile devices, and wearable devices, to name a few. This industry also serves as a test-bed for applications that are being developed to other areas like the ones described previously.

With the intention to create a game situated in the physical world, Thomas, Donoghue, and Squires (2000) developed an extension of the game Quake by combining optical tracking with GPS and magnetic compass for tracking fiducial markers; i-Glasses were used to observe the scenes and a laptop included in a backpack kid to perform the processing of information. Cheok (2010) proposed a game that uses GPS and inertial systems to record the position of players. PacMan and the phantoms are human persons which run on the city while wearing a vision system and a portable computer for perceiving the real world. Mobile augmented reality offers a new way of interacting with applications without requiring in some cases to use a mouse, joystick, game control, or physical device; tracking of the movements performed by the user can be used to manipulate the application. Following this philosophy Lv, Halawani, Feng, Réhman, and Li (2015) proposed an interactive game manipulated by gestures. Gestures are captured by a wearable camera and displayed through a wearable projector. Proofs were performed in the hardware of a hybrid wearable framework, which is composed by a mobile phone and a holder base; also google glasses were used to evaluate the proposal. A game which has gained popularity today is Pokémon Go, a game with mobile augmented reality developed by Niantic (2016). By exploiting the popularity of Pokémon characters and with the intention to generate mobility on gamers, the game shows the presence of a Pokémon which has to be trapped in some public places. It can be observed that mobile augmented reality games change the way a user interacts with a video game, and also how this approach tries to eradicate sedentary behavior.

Factors

- **Content:** For mixing real objects with virtual ones in a game, lighting conditions and background colors need to be defined adequately, alignment of the real image and virtual elements in the displayed scenario has to be considered, and the content has to react adequately to the several ways of interaction allowed in the game.
- **Scalability:** Multiplayer games require the integration of several users into the game; therefore, mobile augmented games need to operate adequately in mobile devices although the number of users is increased. Moreover, information to be transmitted among users needs to be classified for reducing the data processing while more users take part of the game.

- **Usability:** The kind of interactions allowed in every game needs to be established adequately to maintain the usability of the application; also, to provide an improvement on the interaction, the devices to be used not should be intrusive with the users based on the actions they need to perform.
- **User Experience:** An attractive game with a good storyline, and good graphics will be preferred always by users; with the aim of improving the user experience, augmented elements have to be chosen and displayed adequately. Furthermore, make to feel the user being part of the game increases their affinity with it.

Challenges

A realistic game requires to consider quality of graphics, interactions, behaviors, sensations, communication, and a good game storyline. Mobile augmented reality opens the possibility to integrate real objects to the game and generates new issues to be addressed like: real time processing of tracking information for keeping the interactivity of the game; persons and places in the real world have to be integrated congruently to the storyline of the game to generate a more realistic interaction; allow the inclusion of any kind of mobile device available to interact with the game; moreover, frameworks which hide the communication issues raised by the heterogeneity of the devices have to be developed. Multiplayer games can be enhanced by the inclusion of georeferenced sources to allow users located in distinct places can interact into the game.

MOBILE AUGMENTED REALITY IN MEDICINE

Augmented reality has opened a wide range of options; due to the use of mobile devices is possible to feel and experiment with virtual worlds. In medicine, it is used to solve problems related to surgery, medical education, and rehabilitation, among others; examples of this are described below.

This technology is widely used in surgery to create virtual transparency of the patient anatomy, which provides higher accuracy and precision with fewer risks, the possibility of diagnosing the patient's condition during surgery, and the performance of guided surgery within less time (Hyppölä, Martínez, & Laukkanen, 2014). AR can introduce new elements to the operating room; the real objects in the system will be patient and instruments, and virtual objects can be models of instruments and patient models built with images of computed tomography, magnetic resonance imaging or surgery regions. 3D visualization and modelling help surgeons to conduct minimally invasive surgery that provides greater benefits to patients (Nicolau, Soler, Mutter, & Marescaux, 2011). Sielhorst, Obst, Burgkart, Riener, and Navab (2004) describe an extension of a birth simulator for medical training with an augmented reality system. This simulation system comprised direct haptic and auditory feedback, and provided important physiological data including values of blood pressure, heart rates, pain and oxygen supply, necessary for training physicians. This plays an important role in increasing the efficiency of the training, since the physician now concentrates on the vaginal delivery rather than the remote computer screen. In addition, forceps are modeled and an external optical tracking system is integrated in order to provide visual feedback while training with the simulator for complicated procedures such as forceps delivery. Fischer, Neff, Freudenstein, and Bartz (2004) proposed an alternative approach of building a surgical AR system by harnessing existing, commercially available equipment for image guided surgery. They describe the prototype of an augmented reality application, which receives all necessary information from a device for intraoperative navigation. AR may also prevent patient and operator from other risks such as exposure to radiation in some procedures (Fritz et al., 2012). Andersen et al. (2016) designed and implemented a surgical telementoring system called the System for Telementoring with Augmented Reality that uses a virtual transparent display to convey precise locations in the operating field to a trainee surgeon. Today AR is applied in some fields such as orthopedics and neurosurgery, its application is investigated in abdominal surgery where navigation is more complicated due to deformation of the organs.

Medical education is another area where AR technology is used; anatomy has received special attention since students need to remember several components of the human body. With augmented reality, it is possible to visualize 3D anatomical imagery, at the same time that students can manipulate real objects and observe extra detailed information about them. Taking those benefits into account, AR technology provides the necessary elements to perform some of the activities required in traditional dissection classes (Kamphuis, Barsom, Schijven, & Christoph, 2014). Hamza-Lup, Santhanam, Imielinska, Meeks, and Rolland (2007) developed a system that allows real-time visualization of 3D lung dynamics superimposed directly on a manikin or on a patient in the operating room. In that visualization they combine a generic functional lung model with patient-specific data extracted from high-resolution computed tomography. This results in a dynamic, real-time visualization of virtual lungs that is overlaid onto the patient's body. Chien, Chen, and Jeng (2010) use AR technology to create an interactive learning system, which help medical students to understand and memorize the 3D anatomy structure easily with tangible augmented reality support, this AR system can help young medical students to learn the complex anatomy structure better and faster than only with traditional methods. Blum, Kleeberger, Bichlmeier, and Navab (2012) describe the magic mirror ('mirracle') which is an AR system that can be used for undergraduate anatomy education. The trainee stands in front of a TV screen that has a camera and the Kinect attached to it. The camera image of the trainee is flipped horizontally and shown on the TV screen, mimicking a mirror function. Part of an anonymous CT dataset is augmented to the user's body and shown on the TV screen. This creates the illusion that the trainee can look inside his body. A gesture-based user interface allows real time manipulation of the visualization of the CT data. The trainee can scroll through the dataset in sagittal, transverse and coronal slice mode, by using different hand gestures.

Furthermore, in the field of medicine, AR has been widely used in cognitive and motor rehabilitation in patients with brain damage; examples of these cases include a Spatial Augmented Reality system for rehabilitation of hand and arm movement. It is a system which tracks a subject's hand and creates a virtual audio-visual interface for performing rehabilitation-related tasks that involve wrist, elbow, and shoulder movements. It measures range, speed, and smoothness of movements locally and can send the real-time photos and data to the clinic for further assessment (Mousavi Hondori, Khademi, Dodakian, Cramer, & Lopes, 2013). Other tools are computer games for rehabilitating stroke patients affected with upper limb disabilities (De Leon, Bhatt, & Al-Jumaily, 2014), upper-limb stroke rehabilitation (Burke et al., 2010; Jordan & King, 2011), rehabilitation of hand and arm movement (Wang, Hsu, Chiu, & Tsai, 2011; Mousavi et al., 2013), neurological rehabilitation (Poore & Tevfik, 2011) and rehabilitation using multimodal Feedback (Aung & Al-Jumaily, 2011; Vieira, Sousa, Arsénio, & Jorge, 2015).

In conclusion, augmented reality has multiple applications in the medical area; this section described only in surgery, education, and rehabilitation activities, but it is also applied to other medical disciplines.

Factors

Applications developed with augmented reality should consider factors that involve improving the techniques used for surgery, provide better visual tools for teaching Anatomy, and develop software for rehabilitation of injuries or deficiencies in the human body.

- **Availability:** Surgery is one of the most addressed issues in medicine through augmented reality, and also, one of the riskiest activities. When working with applications for helping surgery, the availability of data and information (images, diagrams, procedures, etc.) is required to provide a good operation.
- **Content:** The information displayed should be analyzed carefully and several tests should be performed before to deliver the application. Calculations and graphics should be accurate and displayed in real time to provide a better interaction. The amount of data available regarding the human body is huge; choosing the right information is an important factor to consider.
- **Security:** Augmented reality applications employed in certain activities within medicine require not to be distracting, such that users could continue their work only with the necessary assistance to complete it. Show the right information is essential to ensure a safe interoperability.
- **Usability:** The software should consider easy and intuitive elements for different users who can use the system, such as medical specialists, patients, technical personnel, among others.
- **User Experience:** Frequently, surgeons fall under stress when surgical interventions are performed. Mobile augmented reality applications have the potential to influence in the ambience; by generating a more relaxing ambience the probabilities of success on performing the activities are increased.

Challenges

In the medical area there are situations that have not yet been solved in which researchers will continue investigating to provide solutions. Some of these issues are related to deformations, position changes, and size of the organs of the human body; these behaviors are difficult to compute, simulate, and visualize in real time; certainly there are good challenges for software developers.

MODEL FOR DEVELOPING MOBILE AUGMENTED APPLICATIONS

Mobile augmented reality applications need to identify physical objects that will later be analyzed to provide the additional information required based on the activity to be performed. The authors propose a model to develop augmented reality applications that requires to evaluate their improvements; this proposal is illustrated in Figure 1.

The augmented reality model has to be constructed by considering the goal and the activities required to reach it, physical objects to be manipulated, and the virtual information to be generated. The activities define the interactions allowed with the physical objects based on the devices to be used. The physical objects are tracked to establish the relations with the information to be displayed. 3D models are used to generate a more attractive and lifelike application. All these elements are integrated for generating

Mobile Augmented Reality

Figure 1. Developing process of an augmented reality model

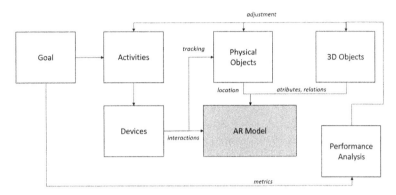

the augmented reality application. To evaluate the performance of the application in order to reach the defined goals, some metrics are defined, which will help to adjust the diverse elements which are part of the model.

FUTURE RESEARCH DIRECTIONS

Research on mobile augmented reality has an open field for generating new proposals. In this chapter areas with major influence of this technology were tackled, but many others remain without significant advances, and good benefits can be obtained by including mobile augmented reality on them; for example archeology, architecture, literature, logistics, psychology, and rescue, to name a few. There exist other issues that should be addressed.

- **Communication:** As new mobile devices are designed, communication protocols need to be adapted, or new ones have to be proposed for allowing the intercommunication among those devices. Scalability and confidence in transmissions are required to ensure quality of service. Zigbee is a promising technology to communicate with wearable devices, but its limited bandwidth compromises the quantity of data that can be transmitted; the development of augmented reality applications must consider those constraints to provide a good service.
- **Devices:** Nowadays, there are under development portable devices like smart glasses, projectors, smart watches, and wearable controls; however, the sensors which are required to capture the user movements are not small enough to be carried in a non-intrusively manner. Tracking and accurate location are necessary in order to provide useful information to the user; embedded systems that help to perform those activities can be proposed.
- **Energy:** While more processing and communication activities are performed by a device, more energy is consumed. It is necessary to balance the activity load among devices according to their capabilities to extend the battery life. Also, long life batteries or low consumption devices that increase the productivity can be developed. Most of augmented reality applications need to process 3D graphics, which require significant processing; this is the reason why energy is an important factor to take into account.

- **Frameworks:** Development of applications which require the interaction with several kind of devices, the sensing of the environment, processing of the acquired information, and the display of data to the user through 3D scenes, is a complex task which requires several areas of expertise. It is necessary to develop frameworks which hide all the complexity to the user, but which facilitate the generation of content and the establishment of interactions easily in order to allow any user, with only experience in their daily activities, to make use of mobile augmented reality.
- **Security:** Last but not least, security of both information and user, are important factors to consider when mobile augmented reality applications are developed. Guidelines are required to define how the content is displayed to the user in order to avoid the loss of attention in the activity being performed. In some situations, the information manipulated through the application is confidential; so, it is necessary to define mechanisms for protecting the information that is transferred among devices.

CONCLUSION

Mobile augmented reality is an emergent technology for the interaction of users with mobile and wearable devices. The creation of new devices that user can carry with them all the time, allows to take advantage of this technology for assisting users in their daily activities and work. In this chapter, some approaches were presented, which provide solutions to be applied in automotive industry, commerce, education, entertainment, and medicine. Attention was centered in factors like availability, content, data protection, privacy, scalability, security, and usability. Challenges concerning security, communication, display of information, battery consumption, quality of data, and frameworks were presented with the intention to contribute to the fast integration of this technology in the daily life of users, at the same time of providing a secure interaction.

In the automotive industry, activities like design, driving, manufacturing, mechanics, and sales can be improved through mobile augmented reality; moreover, factors like content, data protection, privacy, security, and usability should be considered in the development of applications for this sector. Exhibition and shopping are tasks that can be benefited from augmented reality in commerce; also content, data protection, privacy, and security are factors to take into account by applications in that sector. Education is an area which has received special attention; since there are no security risks that applications can cause, it serves as a platform to prove new devices and techniques that helps to improve learning and teaching activities; factors like content, scalability, and usability, are important for good performance of the applications. In the entertainment industry, video games have received special attention; the user experience is the main factor to consider in the video games. Finally, medicine was the last sector covered in this chapter; activities like rehabilitation and surgery are the most addressed due to the benefits provided by augmented reality applications; however, the work with humans requires special care, making of security the main factor to consider when applications are developed for surgery and rehabilitation.

REFERENCES

Abdi, L., Abdallah, F. B., & Meddeb, A. (2015). In-vehicle augmented reality traffic information system: A new type of communication between driver and vehicle. *Procedia Computer Science*, *73*, 242–249. doi:10.1016/j.procs.2015.12.024

Andersen, D., Popescu, V., Cabrera, M. E., Shanghavi, A., Gomez, G., Marley, S., & Wachs, J. P. et al. (2016). Medical telementoring using an augmented reality transparent display. *Surgery*, *159*(6), 1646–1653. doi:10.1016/j.surg.2015.12.016 PMID:26804823

Ashida, K., Takamoto, J., Ibuki, M., Yamamoto, S., Matsui, H., Kumazaki, D., & Suga, A. (2008). *U.S. Patent No. D559254*. Washington, DC: US Patent Office.

Aung, Y. M., & Al-Jumaily, A. (2011). Development of augmented reality rehabilitation games integrated with biofeedback for upper limb. *Proceedings of the 5th International Conference on Rehabilitation Engineering & Assistive Technology*, 1-4.

Bay, H., Ess, A., Tuytelaars, T., & van Gool, L. (2008). SURF: Speed up robust features. *Computer Vision and Image Understanding*, *110*(3), 346–359. doi:10.1016/j.cviu.2007.09.014

Billinghurst, M., Kato, H., & Poupyrev, I. (2008). Article. *ACM SIGGRAPH Asia*, *7*, 1–11.

Blum, T., Kleeberger, V., Bichlmeier, C., & Navab, N. (2012). Mirracle: An augmented reality magic mirror system for anatomy education. *2012 IEEE Virtual Reality Workshops*, 115-116. doi:10.1109/VR.2012.6180909

Bolt, R. A. (1980). Put-that-there: Voice and gesture at the graphics interface. *Computer Graphics*, *14*(3), 262–270. doi:10.1145/965105.807503

Bowers, B. (2001). *Sir Charles Wheatstone FRS 1802-1875*. London: The Institution of Electrical Engineers in Association with the Science Museum.

Burke, J. W., McNeill, M. D. J., Charles, D. K., Morrow, P. J., Crosbie, J. H., & McDonough, M. S. (2010). Augmented reality games for upper-limb stroke rehabilitation. *2010 Second International Conference on Games and Virtual Worlds for Serious Applications*, 75-78. doi:10.1109/VS-GAMES.2012.21

Cheok, A. D. (2010). Human Pacman: A mobile augmented reality entertainment system based on physical, social, and ubiquitous computing. In A. D. Cheok (Ed.), *Art and Technology of Entertainment Computing and Communication* (pp. 19–57). Springer London. doi:10.1007/978-1-84996-137-0_2

Chien, C., Chen, C., & Jeng, T. (2010). An interactive augmented reality system for learning anatomy structure. *Proceedings of the International Multiconference of Engineers and Computer Scientists*, *2010*, 370–375.

Committee, A. S. (2000). *The Link flight trainer: A historic mechanical engineering landmark. Roberson Museum and Science Center*. Binghamton, NY: ASME International, History and Heritage Committee.

De Leon, N. I., Bhatt, S. K., & Al-Jumaily, A. (2014). Augmented reality game based multi-usage rehabilitation therapist for stroke patients. *International Journal on Smart Sensing and Intelligent Systems*, *7*(3), 1044–1058.

DeFanti, T. A., & Sandin, D. J. (1977). *Final Report to the National Endowment of the Arts* (U.S. NEA R60-34-163). Chicago: University of Illinois.

Fischer, J., Neff, M., Freudenstein, D., & Bartz, D. (2004, June). Medical augmented reality based on commercial image guided surgery. In S. Coquillart (Chair), *10th Eurographics Symposium on Virtual Environments*. Symposium conducted at the meeting of the Eurographics Association, Grenoble, France.

Fritz, J., U-Thainual, P., Ungi, T., Flammang, A. J., Fichtinger, G., Iordachita, I. I., & Carrino, J. A. (2012). Augmented reality visualization with use of image overlay technology for MR imaging-guided interventions: Assessment of performance in cadaveric shoulder and hip arthrography at 1.5 T. *Radiology, 265*(1), 254–263. doi:10.1148/radiol.12112640 PMID:22843764

George, P., Thouvenin, I., Frémont, V., & Cherfaoui, V. (2012). DARIA: Driver assistance by augmented reality for intelligent automotive. *Intelligent Vehicles Symposium (IV)*, 1043-1048.

Gowda, A. D., & A S, M. (2016). Smart shopping using augmented reality on android OS. *International Journal of Engineering Research and General Science, 4*(3), 211–217.

Hamza-Lup, F. G., Santhanam, A. P., Imielinska, C., Meeks, S. L., & Rolland, J. P. (2007). Distributed augmented reality with 3-D lung dynamics – a planning tool concept. *IEEE Transactions on Information Technology in Biomedicine, 11*(1), 40–46. doi:10.1109/TITB.2006.880552 PMID:17249402

Heilig, M. (1962). *U.S. Patent No. 3050870*.

Henderson, S. J., & Feiner, S. (2008). Opportunistic controls: leveraging natural affordances as tangible user interfaces for augmented reality. *Proceedings of the 2008 ACM symposium on Virtual reality software and technology* (pp. 211-218). New York: ACM. doi:10.1145/1450579.1450625

Hyppölä, J., Martínez, H., & Laukkanen, S. (2014, March). Experiential learning theory and virtual and augmented reality applications. In R. Fisher (Chair), *4th Global Conference on Experiential Learning in Virtual Worlds*. Symposium conducted at the meeting of Inter-Disciplinary.Net, Prague, Czech Republic.

Jordan, K., & King, M. (2011). Augmented reality assisted upper limb rehabilitation following stroke. In A. Y. C. Nee (Ed.), *Augmented reality: Some emerging application areas* (pp. 155–174). Rijeka, Croatia: InTech. doi:10.5772/25954

Kamphuis, C., Barsom, E., Schijven, M., & Christoph, N. (2014). Augmented reality in medical education? *Perspectives on Medical Education, 3*(4), 300–311. doi:10.1007/s40037-013-0107-7 PMID:24464832

Kaufmann, H., & Meyer, B. (2008). Simulating educational physical experiments in augmented reality. *Proceedings of ACM SIGGRAPH ASIA 2008 Educators Programme*, 1-8. doi:10.1145/1507713.1507717

Kim, S., & Dey, A. K. (2009). Simulated augmented reality windshield display as a cognitive mapping aid for elder driver navigation. *Proceedings of ACM CHI 2009 Conference on Human Factors in Computing Systems*, 133-142. doi:10.1145/1518701.1518724

Kjaerside, K., Kortbek, K. J., Hedegaard, H., & Gronbaek, K. (2005, June). Ardresscode: Augmented dressing room with tag-based motion tracking and real-time clothes simulation. In J. Zara (Chair), *Central European Multimedia and Virtual Reality Conference 2005*. Symposium conducted at the meeting of The Eurographics Association, Prague, Czech Republic.

Konchada, V., Shen, Y., Burke, D., Argun, O. B., Weinhaus, A., Erdman, A. G., & Sweet, R. M. (2011). The Minnesota pelvic trainer: A hybrid vr/physical pelvis for providing virtual mentorship. *Studies in Health Technology and Informatics, 163*, 280–282. PMID:21335805

Lv, Z., Halawani, A., Feng, S., Réhman, S., & Li, H. (2015). Touch-less interactive augmented reality game on vision based wearable device. *Personal and Ubiquitous Computing, 19*(3), 551–567. doi:10.1007/s00779-015-0844-1

MacIntyre, B., Gandy, M., Dow, S., & Bolter, J. D. (2004). DART: A toolkit for rapid design exploration of augmented reality experiences. In S. K. Feiner (Ed.), *Proceedings of the 17th Annual ACM Symposium on User Interface Software and Technology* (pp. 197-206). Santa Fe, NM: ACM. doi:10.1145/1029632.1029669

Merino, C., Pino, S., Meyer, E., Garrido, J. M., & Gallardo, F. (2015). Realidad aumentada para el diseño de secuencias de enseñanza-aprendizaje en química. *Educación en la Química, 26*(2), 94–99.

Mousavi Hondori, H., Khademi, M., Dodakian, L., Cramer, S. C., & Lopes, C. V. (2013). A spatial augmented reality rehab system for post-stroke hand rehabilitation. *Medicine Meets Virtual Reality, 20*, 279–285. PMID:23400171

Niantic. (2016). *Pokémon Go*. Retrieved from the Niantic, Inc. website: http://pokemongo.nianticlabs.com/es/

Nicolau, S., Soler, L., Mutter, D., & Marescaux, J. (2011). Augmented reality in laparoscopic surgical oncology. *Surgical Oncology, 20*(3), 189–201. doi:10.1016/j.suronc.2011.07.002 PMID:21802281

Polzin, R. S., Kipman, A. A., Finocchio, M. J., Geiss, R. M., Stone Perez, K., Tsunoda, K., & Bennett, D. A. (2014). *U.S. Patent No. 8744121*. Washington, DC: US Patent Office.

Poore, J., & Tevfik, B. (2011). *Augmented reality games for neurological rehabilitation*. Retrieved from Colorado State University, Department of Electrical and Computer Engineering website: http://projects-web.engr.colostate.edu/ece-sr-design/AY11/rehabilitation/documents/first_report.pdf

Prensky, M. (2008). *The role of technology in teaching and the classroom*. Retrieved from http://www.marcprensky.com/writing/Prensky-The_Role_of_Technology-ET-11-12-08.pdf

Shapiro, L. (2011). *Embodied Cognition*. New York: Routledge.

Sielhorst, T., Obst, T., Burgkart, R., Riener, R., & Navab, N. (2004). An augmented reality delivery simulator for medical training. In M. Berger, & N. Navab (Chairs), *Workshop AMI-ARCS 2004*. Symposium conducted at the meeting of IRISA, Rennes, France.

Skulmowsky, A., Pradel, S., Kühnert, T., Brunnett, G., & Daniel, R. G. (2016). Embodied learning using a tangible user interface: The effects of haptic perception and selective pointing on a spatial learning task. *Computers & Education, 92-93*, 64–75. doi:10.1016/j.compedu.2015.10.011

Sung, Y. T., Chang, K. E., & Liu, T. C. (2016). The effects of integrating mobile devices with teaching and learning on students learning performance: A meta-analysis and research synthesis. *Computers & Education, 94*, 252–275. doi:10.1016/j.compedu.2015.11.008

Sutherland, I. E. (1965). The ultimate display. In W. A. Kalenich (Ed.), *Proceedings of IFIP Congress* (pp. 506-508). New York: Spartan Books, Macmillan.

Sutherland, I. E. (1968). A head-mounted three dimensional display. *AFIPS Conference Proceedings, 33, 1* (pp. 757-764). San Francisco, CA: The Thompson Book Company.

Sweller, J. (1988). Cognitive load during problem solving: Effects on learning. *Cognitive Science, 12*(2), 257–285. doi:10.1207/s15516709cog1202_4

Sweller, J., van Merrienboer, J. J., & Paas, F. G. (1998). Cognitive architecture and instructional design. *Educational Psychology Review, 10*(3), 251–296. doi:10.1023/A:1022193728205

Thomas, B., Close, B., Donoghue, J., & Squires, J. (2000). ARQuake: An outdoor/indoor augmented reality first person application. *Proceedings of the Fourth International Symposium on Wearable Computers* (pp. 139-146). Atlanta, GA: IEEE. doi:10.1109/ISWC.2000.888480

Ullmer, B., Ishii, H., & Jacob, R. J. K. (2005). Token+constraint systems for tangible interaction with digital information. *ACM Transactions on Computer-Human Interaction, 12*(1), 81–118. doi:10.1145/1057237.1057242

Underkoffler, J. (1999). *The I/O bulb and the luminous room* (Doctoral dissertation). Retrieved from http://tmg-trackr.media.mit.edu/publishedmedia/Papers/300-The%20IO%20Bulb/Published/PDF

Vieira, J., Sousa, M., Arsénio, A., & Jorge, J. (2015). Augmented reality for rehabilitation using multimodal feedback. *Proceedings of the 3rd 2015 Workshop on ICTs for improving Patients Rehabilitation Research Techniques*, 38-41. doi:10.1145/2838944.2838954

Viola, P., & Jones, M. J. (2004). Robust real-time face detection. *International Journal of Computer Vision, 57*(2), 137–154. doi:10.1023/B:VISI.0000013087.49260.fb

Wang, C., Chiang, Y., & Wang, M. (2015). Evaluation of an augmented reality embedded on-line shopping system. *Procedia Manufacturing, 3*, 5624–5630. doi:10.1016/j.promfg.2015.07.766

Wang, H., Hsu, C., Chiu, C., & Tsai, S. (2011). The design and implementation of augmented reality gaming system in hand rehabilitation. *Communications in Information Science and Management Engineering, 1*(8), 37–40. doi:10.5963/CISME0108009

Wickens, C. D., Hollands, J. G., Simon, B., & Parasuraman, R. (2013). *Engineering psychology and human performance*. Pearson.

Zhu, W., Owen, C. B., Li, H., & Lee, J. (2004). Personalized in-store e-commerce with the PromoPad: An augmented reality shopping assistant. *Electronic Journal for E-commerce Tools and Applications, 1*(3), 1–19.

ADDITIONAL READING

Abu, G., & Pandhare, A. (2016). Applications of augmented reality. *International Journal of Trend in Research and Development, 3*(2), 424–427.

Barfield, W. (2015). *Fundamentals of wearable computers and augmented reality*. Florida: CRC Press. doi:10.1201/b18703

Chow, E. H., Thadani, D. R., & Wong, E. Y. (2015). Mobile technologies and augmented reality: Early experiences in helping students learn about academic integrity and ethics. *International Journal of Humanities Social Sciences and Education, 2*(7), 112–120.

Fernandes Alcantara, M., Goncalves Silva, A., & Silva Hounsell, M. (2015). Enriched barcodes applied in mobile robotics and augmented reality. *IEEE Latin America Transactions, 13*(12), 3913–3921. doi:10.1109/TLA.2015.7404927

Ghouaiel, N., Cieutat, J.-M., & Jessel, J.-P. (2014). Mobile augmented reality: Applications and specific technical issues. In L. Chen, S. Kapoor, & R. Bhatia (Eds.), *Intelligent Systems for Science and Information* (pp. 139–151). Cham, Switzerland: Springer International Publishing. doi:10.1007/978-3-319-04702-7_8

Jain, P., Manweiler, J., & Choudhury, R. R. (2015). Overlay: Practical mobile augmented reality. *Proceedings of the 13th Annual International Conference on Mobile Systems, Applications, and Services* (pp. 331-344). NY, USA: ACM.

Kourouthanassis, P., Boletsis, C., Bardaki, C., & Chasanidou, D. (2015). Tourists responses to mobile augmented reality travel guides: The role of emotions on adoption behavior. *Pervasive and Mobile Computing, 18*, 71–87. doi:10.1016/j.pmcj.2014.08.009

Lebeck, K., Kohno, T., & Roesner, F. (2016). How to safely augment reality: Challenges and directions. *Proceedings of the 17th International Workshop on Mobile Computing Systems and Applications* (pp. 45-50). Florida: ACM. doi:10.1145/2873587.2873595

Lin, C. Y., Chai, H. C., Wang, J. Y., Chen, C. J., Liu, Y. H., Chen, C. W., & Huang, Y. M. (2016). Augmented reality in educational activities for children with disabilities. *Displays, 42*, 51–54. doi:10.1016/j.displa.2015.02.004

Parise, S., Guinan, P. J., & Kafka, R. (2016). Solving the crisis of immediacy: How digital technology can transform the customer experience. *Business Horizons, 59*(4), 411–420. doi:10.1016/j.bushor.2016.03.004

Petersen, N., & Stricker, D. (2015). Cognitive augmented reality. *Computers & Graphics, 53*, 82–91. doi:10.1016/j.cag.2015.08.009

KEY TERMS AND DEFINITIONS

Ad Hoc Network: A network where an access point does not exist, and devices communicate among them to transfer information.

AR Card: Special marked card which when visualized through the camera of a mobile device, the device displays 3D information over the card.

E-Textile: It is the name assigned to conductive fabric and materials, which are used to manufacture clothes and allow the interconnection and integration of electronics and clothing.

Haptic Device: Human-computer interface to manipulate virtual objects through the physical contact with the user; actions performed by the user by means of the manipulation of the device are transferred to the virtual world.

Infrastructure Network: A network which allows the connectivity of multiple devices through an access point, which coordinates the information interchange.

Smart Glass: Device composed by a tiny camera that captures the environment, virtual information is displayed in the lenses, and integrated wireless communication allows the interaction with other devices. This is the evolution of head-mounted display.

Wearable Computing: Term used for identify all those devices that a user could carry on the body without feeling it as an intrusive or uncommon device.

Chapter 11
Using Cognitive Psychology to Understand Anticipated User Experience in Computing Products

Emmanuel Eilu
Makerere University Kampala, Uganda

ABSTRACT

User Experience assessment is an evaluation of user's experience with the product, system or service during 'use' (i.e., actual interaction experience) as well as 'anticipated or before use' (i.e., pre-interaction experience). Whereas many user experience researchers may be conversant with explaining a person's experience during use of a product, system or service, they find it difficult to explain experience before a product or service is used (Anticipated Use), which in this chapter is referred to as Anticipated User Experience (AUX). This chapter applies the theory of cognitive psychology and its principles to best explain how Anticipated User Experience occurs and how this experience can be achieved. This chapter goes a long way in informing user experience researchers and practitioners on the relevance of attaining AUX in a computing product and how it can be achieved.

INTRODUCTION

User experience is the delight and fulfilment a user gets before, amid and after utilizing an item or service. User Experience has been under broad review throughout the most recent decade, producing various User Experience models and assessment techniques. However, the greater part of these models do not show how the delight and fulfilment a user gets before utilising the product can be achieved. In this paper, it is referred to as "Anticipated User Experience" (AUX). Understanding Anticipated User Experience (AUX) till now is a challenge to many user experience designers. The greater part of existing user experience theories and models concentrate on comprehending and accomplishing user experience during and after product use. Yet, AUX is an extremely important aspect in any product or service if the acceptance of that particular product or service is to be guaranteed. AUX in simple terms is an anticipated, pictured or imagined pleasure and satisfaction the user feels about the product even before it's made available to the user. AUX plays a great role in product acceptance. There is need to understand UAX and how it can achieved in computing products. One of the theories that can help user experience researchers and practitioners better understand AUX and how it can be achieved in computing products is the Cognitive Psychology theory.

This Chapter

This chapter attempts to use Cognitive Psychology theory and its principles to better understand AUX and how it can be achieved in mobile computing products. This chapter goes a long way in informing designers of computing products on the relevance of attaining AUX in a computing product and how it can be achieved

BACKGROUND

According to Pakanen (2015), the foundations of user experience research is rooted on usability. As defined by ISO 9241-11 (1998), usability is the extent to which a product [a portal] can be used by specified users to achieve specified goals with effectiveness, efficiency and satisfaction in a specified context of use. According to the revised definition of usability in ISO FDIS 9241-210, it is the extent to which a system, product or service can be used by specified users to achieve specified goals with effectiveness, efficiency and satisfaction in a specified context of use. Designing for effectiveness and efficiency in computing products has rather been successful. However designing for satisfaction has somewhat been under looked. However, the swift of ideology from compulsory work-oriented use of a computer (effectiveness and efficiency) to leisure and personal use (satisfaction) has led to an interest in expanding usability research to focus on User Experience (Pakanen, 2015). During the past 20 years, designing for user experience has been admired (Arhippainen, 2003, 2009). According to Kaikkonen (2009), research in the 1990's was focusing on user interface and interaction, and usability was synonymous with this. In recent years, User Experience has been recognized as an important and determinant factor in design, development and evaluation of human-computer interfaces and man-machine interfaces (Adikari et al., 2010). The raise of the term *'user experience'* came up during mobile devices and services error. It showed that the ease of use, efficiency and effectiveness are not adequate to warranty user satisfaction, in addition to cognitive aspects; wider perspective to humans was needed (Kaikkonen, 2009). Everything has influence on user experience including those that the designer and developer cannot influence, like user past experiences, environmental factors, and many others. Consequently, the increasing use of the term shows the need for a wider approach to the user than 'usability' can offer (Kaikkonen, 2009). Product development in user experience is no longer only about applying features and testing their usability, but it is about designing products that are enjoyable and support essential social, cultural and economic needs and values of the user (Väänänen et al., 2009). A part from optimizing human performance like usability does, user experience aims further at optimizing user satisfaction with achieving both pragmatic and hedonic (pleasurable) goals (Bevan, 2010). According to Arhippainen (2003, 2009) user experience initially was included in usability issues, but afterwards, it has been understood that even a product with good usability can cause negative experiences or dissatisfaction and vice versa. User experience explores much more than just usability of product, even if the usability is an important aspect, without doubt- deficiencies in usability of product will affect user experience in a negative way (Arhippainen, 2003, 2009).

However, despite the growing interest in user experience, it has been hard to arrive at a common agreement on the definition, nature and scope of user experience (Law et al., 2009). Different definitions for user experience have emerged (Arhippainen, 2003, 2009). With McMullin (2004) defining user experience as a systematic cycle that attempts to satisfy hopes, dreams, needs, and desires of the systems' user. While Khambete & Athavankar (2010) define user experience as something that involves the dynamics of space, time, objects, the states of the participants and the context in which the experience occurs. However, the most accepted and adopted definition of user experience is the one forwarded by ISO and (Bevan, 2010). ISO FDIS 9241-210 defines user experience as a person's perceptions and responses that result from the use or anticipated use of a product, system or service. Bevan (2010) in his definition equates user experience to users' culture, attitude, views, emotions, beliefs, preferences, perceptions, physical and psychological responses and behavioral accomplishments that occur *before*, during and after using a service or product. Emphasis in the two definition is on *''Use''* and *''Anticipated Use''* in ISO and *''Before Use''* and *''During Use''* from (Bevan, 2010). According to Adikari et al. (2010) User Experience assessment is an evaluation of user's experience with the product, system or service during 'use' (i.e., actual interaction experience) as well as 'anticipated or before use' (i.e., pre-interaction experience such as needs and expectations). Whereas many User Experience scholars do agree with the two definitions, there is still a challenge in understanding experience before use, the experience which is normally referred to as *"anticipated"* or *"desirability"* (Law et al., 2009; Hassenzahl, 2009).

This kind of experience (before use-experience) normally creates a desire to use a product or service. It is this kind of experience that someone gets even before the person uses the product or service. It is like watching an advertisement of a product on Television, the intended user would see it important to vividly imagine use cases with contextual factors and this is likely to evoke realistic experiences out of anticipated use (Law et al., 2009). Hasselzahl and Tractinsky (2006) identified *anticipated use* as the user's internal state which is associated with the user's tendency, expectations, needs, motivation, mood and many others. Law *et al* (2009) calls it *"imagined use"*. In this paper, it is referred to as Anticipated User Experience (AUX). According to Roto et al. (2011), AUX can be achieved through expectations created from previous experiences and other sources of information. In the design domain, Yogasara (2014) observes that the importance of emphasizing AUX in computing products and services has been acknowledged. Therefore, when AUX in a given product or service is emphasized, there are high chances for such a product or service to be accepted and used by the user. According to Yogasara (2014), AUX in any product or service plays an important role in user experience. It influences users' desire for the product even before the product is made available to the user. It is closely associated with the marketing and awareness stage of a product or service, where the product's image is formed in a user's mind either through marketing (advertisements) or a product's appearance (Yogasara, 2014). He admits that that these experiences and expectations before actual interaction are normally shaped by advertisements, brand image, other people's opinions, and prior experiences with similar products. It is an imagined experience. According to Yogasara (2014), AUX contributes to the current behaviour of the intended user. People often make decisions, select their actions, and adjust their behaviour according to anticipated emotions or feelings usually based on a number or a combination of past emotional outcomes and present affects (Baumeister et al., 2007). In other words, AUX has the ability to positively influence perception and change people's attitude towards using a given product or service. Anticipating a certain out-come may trigger certain specific emotions, and it is these emotions that may stimulate behavioral adjustments (Huron, 2006). Human-product or service interaction does not only include instrumental and non-instrumental interactions, but also non-included physical interaction such as recalling, fantasizing about a product, or anticipating product usage (Desmet & Hekkert, 2007). From Desmet & Hekkert's assertion, the potential outcomes of the interaction can also be imagined, anticipated, or fantasized about, which in turn may evoke positive emotional responses. This response leads to the acceptance and usage of a product or service. Based on the above discussion, this chapter therefore defines AUX as *a positive imagination, fantasy and excitement that a user feels about a given product or service before the product is made available to the user.*

As stated earlier, Law et al. (2009) and Hassenzahl (2009) observe that whereas many user experience researchers may be conversant with explaining the experience that can be attained during usage of a service, they find it difficult to explain how a service or product can create good experience before it is used. Yogarasa (2014) for example studied over 10 definitions of user experience and many of the definition defined user experience as satisfaction a user get during actual interaction with the product or service. Only a few, however, touch on the subject of AUX and no further explanations regarding this subject are provided. Also, of the 27 user experience definitions listed on ''all about UX'' website, less than 15% consider the influence of AUX as a major determinant of user experience (Yogasara, 2014). This further indicates a research gap in the area of AUX that needs to be filled. Owing to all discussions on AUX made so far, this Chapter makes 3 major observation that underlie AUX. Firstly, AUX is a cognitive process. It is linked to imagination, fantasy, recalling, perception etc (Hasselzahl &Tractinsky, 2006; Desmet & Hekkert, 2007; Law et al., *2009;* Yogasara, 2014). *S*econdly, for AUX to occur, there should be a stimulating factor (Roto et al., .2011; Yogasara 2014). Thirdly, AUX influences user behaviour (Huron, 2006; Baumeister et al., 2007; Yogasara, 2014). It appears these three observations or characteristics associated with AUX are closely related to Cognitive Psychology and its principles. Cognitive Psychology theory and its principles could be a key to better understand AUX and how it can be achieved. Basing on these observations, this paper goes ahead to probe the following questions

Research Questions

1. What is Cognitive Psychology and how does it help understand AUX.
2. What factors can stimulate AUX cognitive process.
3. To what extent do these factors stimulate AUX cognitive process to lead to behavior change (acceptance).

Two methods were used to probe the above stated questions.

Methodology

Literature Review

Literature review was one of the methods used to address the first and second research questions. Literature from different sources was reviewed namely; journals, conference proceedings, books and research reports. Through a thematic review of literature, this chapter was able to identifying the key concepts in the principles of cognitive psychology and its relationship with AUX. This chapter was able

Using Cognitive Psychology to Understand Anticipated User Experience

to understand what cognitive psychology is and its application on AUX. Through thematic review also, this chapter was able to elicit a number of stimuli that can positively influence AUX cognitive process.

Systematic Review

Systematic Review was used to answer research question three. Systematic Review method aided in providing empirical data on the extent the factors identified through literature can stimulate AUX cognitive process and therefore lead to behaviour change. Systematic Review is a 'rigorous method used to map the evidence base on an unbiased way as possible, and to assess the quality of the evidence and synthesize it' (DFID, 2013). A systematic review answers a defined research question or research questions by collecting and summarising all empirical evidence that fits pre-specified eligibility criteria. Systematic reviews are useful where:

- There is a substantive research question.
- Several empirical studies have been published.

According to Zanker & Mallett (2013), Systematic Review is considered by some to offer 'the most reliable and comprehensive statement about what works and how it worked. Systematic Review has many times been employed in evidence-informed policymaking within the arena of international development (Zanker & Mallett, 2013). In 1984 Cooper (1984) proposed a five stage systematic review process that was followed in this chapter'

- **Problem Formulation:** Statement of objective 3 of this chapter. This chapter used systematic review to investigate the extent to which factors identified in the literature stimulate AUX cognitive process which leads to behavior change.
- **Data Collection:** An unbiased literature search Reports, Journals, Conference proceedings, Books and many others. This effort resulted in 26 citations from which relevant studies were selected for the review. Their potential relevance was examined, and about 8 citations were excluded as irrelevant. The full papers of the remaining 18 citations were assessed to select those primary studies that explain how the elicited stimuli from the literature would positively stimulate AUX cognitive process.
- **Data Evaluation:** Assessing the studies for inclusion in the review. In data evaluation, a simple data extraction table was used to organize the information extracted from each reviewed study (e.g., authors, publication year, abstract, study design and particularly the outcomes of these researches).
- **Analysis and Interpretation:** Narrative Synthesis was the main form of analysis used in this chapter. Narrative synthesis refers to an approach to the systematic review and synthesis of findings from multiple studies that relies primarily on the use of words and text to summarize and explain the findings of the synthesis. Narrative synthesis adopts a textual approach to the process of synthesis to 'tell the story' of the findings from the included studies.
- **Public Presentation:** Discussion and context of findings. Through systematic review, this chapter was able to establish the extent each of the stimuli elicited from the literature influences AUX cognitive process using empirical evidence

MAIN FOCUS OF THE CHAPTER

Results from Literature

Cognitive Psychology and Its Application on AUX

First introduced by K. S. Lashley in the 1950s. Cognitive psychology is the scientific investigation of mental processes that affect behaviour, this includes all our mental abilities such as perceiving, learning, remembering, thinking, reasoning, imagining, understanding etc (Lin Lu & Dosher, 2007). Cognitive psychology is a relatively modern approach to human behaviour that focuses on how we think, with the belief that such cognitive processes affect the way in which we behave (Robert & Sternberg, 2012; Psychologist World, 2016). Cognitive phycology is based on two assumptions as stated in (Lin Lu & Dosher, 2007; Robert & Sternberg, 2012; Psychologist World, 2016). These two assumptions in Cognitive Psychology help to better understand the three observations in AUX. The assumptions state that;

- Behaviour is controlled by our own cognitive processes, as opposed to genetic factors.
- Human behaviour can be explained as a series of responses to external stimuli.

In other words, Psychologist World (2016) observes that cognitive psychology is based on the principle that our behaviour is a direct result of a series of stimuli and the outcome is determined by cognitive processes as represented below in Figure 1.

In summary, the principle of cognitive psychology claims that human behaviour is a direct result of a cognitive process, and the cognitive process is often influenced by a stimuli. Detailed discussion of this principle is presented in the next section.

AUX as a Cognitive Process

As observed earlier, AUX is a cognitive process (Hasselzahl &Tractinsky, 2006; Desmet and Hekkert, 2007; Law et al., *2009;* Yogasara, 2014). In Cognitive Psychology, human behaviour is controlled by human's own cognitive processes. Whereas there is no standard procedure of presenting the cognitive process, McLeod (2015) observes that cognitive psychology recommends the following assumptions, among others, that guide the cognitive process;

Figure 1. The principle of cognitive psychology
(Psychologist World, 2016)

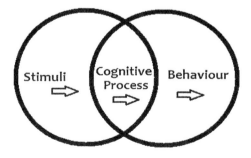

1. A series of cognitive processing systems that transform, or alter the information in systematic ways.
2. Information should be made available from the environment (Stimuli) to influence and to be processed by a series of cognitive processing systems.

The presentation of AUX cognitive process in this section is guided by the above assumptions. There are three processing systems in the entire AUX cognitive process and these include; Reproductive Imagination, Fantasy and Excitement. These processing systems transforms information from the environment in a systematic manner. The AUX cognitive process begins with Reproductive Imagination process, followed by Fantasy process, and finally Excitement process. According to ChangingMinds. org (2016) a renowned psychology website, Imagination and Fantasy (also generally referred to Cognitive arousal) leads to Excitement (also sometimes referred to as Affective arousal). Excitement finally triggers physical arousal or physical behaviour, such as taking physical action or change in behaviour. It should also be noted that reproductive imagination is the key that leads to fantasy (Brenner, 2003; Juan, 2006; Roy, 2011). The three processing systems identified in the cognitive process (Reproductive Imagination, Fantasy and Excitement) specify the processes and structures that form the basis of AUX cognitive process. The detailed AUX cognitive process is described below;

Reproductive Imagination Process in AUX

The entire though process begins with reproductive imagination, which becomes the first phase of AUX cognitive process. Imagination, to begin with, is part of cognition psychology, which in particular is a cognitive process, and deals with the ability to form a mental image of something that is not available at that moment. It is a vital component of creative visualization, positive thinking and affirmations (Maher, 1915; Sasson, 2016). Maher (1915) and Angel (2015) observe that, psychologists over years have identified several forms of imagination. However, imagination is largely perceived in two forms, that is, Reproductive Imagination and Productive Imagination. Reproductive imagination is a term employed to designate the power of forming mental pictures of objects and events as they have been originally experienced before (Maher, 1915; Angell, 2015). For example, a voter may visualize a mobile device that he saw in a demo somewhere. Such an application according to Angell (2015) would illustrate what psychologists mean by reproductive imagery. Productive Imagination on the other hand signifies the power of constructing images of objects not previously perceived (Maher, 1915; Angell, 2015). Reproductive imagination experience is an internal experience a person enjoys from visualizing a recently seen or experienced phenomenon. Reproductive imagination is comprised of three phases, namely crystallization, dialectics and effectiveness (Liang et al., 2012; Liang, 2013).

Fantasy Process in AUX

This is the second phase of the AUX cognitive process. Fantasy is a product of imagination, or Imagination is the key that leads to fantasy (Jaun, 2006). The word fantasy is most often referred in psychology as a "mental apprehension of an object of perception; the faculty by which this is performed" and further as "the fact or habit of deluding oneself by imaginary perceptions or reminiscences" or "a day-dream arising from conscious or unconscious wishes or attitudes" (Brenner, 2003). In psychology, Fantasy is an imaginary thought or image, or a set of images that provides a pleasurable experience for an individual, or as a means of visualizing other possibilities (Psychology Glossary, 2016). The term fantasy and imagination in many cases are used interchangeably. However, Albertus Magnus differentiates the two and considers imagination *"the repository of images and fantasy the active power operating them"*. Other psychology scholars such as Reynolds compares imagination with genius, and fantasy with taste. Fantasy, which is the lower and more restricted term, is defined in relation to imagination, which is a higher and more comprehensive term (Brenner, 2003). Roy (2011) considers fantasy as an imagination in its extremes and an important part of our cognitive thinking, it demonstrations how far our minds can stretch beyond the normal and the natural. Fantasy experience is putting reproductive imagination in to imaginary action while enjoying the whole experience, drawing imaginary use cases of a particular product. This therefore leads to excitement.

Excitement Process in AUX

The outcome of reproductive imagination is fantasy, the end result of fantasy is excitement (Changing-Minds.org, 2016). According to Merriam Webmaster (2015) excitement is a feeling of eager enthusiasm and interest, a quality that causes feelings of eager enthusiasm. According to Patel (2015), when a person is excited, the emotions become more powerful and in many cases affects the person's decision-making abilities. An excited person is more likely to make a decision, any decision including a bad one (Patel, 2015). Excitement in Psychology sometimes can be referred to as Affective or Emotional arousal, and this happen when someone is emotionally charged up and feels passionate about something, this creates physical agitation and 'readiness for action' (Berger, 2011). Excitement in this context, is a powerful argue to transform fantasy to reality.

The Analogy

This paper summaries the three phases in this analogy. Once a user begins to imaging a mobile device which he/she saw in recently in a demo or an advert (Reproductive Imagination), the user is likely to begin creating interesting use cases in the mind (Fantasy), once this is done over and over again, it will result to the argue to use the system in reality (Excitement). The interrelationship of the three though processes is illustrated below in Figure 2;

Figure 2. The processing systems in the AUX cognitive process

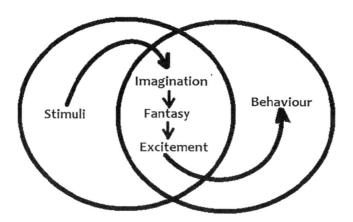

Factors That Stimulate AUX Cognitive Process (Stimuli)

AUX cognitive process is heavily dependent on the nature of environmental stimuli. In this chapter, there was need to establish factors (stimuli) that can influence cognitive processing systems, that is; reproductive imagination, fantasy and excitement. From the literature available, this chapter found out that persuasive techniques used in marketing and in persuasive technology design techniques, technology acceptance theory and other theories could influence the three processing systems. Fogg (2002) defines persuasive design or technology as a technology that is designed to change attitudes, perceptions or behaviours of the users through persuasion and social influence, rather than coercion. According to Fogg (2003), persuasive computing is "any interactive computing approach designed to change people's attitudes or behaviours". Central to the concept of persuasive design is the notion of persuasion, which Fogg explained as "an attempt to shape, reinforce, or change behaviours, feelings, or thoughts about an issue, object, or action". Fogg (2004) and Cialdini (2007) forwarded a number of persuasive strategies when used in designing computing products or services would change attitudes, perceptions or behaviours of the users towards a given product.

- **Demos:** Show the product features and how they work through trial versions, on TV etc.
- **Tailoring:** Provides relevant and personalized information on the product to change a person's attitudes.
- **Reciprocity:** Uses principles of operant conditioning, such as reinforcement and shaping, to change behaviors.
- **Push Messaging/ Reminders:** Gentle and polite reminders at certain times of the day to influence behavior.
- **Social Learning:** Use of peers, role models or acquaintances to influence behavior of others.

On the other hand, Acceptance theory as originally proposed by Davis in 1986 could also influence the three processing systems. According to Park (2009), TAM is one of the well-known models related to technology acceptance and use. The TAM is an adoption of the theory of reasoned action (TRA), specifically tailored for modelling user acceptance of information systems/technology. The TRA assumes that behavioural intentions are prior to every behaviour and that attitude and biased norms are critical for the intentional behaviour (Riedel et al., 2006). The advantage with TAM is that it offers an informative representation of the mechanisms by which design choices influence user acceptance (Park, 2009). According to Park (2009), TAM accounts for more than 60 percent of user acceptance and usage of a given technology. In TAM, perceived usefulness and perceived ease of use are the significant pillars that constitute and determine user acceptance of a given form of technology.

- PU defines the degree to which a person believes that using a particular service or product could improve his or her work performance, say, reduce on the overall time a person takes doing a task or make it very convenient for a person to perform a task.
- PEU on the other hand describes the degree to which a person believes that using a particular service or product is free of effort. The service or product should be easy to learn and use, no much effort is required.

Persuasive and acceptance design requirements were found to be able to create positive imagination, fantasy and build excitement. How besides persuasive and acceptance theories, the literature also reveals that the following requirements can influence AUX.

- Prior Success or Legacy. The legacy of a product matters. The history of success makes a product reliable and this wins trust from the users. According Murthy (2007), the legacy of the product which is associated to its reliability conveys the concept of dependability, successful operation or performance and the absence of failures. It is an external property of great interest to both manufacturer and consumer. Unreliability (or lack of reliability) conveys the opposite (Murthy, 2007).
- Trend and Aesthetics. The inclination of the time influences the design of products. Usually people want to buy up to date items and not ones based on 'last years look' (Ryan, 2004). Aesthetic on the other-hand is the branch of philosophy that deals with the principles of beauty and artistic taste (Business Dictionaries, 2015). People are not only attracted to the latest trend of products but also to beautiful things.
- Social and Cultural Environment. These are a set of beliefs, customs, practices and behavior that exists within a given population (Business Dictionaries, 2015). Some product are aimed at different cultures and countries. A product acceptable in one culture may be looked up one as offensive or less desirable in another. The most critical thing is to align the product or service with these set of beliefs and practices.

RESULTS OF SYSTEMATIC REVIEW

Stimuli and Its Influence on AUX Cognitive Process and Behaviour

Systematic review was used to establish the extent to which the elicited stimuli from the literature would positively stimulate AUX cognitive process (change attitudes, perceptions or behaviours of the users towards a given product). As already stated and illustrated in Figure 2, the end result of AUX cognitive process is behaviour change. This could be in terms of accepting a service or product by the user. In this section, the chapter goes ahead to discuss the relevance of each stimuli to mobile computing products and extend each stimuli influence user behaviour on a computing product. Ten stimuli that influence AUX cognitive process and lead to user behavioural change. For easy understanding, this chapter organises these Ten Stimuli in to two broad categories. Namely; Contextual Stimuli and Product Stimuli

Contextual Stimuli

Verschuren and Hargot (2005) defines surrounding Characteristics as existing prerequisites set by the political, cultural, economic, juridical and or social environment. These are the social, cultural, political factors that play a great role in influencing AUX as discussed below;

Socio-Cultural System

Cultural and religious attributes of the user or group of uses should be highly observed when developing a computing product. There should be no or little conflict between socio-cultural life style of the user and the technology being used. It is right to say that social and cultural acceptance of a particular technology is very vital (Shaukat & Zafar, 2010). For example, a number of manufacturers have a good understanding of colours and colour schemes and how they influence demand of the product in different countries. In China for example, black is always associated with evil, dirt, sin, disasters and bad luck. Black garments in China may not sell as much as white or blue (Ryan, 2004). Social factors are among the major influencers of reproductive imagination. In such situations, Väänänen et al. (2009) recommends designing computing products that are enjoyable and support essential social and cultural needs and values of the users.

Social and Political Learning

According to Cherry (2014), the social learning theory was first proposed by Albert Bandura and has become one of the most influential theory of learning and development. Whereas behavioural theories of learning suggested that all learning was as a result of associations formed by conditioning, reinforcement, and punishment, Bandura's social learning theory proposed that learning can also occur simply by observing what other people are doing. His theory added a social element, arguing that people can learn new information and behaviours by observing other people (Cherry, 2014). Cialdini (2007)] explains that when someone is not sure about a certain course of action, then the person is more likely to look to those around him or her for guidance as far as that particular course of action is concerned. Cialdini (2007) gives a good example on how social validation can significantly influence motivation and persuasion. He explains that, a study found that a fund-raiser who showed homeowners a list of neighbors who had donated to a local charity significantly increased the frequency of contributions; the longer the list, the greater the effect. Cialdini (2007) further quotes Cavett Robert- a sales and motivation consultant who stated that, people are persuaded more by actions or behaviours of other people than any other proof that can be offered. A classic effect of social learning was when Lee's article (circa 2011), reported that and endorsement by Jessica Simpson and aesthetician Nerida Joy helped Beautymint attract 500,000 visitors on day one of its launch (Feldman, 2014). In a survey carried out in Uganda, there was a 32% acceptance rate in respondents who originally did not agree to use their mobile phones for voting, when there was an indication that their political leaders, neighbours, peers or influential would be using their mobile phones for voting (Eilu et al., 2014). Using social leaders, political leaders and technocrats influence wide use of computing products in today's society.

Product Stimuli

Refers to the preferred product/service functionality, features, appearance and other hedonic features. The requirements include; ease of use, perceived usefulness, Demos, push messaging/reminders etc as further discussed below.

Perceived Usefulness

Slatten (2010) describes perceived usefulness as one of the most important factors for influencing user technology acceptance. Perceived usefulness according to Park (2009) is the degree to which a person believes that using a particular system would provide him/her with some benefits. Mobile Phone for example have proved to be a very useful tool in today's world. Mobile phones services have made it easier for people do things, it has provided services such as easy mobile money transfers and mobile bill payment to mention but a few. Through these mobile services, money transfer and bill payment has become easier and convenient in developing countries like Uganda, Kenya and Tanzania. Statistics show that, because of its portability, about 67% of users would prefer buying products using their mobile devices, while 50% of users are more likely not to use an online site if it is not available on their mobile devices even when they like the product available online. About 48% of online customers find a lot of dissatisfaction when they get to a site that is not mobile-friendly (Dholakiya, 2013).

Prior Success and Great Legacy

A good reputation represents a great marketing strategy. A mobile computing product with a good reputation inspires others to use it (Lickerman, 2010). A product with a good reputation has a higher chance of being accepted even by those who have never used it. The Moleskine notebooks for example had built a good reputation over several years as a reliable notebook with a durable cover. Though the books disappeared in the mid-1980s, in 1986 the note books resurfaced and sold out in record numbers because of their earlier reputation (Berget, 2009). Tartan is a type of Scottish cloth which is checked or cross-barred. A variety of colours are used to produce the patterns. Today, tartan patterns are exported all over the world from Scotland and these types of pattern are recognized throughout the world as Scottish. Tartan patterns materials are not only always bought because they are associated with Scotland but they are also purchased for their high quality (Ryan, 2004). For mobile computing product to gain more acceptance, there is need to build legacy or associate it with products that have a solid legacy.

Demos

Reproductive imagination depends directly on the richness of what was seen before, because this is the experience that provides the material from which the products of reproductive imagination, and later fantasy and excitement are constructed. The richer a person's experience, the richer is the material one's imagination has access to (Vygotsky, 2004). Demos are considered a rich source of material for reproductive imagination. A demo is a step by step process which guides the user on how the product works or how it is used. Demos can be done through various means, say on TV, radio, prototypes or sample products. A demo proves to a user that a computing product meets the specified standards and this reduces the level of uncertainty. Demos on TV or radios, or Test prototypes of a computing product given to the user before the user purchases the product are a great confidence and trust builders. Demos are recipe for reproductive imagination, where someone begins to form mental pictures of objects and events seen earlier. A number of researches have been carried to find out the effectiveness of demos on products and the results are stunning. For example, 58% of the respondents in a survey carried out by a research firm Knowledge Networks said they would only buy a product after trying it. In a related survey, there was an increase of 475% when product demonstration in-store were allowed, and further increased sales from the entire product line by as much as 177% on the demonstration day (Lovell, 2015). Khaled (2008) acknowledges that one of the major aim of a demo is to increase self-efficacy and facilitate goal completion of a computing product.

Ease of Use

In designing computing products, unnecessary long steps should be avoided and short simple steps should be made a priority when developing a computing product. Once the tasks in the demo are few and simple, it influences the degree to which a person believes that using a computing product is free of effort. Slatten (2010) believes that users will choose to avoid learning complex process because of the perceived or real difficulty associated with such a process. A survey by usability net revealed that 50% of online visitors to a site would not return if finding information they needed was difficult (Usability Net, 2006).

Tailoring or Customization

According to Fogg (2004), information availed in a computing product will provide better user experience if it is tailored to the individual's needs, interests, personality, usage context, or other factors relevant to the individual. Product customization is gaining ground in many companies. Online product customization for example, raises brand awareness, increases engagement, helps companies identify emerging trends, and adds an additional source of revenue to their online business (Margalit, 2015). Product customization has been taken so seriously, that today, many popular brands are embracing product customization and adding it to their business strategy (Margalit, 2015). Margalit (2015) argues that the decision to buy a customized product is mediated by unconscious factors. There is a tendency for a person to pay more attention to information meant for him or her. When a product is customized, the buyer's touch receptors are activated, creating a sensual stimulation. In a case where touch is not possible, the interaction with the product brings an imaginary path to life, and the imagination of ownership (Margalit, 2015). Cialdini (2007) observes that having a product with a customer's name on can increase chances of winning over the customer. A survey carried out in the United States revealed that clients were willing to pay as high as 20% more to have their products customized (Spaulding & Perry, 2013). A mailing survey found out that there was a 56% response rate for customized mails sent out to people, compared to 30% response rate for non-customized mails (Cialdini, 2007). A survey of more than 1,000 online shoppers conducted by Bain & Company in the United States revealed that more than a quarter of shoppers, 25-30%, are interested in online customization options, even if only 10 percent have tried it until now (Margalit, 2014).

Reciprocity

Reciprocation is a form of reward a user receives after performing a certain task. Computing products that are designed to reward a user attracts more attention from users than others. There is an overwhelming evidence that reciprocity is a powerful determinant of human behaviour. Many scholars including sociologist, ethnologist and economist emphasize the omnipresence of reciprocal behaviour (Falk & Fischbacher, 2000). According to Molm et al. (2007), the appreciation of reciprocity in social life is by no means restricted to exchange theorists, that is ''give and take''. In contrast, as a social construct, reciprocity means that in response to friendly actions or behaviour, people are frequently much nicer and much more cooperative; in reverse, in response to hostile actions they are frequently much nastier and even brutal. The principle of Reciprocity is equally used in a variety of situations, businesses use it in advertising, marketing, and propaganda (Budiu, 2014). Reciprocity appeals to sub conscious part of a human to be good to those who are good. For example, in order to influence voters to use a given mobile device, rewards to users could be considered. The mobile device should have an in built rewarding service, such as free games, application downloads etc. Therefore, rewards like free airtime, free online game access etc. A survey in the United States indicated that an appeal by Disabled American Veterans organization for contributions increased from 18% to 35% when free personalized address labels where later sent. In another example, research found out that people bought twice as many raffle tickets from a stranger if the stranger first gave or promised a can of Coke (Cialdini, 2007). In another case in the United States, a waiter's tips increased by 23% when a waiter gave one mints to a client but quickly returned to offer a second mint (Cialdini, 2007).

Push Messaging/ Reminders

Push messaging notification is, in mobile marketing context, a short text message sent to remind someone to perform a certain act. These messages are send at certain convenient times. Some technologies like suggestive technologies identify these convenient times and then remind users to perform a particular tasks. For example, during the 2 years of observation at a United States hospital, safe behaviour before implementation of the *'gentle reminder'* was 55% in one ward; it increased after 2 months to above 80%; 4 months later, it was 90%. A similar pattern was documented in all wards. This is one recommended way to overcome unsafe behaviour (Erev et al., 2010). Another survey by Localytics revealed that there is a 66% higher click rate on reminders that are sent on weekdays between 12 p.m. to 5 p.m. (VentureBeat, 2014). In another case, there is a higher chance that customers will reconsider and buy a product when constantly reminded about the product (VentureBeat, 2014).

Trend and Aesthetics

A trendy computing product is the newest, the latest, and the greatest product of the day that catches much of people's attention. According to Baumgartner (2012), the brain loves trends because they are fast acting short-lived blasts of newness. Studies have shown that the substantia negra/ventral tegmental area of the brain is stimulated when presented with something new. The study also showed that humans seek novelty in order to receive reward (Baumgartner, 2012). When a person is presented with something new which he/she has not seen before, all cognitive processes involved in learning (working memory, visual processing, problem solving, etc.) are enhanced (Baumgartner, 2012). Fashion goes hand in hand with beauty. The selection of the product's features, colors and shape generates thoughts on 'how it would feel' to own that object (Margalit, 2015). Cialdini (2007) observes that people tend to be attracted to beautiful and trendy things. Physically attracted people are more likely to receive favour than the other less attractive. A study conducted in the USA in 1993 revealed that, physically attractive fund-raisers for the American Heart Association generated nearly twice as many donations (42% versus 23%) as did other requesters (Cialdini, 2007). Studies have also shown that up to about 90% of a product's assessment is based on beauty and color alone. This mental assessment is normally made within 90 seconds of initial viewing (Patel & Puri, Na). A new analysis by the Design Management Institute, a Boston-based nonprofit focused on design management released a report that, in the decade, design-driven companies such as Apple, Coca Cola, Ford, Herman Miller, IBM, Intuit, Newell Rubbermaid and Nike outperformed the Standard & Poor's 500—a stock market index of 500 large publicly traded companies—by 228% (Dunne, 2014)

Using Cognitive Psychology to Understand Anticipated User Experience

In summary, key external stimuli that designers should consider when designing for AUX in computing products are presented. As illustrated in figure 3 below, the stimuli discussed in this chapter influences reproductive imagination first, however, fantasy and excitement are further stimulated through constant reference to the stimuli as illustrated in Figure 3.

AUX Influences User Human Behaviour

Human behaviour refers to a collection of every physical action and observable emotion associated with individuals. In computing products, this could be displayed in form of acceptance and usage of a computing product. One of the assumptions in cognitive psychology is that behaviour is controlled by our own cognitive processes, as opposed to genetic factors. Meaning that the outcome of the cognitive process influences behaviour. However, before 1940s this cognitive psychology assumption on the cognitive process leading to human behaviour was widely contested. The "father" of radical behaviourism (human behaviour) John Watson (1878–1958) objected to this theory and dismissed cognitive process as nothing more than subvocalized speech (Robert & Sternberg, 2012). Until early 1940s, the behaviorist view that human behaviour is directly influenced by the surrounding environment and no thinking process is required- dominated experimental psychology (Riegler & Riegler, 2012). Behaviorist scholars such as B. F. Skinner emerged as the most extreme advocate for behaviorism claiming that mental events such as cognitive processes have no place in the science of psychology. He insisted that behaviour is directly influenced by the environment only, and no thinking process is involved. He argued that though cognitive processes are real, they are unobservable and hence unnecessary to a scientific explanation of behavior (Riegler & Riegler, 2012). However, towards the end of the 1940s and early 1950s there was an abrupt change in research activities, interests, and the scientific beliefs, which marked a definitive break from behaviorism (Riegler & Riegler, 2012; Robert & Sternberg, 2012). The rapid change in research is seen

Figure 3. Processes and structures in cognitive performance that influences AUX

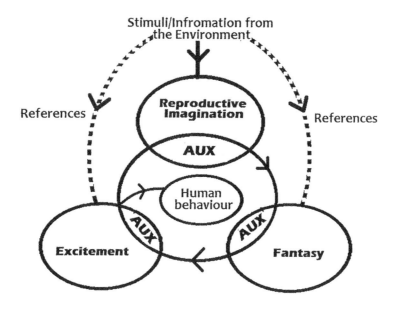

by some as a revolution in which behaviorism was rejected and replaced with cognitive psychology (Riegler & Riegler, 2012). This changes referred to as the "cognitive revolution" which took place in response to behaviorism (Robert & Sternberg, 2012). Cognitivism is the belief that much of human behaviour can be understood in terms of how people think. Cognitive psychologists assume that behaviour is the result of cognitive process. It rejects the notion that psychologists should avoid studying mental processes because they are unobservable (Robert & Sternberg, 2012).

FUTURE RESEARCH DIRECTION

Although little attention has been given to AUX, a number of user experience researchers and practitioners are currently trying to understand AUX and how it can be achieved. There is need to focus more on identifying and categorising AUX stimulants against particular computing products.

CONCLUSION

The term '*user experience*' came up during the explosive wide spread use of mobile devices and services error. User experience shows that the ease of use, efficiency and effectiveness of a product or service are not adequate to guarantee user satisfaction. In order to guarantee satisfaction, user experience research commenced. However, understanding user experience and how it can be achieved has always been quite challenging to researchers. Particularly, Anticipated User Experience (AUX), one of the tenets of user experience. In this chapter, I attempt to address this gap. This chapter applies the theory of cognitive psychology and its principles to best explain how Anticipated User Experience occurs and how it can be achieved. This chapter makes 3 major observation that underlie AUX; AUX is a cognitive process. It is linked to imagination, fantasy and excitement. For AUX to occur, there should be a stimulating factor such as conducive Social Environments, Social Leaning, Reciprocity, Tailoring, Push Messaging, Trends and Aesthetics, Demos, Ease of Use, Perceived Usefulness and Legacy. Thirdly, AUX influences user behaviour displayed inform of acceptance and usage of computing products. This chapter goes a long way in informing designers of computing products on the significance of attaining Anticipated User Experience and how it can be achieved.

REFERENCES

Aker, C. J., & Mbiti, I. M. (2010). *Mobile Phones and Economic Development in Africa*. Retrieved in May 13, 2015, from http://www.cgdev.org/files/1424175_file_Aker_Mobile_wp211_FINAL.pdf

Amin, E. M. (2005). *Social Science Research: Conception, Methodology and Analysis*. Makerere University.

Angell, J. M. (2015). *Psychology: Chapter 8: Imagination*. Retrieved in May 13, 2015, from https://brocku.ca/MeadProject/Angell/Angell_1906/Angell_1906_h.html

Arhippainen, L. (2009). *Studying User Experience: Issues And Problems Of Mobile Services– Case Adamos: User Experience (Im)Possible To Catch?* Doctoral Dissertation.

Battarbee, K., & Mattelmäki, T. (2002) Meaningful Product Relationships. *Proceedings of Design and Emotion Conference.*

Baumeister, R. F., Vohs, K. D., DeWall, C. N., & Zhang, L. (2007). *How emotion shapes behavior*: Feedback, anticipation, and reflection, rather than direct causation. *Personality and Social Psychology Review, 11*(2), 167–203. doi:10.1177/1088868307301033 PMID:18453461

Baumgartner, J. (2012). The Psychology of Dress. *The Psychology of Fashion.* Retrieved in May 13 2015, from https://www.psychologytoday.com/blog/the-psychology-dress/201202/the-psychology-fashion

Berger, J. (2011). Arousal Increases Social Transmission of Information. *Psychological Science.* doi:10.1177/0956797611413294 PMID:21690315

Berget, J. P. (2009). *Why Legacy is Important in Marketing.* Retrieved from http://slymarketing.com/why-legacy-important-marketing/

Betts, G. H. (1916). *Imagination: The mind and its education.* D. Appleton and Company.

Bevan, N. (2010). *What is the difference between the purpose of usability and user experience evaluation methods?* Retrieved in April 10 2016, from http://www.nigelbevan.com/papers/What_is_the_difference_between_usability_and_user_experience_evaluation_methods.pdf

Brenner, A. (2003). *Fantasy.* Academic Press.

Budiu, R. (2014). *The Reciprocity Principle: Give Before You Take in Web Design.* Retrieved in April 10 2016, from https://www.nngroup.com/articles/reciprocity-principle/

Business Dictionaries. (2015). *Socio-cultural environment.* Retrieved in April 10 2016, from http://www.businessdictionary.com/definition/socio-cultural-environment.html

Campbirdge Dictionary. (2015). Retrieved in April 10 2016, from reminder.http://dictionary.cambridge.org/dictionary/english/reminder

Carlile, P.R., & Christensen, C.M. (2004). *The Cycles of Theory Building in Management Research.* Academic Press.

Cartwright, P., & Noone, L. (2006). *Critical imagination*: A pedagogy for engaging pre-service teachers in the university classroom. *College Quarterly, 9*(4), 1.

ChangingMinds.org. (2016). *How we change what others think, feel, believe & do.* Retrieved in Feb 19 2016, from http://changingminds.org/

Chattratichart, J., & Jordan, P. W. (2003). Simulating 'lived' user experience - virtual immersion and inclusive design. In M. Rauterberg, M. Menozzi, & J. Wesson (Eds.), *Proceedings of Human-Computer Interaction - INTERACT'03* (pp. 721–724). Amsterdam: IOS Press.

Cherry, K. (2014). *Social Learning Theory: How People Learn By Observation.* Retrieved in Feb 19 2016, from http://psychology.about.com/od/developmentalpsychology/a/sociallearning.htm

Cialdini, R. B. (2007). *Influence: The Psychology of Persuasion.* New York, NY: HarperCollins Publishers Inc.

Daniels, D. (2015). *Pragmatic Marketing: Secrets of a Winning Product Launch.* Retrieved in Feb 19 2016, from http://pragmaticmarketing.com//resources/6-secrets-of-a-winning-product-launch?p=2

DeMers, J. (2015) *The Importance Of Social Validation In Online Marketing.* Forbes. Retrieved in Feb 19 2016, from http://www.forbes.com/sites/jaysondemers/2015/02/19/the-importance-of-social-validation-in-online-marketing/

Desmet, P. M. A., & Hekkert, P. (2007). Framework of product experience. *International Journal of Design, 1*(1), 57–66.

DeVries, W. A. (1988). *Hegel's theory of mental activity: An introduction to theoretical spirit.* Ithaca, NY: Cornell University Press. Retrieved August 18, 2011, ttp://pubpages.unh.edu/~wad/HTMA/HTMAfrontpage

Dholakiya, P. (2013). *Want Conversions? Start with User-Friendly, Useful Landing Pages.* Retrieved in Feb 19 2016, from https://blog.kissmetrics.com/want-conversions/

Dunne, E. (2014). *Study: Good Design Is Good For Business.* Retrieved in August 21 2015, from http://www.fastcodesign.com/3026287/study-good-design-really-is-good-for-business

Eilu, E., Baguma, R., & Pettersen, J. S. (2014). *Persuasion and Acceptance of Mobile Phones as a Voting Tool in Developing Countries.* Accepted for Presentation at the 4th International Conference on M4D Mobile Communication for Development, Dakar, Senegal.

Erev, I., Rodensky, D., Levi, M. A., Hershler, M.E., Admi, H., Donchin, Y. (2010). *The value of 'gentle reminder' on safe medical behaviour: Quality improvement report.* Academic Press.

Falk, A., & Fischbacher, U. (2000). *A Theory of Reciprocity.* Retrieved in August 21 2015, from http://e-collection.library.ethz.ch/eserv/eth:25511/eth-25511-01.pdf

Feldman, B. (2014). *Social Proof: Your Key to More Magnetic Marketing.* Retrieved in August 21 2015, from https://blog.kissmetrics.com/social-proof/

Fogg, B. J. (2002). *Persuasive Technology: Using Computers to Change What We Think and Do.* Morgan Kaufmann.

Fogg, B.J. (2004). *Captology Understanding How Computers Manipulate People.* Persuasive Technology Lab, Stanford University.

Fogg, B. J. (2008). Mass interpersonal persuasion: An early view of a new phenomenon. In *Proc. Third International Conference on Persuasive Technology, Persuasive 2008.* Berlin: Springer. doi:10.1007/978-3-540-68504-3_3

Forlizzi, J., & Battarbee, K. (2004). Understanding experience in interactive systems. In *Proceedings of the 5th Conference on Designing Interactive Systems: Processes, Practices, Methods, Techniques - DIS'04* (pp. 261-268). New York: ACM Press.

Forlizzi, J., & Ford, S. (2000). *The Building Blocks of Experience: An Early Framework for Interaction Designers.* DIS '00, Brooklyn, NY.

Greenwald, M. (2014). Of the Best Product Demos Ever and why they're so effective. *Forbes*. Retrieved in August 21 2015, from http://www.forbes.com/sites/michellegreenwald/2014/06/04/8-of-the-best-product-demos-ever-and-why-theyre-so-effective/

Hassenzahl, M., & Tractinsky, N. (2006, March-April). User experience – a research agenda. *Behaviour & Information Technology*, 25(2), 91–97. doi:10.1080/01449290500330331

Heikkinen, J., Olsson, T., & Väänänen-Vainio-Mattila, K. (2009). Expectations for user experience in haptic communication with mobile devices. *Proceedings of the 11th International Conference on Human-Computer Interaction with Mobile Devices and Services, MobileHCI'09*. Retrieved in August 21 2015, from http://womeninbusiness.about.com/od/marketingpsychology/a/prineciprocity.htm

Huron, D. (2006). Sweet anticipation: Music and the psychology of expectation. Cambridge, MA: A Bradford Book, The MIT Press.

ISO DIS 9241-210. (2008). *Ergonomics of Human Systems Interaction-Part 210: Human centered design for interactive systems*. ISO Switzerland.

Jones, N. (2002). *Gatner research: Citizens to Vote With Mobile Phone Messages in U.K. Elections*. Publication Date: 8 February 2002 ID Number: FT-15-4730.

Juan, S. (2006). *Which comes first: imagination or fantasy?* Retrieved in June 25 2015, from http://www.theregister.co.uk/2006/05/19/the_odd_body_imagination_fantasy/

Kaikkonen, A. (2009). *Internet On Mobiles: Evolution Of Usability And User Experience*. Doctoral Dissertation. Retrieved in June 25 2015, from URL: http://lib.tkk.fi/Diss/2009/isbn9789522481900/

Khaled, R. (2008). *Culturally-Relevant Persuasive Technology*. PhD Thesis.

Khambete, P., & Athavankar, U. (2010). *Grounded Theory: An Effective Method for User Experience Design Research*. Academic Press.

Krejcie, R. V., & Morgan, D. W. (1970). Determining sample size for research activities. *Educational and Psychological Measurement*, 30(3), 607–610. doi:10.1177/001316447003000308

Law, E. L.C., Roto, V., Hassenzahl, M., Vermeeren, A. P.O.S., Kort, J. (2009). *Understanding, Scoping and Defining User Xperience: A Survey Approach*. CHI 2009 User Experience.

Liang, C. (2013). *The predictive model of imagination stimulation*. Academic Press.

Liang, C., Hsu, Y., Chang, C. C., & Lin, L. J. (2012). In search of an index of imagination for virtual experience designers. *International Journal of Technology and Design Education*. doi:10.1007/s10798-012-9224

Lickerman, A. (2010). *The Value of A Good Reputation; Why we should care about how others perceive us*. Retrieved in May 12 2015, from https://www.psychologytoday.com/blog/happiness-in-world/201004/the-value-good-reputation

Lin Lu, Z., & Dosher, B. A. (2007). Cognitive psychology. *Scholarpedia*, 2(8), 2769. doi:10.4249/scholarpedia.2769

Lovell, S. (2015). *'Try before you buy' – why brands should use product demonstration?*. Retrieved from http://gottabemarketing.co.uk/try-before-you-buy-product-deminstration/

Maher, M. S. J. (1915). *Psychology; Empirical Rational: Imagination*. Retrieved in May 12 2015, from https://www3.nd.edu/~maritain/jmc/etext/psych008.htm

Mäkelä, A., & Fulton Suri, J. (2001). Supporting Users' Creativity: Design to Induce Pleasurable Experiences. *Proceedings of the International Conference on Affective Human Factors Design*, 387-394.

Margalit, L. (2014). *The Psychology of Online Customization*. Retrieved in May 12 2015, from https://www.psychologytoday.com/blog/behind-online-behavior/201503/the-psychology-online-customization

McLeod, S. (2015). *Cognitive Psychology*. Retrieved in September 05 2015, from http://www.simplypsychology.org/cognitive.html

McMullin, J. (2004). *The Experience Cycle model*. Academic Press.

Merriam-Webster Dictionary. (2015). *Full Definition of Reciprocity*. Retrieved in September 05 2015, from http://www.merriam-webster.com/dictionary/reciprocity

Molm, L. D., Schaefer, D. R., & Collett, J. L. (2007). The Value of Reciprocity. *The Value of Reciprocity Social Psychology Quarterly*, *70*(2), 199–217. doi:10.1177/019027250707000208

Murthy, D. N. P. (2007). *Product reliability & warranty: an overview & future research*. Retrieved in September 05 2015, from http://www.scielo.br/scielo.php?script=sci_arttext&pid=S0103-65132007000300003

Oxford Dictionaries. (2015). *Aesthetic*. Retrieved in September 05 2015, from http://www.oxforddictionaries.com/definition/english/aesthetic

Park, S. Y. (2009). An Analysis of the Technology Acceptance Model in Understanding University Students' Behavioral Intention to Use e-Learning. *Journal of Educational Technology & Society*, *12*(3), 150–162.

Patel, N., & Puri, R. (n.d.). *A complete guide to understanding consumer psychology*. Retrieved in September 05 2015, from https://www.quicksprout.com/the-complete-guide-to-understand-customer-psychology-chapter-4/

Payal, V. N. (2014). GSM: Improvement of Authentication and Encryption Algorithms. *International Journal of Computer Science and Mobile Computing*, *3*(7), 393-408.

Perdue, K. (2003). *Imagination. The Chicago school of media theory*. Retrieved October 16, 2012, from http://lucian.uchicago.edu/blogs/mediatheory/keywords/imagination/

Psychologist World. (2016). *Cognitive Approach (Psychology) Introduction to the cognitive approach in psychology*. Author.

Ribot, T. (1906). *Essay on the creative imagination*. Chicago, IL: Open Court. doi:10.1037/13773-000

Riedel, J.C., & Fransoo, V.C.S. (2006). *Modelling dynamics in decision support systems*. Academic Press.

Riegler, B.R., & Riegler, G.L. (2012). *Cognitive Psychology: Applying The Science of the Mind*. Pearson.

Robbins, S. P., & Judge, T. A. (2009). *Organisational decisions* (13th ed.). Prince Hall.

Robert, J., Sternberg, C. J., & Sternberg, K. (2012). *Cognitive Psychology* (6th ed.). Academic Press.

Roto, V., Lee, M., Pihkala, K., Castro, B., Vermeeren, A. P. O. S., & Law, E. (2010). *All about UX: Information for user experience professionals*. Retrieved in September 22 2015, from http://www.allaboutux.org/

Roy, S. (2011). *The Psychology of Fantasy*. Retrieved in September 22 2015, from http://www.futurehealth.org/articles/The-Psychology-of-Fantasy-by-Saberi-Roy-100901-178.html

Ryan, V. (2004). *Colours and Cultures*. Retrieved in September 22 2015, from http://www.technologystudent.com/despro2/colcul1.htm

Segerståhl, K., & Oinas-Kukkonen, H. (2008). Distributed User Experience in Persuasive Technology Environments. Linnanmaa.

Shaukat, M., & Zafar, J. (2010). Impact of Sociological and Organizational Factors on Information Technology Adoption: An Analysis of Selected Pakistani Companies. *European Journal of Soil Science*, *13*(2), 305.

Slatten, A. L. D. (2010). An Application and Extension of the Technology Acceptance Model to Non-profit Certification. Journal for Non-Profit Management, 14.

Spaulding, E., & Perry, C. (2013). *Making it personal: Rules for success in product customization*. Retrieved from http://www.bain.com/publications/articles/making-it-personal-rules-for-success-in-product-customization.aspx

The Free Dictionary. (2015). *Tailoring*. Retrieved in September 22 2015, from http://www.thefreedictionary.com/tailoring

Usability Net. (2006). *The business case for usability*. Retrieved in September 22 2015, from http://www.usabilitynet.org/management/c_business.htm

Väänänen, K.V.M., Roto, V., & Hassenzahl, M. (2009). *Towards Practical User Experience Evaluation Methods*. Academic Press.

Väänänen-Vainio-Mattila, K., & Wäljas, M. (2009). Development of evaluation heuristics for web service user experience. In *Extended Abstracts of the 27th International Conference on Human Factors in Computing Systems - CHI'09* (pp. 3679-3684). New York: ACM Press. doi:10.1145/1520340.1520554

VentureBeat. (2015). *How marketers are tailoring their content strategy for a multi-channel, multi-device world*. Retrieved in June 12 2015, from http://venturebeat.com/2015/08/31/how-marketers-are-tailoring-their-content-strategy-for-a-multi-channel-multi-device-world/

Verschuren, P., & Hartog, R. (2005). *Evaluation in Design-Oriented Research*. Springer.

Vygotsky, S. L. (2004). Imagination and Creativity in Childhood. *Journal of Russian & East European Psychology*, *42*(1).

Wolfe, L (2015). *The Principle of Reciprocity and how it applies to business*. Academic Press.

Yogasara, T. (2014). *Anticipated User Experience in the Early Stages of Product Development.* PhD Thesis.

Yogasara, T., Popovic, V., Kraal, B., & Chamorro-Koc, M. (2011). *General Characteristics of Anticipated User Experience (Aux) with Interactive Products.* Academic Press.

Yogasara, T., Popovic, V., Kraal, B., & Chamorro-Koc, M. (2012). *Anticipating user experience with a desired product.* The AUX Framework.

KEY TERMS AND DEFINITIONS

Anticipated User Experience: A positive imagination, fantasy and excitement that a user feels about a given product or service before the product is made available to the user.

Cognitive Psychology: The scientific investigation of mental processes that affect behaviour, this includes all our mental abilities.

User Experience: Defined as a person's perceptions and responses that result from the use or anticipated use of a product, system or service.

Chapter 12
Mobile Commerce Technologies and Management

Kijpokin Kasemsap
Suan Sunandha Rajabhat University, Thailand

ABSTRACT

This chapter reveals the prospect of mobile commerce (m-commerce); m-commerce and trust; m-commerce, privacy, and security issues; m-commerce adoption and technology acceptance model (TAM); and the significant perspectives on m-commerce. M-commerce is used for business transactions conducted by mobile phones for the promotional and financial activities using the wireless Internet connectivity. M-commerce is the important way to purchase the online items through online services. The main goal of m-commerce is to ensure that customers' shopping experience is well-suited to the smaller screen sizes that they can see on smartphones and tablets. Computer-mediated networks enable these transaction processes through electronic store searches and electronic point-of-sale capabilities. M-commerce brings the new possibility for businesses to sell and promote their products and services toward gaining improved productivity and business growth.

INTRODUCTION

Recently, the emergence of wireless and mobile networks has made possible the admission of electronic commerce (e-commerce) to mobile commerce (m-commerce), which is defined as the buying and selling of commodities, services, and information on the Internet through the utilization of mobile handheld devices (Lee, Hu, & Yeh, 2009). M-commerce technology is widely acknowledged as the next business format and its emergence has changed the business landscape (Faqih&Jaradat, 2015). M-commerce is of growing interest for vendors and customers and with that its importance within the combination of marketing and distributing channels (Möhlenbruch, Dölling, & Ritschel, 2010). The wide utilization of m-commerce includes the applications in banking, shopping, ticketing, entertainment, event management, and education (Gupta, Muttoo, & Pal, 2016).

E-commerce is the business that is transacted by transferring data electronically, especially over the Internet. (Kasemsap, 2016a). Through e-commerce, the cost for the entrepreneurs to sell their products

DOI: 10.4018/978-1-5225-2469-4.ch012

can be saved and diverted to another aspect of their business (Kasemsap, 2016b). Both m-commerce and e-commerce applications have been enhanced by social networking sites, such as Facebook, Twitter, and LinkedIn (Lin & Lu, 2015). Social commerce is recognized as the commerce activities mediated by social media (Curty & Zhang, 2011). In social commerce, people explore the commerce opportunities by engaging in a collaborative online environment (Curty & Zhang, 2011). Organizations should accommodate and energize mobile users and design changes to their social networking sites to facilitate interaction and information sharing (Heinrichs, Lim, & Lim, 2011). Secure, reliable, and economical modes of payment play a critical role in the successful implementation of m-commerce (Lu, Yang, Chau, & Cao, 2011).

Advances in wireless technology have increased the number of people using mobile phones and accelerated the rapid development of mobile service conducted with these devices (Wang, Lin, &Luarn, 2006). M-commerce emerges as the ubiquitous technology among the existing wireless payment mode (Arora, 2016). Smartphones penetrate business and consumer markets, and mobile applications have engendered an innovative market (Keith, Babb, Lowry, Furner, &Abdullat, 2015). In order to attract more m-commerce users, companies and managers need to carefully consider elements (e.g., subjective norm, cost, risk, and enjoyment) that relate more to individuals' roles from both social and consumer perspectives (Zhang, Zhu, & Liu, 2012).

This chapter is based on a literature review of m-commerce. The extensive literature of m-commerce provides a contribution to practitioners and researchers by revealing the applications and implications of m-commerce in order to maximize the business impact of m-commerce in the digital age.

BACKGROUND

Development in the number of the Internet and telecommunications users since the late 1990s has provided companies with new ways to conduct business and exchange information through the development of the e-commerce market (Charbaji, Rebeiz, & Sidani, 2010). M-commerce is the new trend in business transactions (Hu, 2009) and the Internet-enabled smartphones become very popular these days (Hu, Zuo, Kaabouch, & Chen, 2010). Products are recommended to users based on their browsing behavior on the new mobile channel as well as the consumption behavior of heavy users of existing channels, such as catalogs and the Internet (Liu & Liou, 2011). Consumers use smartphones in their shopping experiences, such as browsing for product information and consulting friends before buying online (Ozuem & Mulloo, 2016). When shopping with a virtual catalog, customers can select products which meet their requirement concerning price range (Pierre, 2009).

Mobile devices, such as smartphones, tablets, and e-books readers, are gradually replacing traditional personal computers as the main method of accessing the Internet, organizing the widely deployed 3G/4G telecommunication technologies (Miao & Jayakar, 2016). Through m-commerce, users receive the large volumes of commercial messages (Wang, Hong, Xu, Zhang, & Ling, 2014). The evolution has triggered an increase in the use of mobile phones to conduct m-commerce and mobile shopping on the mobile web (Liou & Liu, 2012). M-commerce makes networks more productive by bringing together voice, data communication, and multimedia services (Wang, 2007). Regarding mobile shopping, the use context is found to have the mediating role between time saving and satisfaction as well as between mobility and satisfaction (Kim, Chung, Lee, & Preis, 2015).

The rapid development of information technology enables the increasing numbers of consumers to seek and reserve products online first and then consume them in brick-and-mortar stores (Xiao & Dong, 2015). Guillén et al. (2016) indicated that the rapid growth of social networks has led many companies to use mobile payment systems as business sales tools. M-commerce environments are characterized by the high complexity, including myriads of technical and organizational aspects (Veijalainen & Weske, 2003). It is necessary to investigate how to design and develop m-commerce applications to ensure the successfulness of their deployment (Binsaleh & Hassan, 2013). Mobile voice service and innovation experience of mobile data services affect subscribers' consumption intention greatly; subscribers' perceived ease of use and brand experience can affect the subscribers' attitude toward mobile data services (Qi, Li, Li, & Shu, 2009).

PRINCIPLES AND PRACTICES OF MOBILE COMMERCE

This section provides the prospect of m-commerce; m-commerce and trust; m-commerce, privacy, and security issues; m-commerce adoption and TAM; and the significant perspectives on m-commerce.

Prospect of Mobile Commerce

The proliferation of mobile devices has created many opportunities for m-commerce (Chen, 2012), and mobile phones serve a wide variety of customers as the multi-sided platforms (Campbell-Kelly, Garcia-Swartz, Lam, & Yang, 2014). While the traditional PC access to the Internet continues to be vital for exploiting the advantages of the Internet, the mobile access appears to attract more people because of flexible accesses to the Internet in the ubiquitous manner (Sumita & Yoshii, 2010). Mobile phones are the global communication and business devices (Eze & Poong, 2016). Many mobile applications can have a much larger impact on the emerging economies than those of the developed world (Patel, 2006).

The advent of m-commerce is a natural phenomenon of the wireless convergence (Moqbel, Yani-De-Soriano, & Yousafzai, 2012). M-commerce applications have two major characteristics: mobility and broad reach (Ngai & Gunasekaran, 2007). In today's digital economy and the extended enterprise paradigm, mobility is on the rise (Godbole, 2006). Regarding mobility, m-commerce users can conduct business through mobile devices in real time. With m-commerce, users can be reached at any time through mobile devices (Ngai & Gunasekaran, 2007). Access to multiple wireless networks leads to the high transaction completion probability even when individual wireless networks do not offer the continuous access (Varshney, 2007). Smartphone users can access the Internet and search information more easily with built-in keyboards, which allow for the rapid data entry regarding m-commerce applications (Okazaki & Mendez, 2013).

Group-oriented m-commerce services are likely to be transaction-oriented with the significant monetary value of business transactions (Varshney, 2008). The challenges in the group-oriented m-commerce services include how to provide group management for m-commerce users who may even experience connectivity problems; how to support transaction performance, such as reliability; how to charge for transactions and other business model and strategies issues for wireless service providers; and user issues including adoption, training, and trust issues (Varshney, 2008). Secure multicast is a key element of m-commerce infrastructure toward supporting the group-oriented m-commerce applications, such as mobile auctions, product recommendation systems, and financial services (Eltoweissy, Jajodia, & Mukkamala, 2008).

In the m-commerce schemes, the users can use their electronic identity to register, such as mobile phone number and e-mail address. Then, they can use their mobile phones to conduct transactions at any place and any time (Han et al., 2016). When m-commerce users move between the stores, the mobile information which includes user identification, stores, and item purchased are stored in the mobile transaction database (Argade & Chavan, 2015). In m-commerce, companies provide location based services to mobile users, who report their locations with a certain level of granularity to maintain a degree of anonymity (Chorppath & Alpcan, 2013). Yu et al. (2011) indicated that mobility management, which includes location management and handoff management, is essential in cellular wireless networks to provide service to mobile users.

Examples from the financial sector include the instant funds transfer (mobile banking) and share trading (mobile brokerage) (Petrova, 2008). Mobile banking is an emerging application of m-commerce that can become the additional revenue source to both banks and telecommunications service providers (Kim, Shin, & Lee, 2009). Mobile banking helps fulfill the personal banking requirements and provides the location-free conveniences. Mobile brokerage services allow mobile phone users to conduct financial transactions using client software at any place and any time (Lin, Lu, Wang, & Wei, 2011). Compared with online brokerage services and telephone-based trading services, mobile brokerage services have unique features, such as ubiquity and mobility (Lin et al., 2011). Investors are able to conduct financial trades without spatial and temporal constraints (Kleijnen, Ruyter, & Wetzels, 2007).

As advertisers increasingly rely on mobile-based data, consumer perceptions regarding the collection and use of such data becomes of great interest to scholars and practitioners (Eastin, Brinson, Doorey, & Wilcox, 2016). Knowing the demographic patterns in m-commerce usage activities can assist companies in improving their mobile advertising strategies (Chong, 2013). Over the years, mobile messaging has become an essential method of communication through the utilization of the Internet and mobile networks (Yow & Mittal, 2008). Perceived network externalities contribute to higher perceived usefulness and perceived ease of use and significantly affect the m-commerce-related short message service (SMS) adoption (Lu, Deng, & Wang, 2010). Smartphone-based mobile advertising is a promising marketing approach, especially in retail and point-of-purchase environments (Atkinson, 2013). Trust in SMS advertising and subjective norms contribute to the intention to use (Zhang & Mao, 2008).

This section is dealing with the prospect of m-commerce and the next section is dealing with m-commerce and trust. With the proliferation in the mobile device usage, most of the business activities are centered around mobiles. Realizing the importance of m-commerce app development, most of the business organizations implement their own mobile app to offer a better service to their customers. Businesses are lining up to build m-commerce app dedicated to access their products and services.

Mobile Commerce and Trust

Trust over the mobile platforms is more critical due to the open nature of wireless networks (Nilashi, Ibrahim, Mirabi, Ebrahimi, & Zare, 2015). The high level of trust enhances the transaction and eliminates perceived risks regarding m-commerce adoption (Li & Yeh, 2010). Despite the popularity and potentials of m-commerce, real m-commerce activities in many developing countries remain low due to the users concern about trust issues (Chong, Chan, & Ooi, 2011). One of the main reasons for low intention is uncertainty that leads to lack of trust that has been challenged and surveyed by the network capabilities

(Siau & Shen, 2003) and interface design (Lee & Benbasat, 2003). The link between interface design and ease of use holds an important approach to motivating the females' utilization of m-commerce (Okazaki & Mendez, 2013).

The website with a greater level of trust is often associated with a higher degree of purchase intentions and higher customer retention rate (van der Heijden, Verhagen, & Creemers, 2003). M-commerce has empowered consumers to easily switch from one website to another at any time (Peng, Quan, & Zhang, 2013). By using websites for promotion and sales, companies are able to sell to markets (Falk & Hagsten, 2015). McKnight and Chervany (2001) claimed a well-designed mobile website with considering the ease of navigation and useful links to other websites give customers adequate information for making purchase decisions. Chae and Kim (2003) stated that limitation in system resources (e.g., small screens in mobile device and reduced multimedia processing capabilities) can constrain the development of trust in m-commerce.

Many predictors of trust have been acknowledged in previous literature, such as perceived site attribute (Wakefield, Stocks, & Wilder, 2004) and website quality (Koufaris & Hampton-Sosa, 2004). With the aid of information technology, a virtual shopping environment is effective in prompting the transactions and has implications of building trust (Li & Yeh, 2010). Trust plays a crucial role in the formation of dependent relationships represented by online transactions (Li, Pieńkowski, van Moorsel, & Smith, 2012). The user's trust behavior toward m-commerce application is composed of three constructs: using behavior, reflection behavior, and correlation behavior (Yan, Dong, Niemi, & Yu, 2013). However, the difficult task for m-commerce is to ensure consumers' trust by making them feel comfortable with wireless transactions (Kapoor, 2016).

This section is dealing with the prospect of m-commerce and trust and the next section is dealing with m-commerce, privacy, and security issues. With the increased usage of smartphones, there are a sizable reduction in the number of people browsing through desktops. Everyone prefers to purchase using a mobile app as they find it more interesting and quick. Consumers are greatly benefited from a mobile app because they need not search for the discounted products and latest collections. Trust is very important to make consumers satisfied with m-commerce application. Most online consumers are intensely aware that various websites are collecting and storing their private information. The more consumers trust companies, the more willing they are to share data regarding m-commerce application.

Mobile Commerce, Privacy, and Security Issues

Mobile devices are at a higher security risk due to the large amount of critical financial and personal data available on them (Adlakha, 2016). In m-commerce, consumers are facing higher security and privacy risks because of the data transaction in a wireless environment (Chong, Darmawan, Ooi, & Lin, 2010). It is essential to consider the maximum security that is an urgent precedence in the m-commerce applications (Nilashi et al., 2015). For example, a secure credit-card mechanism for m-commerce should securely protect the corresponding transactions and personal information (Leu, Huang, & Wang, 2015).

Security is the most important issue for the widespread deployment of m-commerce applications based on mobile agent technology (Muñoz & Maña, 2011). Mobile agent is the technological promising product that can migrate from host to host toward autonomously performing tasks (Chang & Lin, 2006). Most of the existing solutions (e.g., execution tracing, code obfuscation, encrypted code execution, and partial result encapsulation) mainly cover the security threats of mobile agents' code (Shibli, Masood, Ghazi, & Muftic, 2015). For secure communications in public network environments, various three-party authenticated key exchange (3PAKE) protocols are proposed to provide the transaction confidentiality and efficiency in the m-commerce environments (Yang & Chang, 2009). Cryptography is one of the most important elements in providing security for the m-commerce systems (Hu, Hoang, & Khalil, 2011).

Regarding security issues, the secure m-commerce system must have the following properties: confidentiality, authentication, integrity, authorization, availability, and non-repudiation (Lee, Kou, & Hu, 2008). The authentication and authorization of transactions are generally performed by certification authority (Pai &Wu, 2011). For business transactional tasks, both perceived usefulness and perceived security affect the user's intention to utilize m-commerce (Leonard, 2010). Chen (2008) indicated that perceived transaction, perceived transaction rapidity, privacy, and security concerns are the important antecedents of consumer intention to m-commerce.

Big data is a data management system that can reduce the need for storage facilities for data and lower the computational requirements for various sectors (Kasemsap, 2016c). When the adoption of big data is properly aligned to the business, existing governance structures can be easily adjusted to address security, assurance, and general approach to embracing new technologies (Kasemsap, 2017a). Regarding privacy perspectives, consumers' mobile information privacy concerns are largely rooted in the big data-related ecosystem (Cleff, 2007).

Nowadays, digital consumer is no longer entirely anonymous since every form of communication virtually generates the data that can be collected, aggregated, and analyzed (Wessels, 2012). Information gathered for one purpose can be readily retrieved for another, and the possible linkage between mass amounts of aggregated data about an individual makes collected data personally identifiable (Eastin et al., 2016). Okazaki et al. (2009) indicated six determinants toward individuals' information privacy concerns in online settings (i.e., data collection, data control, unauthorized secondary use, improper access, location tracking, and awareness related to these practices).

This section is dealing with the prospect of m-commerce, privacy, and security issues, whereas the next section is dealing with m-commerce adoption and technology acceptance model. Consumers' lack of trust remains the largest single obstacle to growth in the mobile content and commerce industry. M-commerce brings the new challenges in providing information security as information travels through multiple networks often across wireless links. Online businesses must strive to maintain open privacy policies along with strong encryption and authentication processes in order to stay ahead of the attackers. When consumers feel free of risks and have high level of trust in the intention to utilize m-commerce, they actually adopt it.

Mobile Commerce Adoption and Technology Acceptance Model

The technology acceptance model (TAM) has been used for several years to predict the attitudes and behaviors of consumers of mobile services (Molina-Castillo & Meroño-Cerdan, 2014). TAM can be applied to predict consumer's acceptance to the mobile telecommunication innovations in the presence of network externalities (Wang, Lo, & Fang, 2008). TAM with trust is utilized to explore the influence of perceived ease of use and perceived usefulness in mobile shopping on customers' attitudes and behavioral intention to use, mobility, convenience, and information richness, characterized by m-commerce (Chen & Lan, 2014).

Jung et al. (2009) considered TAM, content, and cognitive concentration in a mobile TV service for the potential users who intend to use and suggested that cognitive concentration has the critical impacts on consumer's intention to subscribe the mobile TV service. Luarn and Lin (2005) extended TAM with three factors (i.e., credibility, self-efficiency, and financial cost) and examined consumer's behavioral intention to use the mobile banking. User satisfaction is recognized as one of the useful measures of m-commerce system success (Wang & Liao, 2007). Thus, it is an essential to identify the user requirements for establishing the m-commerce system standards (Büyüközkan, 2009). Brand loyalty is the tendency of customers to keep buying the same brand of a particular product instead of trying other brands (Kasemsap, 2016d). Users' continuance usage and satisfaction on m-commerce can promote brand loyalty (Hew, Lee, Ooi, & Lin, 2016).

The adoption of mobile phones tends to lead to an increase in the demand for mobile content (Eze & Poong, 2016). Investigating mobile technology adoption is the growing interest for technology developers, marketing managers, and researchers (Su & Adams, 2009). Network designers, service providers, and application developers must carefully take the considerations of various users into account to provide better services and attract them to m-commerce (Pedersen, Methlie, & Thorbjornsen, 2002). The importance of perceived value in customer decision making is well-recognized (Pihlström & Brush, 2008).

Perceived compatibility, perceived usefulness, and perceived ease of use are positively related to the behavioral intention to use m-commerce (Halbach & Gong, 2013). Position experience, cognitive style, and computer self-efficacy are the major factors that can predict the suitability of applying mobile devices (e.g., personal digital assistants and tablets) technology for insurance tasks (Lee, Cheng, & Cheng, 2007). Users greatly value the accuracy of the reliability and simplicity of the m-commerce applications (Choi & Stvilia, 2013). Retailers have increased their investment in mobile shopping channels to deliver the content, products, and promotions to customers toward enhancing m-commerce adoption (Liu & Liou, 2011).

Adopters perceive smartphones as both an essential device in which to invest money and a symbolic device to signal their affiliation and timely technology adoption (Kim, Chun, & Lee, 2014). The ability of customer value and the lock-in mechanism are important for maintaining online shoppers through m-commerce (Shih, 2012). Mobile users perceive their devices as a method to quickly assist them in their offline activities, such as rapid communication, price checking, and location-related services (Narang & Arora, 2016). Location-based m-commerce incorporates location-aware technologies, wire-free connectivity, and locationalized web-based services to support the processing of location-referent transactions (Wyse, 2009).

Significant Perspectives on Mobile Commerce

M-commerce is the modern technology which has enhanced the way of conducting business transactions (Sharma, 2016). M-commerce is an emerging discipline involving the use of mobile computing technologies (Faqih &Jaradat, 2015). M-commerce developers and practitioners must understand consumers' perception of m-commerce applications in order to better design and deliver the m-commerce service (Chen, Li, Chen, & Xu, 2011). In m-commerce, the consumer has to deal with the online vendor and the payment provider (Köster, Matt, & Hess, 2016). Consumers order products and services at the online vendor and provide personal information (e.g., financial data) to the mobile payment provider (Köster et al., 2016).

Mobile technologies allow companies to execute various activities toward communicating with current and potential customers (Kapoor, 2016). Modern marketers have the capability to aggregate multiple information sources to establish personal profiles about consumers, which can be used to narrowly target individuals with various forms of marketing communications (Vesanen, 2007). The benefits of m-commerce include the easy purchase process, instant updating of the information, improved customer satisfaction, cost savings, new opportunities for business, and the ability for the customer to use it everywhere (Kasemsap, 2016e). M-commerce offers the convenience of personalized marketing on mobile phones (Kaur & Malhotra, 2016).

Mobile tasks performed on handheld devices significantly challenge the m-commerce developers to adopt new methods and design guidelines that take into account contextual variations in the m-commerce environments (Chan & Fang, 2003). Companies and businesses have opted to adopt m-commerce to offer more widespread services to their different stakeholders (Alfahl, Sanzogni, Houghton, & Sandhu, 2014). The increased mobile usage is a clear example of the growth of mobile services as it offers the significant opportunities as the independent sales channel toward deserving special attention from researchers (Aldas-Manzano, Ruiz-Mafe, & Sanz-Blas, 2009).

Business process management is the development and control of processes used in a company, department, and project to ensure they are effective (Kasemsap, 2017b). With the emergence of business process management, the focus has shifted to the development of electronic services that integrate business processes and that diversify the functionalities available to customers (Chou & Seng, 2012). Due to its inherent characteristics (e.g., ubiquity, personalization, flexibility, and dissemination), m-commerce promises business market potential, great productivity, and high profitability (Siau, Lim, & Shen, 2003). The Internet-enabled mobile handheld devices are one of the most important components of the m-commerce system, making it possible for mobile users to directly interact with m-commerce applications. Enhancing users' ability and their sense of personal control can promote the use of future mobile text-based applications and services (Mahatanankoon & O'Sullivan, 2008).

There are various barriers that obstruct the proliferation of m-commerce, such as mobile devices inefficiency, lack of consumer's trust, incompatible networks, poor network coverage, limited bandwidth, low speed, lack of security, high cost of handsets, and lack of awareness (Gupta, 2016). Poor usability of mobile Internet sites for commerce activities is the major obstacle for the adoption of mobile solutions (Chan & Fang, 2009). Even though providing a mobile site for mobile device users may greatly enhance their news reading experience, many news organizations cannot afford the additional option due to limited funding and staffing resources (Lu, Wang, & Ma, 2013).

FUTURE RESEARCH DIRECTIONS

The classification of the extensive literature in the domains of m-commerce will provide the potential opportunities for future research. M-commerce describes online sales transactions that use wireless electronic devices, such as handheld computers, mobile phones, and laptops. These wireless devices interact with computer networks that have the ability to conduct the online merchandise purchases. Customer satisfaction is significant because it provides marketers and business owners with an effective marketing metric that they can utilize to manage and improve their businesses (Kasemsap, 2016f). Creating positive customer experiences across all stages of the product's life cycle drives higher satisfaction among current customers, which builds customer loyalty (Kasemsap, 2016g). An examination of linkages among m-commerce, customer satisfaction, and customer loyalty would seem to be viable for future research efforts.

Consumer attitude influences the way consumer thinks and it is important for the marketers who study it to understand how consumer behaves toward increasing business profit (Kasemsap, 2017c). The use of social media has created the highly effective communication platforms where any user, virtually anywhere in the world, can freely create the content and disseminate this information in real time to a global audience (Kasemsap, 2017d). Social media offers a chance to redefine the delivery of service to customers, thus changing the way they think about a company's brands while considerably lowering service costs (Kasemsap, 2017e). Web mining techniques can be applied with the effective analysis of the clearly understood business needs and requirements (Kasemsap, 2017f). Business process management is the development and control of processes used in a company, department, and project to ensure they are effective (Kasemsap, 2017g). Considering the relationships among m-commerce, consumer attitude, social media, web mining, and business process management would be beneficial for future research directions.

CONCLUSION

This chapter highlighted the prospect of m-commerce; m-commerce and trust; m-commerce, privacy, and security issues; m-commerce adoption and TAM; and the significant perspectives on m-commerce. M-commerce is used for business transactions conducted by mobile phones for the promotional and financial activities using the wireless Internet connectivity. M-commerce is the important way to purchase the online items through online services. The main goal of m-commerce is to ensure that customers' shopping experience is well-suited to the smaller screen sizesthat they can see on smartphones and tablets. Computer-mediated networks enable these transaction processes through electronic store searches and electronic point-of-sale capabilities.

M-commerce brings the new possibility for businesses to sell and promote their products and services toward gaining improved productivity and business growth. There are a number of business opportunities and grand challenges of bringing forth viable and robust wireless technologies ahead for fully realizing the enormous strength of m-commerce in the digital age and thereby meeting both the basic requirements and advanced expectations of mobile users and providers. The successful future of m-commerce depends on the power of the underlying technology drivers and the attractiveness of m-commerce applications toward obtaining m-commerce adoption.

REFERENCES

Adlakha, M. (2016). Mobile commerce security and its prevention. In S. Madan & J. Arora (Eds.), *Securing transactions and payment systems for m-commerce* (pp. 141–157). Hershey, PA: IGI Global. doi:10.4018/978-1-5225-0236-4.ch007

Aldas-Manzano, J., Ruiz-Mafe, C., & Sanz-Blas, S. (2009). Mobile commerce adoption in Spain: The influence of consumer attitudes and ICT usage behaviour. In B. Unhelkar (Ed.), Handbook of research in mobile business, second edition: Technical, methodological and social perspectives (pp. 282–292). Hershey, PA: IGI Global.

Alfahl, H., Sanzogni, L., Houghton, L., & Sandhu, K. (2014). Mobile commerce adoption in organizations: A literature review and preliminary findings. In I. Lee (Ed.), *Trends in e-business, e-services, and e-commerce: Impact of technology on goods, services, and business transactions* (pp. 47–68). Hershey, PA: IGI Global. doi:10.4018/978-1-4666-4510-3.ch003

Argade, D., & Chavan, H. (2015). Improve accuracy of prediction of users future m-commerce behaviour. *Procedia Computer Science*, *49*, 111–117. doi:10.1016/j.procs.2015.04.234

Arora, J. B. (2016). Regulatory framework of mobile commerce. In S. Madan & J. Arora (Eds.), *Securing transactions and payment systems for m-commerce* (pp. 176–192). Hershey, PA: IGI Global. doi:10.4018/978-1-5225-0236-4.ch009

Atkinson, L. (2013). Smart shoppers? Using QR codes and green smartphone apps to mobilize sustainable consumption in the retail environment. *International Journal of Consumer Studies*, *37*(4), 387–393. doi:10.1111/ijcs.12025

Binsaleh, M., & Hassan, S. (2013). Systems development methodology for mobile commerce applications: Agile vs. traditional. In H. El-Gohary (Ed.), *Transdisciplinary marketing concepts and emergent methods for virtual environments* (pp. 264–278). Hershey, PA: IGI Global. doi:10.4018/978-1-4666-1861-9.ch018

Büyüközkan, G. (2009). Determining the mobile commerce user requirements using an analytic approach. *Computer Standards & Interfaces*, *31*(1), 144–152. doi:10.1016/j.csi.2007.11.006

Campbell-Kelly, M., Garcia-Swartz, D., Lam, R., & Yang, Y. (2014). Economic and business perspectives on smartphones as multi-sided platforms. *Telecommunications Policy*, *39*(8), 717–734. doi:10.1016/j.telpol.2014.11.001

Chae, M. H., & Kim, J. W. (2003). Whats so different about the mobile Internet? *Communications of the ACM*, *46*(12), 240–247. doi:10.1145/953460.953506

Chan, S. S., & Fang, X. (2003). Mobile commerce and usability. In E. Lim & K. Siau (Eds.), *Advances in mobile commerce technologies* (pp. 235–257). Hershey, PA: Idea Group Publishing. doi:10.4018/978-1-59140-052-3.ch011

Chan, S. S., & Fang, X. (2009). Interface design issues for mobile commerce. In D. Taniar (Ed.), *Mobile computing: Concepts, methodologies, tools, and applications* (pp. 526–533). Hershey, PA: IGI Global. doi:10.4018/978-1-60566-054-7.ch045

Chang, C. C., & Lin, I. C. (2006). A new solution for assigning cryptographic keys to control access in mobile agent environments. *Wireless Communications and Mobile Computing*, *6*(1), 137–146. doi:10.1002/wcm.276

Charbaji, R., Rebeiz, K., & Sidani, Y. (2010). Antecedents and consequences of the risk taking behavior of mobile commerce adoption in Lebanon. In H. Rahman (Ed.), *Handbook of research on e-government readiness for information and service exchange: Utilizing progressive information communication technologies* (pp. 354–380). Hershey, PA: IGI Global. doi:10.4018/978-1-60566-671-6.ch018

Chen, L. D. (2008). A model of consumer acceptance of mobile payment. *International Journal of Mobile Communications*, *6*(1), 32–52. doi:10.1504/IJMC.2008.015997

Chen, S. C. (2012). To use or not to use: Understanding the factors affecting continuance intention of mobile banking. *International Journal of Mobile Communications*, *10*(5), 490–507. doi:10.1504/IJMC.2012.048883

Chen, Y., & Lan, Y. (2014). An empirical study of the factors affecting mobile shopping in Taiwan. *International Journal of Technology and Human Interaction*, *10*(1), 19–30. doi:10.4018/ijthi.2014010102

Chen, Z. S., Li, R., Chen, X., & Xu, H. (2011). A survey study on consumer perception of mobile-commerce applications. *Procedia Environmental Sciences*, *11*, 118–124. doi:10.1016/j.proenv.2011.12.019

Choi, W., & Stvilia, B. (2013). Use of mobile wellness applications and perception of quality. *Proceedings of the American Society for Information Science and Technology*, *50*(1), 1–4.

Chong, A. Y. L. (2013). Mobile commerce usage activities: The roles of demographic and motivation variables. *Technological Forecasting and Social Change*, *80*(7), 1350–1359. doi:10.1016/j.techfore.2012.12.011

Chong, A. Y. L., Chan, F. T. S., & Ooi, K. B. (2011). Predicting consumer decisions to adopt m-commerce: Cross country empirical examination between China and Malaysia. *Decision Support Systems*, *53*(1), 34–43. doi:10.1016/j.dss.2011.12.001

Chong, A. Y. L., Darmawan, N., Ooi, K. B., & Lin, B. (2010). Adoption of 3G services among Malaysian consumers: An empirical analysis. *International Journal of Mobile Communications*, *8*(2), 129–149. doi:10.1504/IJMC.2010.031444

Chorppath, A. K., & Alpcan, T. (2013). Trading privacy with incentives in mobile commerce: A game theoretic approach. *Pervasive and Mobile Computing*, *9*(4), 598–612. doi:10.1016/j.pmcj.2012.07.011

Chou, T. H., & Seng, J. L. (2012). Telecommunication e-services orchestration enabling business process management. *Transactions on Emerging Telecommunications Technologies*, *23*(7), 646–659. doi:10.1002/ett.2520

Cleff, E. (2007). Privacy issues in mobile advertising. *International Review of Law Computers & Technology*, *21*(3), 225–236. doi:10.1080/13600860701701421

Curty, R. G., & Zhang, P. (2011). Social commerce: Looking back and forward. *Proceedings of the American Society for Information Science and Technology*, *48*(1), 1–10. doi:10.1002/meet.2011.14504801096

Eastin, M. S., Brinson, N. H., Doorey, A., & Wilcox, G. (2016). Living in a big data world: Predicting mobile commerce activity through privacy concerns. *Computers in Human Behavior*, *58*, 214–220. doi:10.1016/j.chb.2015.12.050

Eltoweissy, M., Jajodia, S., & Mukkamala, R. (2008). Secure multicast for mobile commerce applications: Issues and challenges. In A. Becker (Ed.), *Electronic commerce: Concepts, methodologies, tools, and applications* (pp. 930–951). Hershey, PA: IGI Global. doi:10.4018/978-1-59904-943-4.ch077

Eze, U. C., & Poong, Y. S. (2016). Investigating the moderating roles of age and ethnicity in mobile commerce acceptance. In J. Prescott (Ed.), *Handbook of research on race, gender, and the fight for equality* (pp. 90–112). Hershey, PA: IGI Global. doi:10.4018/978-1-5225-0047-6.ch005

Falk, M., & Hagsten, E. (2015). E-commerce trends and impacts across Europe. *International Journal of Production Economics*, *170*, 357–369. doi:10.1016/j.ijpe.2015.10.003

Faqih, K. M. S., & Jaradat, M. R. M. (2015). Assessing the moderating effect of gender differences and individualism-collectivism at individual-level on the adoption of mobile commerce technology: TAM3 perspective. *Journal of Retailing and Consumer Services*, *22*, 37–52. doi:10.1016/j.jretconser.2014.09.006

Godbole, N. (2006). Relating mobile computing to mobile commerce. In B. Unhelkar (Ed.), *Handbook of research in mobile business: Technical, methodological, and social perspectives* (pp. 463–486). Hershey, PA: IGI Global. doi:10.4018/978-1-59140-817-8.ch033

Guillén, A., Herrera, L. J., Pomares, H., Rojas, I., & Liébana-Cabanillas, F. (2016). Decision support system to determine intention to use mobile payment systems on social networks: A methodological analysis. *International Journal of Intelligent Systems*, *31*(2), 153–172. doi:10.1002/int.21749

Gupta, P. (2016). Exploring barriers affecting the acceptance of mobile commerce. In S. Madan & J. Arora (Eds.), *Securing transactions and payment systems for m-commerce* (pp. 234–250). Hershey, PA: IGI Global. doi:10.4018/978-1-5225-0236-4.ch012

Gupta, R., Muttoo, S. K., & Pal, S. K. (2016). Understanding fraudulent activities through m-commerce transactions. In S. Madan & J. Arora (Eds.), *Securing transactions and payment systems for m-commerce* (pp. 68–93). Hershey, PA: IGI Global. doi:10.4018/978-1-5225-0236-4.ch004

Halbach, M., & Gong, T. (2013). What predicts commercial bank leaders' intention to use mobile commerce?: The roles of leadership behaviors, resistance to change, and technology acceptance model. In M. Khosrow-Pour (Ed.), *E-commerce for organizational development and competitive advantage* (pp. 151–170). Hershey, PA: IGI Global. doi:10.4018/978-1-4666-3622-4.ch008

Han, J., Yang, Y., Huang, X., Yuen, T. H., Li, J., & Cao, L. (2016). Accountable mobile e-commerce scheme via identity-based plaintext-checkable encryption. *Information Sciences*, *345*, 143–155. doi:10.1016/j.ins.2016.01.045

Heinrichs, J. H., Lim, J. S., & Lim, K. S. (2011). Influence of social networking site and user access method on social media evaluation. *Journal of Consumer Behaviour*, *10*(6), 347–355. doi:10.1002/cb.377

Hew, J. J., Lee, V. H., Ooi, K. B., & Lin, B. (2016). Mobile social commerce: The booster for brand loyalty? *Computers in Human Behavior*, *59*, 142–154. doi:10.1016/j.chb.2016.01.027

Hu, J., Hoang, X. D., & Khalil, I. (2011). An embedded DSP hardware encryption module for secure e-commerce transactions. *Security and Communication Networks*, *4*(8), 902–909. doi:10.1002/sec.221

Hu, W. (2009). Fundamentals of mobile commerce systems. In W. Hu (Ed.), *Internet-enabled handheld devices, computing, and programming: Mobile commerce and personal data applications* (pp. 1–25). Hershey, PA: IGI Global. doi:10.4018/978-1-59140-769-0.ch001

Hu, W., Yeh, J., Yang, H., & Lee, C. (2008). Mobile handheld devices for mobile commerce. In A. Becker (Ed.), *Electronic commerce: Concepts, methodologies, tools, and applications* (pp. 152–162). Hershey, PA: IGI Global. doi:10.4018/978-1-59904-943-4.ch015

Hu, W., Zuo, Y., Kaabouch, N., & Chen, L. (2010). A technological perspective of mobile and electronic commerce systems. In M. Khosrow-Pour (Ed.), *E-commerce trends for organizational advancement: New applications and methods* (pp. 16–35). Hershey, PA: IGI Global. doi:10.4018/978-1-60566-964-9.ch002

Jung, Y., Perez-Mira, B., & Wiley-Patton, S. (2009). Consumer adoption of mobile TV: Examining psychological flow and media content. *Computers in Human Behavior, 25*(1), 123–129. doi:10.1016/j.chb.2008.07.011

Kapoor, N. (2016). Consumer perception to mobile commerce. In S. Madan & J. Arora (Eds.), *Securing transactions and payment systems for m-commerce* (pp. 217–233). Hershey, PA: IGI Global. doi:10.4018/978-1-5225-0236-4.ch011

Kasemsap, K. (2016a). Advocating electronic business and electronic commerce in the global marketplace. In S. Dixit & A. Sinha (Eds.), *E-retailing challenges and opportunities in the global marketplace* (pp. 1–24). Hershey, PA: IGI Global. doi:10.4018/978-1-4666-9921-2.ch001

Kasemsap, K. (2016b). Implementing electronic commerce in global marketing. In I. Lee (Ed.), *Encyclopedia of e-commerce development, implementation, and management* (pp. 591–602). Hershey, PA: IGI Global. doi:10.4018/978-1-4666-9787-4.ch043

Kasemsap, K. (2016c). Mastering big data in the digital age. In M. Singh & D. G. (Eds.), Effective big data management and opportunities for implementation (pp. 104–129). Hershey, PA: IGI Global. doi:10.4018/978-1-5225-0182-4.ch008

Kasemsap, K. (2016d). Role of social media in brand promotion: An international marketing perspective. In A. Singh & P. Duhan (Eds.), *Managing public relations and brand image through social media* (pp. 62–88). Hershey, PA: IGI Global. doi:10.4018/978-1-5225-0332-3.ch005

Kasemsap, K. (2016e). Investigating the roles of mobile commerce and mobile payment in global business. In S. Madan & J. Arora (Eds.), *Securing transactions and payment systems for m-commerce* (pp. 1–23). Hershey, PA: IGI Global. doi:10.4018/978-1-5225-0236-4.ch001

Kasemsap, K. (2016f). Promoting service quality and customer satisfaction in global business. In U. Panwar, R. Kumar, & N. Ray (Eds.), *Handbook of research on promotional strategies and consumer influence in the service sector* (pp. 247–276). Hershey, PA: IGI Global. doi:10.4018/978-1-5225-0143-5.ch015

Kasemsap, K. (2016g). Encouraging supply chain networks and customer loyalty in global supply chain. In N. Kamath & S. Saurav (Eds.), *Handbook of research on strategic supply chain management in the retail industry* (pp. 87–112). Hershey, PA: IGI Global. doi:10.4018/978-1-4666-9894-9.ch006

Kasemsap, K. (2017a). Software as a service, Semantic Web, and big data: Theories and applications. In A. Turuk, B. Sahoo, & S. Addya (Eds.), *Resource management and efficiency in cloud computing environments* (pp. 264–285). Hershey, PA: IGI Global. doi:10.4018/978-1-5225-1721-4.ch011

Kasemsap, K. (2017b). Mastering business process management and business intelligence in global business. In M. Tavana, K. Szabat, & K. Puranam (Eds.), *Organizational productivity and performance measurements using predictive modeling and analytics* (pp. 192–212). Hershey, PA: IGI Global. doi:10.4018/978-1-5225-0654-6.ch010

Kasemsap, K. (2017c). Mastering consumer attitude and sustainable consumption in the digital age. In N. Suki (Ed.), *Handbook of research on leveraging consumer psychology for effective customer engagement* (pp. 16–41). Hershey, PA: IGI Global. doi:10.4018/978-1-5225-0746-8.ch002

Kasemsap, K. (2017d). Professional and business applications of social media platforms. In V. Benson, R. Tuninga, & G. Saridakis (Eds.), *Analyzing the strategic role of social networking in firm growth and productivity* (pp. 427–450). Hershey, PA: IGI Global. doi:10.4018/978-1-5225-0559-4.ch021

Kasemsap, K. (2017e). Mastering social media in the modern business world. In N. Rao (Ed.), *Social media listening and monitoring for business applications* (pp. 18–44). Hershey, PA: IGI Global. doi:10.4018/978-1-5225-0846-5.ch002

Kasemsap, K. (2017f). Mastering web mining and information retrieval in the digital age. In A. Kumar (Ed.), *Web usage mining techniques and applications across industries* (pp. 1–28). Hershey, PA: IGI Global. doi:10.4018/978-1-5225-0613-3.ch001

Kasemsap, K. (2017g). Mastering business process management and business intelligence in global business. In M. Tavana, K. Szabat, & K. Puranam (Eds.), *Organizational productivity and performance measurements using predictive modeling and analytics* (pp. 192–212). Hershey, PA: IGI Global. doi:10.4018/978-1-5225-0654-6.ch010

Kaur, R., & Malhotra, H. (2016). SWOT analysis of m-commerce. In S. Madan & J. Arora (Eds.), *Securing transactions and payment systems for m-commerce* (pp. 48–67). Hershey, PA: IGI Global. doi:10.4018/978-1-5225-0236-4.ch003

Keith, M. J., Babb, J. S., Lowry, P. B., Furner, C. P., & Abdullat, A. (2015). The role of mobile-computing self-efficacy in consumer information disclosure. *Information Systems Journal*, *25*(6), 637–667. doi:10.1111/isj.12082

Kim, D., Chun, H., & Lee, H. (2014). Determining the factors that influence college students adoption of smartphones. *Journal of the Association for Information Science and Technology*, *65*(3), 578–588. doi:10.1002/asi.22987

Kim, G., Shin, B., & Lee, H. G. (2009). Understanding dynamics between initial trust and usage intentions of mobile banking. *Information Systems Journal*, *19*(3), 283–311. doi:10.1111/j.1365-2575.2007.00269.x

Kim, M. J., Chung, N., Lee, C. K., & Preis, M. W. (2015). Motivations and use context in mobile tourism shopping: Applying contingency and task–technology fit theories. *International Journal of Tourism Research*, *17*(1), 13–24. doi:10.1002/jtr.1957

Kleijnen, M., Ruyter, K. D., & Wetzels, M. (2007). An assessment of value creation in mobile service delivery and the moderating role of time consciousness. *Journal of Retailing*, *83*(1), 33–46. doi:10.1016/j.jretai.2006.10.004

Köster, A., Matt, C., & Hess, T. (2016). Carefully choose your (payment) partner: How payment provider reputation influences m-commerce transactions. *Electronic Commerce Research and Applications*, *15*, 26–37. doi:10.1016/j.elerap.2015.11.002

Koufaris, M., & Hampton-Sosa, W. (2004). The development of initial trust in an online company by new customers. *Information & Management, 41*(3), 377–397. doi:10.1016/j.im.2003.08.004

Lee, C., Hu, W., & Yeh, J. (2009). Mobile commerce technology. In M. Khosrow-Pour (Ed.), *Encyclopedia of information science and technology* (2nd ed., pp. 2584–2589). Hershey, PA: IGI Global. doi:10.4018/978-1-60566-026-4.ch412

Lee, C., Kou, W., & Hu, W. (2008). Mobile commerce security and payment methods. In A. Becker (Ed.), *Electronic commerce: Concepts, methodologies, tools, and applications* (pp. 292–306). Hershey, PA: IGI Global. doi:10.4018/978-1-59904-943-4.ch027

Lee, C. C., Cheng, H. K., & Cheng, H. H. (2007). An empirical study of mobile commerce in insurance industry: Task–technology fit and individual differences. *Decision Support Systems, 43*(1), 95–110. doi:10.1016/j.dss.2005.05.008

Lee, Y. E., & Benbasat, I. (2003). Interface design for mobile commerce. *Communications of the ACM, 46*(12), 48–52. doi:10.1145/953460.953487

Leonard, L. N. (2010). C2C mobile commerce: Acceptance factors. In I. Lee (Ed.), *Encyclopedia of e-business development and management in the global economy* (pp. 759–767). Hershey, PA: IGI Global. doi:10.4018/978-1-61520-611-7.ch076

Leu, F. Y., Huang, Y. L., & Wang, S. M. (2015). A secure m-commerce system based on credit card transaction. *Electronic Commerce Research and Applications, 14*(5), 351–360. doi:10.1016/j.elerap.2015.05.001

Li, F., Pieńkowski, D., van Moorsel, A., & Smith, C. (2012). A holistic framework for trust in online transactions. *International Journal of Management Reviews, 14*(1), 85–103. doi:10.1111/j.1468-2370.2011.00311.x

Li, Y. M., & Yeh, Y. S. (2010). Increasing trust in mobile commerce through design aesthetics. *Computers in Human Behavior, 26*(4), 673–684. doi:10.1016/j.chb.2010.01.004

Lin, J., Lu, Y., Wang, B., & Wei, K. K. (2011). The role of inter-channel trust transfer in establishing mobile commerce trust. *Electronic Commerce Research and Applications, 10*(6), 615–625. doi:10.1016/j.elerap.2011.07.008

Lin, K. Y., & Lu, H. P. (2015). Predicting mobile social network acceptance based on mobile value and social influence. *Internet Research, 25*(1), 107–130. doi:10.1108/IntR-01-2014-0018

Liou, C. H., & Liu, D. R. (2012). Hybrid recommendations for mobile commerce based on mobile phone features. *Expert Systems: International Journal of Knowledge Engineering and Neural Networks, 29*(2), 108–123.

Liu, D. R., & Liou, C. H. (2011). Mobile commerce product recommendations based on hybrid multiple channels. *Electronic Commerce Research and Applications, 10*(1), 94–104. doi:10.1016/j.elerap.2010.08.004

Lu, Y., Deng, Z., & Wang, B. (2010). Exploring factors affecting Chinese consumers usage of short message service for personal communication. *Information Systems Journal*, *20*(2), 183–208. doi:10.1111/j.1365-2575.2008.00312.x

Lu, Y., Wang, X., & Ma, Y. (2013). Comparing user experience in a news website across three devices: iPhone, iPad, and desktop. *Proceedings of the American Society for Information Science and Technology*, *50*(1), 1–4. doi:10.1002/meet.14505001133

Lu, Y. B., Yang, S. Q., Chau, P. Y. K., & Cao, Y. Z. (2011). Dynamics between the trust transfer progress and intention to use mobile payment services: Across-environment perspective. *Information & Management*, *48*(8), 393–403. doi:10.1016/j.im.2011.09.006

Luarn, P., & Lin, H. H. (2005). Toward an understanding of the behavioral intention to use mobile banking. *Computers in Human Behavior*, *21*(6), 873–891. doi:10.1016/j.chb.2004.03.003

Mahatanankoon, P., & OSullivan, P. (2008). Attitude toward mobile text messaging: An expectancy-based perspective. *Journal of Computer-Mediated Communication*, *13*(4), 973–992. doi:10.1111/j.1083-6101.2008.00427.x

McKnight, D. H., & Chervany, N. L. (2001). *Conceptualizing trust: A typology and e-commerce customer relationships model*. Paper presented at the 34th Annual Hawaii International Conference on System Sciences (HICSS 2001), Maui, HI. doi:10.1109/HICSS.2001.927053

Miao, M., & Jayakar, K. (2016). Mobile payments in Japan, South Korea and China: Cross-border convergence or divergence of business models? *Telecommunications Policy*, *40*(2/3), 182–196. doi:10.1016/j.telpol.2015.11.011

Möhlenbruch, D., Dölling, S., & Ritschel, F. (2010). Interactive customer retention management for mobile commerce. In K. Pousttchi & D. Wiedemann (Eds.), *Handbook of research on mobile marketing management* (pp. 437–456). Hershey, PA: IGI Global. doi:10.4018/978-1-60566-074-5.ch023

Molina-Castillo, F., & Meroño-Cerdan, A. (2014). Drivers of mobile application acceptance by consumers: A meta analytical review. *International Journal of E-Services and Mobile Applications*, *6*(3), 34–47. doi:10.4018/ijesma.2014070103

Moqbel, A., Yani-De-Soriano, M., & Yousafzai, S. (2012). Mobile commerce use among UK mobile users: An experimental approach based on a proposed mobile network utilization framework. In A. Zolait (Ed.), *Knowledge and technology adoption, diffusion, and transfer: International perspectives* (pp. 78–111). Hershey, PA: IGI Global. doi:10.4018/978-1-4666-1752-0.ch007

Muñoz, A., & Maña, A. (2011). TPM-based protection for mobile agents. *Security and Communication Networks*, *4*(1), 45–60. doi:10.1002/sec.158

Narang, B., & Arora, J. B. (2016). Present and future of mobile commerce: Introduction, comparative analysis of m commerce and e commerce, advantages, present and future. In S. Madan & J. Arora (Eds.), *Securing transactions and payment systems for m-commerce* (pp. 293–308). Hershey, PA: IGI Global. doi:10.4018/978-1-5225-0236-4.ch015

Ngai, E. W. T., & Gunasekaran, A. (2007). A review for mobile commerce research and applications. *Decision Support Systems*, *43*(1), 3–15. doi:10.1016/j.dss.2005.05.003

Nilashi, M., Ibrahim, O., Mirabi, V. R., Ebrahimi, L., & Zare, M. (2015). The role of security, design and content factors on customer trust in mobile commerce. *Journal of Retailing and Consumer Services*, *26*, 57–69. doi:10.1016/j.jretconser.2015.05.002

Okazaki, S., Li, H., & Hirose, M. (2009). Consumer privacy concerns and preference for degree of regulatory control. *Journal of Advertising*, *38*(4), 63–77. doi:10.2753/JOA0091-3367380405

Okazaki, S., & Mendez, F. (2013). Exploring convenience in mobile commerce: Moderating effects of gender. *Computers in Human Behavior*, *29*(3), 1234–1242. doi:10.1016/j.chb.2012.10.019

Ozuem, W., & Mulloo, B. N. (2016). Manifested consumption: Mobile storefront. In A. Diab (Ed.), *Self-organized mobile communication technologies and techniques for network optimization* (pp. 356–373). Hershey, PA: IGI Global. doi:10.4018/978-1-5225-0239-5.ch013

Pai, H. T., & Wu, F. (2011). Prevention of wormhole attacks in mobile commerce based on non-infrastructure wireless networks. *Electronic Commerce Research and Applications*, *10*(4), 384–397. doi:10.1016/j.elerap.2010.12.004

Patel, A. (2006). Mobile commerce in emerging economics. In B. Unhelkar (Ed.), *Handbook of research in mobile business: Technical, methodological, and social perspectives* (pp. 429–434). Hershey, PA: IGI Global. doi:10.4018/978-1-59140-817-8.ch030

Pedersen, P. E., Methlie, L. B., & Thorbjornsen, H. (2002). *Understanding mobile commerce end-user adoption: A triangulation perspective and suggestions for an exploratory service evaluation framework.* Paper presented at the 35th Annual Hawaii International Conference on System Sciences (HICSS 2002), Maui, HI. doi:10.1109/HICSS.2002.994011

Peng, J., Quan, J., & Zhang, S. (2013). Mobile phone customer retention strategies and Chinese e-commerce. *Electronic Commerce Research and Applications*, *12*(5), 321–327. doi:10.1016/j.elerap.2013.05.002

Petrova, K. (2008). Mobile commerce applications and adoption. In A. Becker (Ed.), *Electronic commerce: Concepts, methodologies, tools, and applications* (pp. 889–897). Hershey, PA: IGI Global. doi:10.4018/978-1-59904-943-4.ch072

Pierre, S. (2009). Security issues concerning mobile commerce. In D. Taniar (Ed.), *Mobile computing: Concepts, methodologies, tools, and applications* (pp. 2653–2659). Hershey, PA: IGI Global. doi:10.4018/978-1-60566-054-7.ch201

Pihlström, M., & Brush, G. J. (2008). Comparing the perceived value of information and entertainment mobile services. *Psychology and Marketing*, *25*(8), 732–755. doi:10.1002/mar.20236

Qi, J., Li, L., Li, Y., & Shu, H. (2009). An extension of technology acceptance model: Analysis of the adoption of mobile data services in China. *Systems Research and Behavioral Science*, *26*(3), 391–407. doi:10.1002/sres.964

Sharma, M. (2016). Services of mobile commerce. In S. Madan & J. Arora (Eds.), *Securing transactions and payment systems for m-commerce* (pp. 251–274). Hershey, PA: IGI Global. doi:10.4018/978-1-5225-0236-4.ch013

Shibli, M. A., Masood, R., Ghazi, Y., & Muftic, S. (2015). MagicNET: Mobile agents data protection system. *Transactions on Emerging Telecommunications Technologies*, *26*(5), 813–835. doi:10.1002/ett.2742

Shih, H. P. (2012). Cognitive lock-in effects on consumer purchase intentions in the context of B2C web sites. *Psychology and Marketing*, *29*(10), 738–751. doi:10.1002/mar.20560

Siau, K., Lim, E., & Shen, Z. (2003). Mobile commerce: Current states and future trends. In E. Lim & K. Siau (Eds.), *Advances in mobile commerce technologies* (pp. 1–17). Hershey, PA: Idea Group Publishing. doi:10.4018/978-1-59140-052-3.ch001

Siau, K., & Shen, Z. (2003). Building customer trust in mobile commerce. *Communications of the ACM*, *46*(4), 91–94. doi:10.1145/641205.641211

Su, Q., & Adams, C. (2009). Mobile commerce adoption: A novel buyer-user-service payer metric. *Journal of Electronic Commerce in Organizations*, *7*(4), 59–72. doi:10.4018/jeco.2009070106

Sumita, U., & Yoshii, J. (2010). Enhancement of e-commerce via mobile accesses to the Internet. *Electronic Commerce Research and Applications*, *9*(3), 217–227. doi:10.1016/j.elerap.2009.11.006

van der Heijden, H., Verhagen, T., & Creemers, M. (2003). Understanding online purchase intentions: Contributions from technology and trust perspectives. *European Journal of Information Systems*, *12*(1), 41–48. doi:10.1057/palgrave.ejis.3000445

Varshney, U. (2007). Supporting dependable group-oriented mobile transactions: Redundancy-based architecture and performance. *International Journal of Network Management*, *17*(3), 219–229. doi:10.1002/nem.619

Varshney, U. (2008). A middleware framework for managing transactions in group-oriented mobile commerce services. *Decision Support Systems*, *46*(1), 356–365. doi:10.1016/j.dss.2008.07.005

Veijalainen, J., & Weske, M. (2003). Modeling static aspects of mobile electronic commerce environments. In E. Lim & K. Siau (Eds.), *Advances in mobile commerce technologies* (pp. 137–170). Hershey, PA: Idea Group Publishing. doi:10.4018/978-1-59140-052-3.ch007

Vesanen, J. (2007). What is personalization? A conceptual framework. *European Journal of Marketing*, *41*(5/6), 409–418. doi:10.1108/03090560710737534

Wakefield, R. L., Stocks, M. H., & Wilder, W. M. (2004). The role of web site characteristics in initial trust formation. *Journal of Computer Information Systems*, *45*(1), 94–103.

Wang, C. C., Lo, S. K., & Fang, W. (2008). Extending the technology acceptance model to mobile telecommunication innovation: The existence of network externalities. *Journal of Consumer Behaviour*, *7*(2), 101–110. doi:10.1002/cb.240

Wang, J. (2007). Mobile commerce. In D. Taniar (Ed.), *Encyclopedia of mobile computing and commerce* (pp. 455–460). Hershey, PA: IGI Global. doi:10.4018/978-1-59904-002-8.ch075

Wang, X., Hong, Z., Xu, Y., Zhang, C., & Ling, H. (2014). Relevance judgments of mobile commercial information. *Journal of the Association for Information Science and Technology, 65*(7), 1335–1348. doi:10.1002/asi.23060

Wang, Y. S., & Liao, Y. W. (2007). The conceptualization and measurement of m-commerce user satisfaction. *Computers in Human Behavior, 23*(1), 381–398. doi:10.1016/j.chb.2004.10.017

Wang, Y. S., Lin, H. H., & Luarn, P. (2006). Predicting consumer intention to use mobile service. *Information Systems Journal, 16*(2), 157–179. doi:10.1111/j.1365-2575.2006.00213.x

Wessels, B. (2012). Identification and the practices of identity and privacy in everyday digital communication. *New Media & Society, 14*(8), 1251–1268. doi:10.1177/1461444812450679

Wyse, J. E. (2009). Location-aware query resolution for location-based mobile commerce: Performance evaluation and optimization. In D. Taniar (Ed.), *Mobile computing: Concepts, methodologies, tools, and applications* (pp. 3040–3067). Hershey, PA: IGI Global. doi:10.4018/978-1-60566-054-7.ch229

Xiao, S., & Dong, M. (2015). Hidden semi-Markov model-based reputation management system for online to offline (O2O) e-commerce markets. *Decision Support Systems, 77*, 87–99. doi:10.1016/j.dss.2015.05.013

Yan, Z., Dong, Y., Niemi, V., & Yu, G. (2013). Exploring trust of mobile applications based on user behaviors: An empirical study. *Journal of Applied Social Psychology, 43*(3), 638–659. doi:10.1111/j.1559-1816.2013.01044.x

Yang, J. H., & Chang, C. C. (2009). An efficient three-party authenticated key exchange protocol using elliptic curve cryptography for mobile-commerce environments. *Journal of Systems and Software, 82*(9), 1497–1502. doi:10.1016/j.jss.2009.03.075

Yow, K. C., & Mittal, N. (2008). Mobile commerce multimedia messaging peer. In A. Becker (Ed.), *Electronic commerce: Concepts, methodologies, tools, and applications* (pp. 514–523). Hershey, PA: IGI Global. doi:10.4018/978-1-59904-943-4.ch042

Yu, F. R., Wong, V. W. S., Song, J. H., Leung, V. C. M., & Chan, H. C. B. (2011). Next generation mobility management: An introduction. *Wireless Communications and Mobile Computing, 11*(4), 446–458. doi:10.1002/wcm.904

Zhang, J., & Mao, E. (2008). Understanding the acceptance of mobile SMS advertising among young Chinese consumers. *Psychology and Marketing, 25*(8), 787–805. doi:10.1002/mar.20239

Zhang, L., Zhu, J., & Liu, Q. (2012). A meta-analysis of mobile commerce adoption and the moderating effect of culture. *Computers in Human Behavior, 28*(5), 1902–1911. doi:10.1016/j.chb.2012.05.008

ADDITIONAL READING

Abu-Shanab, E., & Ghaleb, O. (2012). Adoption of mobile commerce technology: An involvement of trust and risk concerns. *International Journal of Technology Diffusion, 3*(2), 36–49. doi:10.4018/jtd.2012040104

Anderson, J., & Vakulenko, M. (2014). Upwardly mobile. *Business Strategy Review, 25*(4), 34–39. doi:10.1111/j.1467-8616.2014.01118.x

Anong, S. T., & Kunovskaya, I. (2013). M-finance and consumer redress for the unbanked in South Africa. *International Journal of Consumer Studies, 37*(4), 453–464. doi:10.1111/ijcs.12014

Basoglu, N., Daim, T., & Polat, E. (2014). Exploring adaptivity in service development: The case of mobile platforms. *Journal of Product Innovation Management, 31*(3), 501–515. doi:10.1111/jpim.12110

Boakye, K. G. (2015). Factors influencing mobile data service (MDS) continuance intention: An empirical study. *Computers in Human Behavior, 50*, 125–131. doi:10.1016/j.chb.2015.04.008

Bryson, D., Atwal, G., Chaudhuri, H. R., & Dave, K. (2015). Understanding the antecedents of intention to use mobile Internet banking in India: Opportunities for microfinance institutions. *Strategic Change, 24*(3), 207–224. doi:10.1002/jsc.2005

Chang, C. C., Chang, S. C., & Yang, J. H. (2013). A practical secure and efficient enterprise digital rights management mechanism suitable for mobile environment. *Security and Communication Networks, 6*(8), 972–984. doi:10.1002/sec.647

Chen, J., & Shen, X. L. (2015). Consumers decisions in social commerce context: An empirical investigation. *Decision Support Systems, 79*, 55–64. doi:10.1016/j.dss.2015.07.012

Chong, A. (2013). Understanding mobile commerce continuance intentions: An empirical analysis of Chinese consumers. *Journal of Computer Information Systems, 53*(4), 22–30. doi:10.1080/08874417.2013.11645647

Dan, L., & Jing, Z. (2011). TAM-based study on factors influencing the adoption of mobile payment. *China Communications, 8*(3), 198–204.

Gao, L., & Bai, X. (2014). An empirical study on continuance intention of mobile social networking services. *Asia Pacific Journal of Marketing and Logistics, 26*(2), 168–189. doi:10.1108/APJML-07-2013-0086

Garrett, J. L., Rodermund, R., Anderson, N., Berkowitz, S., & Robb, C. A. (2014). Adoption of mobile payment technology by consumers. *Family and Consumer Sciences Research Journal, 42*(4), 358–368. doi:10.1111/fcsr.12069

He, W. (2013). A survey of security risks of mobile social media through blog mining and an extensive literature search. *Information Management & Computer Security, 21*(5), 381–400. doi:10.1108/IMCS-12-2012-0068

Hew, J. J., Lee, V. H., Ooi, K. B., & Wei, J. (2015). What catalyses mobile apps usage intention: An empirical analysis. *Industrial Management & Data Systems, 115*(7), 1269–1291. doi:10.1108/IMDS-01-2015-0028

Humphreys, L. (2013). Mobile social media: Future challenges and opportunities. *Mobile Media & Communication*, *1*(1), 20–25. doi:10.1177/2050157912459499

Jung, J. C., Ugboma, M. A., & Liow, A. K. (2015). Does Alibabas magic work outside China? *Thunderbird International Business Review*, *57*(6), 505–518. doi:10.1002/tie.21739

Jung, Y. (2014). What a smartphone is to me: Understanding user values in using smartphones. *Information Systems Journal*, *24*(4), 299–321. doi:10.1111/isj.12031

Kim, M. J., & Park, J. (2014). A return on investment assessment model for a mobile user interface project at the predevelopment stage. *Human Factors and Ergonomics in Manufacturing & Service Industries*, *24*(2), 216–225. doi:10.1002/hfm.20368

Lee, J. S., & Lin, K. S. (2013). An innovative electronic group-buying system for mobile commerce. *Electronic Commerce Research and Applications*, *12*(1), 1–13. doi:10.1016/j.elerap.2012.09.005

Mervyn, K., & Allen, D. K. (2012). Sociospatial context and information behavior: Social exclusion and the influence of mobile information technology. *Journal of the American Society for Information Science and Technology*, *63*(6), 1125–1141. doi:10.1002/asi.22626

Mir, I. (2011). Consumer attitude towards m-advertising acceptance: A cross-sectional study. *Journal of Internet Banking & Commerce*, *16*(1), 1–22.

Nicholas, D., Clark, D., Rowlands, I., & Jamali, H. R. (2013). Information on the go: A case study of Europeana mobile users. *Journal of the American Society for Information Science and Technology*, *64*(7), 1311–1322. doi:10.1002/asi.22838

Ozcan, P., & Santos, F. M. (2015). The market that never was: Turf wars and failed alliances in mobile payments. *Strategic Management Journal*, *36*(10), 1486–1512. doi:10.1002/smj.2292

Park, J., Han, S. H., Kim, H. K., Moon, H., & Park, J. (2015). Developing and verifying a questionnaire for evaluating user value of a mobile device. *Human Factors and Ergonomics in Manufacturing & Service Industries*, *25*(6), 724–739. doi:10.1002/hfm.20588

Peng, H., & Liu, W. (2011). Drivers and barriers in the acceptance of mobile payment in China. *Communications in Information Science and Management Engineering*, *1*(5), 73–78.

Sheng, M. L., & Teo, T. S. H. (2012). Product attributes and brand equity in the mobile domain: The mediating role of customer experience. *International Journal of Information Management*, *32*(2), 139–146. doi:10.1016/j.ijinfomgt.2011.11.017

Sigman, B. P., & Boston, B. J. (2013). Digital discernment: An e-commerce web site evaluation tool. *Decision Sciences Journal of Innovative Education*, *11*(1), 29–46. doi:10.1111/j.1540-4609.2012.00367.x

Slade, E. L., Dwivedi, Y. K., Piercy, N. C., & Williams, M. D. (2015). Modeling consumers adoption intentions of remote mobile payments in the United Kingdom: Extending UTAUT with innovativeness, risk, and trust. *Psychology and Marketing*, *32*(8), 860–873. doi:10.1002/mar.20823

Wong, C. H., Tan, G. W. H., Tan, B. I., & Ooi, K. B. (2015). Mobile advertising: The changing landscape of the advertising industry. *Telematics and Informatics*, *32*(4), 720–734. doi:10.1016/j.tele.2015.03.003

Xu, C., Peak, D., & Prybutok, V. (2015). A customer value, satisfaction, and loyalty perspective of mobile application recommendations. *Decision Support Systems, 79*, 171–183. doi:10.1016/j.dss.2015.08.008

Zhou, T., & Li, H. (2014). Understanding mobile SNS continuance usage in China from the perspectives of social influence and privacy concern. *Computers in Human Behavior, 37*, 283–289. doi:10.1016/j.chb.2014.05.008

KEY TERMS AND DEFINITIONS

Electronic Commerce: The business conducted through the use of computers or other electronic appliances without the exchange of paper-based documents.

Internet: The worldwide computer network that provides information on many subjects and enables users to exchange the messages.

Mobile Commerce: The business that is conducted on the Internet through the use of mobile phones or other wireless, handheld electronic devices.

Mobile Payment: The point of sale payment made through the wireless devices, such as a mobile phone and personal digital assistant.

Smartphone: The mobile phone with the highly advanced features.

Technology: The purposeful application of information in the design, production, and utilization of products and services.

Technology Acceptance Model: The most widely utilized model to predict and explain the user acceptance of information technology.

Transaction: The exchange of products or services between a buyer and a seller.

Chapter 13
mMarketing Opportunities for User Collaborative Environments in Smart Cities

Artemis D. Avgerou
Imperial College, UK

Despina A. Karayanni
University of Patras, Greece

Yannis C. Stamatiou
University of Patras, Greece

ABSTRACT

Smart City infrastructures connect people with their devices through wireless communications networks while they offer sensor-based information about the city's status and needs. Connecting people carrying mobile devices equipped with sensors through such an infrastructure leads to the "collective intelligence" or "crowdsourcing" paradigm. This paradigm has been deployed in numerous contexts such as performing large-scale experiments (e.g., monitoring the pollution levels or analyzing mobility patterns of people to derive useful information about rush hours in cities) or gathering and sharing user collected experiences in efforts to increase privacy awareness and personal information protection levels. In this chapter, we will focus on employing this paradigm in the mMarketing/mCommerce domain and discuss how crowdsourcing can create new opportunities for commercial activities as well as expansion of existing ones.

INTRODUCTION

Modern cities aim at improving the quality and daily satisfaction of its populations as well as its economy by deploying ICT infrastructures with their physical facilities (Mulligan and Olsson 2013, Rassia and Pardalos 2014). This integration results in a "smarter" city, in the sense that the city "senses" and "understands" its inhabitants' needs and wants and adjusts or rectifies itself in order to satisfy them mainly

DOI: 10.4018/978-1-5225-2469-4.ch013

through monitoring itself and its habitats' opinions and suggestions and notifying the local governors (e.g. municipality or local government).

Within the current decade, many ICT companies created and invested on technologies that can transform cities into smart cities in the above sense. As evidenced from the ongoing work of the Technical University of Vienna, Austria, in http://www.smart-cities.eu/, the EU has developed a major strategic advantage in the Smart City domain by having numerous EU cities equipped with smart city infrastructures and applications. The standard smart city model describes six basic smart city qualities:

- Smart Economy (Innovative spirit, Entrepreneurship, City image, Productivity, Labour Market, International integration).
- Smart Governance (Political awareness, Public and social services, Efficient and transparent administration).
- Smart Living (Cultural and leisure facilities, Health conditions, Individual security, Housing quality, Education facilities, Touristic attractiveness, Social cohesion).
- Smart Mobility (Local Transport System, (Inter-)national accessibility, ICT-Infrastructure, Sustainability of the transport system).
- Smart Environment (Air quality (no pollution), Ecological awareness, Sustainable resource management).
- Smart People (Education, Lifelong learning, Ethnic plurality, Open-mindedness).

Our work, in this chapter, with respect to mMarketing, addresses the first item above.

Usually, the public and private sectors invest in the crowdsourcing and crowdsensing concepts, within the context of a smart city. These terms refer to the massive participation of people (crowd) who deliver, constantly, information (through sensors but also through, e.g., survey input) about their needs, wants, and their environment to applications which collect and analyze this massive information. The information comes from diverse sources which include smart meters, smart energy, smart phones and tablets, public information systems, social network applications and other corporate and public data sources (e.g. civic – municipality – data).

Therefore, a key element in the smart city context is that individuals become actively engaged in the creation of content and in sharing information on the internet, mainly through social networking media. This development taps into "the wisdom of crowds", because it promotes the creation of all sorts of information content, it enhances the diversity of publicly available information and it increases the number of topics on which users can find information. Also, users' opinions and ideas are shared and truly count – anyone can participate and share his views and thoughts about anything.

Along this on-line, user collaboration paradigm, we discuss in this chapter a new mMarketing model based on creating a collaborative environment of shoppers that enables the interaction among them (implementing the "wisdom of crowds" principle) as well as with the participating businesses and marketers, based on the ICT infrastructures provided by smart cities.

BACKGROUND

People's massive interaction with information systems through their mobile devices is, loosely, termed crowdsourcing or collective intelligence. Please see the work by Avgerou, Nastou, Nastouli, Pardalos, and Stamatiou (2016) for application of the model in eCommerce). These terms refer to the data, information, and actions which emerge as a result of the interactions of large numbers of individuals through their mobile devices. This collaboration is supported by the sensing, computational and communications capabilities of modern mobile devices (e.g. mobile phones and tablets). Crowdsourcing applications rely on gathering and processing numerous streams of data of various types, such as environmental variables, location coordinates, real-time information, and on-site conditions from large numbers of individuals moving in dispersed geographical locations. This massive creation of information and data creates opportunities for new types of application domains.

One target application domain, which relies on crowdsourcing on top of the ICT facilities offered by smart cities, appears to offer new opportunities for creating novel applications and services of interest to the marketing and commerce domain. Our focus in this chapter is the mMarketing domain which has, lately, attracted much attention mainly due to the increasing popularity, the powerful features and decreasing cost of using mobile devices. According to the proposed mMarketing model, an individual connects to and interacts with a site in order to use a service (e.g. ordering tickets, paying electronically, buying goods etc.). Smart cities, on the other hand, are characterized by the existence of Wide Area or Metropolitan wireless networks that connect their citizens at any place within their confines, by dispersed data collection sensor/ad-hoc networks that continuously gather environmental or civic information and make it available to individuals roaming the city.

In this chapter we build upon this smart city/crowdsourcing coexistence and propose a mMarketing model that offers novel business and mMarketing opportunities. In this model, we show how to develop mobile device services that involve individuals and businesses that interact among themselves and the smart city facilities in order to create alerts and useful information about events and commercial activities of broad interest. For instance, individuals may send, in real time, to the service information whether a central parking facility is still empty, if a store has sales and on which products, as well as off-line information about good mCommerce sites.

The proposed collaborative environment falls within the Cloud Computing paradigm and, more specifically, the PaaS (Platform as a Service) Cloud Computing Stack Layer. Although we will refer to this model later in this chapter, roughly the proposed environment is intended to be built and offered as a set of components to other businesses wishing to customized then and implement their own mMarketing policies and ideas based on them. This is justifiable by the inherent technical and organizational complexity of this environment which may require specialized ICT expertise (e.g. wireless technologies, internet services, mobile device sensor-based applications, privacy and security etc.) to be developed and operated, which is not usually available within a company.

Our proposal also includes an equally important aspect of this mMarketing model which addresses the privacy of the participating individuals since the data and information they provide to the mMarketing platform may uncover personal information about themselves (e.g. location information, age, education, income range etc.). According to results of recent studies about people's attitude towards the Internet and Internet-based services such as the mMarketing platform described in this chapter, people keep an ambivalent position. Although they feel positively about the value of such services, they seem to be reluctant participate in their workings due to fears of revealing sensitive information about themselves. To this end, the proposed mMarketing platform encompasses as new, strong authentication technology which respects the privacy of people by allowing them to prove only parts of their identity useful to, for instance, a marketing survey but preserving their anonymity at the same time.

THE APPROACH ELEMENTS

Drivers and Inhibitors of Changes in Marketing

In Figure 1 we see a generic model of change for marketing practices described in the work by Beckhard and Reuben (1987). See, also, the work by Downes and Palmer (2005) for an application of this model in identifying marketing practice changes in the ICT business domain. The model rests on the foundational observation that changes in marketing practices are initiated by drivers for change which exist outside the organization (external environment). However, spontaneous changes in marketing practices of an organization can also occur as a vision of a charismatic CEO who has just acquired the management of the organization (internal environment).

In the work by Downes and Palmer (2005) four types of marketing were considered, which were state-of-the-art at that time: Transaction, Database, Interaction, and Network marketing, still in use today but largely superseded by eMarketing. At the time these four marketing types were considered state of the art, eMarketing, which is the currently dominant marketing practice, was still being considered as a possible type of marketing (Downes & Palmer, 2005). However, eMarketing is, for years now, a well-established marketing type which has recently acquired exponential growth and momentum towards mMarketing due to the widespread use of portable devices and high-speed networks.

In this chapter, we will discuss mMarketing as the dominant future marketing practice and will couple it with another important ICT development, the Smart City (ultimately part of the Internet of Things) which is, essentially, an urban ICT infrastructure connecting a huge number of people and devices spread throughout places.

Figure 1. A generic model of change in marketing practices

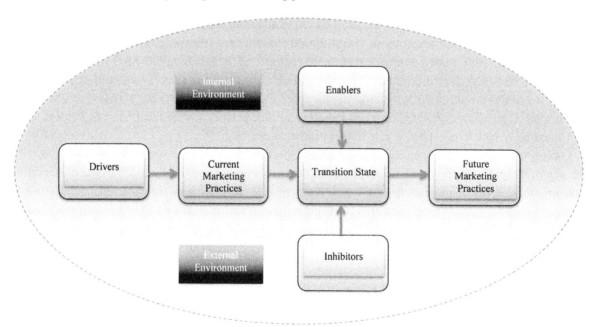

In the context of the marketing practice change model in Figure 1, we can discern the following elements:

- **Current Marketing Practices:** mMarketing is still in its infancy and, mostly, targets users through their mobile devices' applications (e.g. social network applications, on-line purchases etc.) by sending SMS notifications advertising products or, perhaps, asking for people's opinion on a product. Also, targeting massive population targets is a difficult task since one needs to have access to numerous mobile phone numbers or email addresses of people using tablets, which is either costly or hard to accomplish.
- **Drivers:** Drivers refer to the forces which mobilize an organization towards change. Change may refer to shift of attention with respect to industry sector, long-term goal, collaborations as well as marketing strategy. In our case, with respect to marketing strategy change with emphasis on change towards mMarketing, the driving forces can be internal or external. The internal driving forces stem from determined and visionary CEOs as well as CMOs (Chief Marketing Officers) who foresee the potential of new and future technologies
- **Inhibitors:** As explained in the work by Downes and Palmer (2005), one of the major inhibitors towards marketing changes in the past were the costs needed in order to incorporate new technologies into an organization's current marketing practices or integrate novel technological developments into existing legacy ICT systems. Cost was a true concern at that time, which can be broken into the cost of buying ICT equipment for the organization, to implement the organization's marketing practice (e.g. a server containing large amounts of information about large populations) and the cost of buying or developing the necessary software (e.g. buying a professional Data Base Management System to hold customer data to support Database Marketing) or developing a specialized legacy system. We believe that both of these costs can be overcome today through the well-established Cloud-based corporate ICT model (we will discuss it in the next sections). The cloud-based model frees the organization from the burden and high costs of buying, servicing, maintaining and upgrading its own ICT infrastructure (both hardware and software) by paying for the infrastructure much like paying for a utility, such as electricity, on a use and need basis. However, there is another inhibitor which, in our view, presents a challenge as well as opportunity for contemporary ICT-based marketing, and this is the privacy concerns of people when marketing professionals try to gather data about their habits and analyzed them in the context of their marketing practices and strategies. It is true that marketing aims at making customers and this, in turn, entails the analysis of their wants, needs, and behavioural patterns. However, it this analysis is performed carelessly, it leads to violation of people's privacy and leakage of their personal information, leading towards mistrust on their part towards any mMarketing attempt to reach them. On the other hand, conducting completely anonymous surveys may lead to misleading conclusions about target users' profiles and habits since anyone who completes, fully anonymously, an on-line questionnaire may claim anything about his/her profile and characteristics. In this chapter we present a new technology which can lift this inhibitor by empowering users to preserve their anonymity while satisfying the marketers' needs to now certain elements of their mobility and their behavior by proving (using a recently released cryptographic technology called Privacy-ABCs – we describe it later, in another section) parts of their identity characteristics (e.g. profession, place of living, age range etc.) to them. Finally, another inhibitor for successfully applying an mMarketing strategy, beyond privacy concerns of users, is their reluctance to participate

in marketing surveys or offers due to lack of motivation or incentives. Thus, a marketer should design a suitable reward scheme for people who react positively and help the marketer draw useful conclusions. We also discuss such an automatically operating scheme in this chapter too.

- **Enablers:** It is true that over the past decades marketing practices evolved along with available technological means since available technologies were the vehicle towards potential customers. Marketing campaigns have employed, for instance, leaflets, magazine advertisements, TV and radio advertisements, advertisements on popular and frequently visited sites, completion of on-line surveys and many other means to reach people and create customers. The invention of the Web as well as the developments in ICTs which led to the widespread use of new technologies by huge numbers of people shifted the attention of the marketers to the ICTs as the new means of reaching potential customers. The Web and development in ICTs are the enablers of marketing today, and we examine them briefly later in this chapter.
- **Transition State:** Today we are well into the era of eMarketing, where marketing campaigns and strategies target people using laptop or desktop computers mostly. However, in our opinion, if a web page where an ad appears is merely seen through a mobile phone or tablet screen, this is not a mMarketing instance. We believe that mMarketing should target users not in a static, screen-oriented way (as on a laptop, for example) but taking into account their mobility pattern, their dynamic buying behavior, their profiles and location. Furthermore, mMarketing should take into account all these elements respecting the privacy of the target users, which may appear a contradicting goal. In this chapter we amplify on this idea in order to delineate the driving forces which can lead to the new mMarketing era.
- **Future Marketing Practices:** As technology evolves, mMarketing will be the first new marketing type after eMarketing. However, beyond mMarketing as we discuss it in this chapter, future marketing practices may make use of technologies which are considered exotic or unusual for the marketing context. These devices include drones, smart watches (they are used in specialized, mostly medical, applications), wearable Virtual Realty glasses, and other IoT devices such as bio implants and miniature RFID tags. Before mMarketing evolves into something making use of such devices, the marketing industry should study the capabilities of these devices, their power over the potential customer as well as their marketing potential, much like it is done today with customary mobile devices (e.g. mobile phones and tablets).

The Cloud-Based Deployment Model

The rapid increase in the use of Internet services (e.g. social networks, e-shops etc.) combined with the mobility of people today has led to an analogous increase of the need for privacy preserving remote access to services, documents and personal information items from anywhere in the world and with a variety of devices, much like in the proposed mMarketing platform. The concept of "cloud computing" encompasses this need for ubiquitous access to this information and it refers to the use of computing resources, hardware and software, that are delivered as a service over a network, typically the Internet.

The cloud computing model for enabling convenient, on-demand network access to a shared pool of configurable computer resources and services, with minimal management effort or service provider interaction, is defined by the following:

Five essential characteristics:

- On-demand self-service
- Broad network access
- Resource pooling
- Rapid elasticity
- Measured service

Three service models (see Figure 2):

- **SaaS:** Cloud software as a service
- **PaaS:** Cloud platform as a service (the proposed platform follows this)
- **IaaS:** Cloud infrastructure as a service

Four deployment models:

- Private clouds
- Community clouds
- Public clouds
- Hybrid clouds

Two of the most important goals of any Cloud-based service are the following:

- Service availability (24/7 operation)
- Protection of end-users' data

Figure 2. Cloud-service deployment models

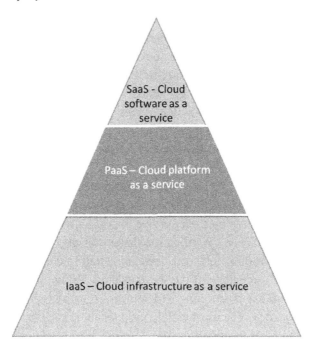

It is evident that mMarketing will be a dominant current and future marketing type and, thus, our belief is that interested stakeholders (e.g. eCommerce operators and companies) will try to address it and use it as a vehicle to implement their marketing policies and strategies. However, a full-featured mMarketing platform, at least as we envisage it and describe it in this chapter, is a complex system engineering task which not all companies are at a position to handle successfully. And if they do, it will be costly to develop, operate day-to-day, administer, and upgrade if necessary. For instance, such a mMarketing application should have advanced privacy-respecting features for the participating users as well as cryptographic algorithms to protect their personal information from disclosure. Building privacy and security critical applications is a demanding and too specialized task to be easy to handle within a non-specialist company.

In view of these considerations, our proposal is to appoint a single Cloud service provider (or a team thereof) who will build and operate this platform, opening it for use by any interested company's marketing department. The department's CMO and the whole marketing team will be able, through the platform, to set-up parametrized marketing surveys, access analytics for the company based on people's activities through the company's access point on the platform, obtain people's opinions about products, analyze mobility patterns of buyers, and perform many other operations as we describe them later in this chapter. The platform provider will license these services as cloud services based on the PaaS model. The services would be offered without charge to people and for some fee to professionals based on the use of the resources (e.g. servers, software, marketing tools etc.).

Key Technological Enablers for User in the Emerging mMarketing Landscape

Technological progress is everywhere and all people are inept at using it in their daily lives. The Smart City is a reality in many places worldwide and the IoT is fast developing. And 4G networks are operating virtually everywhere giving to people unprecedented connection speeds. It is, thus, needless and trivial to say that the required technology to implement mMarketing is already here.

However, we will mention two technological enablers which are key factors in developing mMarketing since they bring "close" the mMarketing platform and the marketer to the individual, following her mobility pattern (user location information) and sensing her environment (sensor devices). Both key factors are implemented into the mobile devices. Since mMarketing aspires to "follow" the individual in her movements within a city environment and "understanding" the surrounding conditions in order to give her timely and appropriate offers, we will dedicate some space below to them.

Location Based Services and People Mobility Tracking

The vast majority of wireless network users today are highly mobile due to the demands placed on them by their professional activities. Regardless of whether users move in their countries or a foreign country, there are a number of services offered by their service providers that depend on the place where they are. For instance, such a service may be a notification that a user may approach some accident scene or a list with hotels with available rooms. Such services are called Location Based Services (LBSs) and today tend to constitute a great percentage of the services provided in wireless communities.

The driving force behind LBS is positioning which means the ability to track the location of mobile users while they are roaming a city. The most widely used technology is the Global Positioning System (GPS). There are, also, some other technologies besides GPS which rely on signals emitted from base stations of a mobile network and not signals coming from satellites. This, however, requires that the user is a subscriber of a mobile telephony provider.

In addition to GPS, Geographic Information Systems (GIS) are also a key technology to LBS. These systems provide data related to a specific geographic location where a user moves. Such data include the morphology of the natural terrain (e.g. mountains, rivers, canyons etc.) as well as man-made structures (e.g. buildings, bridges, highways etc.).

The third component of the technology behind LBS, along with GPS and GIS, is the Location Management Function (LMF). This comprises the interface between GPS and GIS and the wireless network LBSs.

Some types of location based services appear below, which may also be of interest to a marketer, along with examples explaining their functionality:

- Location based information about nearby services and places of interest. As a typical example of services based on this information, consider a mobile user arriving in a city and wishing to find accommodation. LBS (i.e. location based information) would interact with GPS and GIS, through LMF, and send to the mobile equipment of the user a list with hotels having availability of rooms along with the price per night.
- Location based billing for phone calls or data transfer. With this service the user can have different charges per minute for the phone calls or data transfers she/he makes depending on whether she/he is near a home zone, with small charges, or away from such a zone, where the charges may be higher.
- Emergency services. Dialing an emergency number from her/his mobile phone, the user is connected to the appropriate authorities of the place where the user happens to roam when the emergency arose. Due to the importance of this location based service and the requirement for immediate response from the nearest authority, the FCC has forced all wireless carriers in the US to provide a minimum level of accuracy for all users who dial an emergency number from their mobile devices.
- Tracking. This type of LBS includes services as diverse as fleet tracking and e-commerce, for instance. Fleet tracking is used by companies wishing to know the location of their vehicles (e.g. trucks delivering sensitive materials). E-commerce belongs to this type of services since a user may want to know (by enabling this service), for instance, about sales offered by local stores or to be notified if she/he is close to a store that sells things of interest to her/him.

Although the importance of user location information and mobility pattern within a city is hard to deny (we examine location information in some use cases in a following section), this information nevertheless is considered sensitive personal information (much like the IP address of a user's device) and its storage and processing are strictly prohibited by Data Protection Agencies, if this processing leads to the identification of an individual whose activities gave rise to this information.

In accordance to the principle of data minimization, which dictates that sensitive information should be kept to a minimum or suitably distorted so as not to reveal an individual, in the proposed mMarketing platform we foresee mechanisms which decrease the level of raw GPS data granularity to a level which constitutes a balance between revealing the exact mobility pattern of an individual and retaining some value to the marketer gathering this information whenever it does not impinge negatively on the project's objectives. This, further, reduces inference attacks, whereby information which does not lead, by itself, to user identification it can nevertheless be combined with other non-identifying information to form an information set whose elements, jointly, identify the individual. This does not pose a significant problem for the marketer since very accurate position data is usually not necessary to conduct a survey involving users' mobility patterns.

Sensor Devices

Although sensing capabilities similar to the human senses seems far beyond reach for current sensor as well as computational intelligence technology it is, nevertheless, possible to equip mobile devices with the capability to sense a variety of environmental conditions and send useful information which can reveal, to a marketer, correlations between the environmental conditions in a city or store and the buying behavior of people.

Today, it is possible to construct a great variety of accurate, small-sized, mechanical/electronic sensors, some types of which appear below (their names explain their functionality and hint at potential applications):

- Radiation sensors
- Gyroscopes
- Gas sensors
- Temperature sensors
- Mechanical strain/force sensors
- Location (GPS coordinates) sensors and compasses
- Proximity sensors (e.g. RFIDs)
- Microphone and camera
- Magnetic field sensors
- Humidity sensors
- pH sensors
- Speed sensors
- Acceleration (shock) sensors

Smart mobile devices are usually equipped with a small subset of the sensors mentioned above. However, all of these sensors can be found in some IoT devices (e.g. embedded systems and environment monitoring devices), not necessarily the mobile devices normally carried by users.

An important mMarketing goal, in our opinion, would be to identify the existing sensors based on today's technology (as well as envisaged sensors which do not exist yet but could have marketing value if they could be manufactured) and their use in marketing. In other word, it would be useful to define, for each type of sensors, the mMarketing situations in which they could provide useful, for the marketer, information.

Main Inhibiting Factors: The User Privacy and Security Concerns

User data security and user Privacy are fundamental rights of the participating users and should be protected at all cost, in order to ensure massive user participation. Even a single privacy violation incident, may lead to loss of people's trust towards the mMarketing platform and mMarketing in general.

Security and privacy concerns are associated with all type of Cloud or Internet services such as Location Based Services, Virtual and Augmented Reality and network communications. The proposed mMarketing platform and its software and hardware components are exposed to the existing (in the Internet) security, privacy, identity and copyright related threats. These threats range from unauthorized information manipulation and disclosure (violations of personal or corporate confidentiality), to forgery, denial of service, and user profiling. In order to handle these issues, it is imperative to take suitable measures, which include organizational, procedural and technical ones, such as user and data authentication, data and traffic data confidentiality, authorization and access control, digital signatures, privacy and copyright protection where necessary (e.g. in digital products).

In the following, we identify the security threats which are relevant to the proposed mMarketing platform.

Privacy and Security Concerns

Digital traces are left after interacting with any ICT system. Thus, interaction with the proposed mMarketing platform may also raise privacy concerns among participants and, thus, prevent them from participating.

Reasonable privacy and security related concerns include the following:

- The information exchanged between the participants and the mMarketing platform may be intercepted and, later, be revealed to third parties for profit or for other malicious reasons. This will deface the mMarketing platform while it will destroy the trust of people towards it.
- Identity theft may result from disclosure of related, to a participating user, personal information and associated user authentication credentials and may be used for conducting fraudulent actions.
- Spam or unsolicited electronic communication may be imposed on participants from misusing contact information disclosed to illegally acting entities.
- Profiling is possible through recording and analyzing individual actions, transactions, choices, communications, posted messages, buying preferences etc.
- Loss of control of an individual's personal information is also possible if it is disclosed to illegally acting entities. That is, individuals lose control over which information or identity elements to use in specific interactions with the platform (e.g. to fill in a questionnaire for a marketing survey).

Individual's information or pieces thereof associated with actions performed by the individual while interacting with the mMarketing platform may be combined and, later, lead to his/her identification.

Privacy and Security Protection

There is an international consensus on the principles that must be taken into account in handling information and communications of individuals, which are reflected in ethical code of conducts or specific laws. These are as follows:

- A privacy policy should be developed and managed, addressing the proper handling of personal data, consistent with data protection principles and practices and the choices made by individuals they refer to.

- Reasonable security safeguards should be adopted by data controllers to adequately protect personal data in their custody against risks such as loss or unauthorized access, destruction, use, modification or disclosure of data. Security involves both technical and organizational measures.
- Active technical measures include authentication, access control mechanisms and accountability, and passive measures include integrity, cryptographic support, anonymity, pseudonymity, unlinkability and unobservability. Organizational measures include security and privacy protection planning and strategy, security and privacy policy creation and maintenance and disaster recovery and business continuity planning.
- The collection of personal data should be limited only to the necessary for the fulfillment of a specified purpose. Personal data should be collected by lawful and fair means and, where appropriate, with the knowledge or consent of the data subject
- The purposes for the collection of personal data should be identified not later than at the time of data collection. Subsequent use of the data should be limited to the fulfillment of those purposes and not be further processed in ways incompatible with those purposes.
- Personal data should be adequate, relevant and not excessive in relation to the purposes for which they are collected and to the extent necessary for those purposes, should be accurate, complete and kept up-to-date. To assure data integrity, data controllers must take reasonable steps, such as using only reputable sources of data and cross-referencing data against multiple sources, providing consumer access to data, and destroying untimely data or converting it to anonymous form.
- Information concerning development, practices and polices relating to the management of personal data should be made readily accessible to individuals.
- Processors of personal information should be accountable for complying with privacy principles and practices. They should provide means to address improper handling or misuse of personal data. Services and mechanisms should be implemented to ensure that processing of personal data is done according to relevant laws and the choices of individuals they refer to. Auditable controls should be in place to ensure compliance with legislation being in force.
- The use of a reliable mechanism to impose sanctions for noncompliance with privacy practices and principles. The principles of privacy protection can only be effective if there is a mechanism in place to enforce them. The data controller should provide data subject with a means to report alleged violations of privacy policy to judicial authorities. Also provide regulatory authorities access to audit services.

Cryptography, passwords, and especially pseudonyms are considered as very promising measures against many of the mentioned above privacy concerns and requirements. With respect to the later, pseudonyms, we will discuss in the next section a new, privacy-by-design, technology for building privacy preserving authentication credentials for the participating users. These credentials allow the users to preserve their anonymity while proving to the mMarketing platform certain elements of their profile which are important for the surveys and evaluations of marketing policies conducted by marketing specialists.

SOLUTIONS AND RECOMMENDATIONS

In this section we provide the details of our approach of addressing the security and privacy concerns of users while satisfying mMarketing needs

Privacy-ABCs

Many electronic applications and services require some authentication of their users to establish trust relations as well as learn some things about their identity. One widely used mechanism for this is password-based authentication or, more advanced, based on electronic certificates such as X509.

Although electronic certificates can offer sufficient security against theft and misuse of user's identities, while a service can be fully convinced of the identity of the users, these certificates cannot, typically, handle privacy adequately because they reveal all elements of the identity of a person. Any use of such a certificate exposes the identity of the certificate holder to the service, such as the proposed mMarketing platform, requesting authentication of the user. However, this can be an inhibitor (as discussed earlier) for massive user participation in the mMarketing platform and the gathering of valuable information through crowdsourcing.

However, there are many applications and services, including the mMarketing platform proposed in this chapter, where the use of such certificates reveals, unnecessarily, the identity of the holder. For example, in mMarketing surveys a marketer usually needs information about users such as age range, sex, place of living and educational level. Knowing the identities of the participants in a survey is needless and, even, dangerous. Revealing more information than necessary not only harms the privacy of the users but also increases the risk of abuse of information such as identity theft when revealed information is intercepted by malicious parties.

On the other hand, if users are let to respond to a mMarketing survey giving carelessly thought or false data can lead to failure of the (possibly costly to design and implement) survey to provide useful data and conclusions for the marketer who conducts the survey.

Fortunately, a new identity management technology is available based on electronic credentials, which is called *Privacy-ABCs* (Privacy Attribute Based Credentials). This technology has its roots in the two seminal works by Brands (2000) and by Camenish and Lysyanskaya (2001) which led, respectively, to the creation of two largely non-interoperable, proprietary, and expensive to license authentication technologies, Microsoft's Uprove and IBM's Idemix. For an extensive description of this technology and its applications see the work by Rannenberg, Camenisch, and Sabouri (2015).

The Privacy-ABCs technology essentially harmonized the two diferent credentials technologies of Microsoft and IBM, as the main goal of the EU research project ABC4Trust (see the project's Web site https://abc4trust.eu/ and its publicly available deliverables). The project was completed in February 2015, and produced a publicly available (through the Github code repository at https://github.com/p2abcengine/p2abcengine) set of software libraries which allow the development of privacy preserving authentication applications which empower the user to prove things about her towards a service (e.g. a mMarketing survey running on the proposed mMarketing platform) while she retains her anonymity. Thus, the service is convinced of the information supplied by the user (e.g. age range, educational level, profession etc.) without learning anything more like, for instance, the users' identity or other identifying information (e.g. Social Security Number or VAT number).

Figure 3 (taken from the project's deliverables) gives an overview of the different entities in a Privacy-ABCs deployment, like the one envisaged for the proposed mMarketing platform, as well as the interactions between these entities. We start our description of the basic privacy architecture entities by the Issuer. This entity generates and provides credentials containing Attributes to the User. On request, the Issuer generates the credential during the issuance protocol and provides it to the User. Depending on the use case, the credential information may be provided either by the User herself or the Issuer, if

Figure 3. Privacy-ABCs entities and their interactions

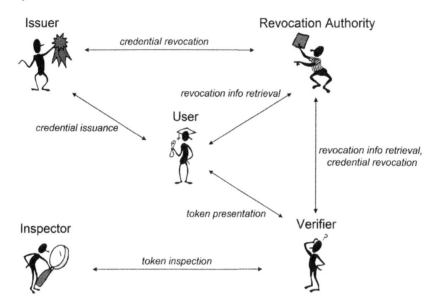

he already holds the respective information in the attribute database. Ideally, the Issuer can provide the information he attests directly, being an authoritative source. In doing so, he should have the right to assign the relevant attribute to various entities. Examples would be assigning attributes of the User to the university for the student status, to the bar association for the attribute of being an advocate, or to the trade register for the company status. Finally, attributes may also be generated "jointly random", e.g. where this may be useful for specific uses or cryptographic processes.

The User is issued the credentials while interacting with the Issuer enabling her to provide proof of certain attributes towards the Verifier. The User acts in different roles. She receives credentials from the Issuer and provides a proof for certain requested attributes towards the verifier. In some cases additional information needed for inspection are provided as well.

The Verifier receives a presentation token from the User allowing him to check that the User has certain attributes. The Verifier usually provides some kind of access restricted service to the User to which the User needs to authenticate and stipulates a policy for access. This will require the user to either reveal or to proof possession of certain attributes Values.

The Inspector reveals the identity or other encrypted attribute values of a User (e.g. lifting anonymity) upon legitimate request. For this, the Inspector has to examine the legitimacy according to the previously declared inspection grounds. We do not consider this feature useful in the context of mMarketing and, thus, we will not discuss it further.

A Revocation Service is responsible for revoking issued credentials. After revocation, the credentials cannot produce valid presentation tokens (i.e. proofs about credentials). The Revocation Service is an optional component of an ABC system. Often, the entity offering the Revocation Service is the same as that offering the Issuer service, which can be assumed to have the most accurate information about users' attributes and credentials.

In an actual deployment, some of the above roles may actually be fulfilled by the same entity or split among many. For example, an Issuer can at the same time play the role of Revocation Authority and/or

Inspector, or an Issuer could later also be the Verifier of tokens derived from credentials that it issued. In our case, the mMarketing platform will have the role of Issuer and Verifiers as well as Revocation Service.

Credentials

A credential is a set of attributes (much like an electronic certificate, an eIdentity or a X509 certificate) issued, and certified (that is, signed electronically), by the credential Issuer (the mMarketing platform in our case) to the stakeholders (end-users of the mMarketing platform).

By issuing a credential, the Issuer certifies for the correctness of the attribute it contains. These attributes are identity elements and information about the User. After issuance, the User can use the credential in order to produce presentation tokens (i.e. proofs) that uncover to other entities, the Verifiers, partial information about the attributes contained in the credential. Although attributes can be of any type (e.g. integers, strings etc.) they must eventually be mapped onto integers in order to be suitably encoded into a credential. This mapping, along with the list and type of encoded credentials, is defined in the credentials specification of the Issuer.

A User can provide certified information to Verifiers in order, for instance, to authenticate herself towards a service, using one or more of her credentials to produce a presentation token which is, then, sent to the Verifier. A presentation token can combine information from any subset of the credentials possessed by the User. Thus, the presentation token can: (i) reveal the values of a subset of the attributes contained in the credentials (e.g., IDcard.firstname = "John"), (ii) show that a credential value satisfies a Boolean predicate, such as an inequality (e.g., IDcard.birthdate >= 10/08/1993), and (iii) the values of two different credentials satisfy a Boolean predicate (e.g., IDcard.lastname = creditcard.lastname). In addition, presentation tokens support a number of advanced features such as pseudonyms, device binding, inspection, and revocation that were described earlier.

A Verifier announces in its presentation policy which credentials from which Issuers it accepts and which information the presentation token must reveal from these credentials. In mMarketing terms, a marketing survey may ask of the participants to reveal e.g. their age, occupation and educational level. Or the survey may let only participants with a specific place of living (to identify how well a product is received in a certain area). The Verifier can cryptographically verify the authenticity of a received presentation token using the credential specifications and issuer parameters of all credentials involved in the token. The Verifier must obtain the credential specifications and issuer parameters in a trusted manner, e.g., by using a traditional PKI to authenticate them or retrieving them from a trusted location.

Presentation tokens based on privacy-ABCs are cryptographically unlinkable and untraceable, meaning that Verifiers cannot tell whether two presentation tokens were derived from the same or from different credentials, and that Issuers cannot trace a presentation token back to the issuance of the underlying credentials.

There are two unique features of Privacy-ABCs, which can be very valuable in the context of the proposed mMarketing platform:

- **Scope-Exclusive Credential Presentation:** As we discussed before, the main goal of Privacy-ABCs is to preserve the anonymity of the users by allowing them to generate different pseudonyms towards different verifiers (e.g. mMarketing surveys, eShops etc.). This ensures full unlinkability among the pseudonyms as well as between the pseudonyms and the identity of the user who generated them. There are situations, however, where this unlinkability property among the generated pseudonyms is not desirable and, thus, it should be assured that all these pseudonyms are linked, together, belonging to the same, but still anonymous, user. For example, in an online, anonymous mMarketing survey, users should not be allowed to participate more than once under different pseudonyms contaminating, thus, the survey's results. In such situations, where all the pseudonyms of a user should be linked as belonging to the same, still anonymous user, the verifier can request a special type of pseudonyms, called scope-exclusive pseudonyms, which are unique for the user's secret key and a given scope (i.e. URL) string. Scope-exclusive pseudonyms for different scope strings remain unlinkable, of course. By using the URL of the mMarketing survey, for instance, as the scope string the verifier can ensure that each user can only create a single pseudonym in order to participate, anonymously, to the survey uncovering only the identity elements which are required by the survey (e.g. age range, educational level, profession etc.) since all the different pseudonyms are linkable as if they were a single pseudonym. As mentioned above, the scope string can be the Web address of the mMarketing platform's front end or the mMarketing survey URL. Thus, each time users visit such a site to perform a transaction, e.g. recommend a product or give their opinion in a survey, they prove only a limited subset of their identity elements which are of absolute interest to the marketer. This anonymous proof of these identity elements, based on scope exclusive Privacy-ABCs credentials, gives more credibility and importance to the users' opinions.

- **Carry-Over Mechanism:** We also have the carry-over feature whereby a credential is issued carrying over attribute values from other credentials of the user, without the issuer learning these values or the identity of the user. Actually, we have an issuer-oblivious transfer of attribute values into newly issued credentials. This feature can lead to applications whereby a user may access multiple web sites, in a chain, where each site can verify that the citizen has accessed the previous, in the chain, based on credentials issues by the previous site towards which, however, the user is still anonymous site. This may be useful in applications where the user can obtain reward points after visiting a series of collaborating sellers, one after the other, proving to each newly visited seller that she has visited the previous one using the issued credentials. Or whenever a user should, first, complete a certain survey and then complete another one, based on the result of the first. Then the user completes the first survey, obtains a credential from this completion and then presents it to the second survey session to prove that she has completed the first survey. The user remains completely anonymous throughout this process.

Credential Presentation

To provide certified information to a Verifier (for authentication or an access decision), the User uses one or more of her credentials to derive a presentation token and sends it to the Verifier. A single presentation token can contain information from any number of credentials. The token can reveal a subset of the attribute values in the credentials (e.g., IDcard.firstname = "John"), prove that a value satisfies a certain predicate (e.g., IDcard.birthdate < 08/10/1993) or that two values satisfy a predicate (e.g., IDcard.lastname = creditcard.lastname).

Apart from revealing information about credential attributes, the presentation token can optionally sign an application-specific message and/or a random nonce to guarantee freshness. Moreover, presentation tokens support a number of advanced features such as pseudonyms, device binding, inspection, and revocation that are described in more details below.

A Verifier announces in its presentation policy which credentials from which Issuers it accepts and which information the presentation token must reveal from these credentials. The Verifier can cryptographically verify the authenticity of a received presentation token using the credential specifications and issuer parameters of all credentials involved in the token. The Verifier must obtain the credential specifications and issuer parameters in a trusted manner, e.g., by using a traditional PKI to authenticate them or retrieving them from a trusted location.

The presentation token created in response to such a presentation policy consists of the presentation token description, containing a mechanism-agnostic description of the revealed information, and the presentation token evidence, containing opaque technology-specific cryptographic data in support of the token description. Presentation tokens based on Privacy-ABCs are in principle cryptographically unlinkable and untraceable, meaning that Verifiers cannot tell whether two presentation tokens were derived from the same or from different credentials, and that Issuers cannot trace a presentation token back to the issuance of the underlying credentials.

It should be noted that presentation tokens are unlinkable on the condition that they do not reveal identifying information (e.g., the User's social security number). In addition, pseudonyms and inspection can be used to create intentional linkability among different presentation tokens and impose their traceability (e.g., for accountability reasons in case of abuse). Finally, Privacy-ABCs can be combined with existing anonymous communication channels (e.g. Tor) to avoid linkability at the lower TCP/IP layers through the IP addresses in the underlying communication channels or the physical characteristics of the device on which the tokens were constructed.

Architecture of the Proposed mMarketing Platform for Smart Cities

In Figure 4 we see the cloud-based architecture of the proposed mMarketing platform (following the PaaS – Platform as a Service – cloud model) which takes advantage of the infrastructure of smart cities as well as the technological enablers mentioned in the previous sections to engage users in mMarketing activities.

By considering the figure as being composed of four "columns", on the left-hand side column we can view the involved mMarketing stakeholders, who are all considered as "End-Users" of the platform and its services:

- **Users Roaming in Smart Cities:** This end-user group is the main one, since the mMarketing platform's operation and success relies on the massive participation of people roaming in a smart city. The platform should take special care so as not to exclude special groups of people such as people with eyesight problems, disabilities as well as elderly people. This can be taken care of in the portal component that appears in the third column in the figure, by employing specially designed user interfaces depending on the characteristics of each user.
- **Organizations Employing mMarketing Strategies:** This is the end-user group whose members plan to user the mMarketing platform in order to implement their mMarketing strategy. This includes eCommerce sellers, service providers, ICT companies and, virtually, any enterprise which wishes to target large number of people easily and have access to numerous interesting information items regarding these people without risking data leakage or violating their anonymity.
- **Market Agencies:** This includes all kinds of market observation agencies including, for instance, national statistical agencies, price observation agencies, marketing policy analysis agencies, the

Figure 4. The mMarketing platform architecture

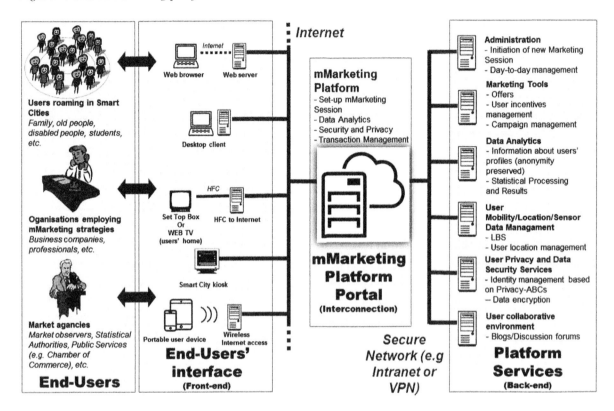

Chamber of Commerce etc. This group is interested in obtaining data analytics and information about mMarketing practices, users' behaviours and mobility patterns within a smart city, users' buying habits and preference etc. in order to derive useful conclusions about the commercial life of a city and the contributions to it by user groups of varying profiles and buying behaviours.

In the second column of the figure we see the means through which end-users connect to the platform. The platform should support all possible ICT means from mobile devices (e.g. tablets, mobile phones, and smart watches) to desktop computers and Internet TVs. Virtually, any IoT device should be connected beyond consumer ICT devices, which includes embedded microcontroller systems taking measurements within a smart city, ad-hoc sensor networks formed by sensor devices spread throughout the city, smart homes, and virtually all devices with wireless connectivity supporting one of the usual wireless communications protocols (e.g. WiFi, Zigbee, Bluetooth etc.). All these devices can be employed for mMarketing projects since they provide to the marketer important fine-grained environmental information within a city including nearby, to the devices, events.

In the third column we have the interface between the end-users and their devices with the platform, the front-end component. This provides all the necessary menus, on-line documentation, helpdesk info, forms, instructions, and authentication mechanisms for signing-up, signing-in, creating an mMarketing session, obtaining identity credentials, participating in mMarketing sessions, obtaining analytics and statistics results, and peer-to-peer communication and collaboration among users.

In the fourth column we see the heart of the mMarketing platform which is comprised of the servers and the mMarketing services they provide to end-users. The proposed platform is envisaged to offer the following services:

- **Administration:** This service is responsible for creating user accounts, letting mMarketing end-users create mMarketing sessions, authenticating users and applying access rights mechanism depending on the role of the end-users.
- **Marketing Tools:** This service let mMarketing professionals create and implement user incentive and reward mechanisms based on the users' participation frequency and intensity in their sessions.
- **Data Analytics:** This service collects, categorizes and analyses data related to the user's behaviours, mobility pattern and profiles. It creates and displays useful data analytics for the mMarketing professionals allowing them to assess the degree of success of their mMarketing sessions.
- **User Mobility/Location/Data Management:** This service is responsible for handling user and user device related data. This data includes users' location, sensor data collected from their mobile devices as well other types of data captured by the devices (e.g. photographs).
- **User Privacy and Data Security Services:** This service is the main strength of the proposed mMarketing platform since it facilitates the massive user participation, as required by the crowdsourcing and collective intelligence principles. Since voluntary participation cannot exist without trust that the mMarketing platform will not be able to link the acquired data with specific user identities, this service preserves full anonymity of the participants while it also allows them to reveal only part of their profile characteristics useful for a mMarketing session (e.g. profession, age range, city of residence etc.). Also, this service applies the usual strong cryptographic algorithms for encrypting user information before it is being used while after its use the information is deleted automatically, in harmony with the "right-to-be-forgotten" initiative.
- **User Collaborative Environment:** This service allows users to form discussion groups where they can share information about offers, mMarketing sessions, interesting commercial places in a city, as well as opinions and views on products they have tried or used.

In Figure 5 we stress the important role of users within the mMarketing platform context. Users perform crowdsensing, that is massive collection and dispatch to the platform of sensory information from their environment: users' location (GPS data from their mobile devices), acceleration, temperature and magnetic field information, acceleration, images from camera, orientation etc. In essence, the users, through their devices, constantly monitor and submit information to the platform about their environment and mobility pattern.

Then the users perform the crowdsourcing operation, which is the provision of information on how they think about elements that marketers consider important. This includes their participation in online surveys and opinion polls, participation in trials of new products, provision of feedback on products they have used and recommendations for improvements as well as their opinion about marketing campaigns they have participated in. In both these processes privacy and personal data protection permeate all the activities of the users horizontally so as to prevent them from losing their trust towards the platform. In addition, an ongoing incentives and reward system is in operation in order to attract as many participants as possible, fulfilling mMarketings objectives.

Figure 5. Crowsourcing and crowdsensing

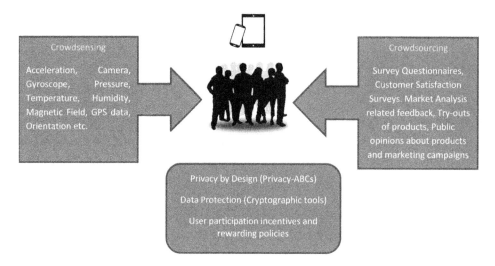

mMarketing Platform Use Cases and Day-to-Day Operation

The main day-to-day business scenario in such an ecosystem is described in what follows, with most important element the protection of the privacy of participating users as well as the maximum, possible, release of information towards the mMarketing platform.

To begin with, all end-users should receive credentials with identity information, based on Privacy-ABCs. All stakeholders receive a credential, called credID, depending on their role (see first column in Figure 4). The proposed credentials and their fields are shown in Figure 6.

A user participant starts by subscribing to the service implementing the collective intelligence model using a registration process that, in the end, gives them a credential with fields containing the user's personal information. This credential (see next subsection) will allow, from now on, the user to interact anonymously with the service and shop network as well as the central service. It also allows the user to uncover specific fields from this credential allowing partial identification towards a service, if required.

Figure 6. Proposed end-users' Privacy-ABCs credentials

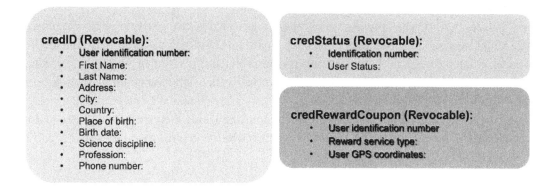

Each time a user submits information to the system, he is rewarded by a credential-coupon, called credRewardCoupon that can be used to prove, anonymously, that the user has provided information related to the city's commercial activities. The collection of such coupons acts as a reward mechanism to attract massive user participation. Users can also assign a credibility value to the information they submit by revealing, e.g., their profession. Then the submitted information carries fields that contain the professions of the users who have reported it.

Other users, now, can access this information by proving, only, their registration to the service in a fully anonymous fashion. To this end, they use their CredID to provide anonymous proof of their eligibility. Then the users, after obtaining the information submitted by others, can assign a usefulness/trustworthiness value to it. This acts as reputation mechanism for the participating users so that users revealing useful/trustworthy information ascend the reputation levels faster than others. The reputation level is reflected by the credStatus that contains a reputation/status level currently assigned to a user.

BC1

Paul is a marketing analyst working at the IT equipment company Orange. His goal is to investigate the market potential of Orange's upcoming new computer, OR11, a model that will be launched in the computer market within the next three months. In order to conduct his investigation, he has developed a survey questionnaire targeting people from 25 to 45 years old. He uses the mMarketing platform to specify the study (i.e. enter the questionnaires in an electronic form) and the target population characteristics. The questionnaire includes questions such as "Do you already have an Orange computer?" and "What is your opinion on the upcoming OR11 model?" and "Would you buy the new OR11 computer?". He also uses the platform's LBS services in order to notify participants whenever they are close to a store offering Orange's products in order to see an early prototype of OR11 where they will be also asked, in real time, by the shop owner about their impressions from the new computer. As an incentive to visit a nearby shop and complete the questionnaire on-line, they are given electronic coupons for a 15% discount for their next buy of an Orange IT product. Based on the platform and the gathered information, Paul is able to produce a detailed study on the market potential for OR11 while he extracts the supporting information from the mMarketing platform along with graphical presentation of the survey results. Using the proposed platform, Paul has access, in an easy and fast way, to numerous participants which will provide him with valuable data for conducting his marketing research on the market potential for the new OR11 computer.

BC2

John is a user participant of the mMarketing platform. He has been in March rewarded with the distinction of "Participant of the Month" due to the number of surveys that he participated in, as well as the quality of the data he has provided. The platform ranks the participants commensurate with their contributions where levels like: Beginner, Average Participant, Professional and Master of Data are attributed to the users in their Privacy-ABCs credentials. As "Participants of the Month" the participants can earn a prize of 20€ credit on their mobile phones, 2Gb free data, free tickets for the theater, and discounts at collaborating restaurants. This action helps to make the participants more eager on participating in eMarketing sessions and at the same time making it fun and rewarding to participate in the platform.

BC3

Steve is a marketing researcher in the advertising industry. He is interested in knowing the gathering places, walking routes, driving routes, train journeys of various demographic groups in order to identify the optimal deployment of advertisements on billboards and on-train ad spaces that his company owns. He would like to track the movements of users as they go about their daily lives. He does not wish, of course, to know individual identities but only broad general demographic profile information of each user. Revealing this information along with GPS coordinates of users, while preserving their anonymity, is possible through the use of Privacy-ABCs.

BC4

Jenny is a marketing researcher in the clothes industry. Her goal is to identify the relationship between the colours of the shop (e.g. colours of the walls and the shelves) and the volume of the customers' purchases over four months in order to determine the optimal colours for the shop surfaces and maximize revenue. Naturally, she is not interested in knowing individual identities but only general information on how colours affect their buying behaviour.

BC5

Jenny is a PhD student working in Marketing Management in the food industry. She is interested in knowing the selling policies that supermarkets follow for placing products on their stores' shelves. The goal is to discover which (perhaps long-term) marketing strategies determine these selling policies. She would, also, like to track the movements of users within stores to understand whether the adopted selling policies affect the customers' behaviour. She is not interested in knowing individual identities but only the general movement and shopping profiles of customers.

BC6

Mark is a marketing researcher in the clothing industry. He is interested in knowing the best level of the shops' temperature in order to maximize the sales total amount in a given period. He would like to track the different perceptions of the potential customers inside shops at any change of temperature. He does not wish to know individual identities bad only the broad perceptions of each user.

BC7

George is a CMO (Chief Marketing Officer) in the toy industry. His goal is to test whether the light colour and its intensity in the store affect the sales volume and how. He intends to use this information in order to propose actions that may lead to an increase in sales in toy stores. Thus, he would like to track the different perceptions of people inside the store, during changes in the light colour or intensity. Of course, he does not wish (or need) to know people's identities but only their broad general perceptions and reactions.

User Incentives and Award Mechanism

Although the protection of the privacy and personal information of participants removes a severe obstacle for user participation, it nevertheless does not suffice to ensure massive people participation. A mMarketing session organized through the proposed scheme should contain an incentives and user reward mechanism to attract users. Possible parameters in determining user participation intensity and loyalty are the following:

- Number of days of active participation in mMarketing sessions: For each participation day, the participant receives 1 point. For example, if she has installed the mobile app in his smartphone on 12/10/2014 and today's date is 15/03/2016, then he earns 520 points.
- Number of sensors the user offers to mMarketing sessions: For each provided sensor, the participant receives 100 points. For instance, if the she allows the use of the accelerometer, the GPS and the gyroscope of his mobile device, he may earn 300 points in total, 100 for each sensor.
- Number of mMarketing sessions in which the user has participated: For each mMarketing session, the participant may earn 50 points. For instance, if she has participated in 25 sessions, she earns 1250 points in total.
- Number of mMarketing surveys in which the user has participated: For each survey in which the user has participated, 200 points are earnt. For instance, if the participant has responded to 3 mMarketing surveys, 600 points are earnt.
- Ratings earnt for proposing ideas for improvements in products or services: For each new idea, the participant earns as many points as the rating of the idea itself. An idea can obtain a rating between 0 and 100, which is converted to a 5-star scale to be displayed to the user. For example, if the user has proposed 4 ideas for product or service improvements, which have received, correspondingly, by the corresponding mMarketing specialist the ratings 65, 79, 24 and 97, then the participant earns 265 points in total.
- Number of times the user has tried out new products: For each product offered for trial in the context of a mMarketing campaign, the participant earns 300 points.
- Number of posts in the user collaborative area: For each posted message the participant receives 5 points.
- Number of received award coupons: For each received coupon (e.g. received after visiting or buying something from a participating store) the participant receives 10 points.

In Table 1 we show the weights of all contributing parameters as well as the points and total scores for two example users. The total rate is the weighted sum of these parameters.

The rating of the user and her rank among all other mMarketing platform users can be seen only by the user through the crowdsourcing tool installed in her device. The crowdsourcing tool, using the appropriate API, obtains periodically from the server the ratings of the user and her ranking, updating them accordingly.

Table 1. The user ranking function

Reward Parameter	Parameter Weight (Is Adjustable)	Example User A		Example User B	
		Value	Points Gathered	Value	Points Gathered
Number of days of active participation in mMarketing sessions	1	From: 12/10/2014 To: 15/03/2016 = 520 days	520	From: 25/2/2015 To: 15/03/2016 = 384 days	384
Number of sensors the user offers to mMarketing sessions	100	3	300	2	200
Number of mMarketing sessions in which the user has participated	50	25	1250	37	1850
Number of mMarketing surveys in which the user has participated	200	3	600	3	400
Ratings earnt for proposing ideas for improvements in products or services	1	4 ideas with ratings: 65+79+24+97 = 265	265	6 ideas with ratings: 23+92+54+86+93+48 = 396	396
Number of times the user has tried out new products	300	2	600	3	900
Number of posts in the user collaborative area	5	10	50	15	75
Number of received award coupons	10	3	30	5	50
User's total Rating			3615		4255

Some Large-Scale and Long-Term Applications of the mMarketing Platform

Diffusing Innovative Products and Services

The diffusion of innovations among people, such as gadgets, internet services and, even, habitshas attracted the attention of a number of researchers from diverse disciplines ranging from sociology and psychology to mathematics and physics.

In our view, innovations may fail to gain much popularity upon their introduction because attempts to introduce them are, usually, rather abrupt and targeted at very large, virtually unrelated groups of individuals, even whole country populations. Things can be worse, if a fault is discovered in the innovation which can lead to loss of trust and desire of people to acquire it (in case of a product) or use it (in case of a service). In other words, marketing campaigns and strategies targeting at large groups of unrelated individuals may fail to diffuse the innovation.

Putting this observation in perspective, we believe that an innovation has better chances of diffusion to large population groups through a step-wise, gradual mMarketing process that targets progressively larger sets of individuals that are as closely related as possible in the beginning at least.

For instance, a new on-line service or mobile device could be offered for trial use first, in closely related groups of individuals, e.g. in a sample of people working in an ICT company or other business sector. Such a controlled exposure of the innovation gives time to the people trying it to act as "early adopters" and initiators. They are expected to communicate their impressions from using the products among themselves reinforcing a positive opinion or discovering, early enough, possible problems with the innovation before they become known to whole populations of potential users.

The mMarketing platform acts as the vehicle which connects these people through the collaborative user area as well as the anonymous surveys which can be contacted gathering their opinions and feedback from using the innovation. If problems arise, they are solved fast and without large-scale damage for the innovation creator. Then the group of trial users is enhanced e.g. to more companies or perhaps universities, reiterating the above process through the mMarketing platform. All the gathered analytics, surveys, opinions (in a single place) are in the hands of marketers who can decide the best time the innovation can be introduced to large populations or to all people of a country, for instance.

This gradual adoption effort, which can be coordinated by the proposed mMarketing platform, has a number of positive side effects. First of all, it allows for a thorough, in-field evaluation of the innovation in a controlled environment which offers full analytics and user survey reports. In addition, time is given to the trial users to develop opinions and views about the innovation that will contribute to its improvements before the next trial round or final introduction.

We believe that following the "gradual introduction" approach outlined above, eventually the innovation (if deserving) will be adopted by the majority of people in a large target population, something which could not be achieved if one attempted to impose, all at once (e.g. by large-scale advertising), its use on the population. The mathematical theory behind this idea relies on the work of (Young, 2003), who proposed a mathematical model based on game and graph theory to explain the diffusion of innovations in closely-related groups of individuals (e.g. friends in social networks).

Market Segmentation

In creating a successful business in any sector, the marketer should be able to address the target population and communicate her message in a convincing manner, so as to create customers. However, communicating a message convincingly requires a deep understanding of the target population's needs, concerns, desires and even ethical values (Kotler & Keller, 2015). This, in turn, necessitates knowing the profiles of the target population as well as attracting a massive audience to which the message will be communicated appropriately

Thus, the marketer should design marketing strategies based on questions such as "What are the characteristics of the potential customers?", "What are their purchasing habits?", "What products to they like?", "How do they move about within a city's confines and what do they look for and where?" etc. The marketing science has concluded that it is not possible to address at once, all people's needs as diverse they may be. Thus, companies usually divide populations into groups or segments, according to some specified characteristics, which appear to be the most promising for the product or service they plan to launch (Kotler & Keller, 2015).

In this context, Kotler and Armstrong define market segmentation as the process of "dividing a market into distinct groups of buyers who have distinct needs, characteristics, or behaviour and who might require separate products or marketing mixes" (Armstrong & Kotler, 2005). Characteristics of interest include the following (we will not go into details since they are well known in the market segmentation domain): age group, profession, educational level, sex, nationality, buying preferences etc.

It is, therefore, evident from the above that the key elements in any segmentation effort are the following:

- Analysis of large target populations in order to derive statistically valid analysis results about the population members needs and desires.

- Knowledge of the profiles of the individuals of the target population, at least with respect to the characteristics of interest.

Both of these elements are offered by the mMarketing platform proposed in this chapter. First, crowdsourcing allows the marketer to address large populations. Then, the marketer knows many of characteristics of interest (e.g. age, sex, education, profession, family size etc.) through the population members' credentials. Finally, all these are possible respecting the privacy of the participants of the mMarketing platform. The other alternative would be to buy people's personal information by data brokers, which can be costly, can miss important population segment since brokers usually obtain data from single sources whose members may not be as diverse as necessary, and finally it is unethical as a marketing strategy. Not to mention that there is absolutely no reward for the individuals whose data is exploited.

Thus, the marketer can organize surveys and opinion gatherings through the mMarketing platform addressing massive individual groups, who voluntarily (without fear of privacy loss and with the incentives and awards offered by the mMarketing platform) reveal about themselves the required, by the marketer, identity elements.

FUTURE RESEARCH DIRECTIONS

In this section we discuss the implication of the proposed mMarketing model in the evolving personal data ecosystem.

Currently, we are witnessing a rapidly evolving personal information market whose main target is people's personal data, including their habits as well as life style and mobility patterns (Newman, 2014). Over the past years, privacy breach incidents have shown that a number of Internet service providers capitalize on people's data by selling their customers' data or gathering such data from various sources for profit.

The bait for people is, most often, a useful free service that asks people to reveal some personal information or a survey which, although it may not directly ask for identifying information (e.g. name, Social Security Number), it nevertheless asks for data of interest to the personal information market. Also, these service providers use this information to track users' behavior, profile them, and then build a personal information database which, in the personal information market, has a value that far exceeds the costs incurred in building and maintaining it.

Many users are attracted to the free service offers without second thoughts about their privacy, often releasing personal information to the service providers who, actually, act as information brokers in the evolving personal information market. This, in turn, leads to a financial asymmetry since the owners of the (unknowingly) marketed information, i.e. the users of the services, are not given any compensation while, at the same time, they essentially lose control over it. Furthermore, this profiling action leads to unfair treatment of people during their online purchases, as investigations have shown. This is because people's data is, often, used by product providers to manipulate the prices of the same product so as to take into account the financial or other status of the targeted potential customers. This is known as price discrimination which essentially leads to amplifying inequalities among people.

Based on the above considerations, the interaction between users and internet service providers does not appear to offer a "win-win" financial relationship. The users appear to be in a disadvantage position with respect to the control over their personal information.

This situation is worse when mobile services are considered since mobile phones may reveal more information than desktops or laptops as they are close to their owners almost all day. Thus, users tend to perform all their transactions through their mobile phones (giving personal information) while the mobile phones capture information about their mobility patterns as well as environmental conditions, revealing information about their movements and positions. Thus, when it comes to mMarketing, which actually needs such data in order to make inferences about various marketing success parameters and sales goals, then protecting the privacy of users is of utmost importance, if people can even consider releasing such data. And even if they do so, under anonymity, they do not receive any compensation for giving away their personal data.

The proposed mMarketing platform attempts to offer a win-win situation for both the users and the mMarketing professionals in which users can remain completely anonymous, the marketer can learn important information about their habits (of interesting to a marketing campaign or product acceptance survey for instance), and the users receive rewards commensurate with their participation. Privacy protection frees users from their concerns and reservations about participating in mMarketing activities while it also relieves mMarketing departments from the necessity to buy (often expensive) people's personal data from data brokers which amplifies, further, the privacy concerns of people and their reluctance to share their data with marketers.

In addition, if marketing practices move away from buying personal data (for use in their marketing surveys) from third parties, then personal information collecting services and information brokers will be forced to rethink the unfair model on which they base their current operations. They will have to either stop basing their profits on people's lack of privacy awareness or provide financial returns to those who give, upon consent, personal information. Especially with respect to this latter issue, see (Li, Li, Miklau, & Suciu, 2013), for an interesting personal data pricing model which can be employed in customer-award based marketing policies and strategies which require personal data from people (respecting their privacy of course).

We feel that our proposal, beyond its applicability in the marketing domain, can also be a step towards reducing the asymmetries that exist in the financial aspects of the personal data ecosystem.

CONCLUSION

In this chapter we presented a generic privacy mMarketing model and the corresponding platform for implementing it, with key element the protection of the privacy of the participants. Privacy protection can attract massive participation, in the sense of crowdsourcing. The enabling medium is the smart city as it provides the necessary infrastructure to interconnect people and devices moving within it confines. The proposed mMarketing platform aims at attracting large masses of people who will participate in information exchanges about product they have bought or tried, in opinion sharing about products they have used and liked or disliked, and writing evaluations of services they have users. In the spirit of works, such as the one by Brady and Fellenz (2008), which investigated the role of ICTs in Contemporary Marketing Practices, it would be interesting to conduct a similar research for the role of crowdsourcing/crowdsensing technologies, as well as the evolving Internet of Things (IoT), in modern marketing practices.

The main element of our approach is the deployment of a new privacy enhancing technology, the Privacy-ABCS. We believe that the privacy enhancing characteristics of this technology will increase trust of people towards mMarketing and encourages their participation in marketing-related activities without risking their privacy and personal information. At the same time, they will be able to prove about themselves elements of their identities of interest to marketing surveys and the marketers, allowing the extraction of conclusions useful to the marketing domain. This can create new marketing and business opportunities for eCommerce sites by exploiting the authoritative opinions in increasing their marketing revenues.

REFERENCES

Armstrong, G., & Kotler, P. (2005). *Marketing: An Introduction* (7th ed.). Upper Saddle River, NJ: Prentice Hall.

Avgerou, A. D., Nastou, P. E., Nastouli, D., Pardalos, P., & Stamatiou, Y. C. (2016). On the Deployment of Citizens Privacy Preserving Collective Intelligent eBusiness Models in Smart Cities. *International Journal of Security and Its Applications, 10*(2), 171–184. doi:10.14257/ijsia.2016.10.2.16

Beckhard, R., & Reuben, T. H. (1987). *Organizational Transition: Managing Complex Change*. Reading, MA: Addison-Wesley.

Brady, M., Fellenz, M. R., & Brookes, R. (2008). Researching the role of information and communications technology (ICT) in contemporary marketing practices. *Journal of Business and Industrial Marketing, 23*(2), 108–114. doi:10.1108/08858620810850227

Brands, S. (2000). *Rethinking Public Key Infrastructures and Digital Certificates: Building in Privacy*. MIT Press.

Camenisch, J., & Lysyanskaya, A. (2001). An Efficient non-transferable anonymous multi-show credential system with optional anonymity revocation. (2001). In *Proceedings of the 30th Annual International Conference on the Theory and Applications of Cryptographic Techniques (EUROCRYPT 2001)* (vol. 2045, pp. 93-118). LNCS, Springer.

Downes, J., & Palmer, R. A. (2005). The Influences on Contemporary Marketing Practices In High-Technology Companies: Research Programme and Preliminary Findings. In *IMP Conference, 2005, Rotterdam, Netherlands*. (Unpublished draft proceeding)

Kotler, P., & Keller, K. L. (2015). *Marketing Management* (15th ed.). Pearson Education Limited.

Li, C., Li, D. Y., Miklau, G., & Suciu, D. (2013). A theory of pricing private data. In *Proceeding of the 16th International Conference on Database Theory (ICDT '13)* (pp. 33-44). New York: ACM Press.

Mulligan, C. E. A., & Olsson, M. (2013). Architectural Implications of Smart City Business Models: An Evolutionary Perspective. *IEEE Communications Magazine, Vol, 51*, 80–85. doi:10.1109/MCOM.2013.6525599

Newman, N. (2014). The Costs of Lost Privacy: Consumer Harm and Rising Economic Inequality in the Age of Google. *William Mitchell Law Review, 40*(2), 849–889.

Rannenberg, K., Camenisch, J., & Sabouri, A. (Eds.). (2015). *Attribute-based Credentials for Trust: Identity in the Information Society*. Springer International Publishing. doi:10.1007/978-3-319-14439-9

Rassia, S. Th., & Pardalos, P. M. (Eds.). (2014). *Cities for Smart Environmental and Energy Futures*. Springer-Verlag Berlin Heidelberg. doi:10.1007/978-3-642-37661-0

Young, H. P. (2003). The Diffusion of Innovations in Social Networks. In *Proc. The Economy as a Complex Evolving System* (Vol. 3, pp. 267–282). Oxford University Press.

Chapter 14
The Effect of Cultural Values in Mobile Payment Preference

Jashim Khan
University of Surrey, UK

Gary James Rivers
University of Surrey, UK

Jean-Éric Pelet
ESCE International Business School, Paris, France

Na Zuo
Digital Economy Consulting Centre, New Zealand

ABSTRACT

The purpose of this study is to compare French and New Zealand consumers' perceptions of mobile payments (m-payments) relative to other options to identify the preferred mode of payment and related spending behaviour. Evidence suggests that payment modes can influence spending behaviours and therefore this is important to commerce to promote payment modes that facilitate transactions. Using the Perceptions of Payment Mode (PPM) scale (Khan et al., 2015), this study was able to identify cultural differences on perceptions of cash payments, though both countries' consumers held negative perceptions of, and emotions towards, m-payments relative to other options. The empirical results are useful in understanding cultural aspects of payment modes and for companies to recognise consumers' associations with these modes to enhance relations, services and the use of m-payments.

INTRODUCTION

In 1997, Kevin Duffey coined the term mobile commerce (m-commerce) to signify electronic commerce via mobile wireless devices: that is, to deliver electronic commerce capabilities directly into the consumer's hand, anywhere, anytime. The notion of mobile commerce is embedded with the concept of 'online shopping', information in consumer's hand 24/7. According to Cosseboom (2014) "mobile commerce is worth US$230 billion, with Asia representing almost half of the market, and has been forecast to reach US$700 billion in 2017" (p.2). Walmart, a giant retailer worldwide, estimated that 40% of all visits to their internet shopping portal in 2012 was from a mobile device (Niranjanamurthy & Kavyashree, 2013). French participation in m-commerce accounts for 22% of all online transactions with 37% on being conducted on smartphones and 63% via tablets (Criteo, 2015). On the other hand,

DOI: 10.4018/978-1-5225-2469-4.ch014

New Zealand has the highest penetration per capita of m-commerce and about to reach US $ 10 billion by 2019 (Jeremiah, 2014). Since 2011, Google Wallet Mobile App in conjunction with Vodafone, O2, and Orange had accelerated m-commerce worldwide. Smartphone customers are the main participants in m-commerce (Criteo, 2015).

The earliest use of mobile devices to pay for purchases was reported in Kenya where money transfer was mainly done through the use of mobile phones. The locals called it M-Pesa (Swahili word for money). M-Pesa is a mobile phone-based money transfer, financing and microfinancing service, launched in 2007 by Vodafone for Safaricom and Vodacom, the largest mobile network operators in Kenya and Tanzania (Saylor, 2012). The idea behind M-Pesa was to create a service involving micro finance where borrowers receive and pay loans using the network of Safaricom airtime resellers (Hughes & Lonie, 2010). The service gained consumer acceptance as it allowed migrant workers sending remittances home across the country and making payments (Hughes & Lonie, 2010). This was an initiative of Safaricom, a pioneer in mobile money transfers via ATM machines. The acceptance of this system was rapid, as mobile ATMs can be at any location and can transmit transactional information wirelessly. Additionally, these machines typically have internal heating and air conditioning units that help keep them functional despite the high temperature of the country in summer.

Another case of mobile-payments usage is in Hungary, where Vodafone allowed cash or bank card payments for monthly phone bills using mobile devices. As Hungarian consumers are not comfortable with direct debits and this is not a standard practice, Vodafone pioneered to install a post-payment service that subscribers can use and access to pay bills and purchase tickets. Mobile tickets are sent to mobile phones using a variety of technologies. Users are then able to use their tickets immediately, by presenting their mobile phone at the ticket check-in as a digital boarding pass. Mobile ticketing technologies were also used for the distribution of vouchers, coupons and loyalty cards. These items were represented by a virtual token that is sent to the mobile phone. A customer presenting a mobile phone with one of these tokens at the point of sale receives the same benefits as if they had the traditional tickets.

Other innovative applications and examples can be found. For instance, Uber, the car-pooling company, relies on mobile messaging and GPS (global positioning system) to track customers in China and payments can be made using a mobile account and smartphone.

Currently, mobile content for mobile devices can be purchased and delivered online and over networks, including the sale of ring-tones, wallpapers and games, amongst other apps. The convergence of mobile hardware to include phones, portable audio and video players into a single device is increasing the purchase and delivery of full-length music tracks and video. The download speeds available with 4G networks make it possible to buy a movie on a mobile device in seconds. Banks and other financial institutions use mobile commerce to allow their customers to access account information and make transactions, such as purchasing stocks or remitting money. These services are often referred to as mobile or m-banking. Catalog merchants can accept orders from customers electronically, via customers' mobile devices. In some cases, merchants can conveniently deliver catalogs electronically, rather than mailing paper versions to the customer. Consumers making mobile purchases can also receive value-add upselling services and offers. Some merchants provide mobile websites that are customized for the mobile devices' smaller screens and limit the user interface of a mobile device. Payments can be made directly inside of an application running on a popular smartphone operating systems, such as Google's Android or Apple's iOS X. In-app purchases are used to buy virtual goods, news and other mobile content and are billed by mobile carriers rather than the app stores themselves. Analyst firm Gartner expects in-application purchases to drive 41% of all apps soon. Ericsson's IPX mobile commerce system, similarly, offers payment options such as try-before-you-buy rentals and subscriptions.

While cash and cheques are still prevalent in some parts of the world, mobile payments are gaining consumer acceptance in many economies due to the high penetration of mobile phone technology (Herzberg, 2003). In some countries, more advanced smart payment systems are in operation. For instance, in Hong Kong, New Zealand and Australia, a contactless and rechargeable smart card allows consumers to pay their bus and train fares, buy snacks at vending machines and cafes, pay parking fees and also pay for access to sporting facilities (Yoon, 2001). For more than a decade, there have been several attempts to integrate 'smart card technology' into 'mobile devices' to enable mobile payments for business to consumer (B2C) payment transaction processing. In the era of third generation (3G) mobile networks, mobile payments are imminent. Many of the European and Asian countries, including Korea, Singapore and Japan have adopted these technologies (Pousttchi et al., 2009). In Japan, it is possible to pay for a vending machine snack by simply dialing a number on one's mobile phone and having the amount charged to one's phone bill. In recent times, mobile phones are increasingly used to make purchases by flashing the mobile phone in front of a scanner at 'manned' or 'unmanned' points of sale (POS).

For the purpose of this chapter, a mobile payment is defined as a type of payment transaction processing in which the payer uses mobile communication techniques in conjunction with mobile devices for initiation, authorization and confirmation of an exchange of financial value in return for goods and services (Pousttchi, 2008; Flattraaker, 2008). Au and Kauffman (2008, pp.141) suggest that "mobile payment is a type of electronic payment transaction in which at least the payer employs mobile telephony device for the realization of payment". Though cards have a degree of mobility they require technology external to the card to function; this is not the case with mobile wireless based systems. Two forms of mobile payments are available: the mobile credit card and the mobile wallet. A mobile wallet is in essence a smart card application stored in a mobile device that functions in a similar manner to debit cards and has bank accounts and security authentication tools (Flatraaker, 2008). On the other hand, a mobile credit card (using the mobile handset) functions as a credit card and permits online purchasing (Dahlberg et al., 2006). In summary, Consumers can use many forms of payment in mobile commerce, including contactless payment mode for in-person transactions through a mobile phone as well as contactless credit and debit cards; premium-rate telephone numbers, which charges to the consumer's telephone bill; mobile-operator billing allows charges to be added to the consumer's mobile telephone bill, credit cards and debit cards, such as credit cards stored in a phone's SIM card; Google Wallet and stored credit card or debit card information in the cloud.

Given increasing acceptance of mobile payments in lieu of contactless cards and cash (notes and coins) to pay for transactions has led researchers to speculate that the mode of payment used could impact consumers' perceptions and thus purchase behavior. There are studies that support this notion but in the context of cash vs. credit card payment modes (Hirschman, 1979; Feinberg, 1986; Raghubir and Srivastava, 2008), and the effect of debit card use (Thomas, Desai and Seenivasan, 2011). These studies compare the cash mode to a variety of electronic fund transfer system (EFTS) and to do so examine transactions across a variety of situations and use a variety of data gathering modes. The increased spending and payment mode link has been attributed to psychological pain when parting with cash that electronic payment decouples payment and consumption experience at the time of payment (Prelec and Loewenstein, 1998; Raghubir and Srivastava, 2008; Thomas, Desai and Seenivasan, 2011). Why parting with cash causes psychological pain and how decoupling operates to influence spending behaviour requires further explanation. The assumption is that the tangibility or physicality of notes and coins creates conscious or unconscious awareness that stimultaes sensory inputs that something of value is being exchanged. This is, in part, intensified by the consumers' ability to process transactional information using perceptual senses such as sight and touch that translates into an embodied cognition of the amount spent in a single episode of purchase. The physical nature of the tokens affect cognition and emotions that cash has specific sensory qualities in that we can touch and see the value of what is being offered and this is not the case with a card (debit/credit/ contactless card) or indeed any electronic transfer such as mobile wallet. Under an electronic payment condition, consumers may not, at that specific point, be mentally (or emotionally) 'tuned in' to the actual amount of money being spent.

While these studies have provided useful information we think that a logical next step is to examine whether cross-cultural differences manifest when comparing payment modes across social groups, purchase situations and replications. This addresses Khan and Craig-Lees (2009, p.2) earlier calls for cross-cultural comparisons of mobile payments and spending behaviours. Our understanding is that the historical and sensory associations with payment mode effect how we account for our daily expenses and savings in mental accounts. The notion is that people who have used cash-based payment mode over time, have formed a complex psychological relationship with such tokens and that the agreed value of the token becomes imbued in the actual token- i.e. that representative value is physically and viscerally experienced. Further we assume that nature of this psychological relationship affects perceptions and thus judgement when paying for transactions and that these perceptions vary across the payment modes used. There is, therefore, a need for research into perceptions of payment mode cross-culturally.

This research makes several unique and important contributions. First, our findings suggest that historical associations with payment modes produce different cognitive and emotional responses that alter spending behaviour. Specifically, people spend more when they use mobile payment mode and spend less when they use debit cards and this effect is more prominent when they use cash-based payment mode. When paying with a mobile wallet, consumers are required to provide biometric data (or flash their contactless card in a scanner), where it is plausible that at the transaction point, and a 'decoupling effect' occurs (Prelec and Loewenstein, 1989). The case for a cash-based payment mode is different, as to complete a commercial exchange one has to part with the physical cash. The sensory perceptions of payment mode suggest that consumers are more aware of the price of the good if they pay with a cheque or cash than mobile-wallet.

Research Context

The study adopts Hofstede's cultural dimensions framework to compare prevalence of m-payment in New Zealand and France. Hofstede (2011) defines national culture as the effects of a society's culture on the values of its members, and how these values relate to behavior. Hofstede's work established a tradition to compare various culture across national dimensions to compare and identify similarity in national values and social beliefs. Hofstede (2011) added a sixth dimension, indulgence versus self-restraint along with original formulation namely: power distance, individualism vs. collectivism, uncertainty avoidance, masculinity vs. femininity, and long-termism. Hofstede (2011) defines: power distance as the extent to which the less powerful members of organizations and institutions accept and expect that power is distributed unequally. Individualism vs. collectivism as the degree to which people in a society are integrated into groups. Uncertainty avoidance as a society's tolerance for ambiguity. Masculinity is defined as a preference in society for achievement, heroism, assertiveness and material rewards for success. Its counterpart represents a preference for cooperation, modesty, caring for the weak and quality of life. Long-termism associates the connection of the past with the current and future actions/challenges. A lower degree of this index (short-term) indicates that traditions are honored and kept, while steadfastness is valued. Indulgence is defined as a measure of happiness; whether or not simple joys are fulfilled. Indulgent society allows relatively free gratification of basic and natural human desires related to enjoying life and having fun.

The study selected two individualistic cultures: namely, France and New Zealand (see Table 1 for Hofstede's score). The rationale for choosing these countries to study perceptions of mobile payments is based on the fact that individualistic societies tend to accept new technologies faster than collectivist societies. Laukanen (2015), for instance, shows that uncertainty avoidance has a highly significant effect on Internet banking adoption, the strongest influence related to image and risk barriers. Given France scores high on uncertainty avoidance compared to New Zealand, the resistance for adoption of mobile payments may be more likely in France. However, it is also noted that France scores high on individualism and this may be a plausible moderator to acceptance of mobile payment. Kivijärvi and Laukanen (2007) suggest that trust in the Internet banking context has cultural factors, possibly facilitating anticipation of cultural differences in consumer behaviour.

The purpose of this study is to compare and contrast perceptions of mobile payment on spending behaviour in a cross-cultural comparison. There are research that suggest the connections between payment mode effect and purchase behavior, scholars have demonstrated the effect of credit card use compared to cash (Hirschman, 1979; Feinberg, 1986; Raghubir and Srivastava, 2008), and the effect of debit card use (Thomas, Desai and Seenivasan, 2011), researchers have yet to uncover the effect of mobile payment on spending behaviour. There is no research to directly analyse how individuals' perceptions of payment mode influence their choice of payment mode. Therefore, there is a need for research whether cognitive and emotional associations with m-payment mode influences choice of payment mode (cash, card and m-payment), which is the subject of the study described in this paper.

The remainder of the article is structured as follows. First we provide a review of the specific literature about perceptions of payment mode and money management practices to present our research hypotheses. Next, we specify the research method followed and detail the results of our exploratory analysis,

Table 1. Dimensions of national culture (France vs. New Zealand)

	France	NZ	Comments
Power	68	22	Extent to which member of a society expect and accept that power is distributed unequally. New Zealand scores very low on 'Power' (22).
Individualism	71	79	New Zealand, with a score of 79 on this dimension, is an Individualist culture
Masculinity	43	58	New Zealand scores 58 on this dimension and is considered a "Masculine" society. With a score of 43, France has a somewhat Feminine culture
Uncertainty	86	49	At 86, French culture scores high on Uncertainty Avoidance. Structure and planning are required.
Long-Termism	63	33	With a low score of 33 in this dimension, New Zealand is shown to be a normative country. France scores high (63) in this dimension, making it pragmatic
Indulgence	48	75	France scores somewhat in the middle (48) where it concerns Indulgence versus Restraint. New Zealand classified by a high score in Indulgence exhibit a willingness to realise their impulses and desires with regard to enjoying life and having fun.

Source: https://geert-hofstede.com/national-culture.html

followed by the discussion, and practical and managerial implications. Finally, we present a conclusion and the limitations of this study.

Perceptions of Payment Mode

Laukanen (2016) suggests that all innovations meet consumer resistance, and overcoming this opposition must occur prior to product adoption. The case of Internet banking, as used in his study, and mobile payments, as used in this study, share similar characteristics. Hallikainen and Laukanen (2016) examines trust in online banking and payment options where the author argues that trust varies greatly between countries from high vs. low context cultures (China and Finland). Khan et al (2015) argue that perceptions of payment mode not only vary across countries but also within the same cultures. In the same study they demonstrated that perceptions of cash and debit card vary because of the tangibility of the payment mode. One would assume that those cultures where individualism prevails people may have similar values towards spending and saving. Khan (2012) suggests that payment mode affects cognition and emotions and hence behaviour. Therefore, an interesting avenue for research remains to explore whether perceptions of payment mode, for example cash and m-payment mode differs across cultures.

The comparison between France and New Zealand is unique as language differs; both countries ranks high on individualism and are mostly similar in masculinity but power distance, uncertainty avoidance, long-termism and indulgence differs (Hofstede, 2001). Khan (2012) argues that possession of money, irrespective of form, generates positive emotions such as feelings of happiness, whereas spending causes sadness and regret after purchase irrespective of culture. Since paying via cash is an immediate depletion of wealth, an individual sees money outflow and they experience a negative emotion: psychological pain. We argue that possessing cash should encourage saving behavior, since the possessor should try to extend the length of possessing the object that generates positive emotions. We argue that pleasure derived from possessing cash could regulate spending behavior by stimulating an individual emotionally to extend the length time associated with the source of the pleasure. The assumption is that emotional associations with notes and coins as physical representations of monetary value may differ across culture as different cultures value possessions differently. The premise is that physicality of payment mode (cash vs. m-payment) also intensifies cognitive ability to process transactional information using perceptual senses such as sight and touch, translates into an immediate experience. Therefore, we hypothesise that:

H1: The emotional associations with payment modes (cash and m-payment) will differ between France and New Zealand.

Money Management and Mental Spending

Researchers have used mental accounting theory to explain how payment mode influences our money management practices. Thaler (1980) suggests mental accounting as a "set of cognitive operations used by individuals and households to organise, evaluate and keep track of financial activities" (p.40). At the transaction level, people are said to tend to "open" an account (likely pre-existing) mentally for each transaction and base their decisions on evaluation of the perceived benefit of consumption and the associated costs. These mental accounts help reduce the cognitive load on the decision makers. Over time people develop mental filters as a short cut to evaluate financial decision-making. The assumption is that mental frames/filters influence the experience of paying by card and cash. A card-based payment mode (one potential account) decouples the payment from consumption – it removes the transparency so the sense that something of value has been transferred is dulled. As Soman (2001) suggests, cash use assists remembering how much money is being 'spent' and how much remains whereas the use of a card does not. Laukanen (2016) argues that money management practices via Internet banking and mobile banking will differ across high context versus low context countries. Our assumption is that money mamngement practices are culture specific and may even differ across countries that are culturally similar (e.g. on Hofstede's dimension, as per Table 1). A market survey in 2016 identified consumers in France are using mobile phones and devices to shop online, but that mobile payment methods aren't as popular (Pymnts, 2016). ANZ bank internal reporting reveals that more than 50% of ANZ Visa transactions are contactless and that more retailers and consumers are adopting mobile payment technology (Boden, 2016). Khan (2012) suggests that people's cognition and emotions effect the choice of payment mode. So, if perceptions of payment mode (cash as opposed to cards and a mobile payment method) differ across cultures then this may have an influence on the choice of payment mode. Therefore, we hypothesise that:

H2a: The emotional associations that people have with cash based payment mode affect their choice of payment mode.
H2b: The emotional associations that people have with card based payment mode affect their choice of payment mode.
H2c: The emotional associations that people have with m- payment mode affect their choice of payment mode.

METHOD

Sampling and Data Collection

Three hundred and eighty-five participants from New Zealand and 300 from France took part in this study with a 90% response rate. Participants in New Zealand were recruited by invitation in the front entry door of a major grocery retailer (Pack n Save) in Christchurch central business district. A non-probability, criterion-based purposive, sample was used allowing researchers select participants who had experience with the central phenomenon being explored (Hair et al. 2009): namely, .payment modes and their effect on spending across cultures. Prior to beginning the survey, participants were presented with an information sheet guaranteeing their anonymity and explaining the goals of the study. This included definitions of cash, debit and credit cards and mobile payments.

French participants were recruited with a responsive online platform that collected demographic data and responses from the PPM scale (see Khan et al. 2015). As with the New Zealand sample, participants were presented with an information sheet prior to starting the survey guaranteeing anonymity and explaining the goals of the study. Included were definitions of cash, debit and credit cards and m-payments. Within the 23 items of PPM scale were questions on number of financial card ownership, payment mode preferences (e.g. cash, card, m-payment), and asking participants to recall and write the number of transactions per week using cash, debit card and credit card and m-payments.

Data were gathered in March- August, 2013 in New Zealand and February –Jult 2014 France. Sixty-one questionnaires were discarded, as some New Zealand participants dropped out during the data collection process and 5 French questionnaires were incompleted.

Of a total 619 respondents, 53% represented themselves as New Zealander and 47% were French. The majority of French participants were female (70%) and the New Zealand sample was relatively evenly distributed with 47% male and 53% female. The majority of French respondents were students under 25 years of age (70%) whereas the New Zealand sample comprised 30% under the age of 25 years; 20% between 26 years and 34 years; another 20% between the age of 35 years and 44 years and remaining participants were above 45 years of age. The New Zealand sample had 38% with secondary high education; 23% went to polytechnics and 38% had university qualifications whereas 80% of French participants had university qualifications.

Dependent and Independent Measures

In our analysis, self-reporting of preferred payment mode (cash, cards and m-payment) in weekly shopping was recorded as the dependent measure to test hypotheses H2a-c. The independent variables consisted of emotions to cash, card and m-payment, as derived from the PPM Scale which was adopted to suit the mobile payment condition. In this chapter we defined positive emotions to payment mode such as security, confidence, assurance and a relaxed mind state as measures of perceptions of payment mode; this included both cash and m-payments. The notion is that positive associations to payment mode encourage saving behaviour, since the possessor tries to extend the length of possessing the object that generates those positive feelings and emotions. The m-payment mode generates lesser degrees of positive emotional attachments; lack of physicality decouples the positive emotional association and the feeling of ownership, thus leading to increased spending. All items were measured on 5-point Likert type scale, where 1 represented strong disagreement and 5 represented strong agreement.

Survey Instrument and Scale Validation

We assessed the reliability and validity of constructs by confirmatory factor analysis (CFA) using AMOS version 20. Most of the fit indices were within the acceptable range. In addition, we found all the individual factor loadings to be highly significant, giving support to convergent validity (Anderson and Gerbing, 1988). We calculated the Cronbach alpha coefficient, composite factor reliability and average variance extracted for each of the scale and the values were in the desirable range. These findings substantiate discriminant and convergent validities of our scales (See Table 2).

Assessing Method Bias

The common method bias (CMB) is a problem because it is an important source of measurement error. Podsakoff, MacKenzie and Podsakoff (2003) advises to minimise and control for method bias via the

Table 2. Internal consistency measures

	Cron (α)	Mean	S.D.	M-Payment	Cash
M-Payment	.80	3.2	1.0	1	
Cash	.78	3.5	1.2	.68**	1

Note: ** Correlations are significant at the 0.01 level (2-tailed). CASH = Perceptions of cash; M-PAYMENT= Perceptions of M-Payment procedural design of the study and statistical controls. Data on the scale items came from a homogeneous French sample whereas the New Zealand sample was hetrogeneous and more reprentative of its population. Psychological and methodological separations were introduced in the design of the study. A psychological separation was created via the use of a cover story and obscuring the main objective of the study. The response format was manipulated via changes in the ordering of the items in both settings. To minimize social desirability we included an additional "uncertain" column in the five point scale (Worcester and Burns 1975; Garland 1991). Statistical remedies included a Harman's single factor test in the EFA. The unrotated single factor accounted for only 29% of variance. If common method bias is an issue, a single factor should instead account for the majority of the variance. Following Podsakoff et al. (2003) we also used a common latent factor (CLF) method to test for method bias. This latent factor was added to our AMOS CFA model, connecting it to all observed items in the model. Comparing the standardized regression weights from this model to the standardized regression weights of a model without the CLF showed a difference of less than .20, with a range between zero and .030. Therefore, we can assume that method bias is not significant in this analysis (see Appendix 2).

A formal test of response bias was adapted following the procedure suggested by Oppenheim (1966). The t-tests revealed no significant difference between the French (M=2.91, SD=.95) and New Zealand respondents (M=2.97, S.D=.85) in either time period; t (253) =-.516, p=.606, therefore minimizing the risk of response bias.

Data Analysis Techniques

To achieve the stated objective, we asked participants about perceptions of various payment mode adapted from PPM scale. Next we asked a battery of questions in regards participants' preference, attitude, and use of cash, mobile payment and debit and credit card. We were particularly interested about whether respondents' feelings and emotions about payment mode differ across cash, cards and mobile payments. The analysis process follows the "constant comparative technique". Data analysis for perception questions, in addition to descriptive statistics, cross-tabs and validity tests, employed One-Way analysis of variance (ANOVA) and discriminant analysis to test hypotheses 2a-c.

We used discriminant analysis to estimate the relationship between a single non-metric dependent variable (choice of payment mode) and a set of metric independent variables, while controlling for country association. The independent variables were five dimensions associated with PPM scale: these were positive emotions to (1) cash, (2) cards, (3) m-payments, (4) social and personal gratification, and (5) money management. The underpinning assumption is that perceptions of payment mode should discriminate consumers' choice of payment mode.

RESULT AND DISCUSSION

The results indicate that French and New Zealand respondents differ on perceptions of cash-based payment mode. More specifically, New Zealand respondents have less positive associations (Mean=2.9, S.D. =1.1) with cash than the French (Mean=3.2, S.D. =1.0). Of interest, respondents from both countries exhibit negative perceptions towards m-payments with New Zealand (Mean = 2.8, S.D. =1.1) and French (Mean=2.8, S.D. = 1.0) responses being highly comparable. The explanation for our findings centres on the primacy of payment mode used in the two countries. For example, according to Statistics

The Effect of Cultural Values in Mobile Payment Preference

New Zealand "approximately two-thirds of total spending in New Zealand is done electronically and the European Central Bank show that the highest rate of electronic card spending is in Denmark, with 71% of its total transactions made electronically". Another plausible explanation for our finding is that New Zealand scored high worldwide for the use of eftpos bank card transactions where consumers benefit from no charges for use. Given coutries differ in payment mode use and preceptions of payment mode. In Table 3, we have presented our findings on the hypothesis H1: whether perceptions of payment modes (cash and m-payments) will differ between France and New Zealand. The One-way ANOVA between subject groups was chosen as an ananlytical technique. Levene's test was used to assess the homogeneity of variances F= 0.418, p= 0.519, indicating the variance assumption is satisfied. On average, New Zealand participants seemed to display less association with cash than the French. The results from the One-Way ANOVA shows the analytical difference is significant F= 7.3, p<.01; and support for H1. We did find difference in perceptions of mobile payment between New Zealand and France.

Payment Mode Choice

Both simultaneous and stepwise estimation techniques are used for discriminant function estimation. This study used the stepwise procedure. A minimum F value of 1.00 (default value) was used for entry. The results indicated that positive emotions to cash, card and m-payment, entered the model and were a significant discriminator between the choices of payment mode. The discriminant function was highly significant (p<.001) with the three independent variables indicated, as shown in Table 3. Wilks' Lambda suggests the ratio of within-groups sums of squares to the total sum of squares. A lambda of 1.00 means that all variance is explained by factors, other than differences between those means. Here, the lambdas for emotions to card is greater than .50, and emotions to cash and m-payment just missed, thus the group means appear to differ on emotional and cognitive associations with card-based payment mode (60%), Cash (48%) and m-payment (46%). The underlying rationale for low indices for cash and m-payment is that in both France and New Zealand debit and credit cards are prevalent.

The standardized canonical discrimination function coefficient and other values are shown in Table 4, which reports the overall Wilks' lambda .460, x2 (N=619) = 65.63, p<.001. The standardized canonical discriminant function coefficient indicates the strength of the relationship for the predictor variables in the function. For the discriminant function, emotions to card and m-payment have a relatively large positive coefficient and cash has negative coefficient. This suggests emotional association to card and m-payment are the main discriminators in choice of payment mode. This result provides evidence to support hypotheses 2a-c. The loading obtained for choice of payment mode (function1) is shown in Table 4.

Table 3. Summary of two-group stepwise discriminant analysis

Steps	Entered	Wilks' Lambda Value	Sig.
1	Emotions Card	.60	<.001
2	Emotions Cash	.48	<.001
3	Emotions M-Payment	.46	<.001

Table 4. Canonical discriminant functions

Independent Variables	Standardised Canonical Discriminant Function Coefficient					Loadings	
Emotions Card	.758					.747	
Emotions Cash	-.607					.624	
M-Payment	.236					.229	
Function	Eigenvalue	% of Variance	Cumulative %	Canonical Correlation	Wilk's Lambda	Chi-Square	Sig.
1b	1.174	100	100	.735	.460	65.63	<.001

a Pooled within-groups correlation between discriminating variables and standardised canonical discriminant functions

b indicates the first canonical discriminant function used in the analysis. 1= choice of cash, 2= choice of card, 3=choice of m-payment

Classification Matrices and Accuracy

Discriminant analysis uses classification matrices as an alternative to r2, as used in multiple regression analysis to assess predictive accuracy. In this process, each observation must be assessed as to whether it was correctly classified, represented by the hit-ratio (% correctly classified), and is analogous to r2 in a regression analysis. In the current study, a hit-ratio of about 89% of original grouped cases and 85% of cross-validated grouped cases was achieved.. A cross validation allows each case to be classified by the discriminant functions derived from all cases other than that case. The overall percentage of cases correctly classified is 90%, though this value is affected by chance agreement. Accordingly, we computed Kappa, an index that corrects for chance agreements, to assess the accuracy in prediction of group membership. The kappa statistic is .70, p<.001, indicating an excellent accuracy in prediction. Kappa ranges in value from -1 to +1, where a value of 1 for kappa indicates perfect prediction, while a value of 0 indicates chance level prediction.

CONCLUSION, LIMITATIONS, AND FUTURE RESEARCH

Using the PPM scale, this study demonstrated that perceptions of, and emotions towards, payment modes varied between French and New Zealand consumers, though both countries' consumers equally associated m-payments negatively relative to others payment modes. The instrument effectively captured consumers' perceptual and emotional associations to predict preferred payment mode, which has implications for spending behaviours. This advances a previous attempt by Thomas et al. (2010) who tried to link emotions to payment mode using happy-sad face scales and word lists identifying negative associations. It further advances the Zellermayer (1996) study by specifying and validating the positive emotional associations with cash that seem to act as a visceral regulator of saving behaviour: namely, positive emotional associations with cash appear to regulate spending behaviour.

The implications for commerce are to better understand consumers in different cultural contexts and the associations they have towards payment modes. This is particularly important in m-commerce where m-payments are likely to be preferred by businesses though, as in this study, may be least preferred by consumers. In line with this research, retailers and other businesses could use the PPM scale to target consumer groups and gain insights into their associations of preferred payment mode and determine bespoke tactics to enhance perceptions and emotional experiences when using m-payments.

The study is limited in both scope and context. Scope is an issue, because of the sampling process and size. Participants were selected via a non-probability, purposeful sampling processes in both countries obscuring representativeness regardless of sample sizes. Of note, the French participants were mostly under 25 years of age while the New Zealand sample was mainly older. Although we have estimated and found no method and response bias in both country samples, the online platform used in France versus the New Zealand shopping centre recruitement approaches should be noted. Consequently, the stated limitations pose a threat to the generalizability of this study.

An avenue for extending present research is to examine perceptions of prepaid cards (although they represent owned money allocated for a specific purpose) as there may be a different level of awareness that the money is 'spent'. Do perceptions vary across prepaid cards relative to cash, debit card, credit card and m-payment modes and their associated effects on spending? These questions remain unanswered.

Another avenue to explore in future research would be extending the present research in comparison of payment mode perceptions in both individualist and collectivist countries to ascertain implication for m-commerce.

REFERENCES

Anderson, J. C., & Gerbing, D. W. (1988). Structural equation modeling inpractice: A review and recommended two-step approach. *Psychological Bulletin*, *103*(3), 411–423. doi:10.1037/0033-2909.103.3.411

Au, Y., & Kauffman, R. (2008). The economics of mobile payments: Understanding stakeholder issues for an emerging financial technology application. *Electronic Commerce Research and Applications*, *7*(2), 141–164. doi:10.1016/j.elerap.2006.12.004

Baron, M. R., & Kenny, D. A. (1986). The moderator-mediator variable distinction in social psychological research – conceptual, strategic, and statistical considerations. *Journal of Personality and Social Psychology*, *51*(6), 1173–1182. doi:10.1037/0022-3514.51.6.1173 PMID:3806354

Belk, W. R., & Wallendorf, M. (1990). The sacred meanings of money. *Journal of Economic Psychology*, *11*(1), 35–67. doi:10.1016/0167-4870(90)90046-C

Boden, R. (2016). *Apple pay launches in New Zealand*. Retrieved September 13, 2016, from http://www.nfcworld.com/2016/10/13/347807/apple-pay-launches-new-zealand/

Cosseboom, L. (2014). *Asia is dominating the mCommerce market, puts US and Europe to shame*. Retrieved September 13, 2016, from https://www.techinasia.com/digi-capital-reports-asia-as-dominant-player-mcommerce-industry-insights

Criteo. (2015). *State of mobile commerce apps and cross-device lead mobile business*. Retrieved September 13, 2016, from http://www.criteo.com/media/2501/criteo-state-of-mobile-commerce-report-q2-2015-ppt.pdf

Dahlberg, T., Mallat, N., Ondrus, J., & Zmijewska, A. (2006). *M-payment market and research-past, present and future*. Helsinki Mobility Roundtable, Helsinki, Finland.

Feinberg, R. A. (1986). Credit cards as spending facilitating stimuli: A conditioning interpretation. *The Journal of Consumer Research*, *13*(12), 348–356. doi:10.1086/209074

Flatraaker, D. (2008). Mobile, Internet and electronic payments: The key to unlocking the full potential of the internal payments market. *Journal of Payments Strategy &Systems*, *3*(1), 60–70.

Garland, R. (1991). The Mid-Point on a Rating Scale: Is it Desirable? *Marketing Bulletin*, *2*(3), 66–70.

Hair, J. F., Black, W. C., Babin, B. J., Anderson, R. E., & Tatham, R. L. (2009). *Multivariate Data Analysis* (7th ed.). Upper Saddle River, NJ: Pearson Education Inc.

Hallikainen, H., & Laukkanen, T. (2016). Consumer Trust towards an Online Vendor in High-vs. Low-Context Cultures. In *2016 49th Hawaii International Conference on System Sciences (HICSS)* (pp. 3536-3545). IEEE.

Herzberg, A. (2003). Payments and banking with mobile personal devices. *Communications of the ACM*, *46*(5), 53–58. doi:10.1145/769800.769801

Hirschman, E. C. (1979). Differences in Consumer Purchase Behavior by Credit Card Payment System. *The Journal of Consumer Research*, *6*(6), 58–66. doi:10.1086/208748

Hofstede, G. (2011). Dimensionalising Cultures: The Hofstede Model in Context. *Online Readings in Psychology and Culture*, *2*(1). doi:10.9707/2307-0919.1014

Hughes, N., & Lonie, S. (2010). M-PESA: Mobile money for the "unbanked". *Innovations: Technology, Governance, Globalization*, *2*(1–2), 63–81.

Jeremiah, D. (2014). *Frost and Sullivan: New Zealand's mobile commerce market to reach nearly $10 billion by 2019*. Retrieved September 13, 2016, from http://ww2.frost.com/news/press-releases/frost-sullivan-new-zealands-mobile-commerce-market-reach-nearly-10-billion-2019/

Khan, J., Belk, R. W., & Craig-Lees, M. (2015). Measuring consumer perceptions of payment mode. *Journal of Economic Psychology*, *47*, 34–49. doi:10.1016/j.joep.2015.01.006

Khan, J., Belk, W. R., & Craig-Lees, M. (2012). *Cash and Cards: Perceptions of Payment Mode*. Paper presented at Australian & New Zealand Marketing Academy Conference (ANZMAC) the Ehrenberg-Bass Institute, University of South Australia.

Khan, J., & Craig-Lees, M. (2009). ''Cashless'' transactions: Perceptions of money in mobile payment. *International Business and Economics Review*, *1*(1), 23–32.

Kivijärvi, M., Laukkanen, T., & Cruz, P. (2007). Consumer trust in electronic service consumption: A cross-cultural comparison between Finland and Portugal. *Journal of Euromarketing*, *16*(3), 51–65. doi:10.1300/J037v16n03_05

Laukkanen, T. (2015). How uncertainty avoidance affects innovation resistance in mobile banking: The moderating role of age and gender. In *System Sciences (HICSS), 2015 48th Hawaii International Conference on* (pp. 3601-3610). IEEE.

Laukkanen, T. (2016). Consumer adoption versus rejection decisions in seemingly similar service innovations: The case of the Internet and mobile banking. *Journal of Business Research*, *69*(7), 2432–2439. doi:10.1016/j.jbusres.2016.01.013

Mishra, H., Mishra, A., & Nayakankuppam, D. (2006). Money: A bias for the whole. *The Journal of Consumer Research*, *32*(4), 541–549. doi:10.1086/500484

Niranjanamurthy, M., Kavyashree, N., Jagannath, S., & Chahar, D. (2013). Analysis of e-commerce and m-commerce: Advantages, limitations and security issues. *International Journal of Advanced Research in Computer and Communication Engineering*, *2*(6).

Podsakoff, P. M., MacKenzie, S. B., Lee, J. Y., & Podsakoff, N. P. (2003). Common method biases in behavioral research: A critical review of the literature and recommended remedies. *The Journal of Applied Psychology*, *88*(5), 879–903. doi:10.1037/0021-9010.88.5.879 PMID:14516251

Pousttchi, K., Schiessler, M., & Wiedemann, D. (2009). Proposing a comprehensive framework for analysis and engineering of mobile payment business models. *Information Systems and E-Business Management*, *7*(3), 363–393. doi:10.1007/s10257-008-0098-9

Preacher, J., Derek, K., Rucker, D., & Hayes, A. F. (2007). Addressing moderated mediation hypotheses: Theory, methods, and prescriptions. *Multivariate Behavioral Research*, *42*(1), 185–227. doi:10.1080/00273170701341316 PMID:26821081

Prelec & Loewenstein. (1998). The red and the black: mental accounting of savings and debt. *Marketing Science*, *17*, 4–28.

Prelec, D., & Simester, D. (2001). Always leave home without it: A further investigation of the credit-card Effect on willingness to pay. *Marketing Letters*, *12*(1), 5–12. doi:10.1023/A:1008196717017

Pymnts. (2016). *Consumers in France aren't using mobile devices for payment*. Retrieved September 8, 2016, from http://www.pymnts.com/news/mobile-payments/2016/france-consumers-mobile-devices-payment/

Raghubir, P., & Srivastava, J. (2008). Monopoly money: The effect of payment coupling and form on spending behavior. *Journal of Experimental Psychology. Applied*, *14*(3), 213–225. doi:10.1037/1076-898X.14.3.213 PMID:18808275

Saylor, M. (2012). The Mobile Wave: How Mobile Intelligence Will Change Everything. Perseus Books.

Soman, D. (2001). Effects of payment mechanism on spending behavior: The role of rehearsal and immediacy of payments. *The Journal of Consumer Research*, *27*(March), 460–474. doi:10.1086/319621

Soman, D. (2003). The effect of payment transparency on consumption: Quasi-experiments from the Field. *Marketing Letters*, *14*(3), 5–12. doi:10.1023/A:1027444717586

Thaler, R. (1980). Towards a positive theory of consumer choice. *Journal of Economic Behavior & Organization*, *1*(1), 39–60. doi:10.1016/0167-2681(80)90051-7

Thomas, M., Desai, K. K., & Seenavasin, S. (2011). How credit card payments increase unhealthy food purchases: Visceral regulation of vices. *The Journal of Consumer Research*, *38*(June), 126–139. doi:10.1086/657331

Worcester, R. M., & Burns, T. R. (1975). A statistical examination of the relative precision of verbal scales. *Journal of the Market Research Society*, *17*(3), 181–197.

Yoon, S. (2001). Tangled octopus. *Far Eastern Economic Review*, *164*(6), 40.

Zellermayer, O. (1996). *The pain of paying* (Unpublished Ph.D. Dissertation). Carnegie Mellon University, Pittsburgh, PA.

KEY TERMS AND DEFINITIONS

Cash: Means paper currency and coins that are liquid and can be exchanged anywhere anytime.

Credit Card: A credit account will have a Visa or Master card platform- and in some instances so will a line of credit. Credit card links to any form of credit.

Debit Card: Money in savings or in cheque account not borrowed money in any form i.e., loan of cash, credit (via credit card or line of credit) or overdraft. To understand the difference between a card that allows you access to your own money whilst allowing to also access a 'credit' amount that is available via a credit account or line of credit. A debit card can be used in a retail setting and you gain access via the use of a pin number.

Emotional Association With Payment Mode: The emotional associations with cash and coins as physical representations of monetary value is based on the premise that physicality of payment mode (cash) intensifies cognitive ability and stimulates emotions to process transactional information using perceptual senses such as sight and touch, translates into an immediate experience.

Hofstede Cultural Dimension: A theory grounded on four basic tenents that may explain peoples' behaviour globally. These are how different societies cope with inequalities, uncertainty and the relationship one has with their primary group and values inherent to male and female chacteristics. Later Hofstede added two more dimensions such as long-termism and indulgence to explain differences in cultural and individual behaviour within a society.

Mental Account: Thaler (1980) suggests mental accounting as a "set of cognitive operations used by individuals and households to organise, evaluate and keep track of financial activities" (p.40). The physicality of payment mode influence how people organsie and evaluate purchase and process cost benefit of purchase.

Mobile Commerce: Mobile commerce (m-commerce) incorporates electronic commerce via mobile wireless devices: that is, to deliver electronic commerce capabilities directly into the consumer's hand, anywhere, anytime. The notion of mobile commerce is embedded with the concept of 'online shopping', information in consumer's hand 24/7.

Mobile-Payment (M-Payment): It is payment services performed from or via a mobile device. In other words, instead of paying with cash, cheque, or credit cards, a consumer can use a mobile phone to pay for a wide range of product or services and digital goods. Mobile payment also referred to as mobile money, mobile money transfer, and mobile wallet. For the purpose of this chapter, a mobile payment is defined as a type of payment transaction processing in which the payer uses mobile communication techniques in conjunction with mobile devices for initiation, authorization and confirmation of an exchange of financial value in return for goods and services.

Payment Mode: Payment Mode: Payment is made, such as cash, check, or debit, credit or pre-paid card.

The Effect of Cultural Values in Mobile Payment Preference

APPENDIX: ASSESSING COMMON METHOD BIAS

Table 5.

Standardized Regression Weights: (Without CLF)				Standardized Regression Weights: (with CLF)				
			Estimate				Estimate	Difference
q8	<---	b	0.554	q8	<---	mm	0.554	0
q10	<---	b	0.624	q10	<---	mm	0.624	0
q12	<---	b	0.787	q12	<---	mm	0.787	0
q15	<---	c	0.74	q15	<---	spent	0.74	0
q7	<---	c	0.721	q7	<---	spent	0.721	0
q25	<---	c	0.81	q25	<---	spent	0.81	0
q31	<---	d	0.872	q33	<---	m-pay	0.84	0.032
q32	<---	d	0.927	q32	<---	m-pay	0.927	0
q33	<---	d	0.84	q31	<---	m-pay	0.872	-0.032
q28	<---	e	0.796	q28	<---	cash	0.797	-0.001
q20	<---	e	0.859	q20	<---	cash	0.858	0.001
q23	<---	e	0.775	q23	<---	cash	0.775	0

Chapter 15
Consumer Behavior in M-Commerce:
Literature Review and Research Agenda

Saïd Aboubaker Ettis
Gulf College – Muscat, Oman

Afef Ben Zin El Abidine
ISET Kairouan, Tunisia

ABSTRACT

Our work has a two-fold general objectives: on the one hand, we wish to describe the mobile commerce environment and, on the other hand, to establish the determinants of the mobile consumer behavior. To achieve our objectives, first, we describe the concept of mobile commerce, constraints and benefits. Second, we study the determinants of mobile commerce adoption. Third, we focus on the determinants of mobile consumer satisfaction and loyalty. Finally, we summarize future avenues of investigation in a research agenda.

INTRODUCTION

In 2014, Worldwide m-commerce revenues approximate the 184 billion U.S. dollars and are projected to reach 669 billion U.S. dollars in 2018 (Statista, 2016). M-commerce is now over 30% of all US e-commerce. Japan and the United Kingdom are now selling more via mobile than through desktop (Criteo, 2016). Growth in the consumer use of mobile devices to shop has made m-commerce an attractive area for research. Though a lot of research is done on m-commerce, the topic is still rising potential issues for further research. This paper reviews the literature on m-commerce and proposes a framework which analyses the progress in mobile commerce research and provides future research directions.

Our work has a two-fold general objective: on the one hand, we wish to describe the mobile commerce environment and, on the other hand, to establish the determinants of the mobile consumer behavior. To achieve our objective, the first section describes the concept of mobile commerce (m-commerce), con-

DOI: 10.4018/978-1-5225-2469-4.ch015

straints and benefits. Likewise, the second section studies the determinants of m-commerce adoption. The third section focuses on the determinants of mobile consumer satisfaction and loyalty. Finally, the fourth section contains a research agenda.

MOBILE COMMERCE: IS IT AN EXTENSION OF E-COMMERCE?

The name "M-Commerce" arises from the mobile nature of the wireless environment that supports mobile electronic business transactions. Devices, such as smartphones, Personal Digital Assistants (PDAs), pagers, notebooks, and even automobiles, can already access the Internet wirelessly and utilize its various capabilities, such as e-mail and Web browsing, (Coursaris and Hassanein, 2002). Previous studies have defined m-commerce as an extension of e-commerce (Niranjanamurthy et al., 2013; Ngai and Gunasekaran, 2007; Varshney and Vetter, 2002; Wei et al., 2009). In these works, the m-commerce is comparable to e-commerce, except the transactions in m-commerce are conducted wirelessly using a mobile device (Lee and Benbasat, 2004). Hence, m-commerce refers to commerce activities via mobile devices. According to Coursaris and Hassanein (2002), differences between the m-commerce and the e-commerce rely on the mode of communication, the types of Internet access devices, the development languages, communication protocols, and the enabling technologies used to support each environment.

On the other hand, Feng, Hoegler, and Stucky (2006) claimed that m-commerce is much more than merely being an extension of e-commerce. According to them, considering m-commerce as an extension of e-commerce is too reductionist as it is exclusively based on the medium and device. The authors stated that m-commerce has different interactions with users, usage patterns, and value chain, thus offering new business models that are not available to e-commerce (e.g. location-based marketing). In the same vein, for Paavilainen (2002), "mobile e-commerce" is a deceptive term because the business models and the value chain are not the same as e-commerce. M-commerce has distinctive attributes that provide consumers with values unavailable in conventional wired e-commerce, including ubiquity, personalization, flexibility, and dissemination (Chong, 2013; Mahatanankoon et al., 2005; Siau et al., 2001). Furthermore, in contrast to e-commerce, m-commerce has "revolutionized consumers' self-perceptions by empowering them to voice their beliefs and preferences continuously and instantaneously" (Khansa et al., 2012, p. 45). M-commerce extends not only the benefits of the web but also allows for unique services and additional benefits when compared to traditional e-commerce applications (Mahatanankoon et al., 2005). M-commerce is "any transaction, involving the transfer of ownership or rights to use goods and services, which is initiated and/or completed by using mobiles access to wireless networks with the help of mobile devices" (Chan and Chong, 2013, p.443).

MOBILE COMMERCE: BENEFITS AND CONSTRAINTS

Benefits for Users

For users, the benefits delivered by mobile technologies might be characterized as "anytime and anywhere computing" (For Au and Kauffman, 2008). The ubiquity, temporal and spatial dimensions of mobility allow users access to information, communication, and services anywhere and anytime (Kim et al., 2010a). The m-commerce offers the opportunities for broadband internet access without geographic locations constraints, personalized services, and location-based services (Chong et al., 2011; Chong, 2013). The use of wireless device enables the user to receive information and conduct transactions anywhere and anytime (Niranjanamurthy et al., 2013). Users spend more time on the mobile phone than a computer. So, consumers might conveniently search for products and services using mobile phones. For instance, they might check train schedule, timing, and book on their mobile phones even at the last moment. Even more, a customer can check the web price of a product while shopping in a supermarket. Despite these benefits, consumers' actual m-commerce usage activities have remained low. The mobile devices are still used for only entertaining activities such as listening to music, social media interaction, and content browsing (Chong, 2013).

Benefits for Organizations

For organizations, the distinctive benefits of m-commerce are offered by the two most important attributes including mobility and reachability, which provide m-commerce with advantages over e-commerce (Kim et al., 2010a). Mobility implies that users can carry smartphones or other mobile devices to buy regardless of time and place. Mobility and reachability of the mobile devices make it possible for organizations to contacted consumers anytime and anywhere. Companies could offer services and goods based on the current location of the customers and provide up-to-date information about products, sales promotions, stores, e-couponing opportunities, etc. So, the location-based information meets the consumer's requirement. The mobility and reachability features facilitate personalization. Personalization will take the form of Location Based Advertising, a customized message, which is usually dedicated to a specific segment group or user based upon time and location (Singh, 2014). Also, m-commerce allows businesses to deliver unique services and serves as a new business model for businesses (Hew, 2017). For several researchers, the use of users' location information for satisfying consumers' needs under different situations in real-time, is an essential feature of m-commerce (Zou and Huang, 2015; Ha et al., 2012; Faqih and Jaradat, 2015; Krotov et al., 2015).

Mobile Commerce Constraints

However, consumers face a lot of obstacles while using m-commerce. Subsequently, consumer attitudes towards m-commerce change rapidly (Sari and Bayram, 2015). These obstacles may be technological, environmental, or consumer related. As stated by Sari and Bayram (2015), some technological constraints are related to the use of mobile phones. Mostly, mobile phones or tablets have smaller screens, which are difficult to read and to use with only one hand. Also, the usage of these devices is affected by noise level, weather, and brightness. Often, users are flustered by the difficulties such as the lack of ease of use and poor interface quality. There are other major obstacles for users using mobile devices, such as the wireless network connection speed, low display resolutions, and limited processing capability (Keengwe, 2014).

The environmental obstacles are related to the lack of standards, uncompleted infrastructure, cost and speed issues (Sari and Bayram, 2015). Security is another serious issue in m-commerce (Ghosh & Swaminatha, 2001). The risk of losing the device and data, unauthorized access, identity theft, and the chance of losing money through mobile payment are seen as the main obstacle to consumer adoption of M-commerce.

The consumer-related variables may be demographic, psychographic, social, and cultural (Sari and Bayram, 2015; Kalliny and Minor, 2006). Factors such as gender, age, region, qualification, income, and profession are seen as critical factors in consumer's perception of m-commerce (Kanwalvir and Aggarwal, 2013). Other major consumer-related concerns are low perceived relative advantage, complexity, the intrusiveness of unwanted advertising messages, and consumer privacy issues (Carter, 2008). Various research has considered consumer characteristics influencing the adoption process of m-commerce such as consumer innovativeness, consumer knowledge of technological innovations, consumer past adoption of technologies, and social influences (Leong et al., 2011; Yang, 2005; Lu et al., 2005).

AN OVERVIEW OF MOBILE CONSUMER BEHAVIOR DETERMINANTS

Our literature review of mobile consumer behavior revealed that most of the existing research studied determinants of consumer adoption of m-commerce. But these studies are fragmented, context specific (depending on countries, applications, devices, etc.), and showing often contradictory results. M-commerce adoption is still an emerging issue of research and results duplication and generalization are further needed. In this section, we draw from existing literature to present well-known factors that influence consumer behavior adoption of m-commerce. This is followed by the development of research propositions.

Prominence of Technological Variables as Determinants of M-Commerce Adoption

Previous m-commerce adoption investigations have frequently used the recognized Technology Acceptance Model (TAM, Davis, 1989), Theory of Planned Behavior (TPB, Ajzen, 1991), Innovation Diffusion Theory (IDT, Rogers, 1995), and Unified Theory of Acceptance and Use of Technology (UTAUT, Venkatesh et al., 2003). In these works, several variables are studied.

Perceived Usefulness: Performance Expectancy

Perceived usefulness is defined as the extent to which a person believes that using the technology will enhance his or her job performance (Davis, 1989). Note that in UTAUT, performance expectancy is similar to the TAM's perceived usefulness. Performance expectancy is defined as the degree to which a user believes that using the system will help him or her to attain gains in job performance (Venkatesh et al., 2003). Perceived usefulness was found to be the most important factor for predicting behavior intention of mobile device users who have not adopted mobile commerce yet (Khalifa and Shen, 2008a). Perceived usefulness was found to have a positive effect on the individual's intention to adopt m-payment (Kim et al., 2010a; Schierz et al., 2010) and m-banking (Riquelme and Rios, 2010). Yu (2012) empirically concluded that consumer intention to adopt m-banking was significantly influenced by performance expectancy. Perceived usefulness positively influences Singapore consumers' attitude toward using m-commerce (Yang, 2005) and positively related to the perceived value of mobile shopping for fashion products among Korean consumers (Ko et al., 2009). Park et al. (2007) and Chong (2013) found performance expectancy as a key determinant of the adoption of m-commerce in China. Other researchers, found that perceived usefulness has no effect on adoption of m-payment (Daştan and Gürler, 2016) and on intention to use m-banking (Akturan and Tezcan, 2012). In view of these inconsistent findings, in this research we consider that a significant positive effect is more plausible than negative effect, therefore:

P1: Perceived usefulness (performance expectancy) positively influences the adoption of m-commerce.

Perceived Ease: Effort Expectancy

Perceived ease of use is defined as the extent which a person believes that using the technology will be free of effort (Davis, 1989). Like perceived ease of use in TAM, effort expectancy in UTAUT refers to the degree of ease associated with the use of a system (Venkatesh et al., 2003). Ease of use plays an important role in forming a favorable attitude to mobile technology adoption (Bruner & Kumar, 2005). Perceived ease of use was found to have a positive effect on the individual's intention to adopt m-payment (Li et al., 2014; Kim et al., 2010a). Park et al. (2007) found performance expectancy as important factors that influenced the adoption of m-commerce in China. Likewise, ease of use was significantly and positively related to perceived value of mobile shopping for fashion products among Korean consumers (Ko et al., 2009). However, Khalifa and Shen (2008a) did not find a significant effect of perceived ease of use on the intention to adopt m-commerce. Similarly, in Singaporean context, Yang (2005) found that the relationship between perceived ease of use and attitude toward using m-commerce was not significant. Yu (2012) empirically concluded that consumer intention to adopt m-banking was not significantly influenced by effort expectancy. So, like usefulness, the literature contains inconsistent findings of the impact of ease of use on m-commerce adoption, depending on products or context. In this research, we postulate a significant positive effect, thus:

P2: Perceived ease of use (effort expectancy) positively influences the adoption of m-commerce.

Subjective Norms: Social Influence

Subjective norms refer to perceived pressures to perform a behavior, according to what others say or do is important (Fishbein and Ajzen, 1975). Venkatesh et al. (2003) used social influence to represent the subjective norm in the Theory of Planned Behavior and Theory of Reasoned Actions. They defined social influence as the degree to which an individual perceives that important others believe he/she should use the technology (Venkatesh et al. 2003). Prior studies argue for the importance of social influence as a determinant of the adoption of m-commerce (Lu, 2014; Gitau and Nzuki, 2014; Teo and Pok, 2003; Kurnia et al., 2006). Empirical examinations indicated aqzézz 2012; Kim et al., 2011; Sadia, 2011; Shin et al., 2010; Wei et al., 2009; Khalifa and Shen, 2008a). Chan and Chong (2013) empirically revealed that individual decisions to adopt entertainment-related m-commerce activities were influenced by friends, family members, and peers. Similarly, empirical evidence indicated that subject norm was a significant factor for individual intention to adopt m-banking for Singaporean (Riquelme and Rios, 2010), Malaysian (Amin et al., 2008), Thai (Sripalawat et al., 2011), Brazilian (Puschel et al., 2010), and Taiwanese consumers (Yu, 2012). Hence, the following proposition is posited:

P3: Subjective norms (social influence) positively influences the adoption of m-commerce.

Perceived Behavioral Control

In extensions of the Theory of Reasoned Actions, the Theory of Planned Behavior, defines perceived behavioral control as "beliefs regarding access to the resources and opportunities needed to perform a behavior, or alternatively, to the intention and external factors that may impede the performance of the behavior" (Ajzen, 1985, p.34). A number of researchers have highlighted the importance of perceived behavioral control by demonstrating its influence on adoption of m-commerce. Mishra (2014) conducted that perceived behavioral control have a positive effect on Indian consumers' intention to adopt m-commerce. The empirical results of Khalifa and Shen (2008a) indicated an important role of self-efficacy, as an integral component of perceived behavioral control, in influencing intentions to adopt m-commerce among Hong Kong cellular phone users who had not adopted m-commerce yet. In the same context, according to Khalifa and Shen (2008b), the effect of perceived behavioral control, on the other hand, is found to be insignificant. Others investigated m-commerce for a specific application. In Brazilian context, Puschel et al. (2010) found that perceived behavioral control, significantly impacted intention to use mobile banking. Kim et al. (2010b) showed that perceived behavioral control positively influences behavioral intention of current users to adopt multimedia messaging services in South Korea. Consistent with prior studies, we propose that:

P4: Perceived behavioral control positively influences the adoption of m-commerce.

Facilitating Conditions

Both businesses and consumers who engage in m-commerce need legal and regulatory protections and infrastructure to support business transactions (Kim et al., 2011). M-commerce cannot occur if its environment prevents it or if the facilitating conditions make the behavior difficult (Kim et al., 2011). Venkatesh et al. (2003) defined facilitating conditions as the degree to which an individual believes that an organizational and technical infrastructure exists to support technology use. The empirical research investigating the relationship between facilitating conditions and adoption of m-commerce are scarce. The existent works postulate for a significant effect of the facilitating conditions on intention to adopt m-commerce. Based on neural network analyses, Chong (2013) found that facilitating conditions are an important predictor of m-commerce adoption in the Chinese market. Yu (2012) illustrated that facilitating conditions significantly affect Taiwanese behavior of using mobile banking. Consequently, the following proposition is recommended:

P5: Facilitating conditions positively influence the adoption of m-commerce.

Compatibility

Compatibility is originally derived from the Innovation Diffusion Theory by Rogers (1995). Compatibility is the degree to which an innovation is consistent with existing values, beliefs, needs, and experiences of the potential adopters (Rogers, 1995). Greater compatibility allows innovation to be interpreted in a more familiar context (Ilie et al., 2005). Wu and Wang (2005) indicate that compatibility influence favorable attitudes toward m-commerce. Lin (2010) provides further evidence that compatibility influence attitudes and behavioral intention to adopt mobile devices for m-business purposes in Taiwan. Similarly, an empirical study by Tanakinjal et al. (2010) suggest that compatibility, determine user's intentions to adopt mobile marketing in Malaysia. Chung and Holdsworth (2012) demonstrate that compatibility determined behavioral intent to adopt m-commerce among the Y Generation in Kazakhstan, Morocco, and Singapore. In m-banking context, Koennig-Lewis et al. (2010) uncovered that compatibility affected consumer intention to adopt mobile banking in Germany. Puschel et al. (2010) found that compatibility significantly affects attitude toward mobile banking adoption in Brazil. Moreover, in Taiwan, customer perceptions about the compatibility of mobile banking with their values, experiences, and needs appear to have a positive effect on attitude toward adopting mobile banking (Lin, 2011). Schierz et al. (2010) showed that compatibility has a significant effect on adoption of Mobile Payment Systems. However, in the work of Kim et al. (2010a), the compatibility of m-payment was not the primary reason for consumer's decision to adopt it. Kim et al. (2010a) found that compatibility has an insignificant effect on user's intention to use m-payment through perceived ease of use and perceived usefulness. Corresponding to all these researchers, the following proposition is suggested:

P6: Perceived compatibility positively influences the adoption of m-commerce.

Consumer Traits as Determinants of M-Commerce Adoption: Rarely Studied Variables

As stated previously, most of the existing research in the area of consumer adoption of m-commerce, examined the role of technological factors. However, little research has examined the effect of consumer traits on m-commerce adoption. Consumer traits have been found to be an important determinant of consumer behavior. In addition, compared to e-commerce it has been shown that consumer traits play a crucial role in enhancing e-commerce adoption and use (Dennis et al., 2009). The most common traits predicting m-commerce adoption will be depicted here.

Demographics

The most commonly examined demographic variables are gender, education, income, and age. In m-banking context, some research indicated demographic user profile of m-banking. The m-banking users were relatively young (Joshua and Koshy, 2011), aged 50 or over (Suoranta and Mattila, 2004), or aged between 30 and 49 (Laukkanen and Pasanen, 2008). Age and gender influence mobile banking adoption, while education, occupation, household income, and size of the household are not significant differentiating variables (Laukkanen and Pasanen, 2008). M-banking users were predominantly males (Puschel et al.,2010; Koennig-Lewis et al., 2010; Joshua and Koshy, 2011). In the same context, it was discovered that gender significantly moderated the effects of performance expectancy and perceived financial cost on behavioral intention, and the age moderated the effects of facilitating conditions and perceived self-efficacy on actual adoption behavior of m-banking in Taiwan (Yu, 2012). The influence of social norm on intention to adopt m-banking in Singapore was stronger among women than among men (Riquelme and Rios, 2010). In m-shopping context, studies have suggested that younger users tend to adopt m-commerce more than older users (Dai and Palvia, 2008). Younger users engage in m-commerce delivery and entertainment activities more often than older users. Educational level has positive significant relationships with both transactions and location-based services m-commerce activities. Gender has no significant relationship with any of these m-commerce usage activities (Chan and Chong, 2013). It was also established that gender and age to have a causal relationship with the adoption of m-commerce (Huitt, 2003; Yang and Lee, 2010). Thus,

Consumer Behavior in M-Commerce

P7: Demographic variables influence the adoption of m-commerce.
P8: Demographic variables moderate the effects of technological variables (such as usefulness, ease of use, perceived behavioral control…) on the adoption of m-commerce.

Innovativeness

Innovativeness refers to the willingness of an individual to try out any new technology (Agarwal & Prasad, 1998). Innovativeness is conceptualized as a trait. Innovative individuals are communicative, curious, dynamic, venturesome, and stimulation–seeking (Kim et al., 2010a). Consumers who are more innovative should be more confident in their beliefs about m-commerce. Malik et al. (2013) and Yang (2005) found out that there is a relationship between innovativeness and the adoption of m-commerce. Innovativeness has a direct and positive influence on the intention to engage in m-shopping (Aldàs-Manzano et al., 2009). Innovativeness is, therefore, an antecedent of m-commerce adoption. Besides et al. (2010a) demonstrate that the effect of perceived usefulness on the intention to use m-payment is more important for early adopters than for late adopters. The effect of perceived ease of use on the intention to use m-payment is more important for late adopters than for early adopters. This means that innovativeness might be considered as a moderating variable.

P9: Innovativeness positively influences the adoption of m-commerce.
P10: Innovativeness moderates the effects of technological variables (such as usefulness, ease of use, perceived behavioral control…) on the adoption of m-commerce.

Self-Efficacy

Self-efficacy as a personal trait refers to the perceived capabilities of an individual to mobilize the motivation, cognitive resources, and courses of action needed to meet given situational demanding (Mort and Drennan, 2007). Self-efficacy, as applied to m-commerce, refers to consumers' judgment of their own capabilities to purchase products using mobile devices (Malik et al., 2013). The relationship between self-efficacy and a system's usefulness, ease of use, and enjoyment has been examined in previous studies (Lu and Su, 2009; Islam et al. 2011; Yang, 2010; Huang and Liaw, 2005). Therefore, self-efficacy may impact the m-commerce adoption. Therefore, we expect self-efficacy to play a significant role in consumer's intention to use m-commerce.

P11: Self-efficacy positively influences the adoption of m-commerce.
P12: Self-efficacy moderates the effects of technological variables (such as usefulness, ease of use, perceived behavioral control…) on the adoption of m-commerce.

Involvement

Involvement as a personal trait is defined as a person's perceived relevance of the object based on inherent needs, values, and interests (Zaichkowsky, 1985). In the same vein, individual involvement in mobile purchasing was defined by San Martín-Gutiérrez et al. (2012) as a stable and long-lasting state of personal relevance and interest in the (mobile) channel used for the purchase. Involvement with mobile devices and/or m-commerce have been considered as a factor affecting the use of mobile devices for transactions (Malik et al., 2013; Mort and Drennan, 2007). Malik et al. (2013) have shown that there is a significantly positive relationship between involvement with a mobile phone and consumer's intention to adopt m-commerce. Further work is needed seeing the limited number of researchers in this area. As the other consumer traits variables, we expect involvement to play a significant role in consumer's intention to use m-commerce.

P13: Involvement positively influences the adoption of m-commerce.
P14: Involvement moderates the effects of technological variables (such as usefulness, ease of use, perceived behavioral control…) on the adoption of m-commerce.

Definitely, there are several other consumer traits that might have the same effects on the adoption of m-commerce but in line with space limitations, we summarize the main traits as cultural differences, personality traits, need for cognition, optimal stimulation level, risk aversion, familiarity with m-commerce, and sensitivity to personal information protection.

DETERMINANTS OF SATISFACTION TOWARD MOBILE VENDOR

Mobile user satisfaction can be defined as "a summary affective response of varying intensity that follows mobile commerce activities, and is stimulated by several focal aspects, such as information quality, system quality, and service quality" (Wang and Liao, 2007, p. 384). It implies confirming expectations and experiencing a positive affective state grounded on the result of preserving the relation in the case of m-commerce (Yeh and Li, 2009).

Vendor website quality, mobile technology quality, and vendor quality are considered as the main antecedents of m-satisfaction (Yeh and Li, 2009). Some researchers have also investigated the effect of personal variables such as impulsiveness, involvement and socio-demographic on the satisfaction of mobile shoppers.

Vendor Website Quality

Previous studies have shown that interactivity, customization, content quality and appearance are the determinants of m-satisfaction related to vendor website quality (Yeh and Li, 2009; Choi et al. 2008; Wang and Liao, 2007). Each variable will be presented below.

Interactivity

Using information such as the users' identity and location, the mobile service provider can provide contextually and timely services at the point of need. Lee (2005) advocated that the main characteristics of interactivity are ubiquitous connectivity and contextual offers and defined it as the continuation of mobile commerce activities regardless of users' time and location. He suggested that interactivity in m-commerce environment should be regarded as a crucial factor for customer relationship building instead of being regarded as pure media features. According to Yeh and Li (2009), customer satisfaction was significantly influenced by interactivity of mobile website as showed by a survey conducted on a sample of students which m-commerce experience was on SMS (30%), Ringtone/MP3 download (27.4%), Wallpaper/ photo download (25%), MMS (5.6%) and E-mail/news (12%). We, therefore, propose that:

P15: Mobile website interactivity will positively influence mobile consumer satisfaction.

Customization

Mobile devices are usually more personal and individualized than stationary devices because people rarely share mobile phones in the same way that they share desktop computers (Chae and Kim, 2003). Consequently, the mobile device transmits its user's identity, an essential condition for customization (Lee and Benbasat, 2004). Lee and Benbasat (2004, p. 82) defined customization as "the site's ability to tailor itself (tailoring) or to be tailored by each user". Venkatesh et al. (2003) further suggested that customization's impact can be extended to enhance the mobile interface design and to improve mobile usability, thus raising the level of satisfaction. Thus:

P16: Mobile website customization will positively influence mobile consumer satisfaction.

Content Quality

Content quality refers to the usefulness, accuracy, availability and updating of the information of the mobile website (Wang and Liao, 2007). "In e-commerce, given the users' static environment, only the currency of content is considered. In contrast, m-commerce makes it possible to reflect the users' dynamic environment into its content by virtue of context-aware applications" (Lee and Benbasat, 2004, p.95). Choi et al. (2008) found that content quality, as an exclusive factor to m-commerce, is the degree to which a person believes that the mobile portal will give him products or service of good quality and would be trustable to order them influence customer satisfaction significantly. He suggested that the following difficulties would make content quality more important for m-commerce than e-commerce. First, seeing its limited screen size, it is hard for a mobile device to show enough pre-information about content in text form and difficulty in controlling the device. Second, customers are not willing to use pre-listening service due to the additional costs. Third, all the content from various providers cannot be individually verified. Last, it is difficult to obtain directly notes after using the services in m-commerce because of the immaturity of the system for reviewing experienced services. Therefore:

P17: Mobile website content quality will positively influence mobile consumer satisfaction.

Mobile Technology Quality

Previous studies have showed that usefulness and ease-of-use, the two vital elements in the Technology Acceptance Model (Davis, 1989), have a special status in m-commerce environment as they affect the intention to adopt m-commerce (Pelet et papadopoulou, 2014; Chong, 2013; Kim et al. 2010a; Manzano et al. 2009; Mao et al. 2005) as well as mobile consumer satisfaction (Hung et al., 2012; Lu, 2014).

Usefulness

While m-commerce has many advantages such as allowing the consumer to shop anytime and anywhere, an important question is whether consumers perceive m-commerce as sufficiently useful. Users will only appreciate m-commerce if they consider m-commerce's offerings being more useful than its alternative such as e-commerce (Chong, 2013). Previous works show that usefulness may positively affect mobile banking users' satisfaction (Lin et al., 2014) and mobile shopper's satisfaction (Hung et al., 2012). However, Yeh and Li (2009) found that usefulness doesn't contribute to the satisfaction if the main usage of mobile phone is entertainment. For this kind of m-commerce activity, younger people don't perceive usefulness as a dominant variable as it was showed in the study of Chan and Chong (2013). Thus:

P18: Usefulness of m-commerce technology will positively influence mobile consumer satisfaction.

Ease of Use

Today using mobile devices is familiar. However, m-commerce applications may be unacquainted to many consumers. As shown previously, some studies found that perceived ease of use has a significant relationship with all m-commerce activities' adoption. Nevertheless, other investigations argued that perceived ease of use doesn't contribute to the intention to use m-commerce. One explanation for this is that users are now somewhat familiar with m-commerce devices and applications; hence they do not find that m-commerce is difficult to use. Another explanation relies on the indirect effect of ease of use on behavioral intention through perceived usefulness as showed by Wu and Wang (2005). Yet, ease of use is considered as the main antecedent of mobile consumer satisfaction (Yeh and Li, 2009). Using a sample of 116 respondents which were spread across 13 popular m-commerce categories, including send/receive emails (18.1%), listen to/ download music (15.5%), download graphics/animation (12.9%), online shopping (9.5%), read and receive news (8.6%), play online games (7.8%), stock trading (6.0%), book travel tickets (5.2%), online fortune telling (4.3%), friend finding (4.3%), buy books (3.4%), take part in Internet auctions (2.6%), and routine bank services (1.7%), Wang and Liao (2007) argued that ease of use is a dimension of user satisfaction with m-commerce systems. Therefore:

P19: Ease of use of m-commerce technology will positively influence mobile consumer satisfaction.

Mobile Vendor Quality

Service quality and perceived value were found to be important factors affecting satisfaction. Each factor will be presented below.

Service Quality

Service quality, as a dimension of m-consumer satisfaction, was measured by the goodness of after-sales services, payment procedures and FAQ services (Wang and Liao, 2007). Shin and Kim (2008) stated that service quality is a consumer's overall impression of the relative efficiency of the service provider, and they found that it is significantly related to customer satisfaction. In a mobile telecommunications services study, service quality measured by call quality, value-added services, and customer support create customer satisfaction for mobile subscribers (Kim et al., 2004). Moreover, in the mobile context, the results of a study conducted by Deng et al. (2010) through a sample of 622 users of the mobile instant message in China indicate that perceived service quality is found to have the greatest effect on customer satisfaction. According to Choi et al., (2008), transaction process which is related to the degree to which a person can operate a systematically efficient and clear order process lead to mobile customer satisfaction. Indeed, they suggested that m-commerce burdens customers with both aspects of cost and time more than e-commerce because additional charges such as connection fee per unit time are required and connection speed is much slower than e-commerce; that is, slower speed brings about more additional charges, which makes customers perceive greater dissatisfaction. Therefore:

P20: Service quality of m-vendor will positively influence mobile consumer satisfaction.

Perceived Value

Perceived value is the evaluation of the benefits of a product or a service by customers based on their advance sacrifices and perceived performance when they use mobile services (Kuo et al., 2009). A study on a sample of 622 users of the mobile instant message in China shows that functional value and emotional value have a significant effect on customer satisfaction, while social value and monetary value are found to have no significant effects (Deng et al., 2010). Furthermore, it was found that perceived value positively influenced customer satisfaction in the context of value-added services (ringtone, multimedia message service, picture download, music download and auto answering message) (Kuo et al., 2009) and in the context of popular m-commerce categories (e.g sending/receiving emails, booking cinema/ theatre tickets, restaurant table reservations) (Lin and Wang, 2006). This leads to:

P21: Perceived value of m-vendor will positively influence mobile consumer satisfaction.

Personal Variables

In previous studies, it was found that impulsiveness, involvement, and socio-demographic characteristics were found to have an impact on mobile shopper satisfaction.

Impulsiveness

A smartphone user can purchase products and services easily and immediately, which make this personal device an ideal vehicle for impulsive purchases (San Martin and Lopez Catalan, 2013). Therefore, it seems realistic that impulsiveness may play a more important role in m-commerce than in e-commerce (San Martin and Lopez Catalan, 2013). Previous works have examined impulsiveness in online shopping (Wells et al., 2011; Zhang et al., 2006). Nevertheless, the effect of impulsiveness in m-commerce remains relatively unexplored (Wilska, 2003). According to San Martin and Lopez Catalan (2013) impulsiveness, which evidences a failure to reflect on the decision process, has a negative impact on a subsequent evaluation of the purchase in terms of satisfaction. They suggested that, once the product is purchased, impulse buyers may re-think about their purchase decision. Probably, impulsive purchases cause a cognitive dissonance and that is why this kind of purchases would tend to be regarded as inappropriate in terms of satisfaction (San Martin and Lopez Catalan, 2013). Thus:

P22: Impulsiveness will negatively influence mobile consumer satisfaction.

Involvement

San Martín-Gutiérrez et al. (2012) found that perceived risk was an inhibitor of involvement in m-purchasing while marketing permission (consumer consent to provide the requesting firm personal data) and personal innovativeness with information technology behave as facilitators of involvement. According to San Martin and Lopez Catalan (2013) involvement with the m-purchase describes a state of personal interest in the mobile phone as a buying-selling medium and predisposes individuals positively towards expectation fulfillment, a positive affective state subsequent to the purchase, and a positive purchase reaction. Hence, highly-involved shoppers are motivated to enhance the quality of their shopping experience and this may lead to a high level of satisfaction (Oliver and Bearden, 1983). Therefore:

P23: Involvement will positively influence mobile consumer satisfaction.

Socio-Demographic Variables

The literature considers personal traits as variables that moderate the relation between a stimulus and a response (San Martín-Gutiérrez et al., 2012). Thus, Goldsmith and Flynn (2004) found that neither age nor income but gender is related to online purchasing of clothes. Yang (2005) affirms that the relation between the perceived utility of purchases of the mobile telephone and ease of use is stronger for men than for women. Likewise, in the context of mobile purchasing, perceived risk acts as an inhibitor of the emergence of involvement in the case of women and shows no significant effect for men (San Martín-Gutiérrez et al., 2012). The results of a study conducted by Deng et al. (2010) show also that gender and age have significant moderating effects on the relationship between trust and mobile customer satisfaction and on the relationship between emotional value and mobile customer satisfaction. Trust is a more important factor for females in obtaining satisfaction with a mobile instant message in China while user enjoyment and interesting experience are stronger drivers of satisfaction for males (Deng et al., 2010). Age also significantly moderates the effects of trust and emotional value on customer satisfaction (Deng et al., 2010). That is, the effect of trust on customer satisfaction is more important for older users than younger ones. This leads to:

P24: The relationship between mobile consumer satisfaction and its determinants will be moderated by socio-demographic variables.

DETERMINANTS OF LOYALTY TOWARD MOBILE VENDOR

The literature indicates that trust, satisfaction, and switching cost are the determinants of loyalty in the context of m-commerce. Loyalty in mobile context has been defined according to the behavioral approach as customers' intention to continually use the mobile service with their present service providers, as well as their inclinations to recommend this service to other persons (Deng et al., 2010). However, following Lin and Wang (2006), loyalty was defined as the customer's favorable attitude toward an m-commerce website, resulting in repeat purchasing behavior. This definition emphasizes the two, behavioral and attitudinal, aspects of loyalty.

Trust

The idea of conducting commerce and sharing personal information over wireless, hand-held devices might be uncomfortable for consumers (Varnali and Toker, 2010). It is also considered that the mobile network is more vulnerable to attacks than a standard physical network or even a web-based network (Davis et al., 2011). According to the literature in m-commerce, there are two different phases of trust: pre-use trust, which belongs to the pre-purchase stage, and post-use trust, which belongs to the post-purchase stage. Pre-use trust means trust before the use of a technology or a service and may impact the intention and behavior of consumer's purchasing decisions; conversely, post-use trust means trust after using the technology and is enhanced by customer satisfaction and which in turn influences future usage, including repurchase decisions (Lin et al., 2014; Siau and Shen, 2003).

Prior studies in m-commerce have viewed trust as either trusting intentions which reflect the willingness of a consumer to take risks in order to fulfill his or her needs (Lin et al., 2014) or trusting beliefs (Lin and Wang, 2006). Trust can be defined as consumers' perceptions of particular attributes of m-vendors, including the ability, integrity and benevolence of the vendors when handling the consumers' transactions (Lin and Wang, 2006; Lin, 2011; Davis et al., 2011). Various factors may influence the process of engendering mobile trust: mobile vendor characteristics such as familiarity, reputation, information quality, third party recognition and attractive rewards, along with mobile technology characteristics such as feasibility, all play a role in the initial trust formation (Siau and Shen, 2003). In the context of mobile advertising, it was empirically found that the reputation of the vendor, disposition to trust and structural assurance have the highest impact on consumer trust, followed by perceived ease of use, third party assurance and perceived privacy (Davis et al., 2011). Li and Yeh (2010) found empirical support for the positive influence of design aesthetics, customization, perceived usefulness and ease of use on the trust toward the mobile website.

In the context of mobile banking, Lin (2011) found that mobile trust through perceived competence and integrity has a significant effect on attitude. If customers believe the mobile banking firm is able to develop effective service delivery strategies and provide adequate protection from fraud and violation of privacy, then adoption, as well as continue-to-use intentions, will increase. The results of a study conducted by Hung et al. (2012) about 244 users of the mobile phone who experience with m-shopping vary between 1 to 4 years show that the repurchasing activities of m-shoppers are primarily dependent on their trust followed by their satisfaction with the purchased products or services. Moreover, results of an empirical investigation across popular m-commerce categories show trust has a positive influence on the mobile shopper loyalty (Lin and Wang, 2006). Thus:

P25: Trust will positively influence mobile consumer loyalty.

Switching Cost

The switching cost refers to the obstacles of switching to another industry when the customer is dissatisfied (Fornell, 1992). In the context of mobile telecommunication services, "switching cost was defined as loss cost, adaptation cost, and move-in cost. Loss cost refers to the perception of loss of social status or performance, when cancelling a service contract with an existing carrier; adaptation cost refers to the perceived cost of adaptation, such as search cost and learning cost; and move-in cost refers to the economic cost involved in switching to a new carrier, such as the purchase of a new device and the subscriber fee" (Kim et al., 2004, p. 149).

Previous studies indicated that switching cost was a significant factor in predicting customer loyalty (Aydin et al. 2005). Regarding the use of mobile instant message service, a high switching cost for changing to a new provider leads to a higher customer loyalty (Deng et al., 2010). Therefore:

P26: Switching cost will positively influence mobile consumer loyalty.

Satisfaction

Customer satisfaction and loyalty in mobile context has rarely been studied (Varnali and Toker, 2010). Satisfied customers are likely to remain with their existing providers and maintain their subscription (Kim et al., 2004). Moreover, it was found that increasing the degree of customer satisfaction through improved service quality and customer value is an effective tool to maintain customer loyalty (Deng et al., 2010). Chae et al. (2002) found evidence suggesting that the dimensions of information quality (connection quality, content quality, interaction quality and contextual quality) have a significant impact on mobile user satisfaction, which, in turn, was shown to be related to customer loyalty. Likewise, Choi et al., (2008) verified empirically that two factors (Content reliability and transaction process) had an effect on mobile user satisfaction which in turn has significant effects on customer loyalty and customer complaint. Therefore:

P27: Satisfaction will positively influence mobile consumer loyalty.

RESEARCH AGENDA

There is a substantial body of literature examining m-consumer behavior. Our analysis of the literature reveals that many factors influence m-shopping adoption and m-consumer behavior.

This paper attempts to present the crucial factors explaining m-consumer behavior and formulate several research propositions. Future studies should strive not only to gain a deeper understanding of consumer's adoption of m-commerce but also to discover relationships between m-commerce use antecedents and consumer behavior. We summarize the main research topics as following:

- Considerable research in the marketing and IS area show that perceived customer value is an important factor in users' adoption of technologies. However, surprisingly, little work is done in the m-commerce field. Perceived value is a customer's overall assessment of the utility of a product based on perceptions of what is received (benefits) and what is given (costs) (Zeithaml, 1988). Consumer's acceptance of m-commerce should be affected by the balance between costs and benefits of the transaction. The important function of perceived value as mediating the effect of m-commerce attributes on the consumers' intention to adopt m-commerce need to examined in future research.

- TAM and subsequent models are taking a cognitive approach ignoring the experiential aspects of using technologies. Nevertheless, if we are to increase adoption of m-commerce, attention must also be turned to the emotional factors related to the experience of m-shopping. Researchers have to focus on the influence of variables such as emotions, hedonic motivations, enjoyment, and flow.
- Gen Y, born after 1980, is the first generation to grow up with technology and the Internet. Further understanding of how friendly Gen Y is with m-commerce, how m-marketing can increase value for these consumers, how m-marketing can enhance their satisfaction and loyalty is important.
- There have been limited studies on the direct and moderation relationships between consumer traits and m-commerce adoption and between these traits and m-consumer behavior. Examining these subjects will provide new insights into whether there are correlations between personal traits of consumers and their m-commerce usage activities.
- Although previous research found significant relationships between the aesthetic beauty of websites and consumer attitude and behavioral intentions in e-commerce, works examining the relationship between the design of the mobile user interface and consumer satisfaction or loyalty are rare (Varnali and Toker, 2010). Cyr et al. (2006) found that design aesthetics of a mobile service have a significant impact on perceived usefulness, ease of use, and enjoyment, which influence consumers' loyalty intentions. In contrast, Magura (2003) found that mobile website design is as of little concern for m-commerce users. Certainly, further research on the influence of design and aesthetics aspects of the mobile user interface on behavioral intentions and attitude would be of great value.

CONCLUSION

The study has aimed to analysis the m-commerce literature and to depict relevant avenues of research. Two big m-commerce streams of research are identified. The first tries to discover the antecedents of consumer adoption of m-commerce, and the second attempts to understand the factors influencing post-adoption behavior such as satisfaction, trust, and loyalty. The review has identified in each section some significant gaps, and some research propositions have been offered that will help research to close these gaps. Future avenues of investigation were also exposed in a research agenda. Topics related to m-commerce were identified: perceived customer value, emotional factors, Gen Y, consumer traits, design aesthetics, and m-commerce interfaces. Whereas, there are much more issues than actually mentioned. M-commerce is still in its infancy and therefore much work is very needed.

REFERENCES

Agarwal, R., & Prasad, J. (1998). A conceptual and operational definition of personal innovativeness in the domain of information technology. *Information Systems Research*, *9*(2), 204–215. doi:10.1287/isre.9.2.204

Ajzen, I. (1985). From Intention to Actions: A Theory of Planned Behavior. In J. Kuhl & J. Beckman (Eds.), *Action Control: From Cognition to Behavior* (pp. 11–39). New York: Springer-Verlag. doi:10.1007/978-3-642-69746-3_2

Ajzen, I. (1991). The theory of planned behavior. *Organizational Behavior and Human Decision Processes, 50*(2), 179–211. doi:10.1016/0749-5978(91)90020-T

Akturan, U., & Tezcan, N. (2012). Mobile Banking Adoption of the Youth Market: Perceptions and Intentions. *Marketing Intelligence & Planning, 30*(4), 4–4. doi:10.1108/02634501211231928

Aldas-Manzano, J., Ruiz-Mafé, C., & Sanz-Blas, S. (2009). Exploring individual personality factors as drivers of M-shopping acceptance. *Industrial Management & Data Systems, 109*(6), 739–757. doi:10.1108/02635570910968018

Amin, H., Hamid, M. R. A., Lada, S., & Anis, Z. (2008). The adoption of mobile banking in Malaysia: The case of Bank Islam Malaysia Berhad. *International Journal of Business and Society, 9*(2), 43–53.

Au, Y. A., & Kauffman, R. J. (2008). The economics of mobile payments: Understanding stakeholder issues for an emerging financial technology application. *Electronic Commerce Research and Applications, 7*(2), 141–164. doi:10.1016/j.elerap.2006.12.004

Aydin, S., Özer, G., & Arasil, Ö. (2005). Customer loyalty and the effect of switching costs as a moderator variable: A case in the Turkish mobile phone market. *Marketing Intelligence & Planning, 23*(1), 89–103. doi:10.1108/02634500510577492

Bruner, G. C. II, & Kumar, A. (2005). Explaining consumer acceptance of handheld Internet Devices. *Journal of Business Research, 58*(5), 553–558. doi:10.1016/j.jbusres.2003.08.002

Carter, E. (2008). Mobile marketing and generation Y African–American mobile consumers: The issues and opportunities. *International Journal of Mobile Marketing, 3*(1), 62–66.

Chae, M., & Kim, J. (2003). Whats So Different About the Mobile Internet? *Communications of the ACM, 46*(12), 240–247. doi:10.1145/953460.953506

Chae, M., Kim, J., Kim, H., & Ryu, H. (2002). Information Quality for Mobile Internet Services: A Theoretical Model with Empirical Validation. *Electronic Markets, 12*(1), 38–46. doi:10.1080/101967802753433254

Chan, F. T. S., & Chong, A. Y. L. (2013). Analysis of the determinants of consumers m-commerce usage activities. *Online Information Review, 37*(3), 443–461. doi:10.1108/OIR-01-2012-0012

Choi, J., Seol, H., Lee, S., Cho, H., & Park, Y. (2008). Customer satisfaction factors of mobile commerce in Korea. *Internet Research, 183*(3), 313–335. doi:10.1108/10662240810883335

Chong, A. Y. L. (2013). A two-staged SEM-neural network approach for understanding and predicting the determinants of m-commerce adoption. *Expert Systems with Applications, 40*(4), 1240–1247. doi:10.1016/j.eswa.2012.08.067

Chong, A. Y. L., Chan, F. T. S., & Ooi, K. B. (2012). Predicting consumer decisions to adopt mobile commerce: Cross country empirical examination between China and Malaysia. *Decision Support Systems, 53*(1), 34–43. doi:10.1016/j.dss.2011.12.001

Chong, J. L., Chong, A. Y. L., Ooi, K. B., & Lin, B. (2011). An empirical analysis of the adoption of m-learning in Malaysia. *International Journal of Mobile Communications, 9*(1), 1–18. doi:10.1504/IJMC.2011.037952

Chung, K. C., & Holdsworth, D. (2012). Culture and behavioural intent to adopt mobile commerce among the Y Generation: Comparative analyses between Kazakhstan, Morocco and Singapore. *Young Consumers. Insight and Ideas for Responsible Marketers*, *13*(3), 3. doi:10.1108/17473611211261629

Coursaris, C., & Hassanein, K. (2002). Understanding m-Commerce: A Consumer-Centric Model. *Quarterly Journal of Electronic Commerce*, *3*(3), 247–271.

Criteo. (2016). *The State of Mobile Commerce (2016).* Retrieved from http://www.criteo.com/resources/mobile-commerce-report

Cyr, D., Head, M., & Ivanov, A. (2006). Design aesthetics leading to m-loyalty in mobile commerce. *Information & Management*, *43*(8), 950–963. doi:10.1016/j.im.2006.08.009

Dai, H., & Palvia, P. (2008). Factors affecting mobile commerce adoption: A cross-cultural study in China and the United States. *The Data Base for Advances in Information Systems*, *40*(4), 43–61. doi:10.1145/1644953.1644958

Daştan, İ., & Gürler, C. (2016). Factors Affecting the Adoption of Mobile Payment Systems: An Empirical Analysis. *Emerging Markets Journal*, *6*, 1. Retrieved from http://emaj.pitt.edu

Davis, F. D. (1989). Perceived usefulness, perceived ease of use, and user acceptance of information technology. *Management Information Systems Quarterly*, *13*(3), 319–339. doi:10.2307/249008

Davis, R., Sajtos, L., & Chaudhri, A. A. (2011). Do Consumers Trust Mobile Service Advertising? *Contemporary Management Research*, *7*(4), 245–270. doi:10.7903/cmr.9696

Deng, Z., Lu, Y., Wei, K. K., & Zhang, J. (2010). Understanding customer satisfaction and loyalty: An empirical study of mobile instant messages in China. *International Journal of Information Management*, *30*(4), 289–300. doi:10.1016/j.ijinfomgt.2009.10.001

Dennis, C., Merrilees, B., Jayawardhena, C., & Wright, L. (2009). E-consumer behavior. *European Journal of Marketing*, *43*(9/10), 1121–1139. doi:10.1108/03090560910976393

Faqih, K. M. S., & Jaradat, M. I. R. M. (2015). Assessing the moderating effect of gender differences and individualism-collectivism at individual-level on the adoption of mobile commerce technology: TAM3 perspective. *Journal of Retailing and Consumer Services*, *22*, 37–52. doi:10.1016/j.jretconser.2014.09.006

Feng, H., Hoegler, T., & Stucky, W. (2006). *Exploring the Critical Success Factors for Mobile Commerce*. Paper presented at the International Conference on Mobile Business, Copenhagen, Denmark. doi:10.1109/ICMB.2006.15

Fishbein, M., & Ajzen, I. (1975). *Belief, Attitude, Intentions and Behavior: An Introduction to Theory and Research*. Boston, MA: Addison-Wesley.

Fornell, C. (1992). A national customer satisfaction barometer: The Swedish experience. *Journal of Marketing*, *56*(1), 1, 6–21. doi:10.2307/1252129

Ghosh, A. K., & Swaminatha, T. M. (2001). Software security and privacy risks in mobile e-commerce. *Communications of the ACM*, *44*(2), 51–57. doi:10.1145/359205.359227

Gitau, L., & Nzuki, D. (2014). Analysis of Determinants of M-Commerce Adoption by Online Consumers. *International Journal of Business Human Technology*, *4*, 3.

Goldsmith, R. E., & Flynn, L. R. (2004). Psychological and behavioural drivers of online clothing purchase. *Journal of Fashion Marketing and Management*, *8*(1), 84–95. doi:10.1108/13612020410518718

Ha, K. H., Canedoli, A., Baur, A. W., & Bick, M. (2012). Mobile banking-insights on its increasing relevance and most common drivers of adoption. *Electronic Markets*, 1–11.

Hew, J. J. (2017). Hall of fame for mobile commerce and its applications: A bibliometric evaluation of a decade and a half 2000–2015. *Telematics and Informatics*, *34*(1), 43–66. doi:10.1016/j.tele.2016.04.003

Huang, H. M., & Liaw, S. S. (2005). Exploring users attitudes and intentions toward the web as a survey tool. *Computers in Human Behavior*, *21*(5), 729–743. doi:10.1016/j.chb.2004.02.020

Huitt, W. (2003). The affective system: Educational Psychology Interactive. Valdosta, GA: Valdosta State University. Retrieved from http://www.edpsycinteractive.org/topics/affect/affsys.html

Hung, M. C., Yang, S. T., & Hsieh, T. C. (2012). An examination of the determinants of mobile shopping continuance. *International Journal of Electronic Business Management*, *10*(1), 29–37.

Ilie, V., van Slyke, C., Green, G., & Lou, H. (2005). Gender differences in perceptions and use communication technologies: A diffusion of innovation approach. *Information Resources Management Journal*, *18*(3), 13–31. doi:10.4018/irmj.2005070102

Islam, M. A., Khan, M. A., Ramayah, T., & Hossain, M. M. (2011). The Adoption of Mobile Commerce Service among Employed Mobile Phone Users in Bangladesh: Self-efficacy as A Moderator. *International Business Research*, *4*(2), 80–89. doi:10.5539/ibr.v4n2p80

Joshua, A. J., & Koshy, M. P. (2011). Usage patterns of electronic banking services by urban educated customers: Glimpses from India. *Journal of Internet Banking and Commerce*, *16*(1), 1–12.

Kalliny, M., & Minor, M. (2006). The antecedents of m-commerce adoption. *Journal of Strategic E-Commerce*, *4*(1/2), 81–98.

Kanwalvir, K., & Aggarwal, H. (2013). Critical Factors in Consumers Perception towards Mobile Commerce in E-Governance Implementation: An Indian Perspective. *International Journal of Engineering and Advanced Technology*, *2*, 3.

Keengwe, J. (2014). *Promoting Active Learning through the Integration of Mobile and Ubiquitous Technologies*. IGI Global.

Khalifa, M., & Shen, K. N. (2008a). Drivers for Transactional B2C M-Commerce Adoption: Extended Theory of Planned Behavior. *Journal of Computer Information Systems*, *48*(3), 111–117.

Khalifa, M., & Shen, K. N. (2008b). Explaining the adoption of transactional B2C mobile commerce. *Journal of Enterprise Information Management*, *21*(2), 110–124. doi:10.1108/17410390810851372

Khansa, L., Zobel, C. W., & Goicochea, G. (2012). Creating a taxonomy for mobile commerce innovations using social network and cluster analyses. *International Journal of Electronic Commerce*, *16*(4), 19–52. doi:10.2753/JEC1086-4415160402

Kim, C., Mirusmonov, M., & Lee, I. (2010a). An empirical examination of factors influencing the intention to use mobile payment. *Computers in Human Behavior, 26*(3), 310–322. doi:10.1016/j.chb.2009.10.013

Kim, C., Tao, W., Shin, N., & Kim, K. (2010b). An empirical study of customers perceptions of security and trust in e-payment systems. *Electronic Commerce Research and Applications, 9*(1), 84–95. doi:10.1016/j.elerap.2009.04.014

Kim, K. K., Shin, H. K., & Kim, B. (2011). The role of psychological traits and social factors in using new mobile communication services. *Electronic Commerce Research and Applications, 10*(4), 408–417. doi:10.1016/j.elerap.2010.11.004

Kim, M. K., Park, M. C., & Jeong, D. H. (2004). The effects of customer satisfaction and switching barrier on customer loyalty in Korean mobile telecommunication services. *Telecommunications Policy, 28*(2), 145–159. doi:10.1016/j.telpol.2003.12.003

Ko, E., Kim, E. Y., & Lee, E. K. (2009). Modeling consumer adoption of mobile shopping for fashion products in Korea. *Psychology and Marketing, 26*(7), 669–687. doi:10.1002/mar.20294

Koening-Lewis, N., Palmer, A., & Moll, A. (2010). Predicting young consumers take up of mobile banking services. *International Journal of Bank Marketing, 28*(5), 410–432. doi:10.1108/02652321011064917

Krotov, V., Junglas, I., & Steel, D. (2015). The mobile agility framework: an exploratory study of mobile technology enhancing organizational agility. *Journal of Theory and Applications in Electronic Commerce Research*, 10.

Kuo, Y. F., Wu, C. M., & Deng, W. J. (2009). The relationships among service quality, perceived value, customer satisfaction, and post-purchase intention in mobile value-added services. *Computers in Human Behavior, 25*(4), 887–896. doi:10.1016/j.chb.2009.03.003

Kurnia, S., Smith, S. P., & Lee, H. (2006). Consumers' perception of mobile internet in Australia. *e-Business Review (Federal Reserve Bank of Philadelphia), 5*(1), 19–32.

Laukkonen, T., & Passanen, M. (2008). Mobile banking innovations and early adopters: How they differ from another online user? *Journal of Financial Services Marketing, 23*(2), 86–94. doi:10.1057/palgrave.fsm.4760077

Lee, T. M. (2005). The impact of perceptions of interactivity on customer trust and transaction intentions in mobile commerce. *Journal of Electronic Commerce Research, 6*(3), 165–180.

Lee, Y. E., & Benbasat, I. (2004). A Framework for the Study of Customer Interface Design for Mobile Commerce. *International Journal of Electronic Commerce, 8*(3), 79–102.

Leong, L. Y., Ooi, K. B., Chong, A. Y. L., & Lin, B. (2011). Influence of individual characteristics, perceived usefulness and perceived ease of use on mobile entertainment adoption in Malaysia – an SEM approach. *International Journal of Mobile Communications, 9*(44), 359–382. doi:10.1504/IJMC.2011.041141

Li, H., Liu, Y., & Heikkilä, J. (2014). Understanding the factors driving nfc-enabled mobile payment adoption: an empirical investigation. *PACIS Proceedings*. Retrieved from http://aisel.aisnet.org/pacis2014/231

Li, Y. M., & Yeh, Y. S. (2010). Increasing trust in mobile commerce through design aesthetics. *Computers in Human Behavior, 26*(4), 673–684. doi:10.1016/j.chb.2010.01.004

Lin, H. F. (2010). An empirical investigation of mobile banking adoption: The effect of innovation attributes and knowledge-based trust. *International Journal of Information Management, 30*(6), 33–45.

Lin, H. F. (2011). An empirical investigation of mobile banking adoption: The effect of innovation attributes and knowledge-based trust. *International Journal of Information Management, 31*(3), 252–260. doi:10.1016/j.ijinfomgt.2010.07.006

Lin, H. H., & Wang, Y. S. (2006). An examination of the determinants of customer loyalty in mobile commerce contexts. *Information & Management, 43*(3), 271–282. doi:10.1016/j.im.2005.08.001

Lin, J., Wang, B., Wang, N., & Lu, Y. (2014). Understanding the evolution of consumer trust in mobile commerce: A longitudinal study. *Information Technology Management, 15*(1), 37–49. doi:10.1007/s10799-013-0172-y

Lu, H. P., & Su, Y. J. (2009). Factors affecting purchase intention on mobile shopping web sites. *Internet Research, 19*(4), 442–458. doi:10.1108/10662240910981399

Lu, J. (2014). Are personal innovativeness and social influence critical to continue with mobile commerce? *Internet Research, 24*(2), 134–159. doi:10.1108/IntR-05-2012-0100

Lu, J., Yao, J. E., & Yu, C. S. (2005). Personal innovativeness, social influences and adoption of wireless internet services via mobile technology. *The Journal of Strategic Information Systems, 14*(3), 245–268. doi:10.1016/j.jsis.2005.07.003

Magura, B. (2003). What hooks m-commerce customers? *MIT Sloan Management Review, 44*(3), 9–16.

Mahatanankoon, P., Wen, H. J., & Lim, B. (2005). Consumer-based m-commerce: Exploring consumer perception of mobile applications. *Computer Standards & Interfaces, 27*(4), 347–357. doi:10.1016/j.csi.2004.10.003

Malik, A., Kumra, R., & Srivastava, V. (2013). Determinants of Consumer Acceptance of M-Commerce. *South Asian Journal of Management, 20*(2), 102–126.

Manzano, J. A., Mafe, C. R., & Blas, S. S. (2009). Exploring individual personality factors as drivers of M-shopping acceptance. *Industrial Management & Data Systems, 109*(6), 739–757. doi:10.1108/02635570910968018

Mao, E., Srite, M., Thatcher, J. B., & Yaprak, O. (2005). A Research Model for Mobile Phone Service Behaviors: Empirical Validation in the U.S. and Turkey. *Journal of Global Information Technology Management, 8*(4), 4, 7–28. doi:10.1080/1097198X.2005.10856406

Mishra, S. (2014). Adoption of M-commerce in India: Applying Theory of Planned Behaviour Model. *Journal of Internet Banking and Commerce, 19*, 1. Retrieved from http://www.arraydev.com/commerce/jibc/

Mort, G. S., & Drennan, J. (2007). Mobile communications: A study of factors influencing consumer use of m-services. *Journal of Advertising Research, 47*(3), 302–312. doi:10.2501/S0021849907070328

Ngai, E. W. T., & Gunasekaran, A. (2007). A review for mobile commerce research and applications. *Decision Support Systems*, *43*, 1, 3–15. doi:10.1016/j.dss.2005.05.002

Niranjanamurthy, M., Kavyashree, N., Jagannath, M. S., & Chahar, D. (2013). Analysis of e-commerce and m-commerce: Advantages, limitations and security issues. *International Journal of Advanced Research in Computer and Communication Engineering*, *2*(6), 2360–2370.

Oliver, R. L., & Bearden, W. O. (1983). The role of involvement in satisfaction processes. In R. P. Bagozzi & A. M. Tybout (Eds.), *Advances in Consumer Research* (pp. 250–255). Ann Arbor, MI: Association for Consumer Research.

Paavilainen, J. (2002). *Mobile business strategies: understanding the technologies and opportunities*. Addison-Wesley.

Park, J. K., Yang, S. J., & Lehto, X. (2007). Adoption of mobile technologies for Chinese consumers. *Journal of Electronic Commerce Research*, *8*(3), 196–206.

Pelet, J. É., & Papadopoulou, P. (2014). Consumer Behavior in the Mobile Environment: An Exploratory Study of M-Commerce and Social Media. *International Journal of Technology and Human Interaction*, *10*(4), 36–48. doi:10.4018/ijthi.2014100103

Puschel, J., Mazzon, J. A., & Hernandez, J. M. C. (2010). Mobile banking: Proposition of an integrated adoption intention framework. *International Journal of Bank Marketing*, *28*(5), 389–409. doi:10.1108/02652321011064908

Riquelme, H., & Rios, R. E. (2010). The moderating effect of gender in the adoption of mobile banking. *International Journal of Bank Marketing*, *28*(5), 328–341. doi:10.1108/02652321011064872

Rogers, E. M. (1995). *Diffusion of Innovation*. New York: Free Press.

Sadia, S. (2011). User acceptance decision towards mobile commerce technology: A study of user decision about acceptance of mobile commerce technology. *Interdisciplinary Journal of Contemporary Research in Business*, *2*(12), 535–547.

San Martín-Gutiérrez, S., López-Catalán, B., & Ramón-Jerónimo, M. (2012). Determinants of involvement in mobile commerce. The moderating role of gender. *EsicMarket*, *141*, 69–101.

Sari, A., & Bayram, P. (2015). Challenges of Internal and External Variables of Consumer Behaviour towards Mobile Commerce. *International Journal of Communications Network and System Sciences*, *8*(13), 578–596. doi:10.4236/ijcns.2015.813052

Schierz, P. G., Schilke, O., & Wirtz, B. W. (2010). Understanding consumer acceptance of mobile payment services: An empirical analysis. *Electronic Commerce Research and Applications*, *9*(3), 209–216. doi:10.1016/j.elerap.2009.07.005

Shin, D. H., & Kim, W. Y. (2008). Forecasting customer switching intention in mobile service: An exploratory study of predictive factors in mobile number portability. *Technological Forecasting and Social Change*, *75*(6), 854–874. doi:10.1016/j.techfore.2007.05.001

Shin, Y. M., Lee, S. C., Shin, B., & Lee, H. G. (2010). Examining influencing factors of post-adoption usage of mobile internet: Focus on the user perception of supplier-side attributes. *Information Systems Frontiers, 12*(5), 595–606. doi:10.1007/s10796-009-9184-x

Siau, K., Lim, E. P., & Shen, Z. (2001). Mobile commerce: Promises, challenges, and research agenda. *Journal of Database Management, 12*(3), 3, 4–13. doi:10.4018/jdm.2001070101

Siau, K., & Shen, Z. (2003). Building customer trust in mobile commerce. *Communications of the ACM, 46*(4), 91–94. doi:10.1145/641205.641211

Singh, V. R. (2014). An Overview of Mobile Commerce in India. *International Journal of Management Research and Review, 4*(3), 354–366.

Sonia, S. M.-G., & Blanca, L.-C. (2013). How can a mobile vendor get satisfied customers? *Industrial Management & Data Systems, 113*(2), 156–170. doi:10.1108/02635571311303514

Sripalawat, J., Thongmak, M., & Ngramyarn, A. (2011). M-banking in metropolitan Bangkok and a comparison with other countries. *Journal of Computer Information Systems, 51*(3), 67–76.

Statista. (2016). *Global mobile retail commerce revenue from 2012 to 2018 in billion U.S. dollars*. Retrieved from https://www.statista.com/statistics/324636/mobile-retail-commerce-revenue-worldwide/

Suoranta, M., & Mattila, M. (2004). Mobile banking and consumer behavior: New insights into the diffusion pattern. *Journal of Financial Services Marketing, 8*(4), 354–366. doi:10.1057/palgrave.fsm.4770132

Tanakinjal, G. H., Deans, K. R., & Gray, B. J. (2010). Third screen communication and the adoption of mobile marketing: A Malaysia perspective. *International Journal of Marketing Studies, 2*(1), 36–47.

Teo, T. S. H., & Pok, S. H. (2003). Adoption of WAP-enabled mobile phones among Internet users. *Omega: The International Journal of Management Science, 31*(6), 483–498. doi:10.1016/j.omega.2003.08.005

Varnali, K., & Toker, A. (2010). Mobile marketing research: The-state-of-the-art. *International Journal of Information Management, 30*(2), 144–151. doi:10.1016/j.ijinfomgt.2009.08.009

Varshney, U., & Vetter, R. (2002). Mobile Commerce: Framework, Applications and Networking Support. *Mobile Networks and Applications, 7*(3), 185–198. doi:10.1023/A:1014570512129

Venkatesh, V., Ramesh, V., & Massey, A. P. (2003). Understanding usability in mobile commerce. *Communications of the ACM, 46*(12), 53–56. doi:10.1145/953460.953488

Wang, Y. S., & Liao, Y. W. (2007). The conceptualization and measurement of m-commerce user satisfaction. *Computers in Human Behavior, 23*(1), 381–398. doi:10.1016/j.chb.2004.10.017

Wei, T. T., Marthandan, G., Chong, A. Y. L., Ooi, K. B., & Arumugam, S. (2009). What drives Malaysian m-commerce adoption? An empirical analysis. *Industrial Management & Data Systems, 109*(3), 370–388. doi:10.1108/02635570910939399

Wells, J. D., Parboteeah, V., & Valacich, J. S. (2011). Online Impulse Buying: Understanding the Interplay between Consumer Impulsiveness and Website Quality. *Journal of the Association for Information Systems, 12*(1), 32–56.

Wilska, T. A. (2003). Mobile phone use as part of young peoples consumption styles. *Journal of Consumer Policy*, *26*(4), 441–463. doi:10.1023/A:1026331016172

Wu, J. H., & Wang, S. C. (2005). What drives mobile commerce? An empirical evaluation of the revised technology acceptance model. *Information & Management*, *42*(5), 719–729. doi:10.1016/j.im.2004.07.001

Yang, K. (2010). Determinants of US consumer mobile shopping services adoption: Implications for designing mobile shopping services. *Journal of Consumer Marketing*, *27*(3), 262–270. doi:10.1108/07363761011038338

Yang, K., & Lee, H. J. (2010). Gender differences in using mobile data services: Utilitarian and hedonic value approaches. *Journal of Research in Interactive Marketing*, *4*(2), 142–156. doi:10.1108/17505931011051678

Yang, K. C. C. (2005). Exploring factors affecting the adoption of mobile commerce in Singapore. *Telematics and Informatics*, *22*(3), 257–277. doi:10.1016/j.tele.2004.11.003

Yeh, Y. S., & Li, Y. M. (2009). Building trust in m-commerce: Contributions from quality and satisfaction. *Online Information Review*, *33*(6), 1066–1086. doi:10.1108/14684520911011016

Yu, C. S. (2012). Factors affecting individuals to adopt mobile banking: Empirical evidence from the utaut model. *Journal of Electronic Commerce Research*, *13*, 2.

Zaichkowsky, J. L. (1985). Measuring the Involvement Construct. *The Journal of Consumer Research*, *12*(3), 341–352. doi:10.1086/208520

Zeithaml, V. A. (1988). Consumer perceptions of price, quality and value: A means-end model and synthesis of evidence. *Journal of Marketing*, *52*(3), 2–22. doi:10.2307/1251446

Zhang, X., Prybutok, V. R., & Koh, C. E. (2006). The role of impulsiveness in a TAM based online purchasing behavior model. *Information Resources Management Journal*, *19*(2), 54–68. doi:10.4018/irmj.2006040104

Zou, X., & Huang, K. W. (2015). Leveraging location-based services for couponing and infomediation. *Decision Support Systems*, *78*, 93–103. doi:10.1016/j.dss.2015.05.007

Chapter 16
Designing Website Interfaces for M-Commerce With Consideration for Adult Consumers

Jean-Éric Pelet
ESCE International Business School, France

Basma Taieb
Cergy Pontoise, France

ABSTRACT

This chapter analyzes the interaction effects between the principal design cues of a mobile commerce website, such as background/foreground colors, font text and layout. Three experiments have been conducted based on visits to a fictitious m-commerce website. Experiment 1 manipulates the levels of color contrast: positive contrast (light text on a dark background) versus negative contrast (dark text on a light background). In experiment 2, contrast and font have been manipulated with a complete factorial plan: 2 x 2 (negative vs positive contrast x serif font vs sans serif font). Finally, contrast and layout have been manipulated in a third experimental 2 x 2 plan (negative vs positive contrast x dense vs airy layout). This research involved 219 French participants. Results show significant effects of the positive contrast (light text on a dark background) of the mobile website design on the purchase and revisit intentions of adults. Discussions about the interaction effects of design elements, limitations and directions for future research follow.

INTRODUCTION

Mobile commerce (m-commerce), which now accounts for 1.6% of the total retail segment, is expected to reach a market share of 2.7% by 2019 (Criteo, 2015). This way of conducting commercial transactions refers to the one- or two-way exchange of value facilitated by a mobile consumer electronic device (e.g. smartphone or tablet), which is enabled by wireless technologies and communication networks (Mobile

DOI: 10.4018/978-1-5225-2469-4.ch016

Designing Website Interfaces for M-Commerce With Consideration for Adult Consumers

Marketing Association, 2013). For example, in France m-commerce accounts for 22% of all online transactions with 37% on smartphones and 63% on tablets, according to Criteo (2015). Mobile devices are likely to continue to increase not only in sessions on online retail websites but also in conversion rates, as a larger and larger percentage of consumers continue to become more comfortable completing transactions with their handheld devices (Criteo, 2015). The shopping process of a mobile consumer usually starts by collecting information about the product and analyzing some parameters such as finding out the opinions of other users by using social media. Indeed, the latter have quickly become the most popular destinations on the Internet (Gil-Or, 2010). Then, consumers usually check the purchase price. This means comparing the prices in various online stores, through applications or price comparison websites. Shoppers use their mobile devices to visit comparison-shops, read product reviews and purchase products, browsing social media as a point of departure (Refuel Agency, 2015). They then check the possibility of obtaining discounts while purchasing products and finalize this shopping session with the online purchase of the product after its selection has been made following the analysis of its advantages and price attractiveness (Mobile Institute, 2015). Among mobile consumers, another activity appears to have become increasingly practiced, namely "showrooming", where people use smartphones while inside traditional shops and search for information about the products on the Internet at the same time, comparing the online and shop prices; in cases where a more attractive offer is found on the Internet, consumers will leave the shop and make their purchase online (Pralat, 2013).

Several research projects on m-commerce in the recent past focused on the adoption, acceptance and use of mobile shopping, and the utilitarian and hedonic factors that might influence it (Li, Dong, & Chen, 2012; Lu & Su, 2009; Wu & Wang, 2005; Holmes, Byrne, & Rowley, 2013). Several attempts were aimed at presenting conceptual reviews of mobile marketing and mobile retailing (e.g. Shankar, Venkatesh, Hofacker, & Naik, 2010; Varnali & Toker, 2010). However, in the context of smartphones, very few recent studies examine consumer attitudes towards smartphone marketing (Persaud & Azhar, 2012). For this reason, the authors consider that e-commerce studies can be applicable in the m-commerce context, considering analogical conditions such as indoor conditions for example.

However, results of previous studies (see for example Pelet & Uden, 2015, Pelet & Papadopoulou, 2015, Pelet & Lecat, 2014) reinforce the importance of taking into account the environment of the m-commerce website where consumers spend time shopping, commenting, sharing, marking, buying, and selling. In particular, the color, font and layout scheme appear to be important aspects of the website interface which should be considered in order to better understand consumer behavior. Holmes et al. (2013) also reported that when shopping on a smartphone, consumers value its convenience and accessibility.

Furthermore, older users represent an important consumer sector with real growth potential (Piqueras-Fiszman, Ares, Alcaide-Marzal, & Diego-Más, 2010). A major factor to be considered remains the quality of the navigation experience in relation to the progressive aging of the population using the Internet, especially when people browse m-commerce websites on their mobile devices. Eyesight deteriorates over the age of 45 (Mutti & Zadnik, 2000). Hence, understanding age-related differences in website navigation via mobile devices is a key issue for designers (Bergstrom, Olmsted-Hawala, & Jans, 2013).

The importance of color, text and layout in website design and its effect on trust, purchase and revisit intentions implied by the results of previous studies (see for example Pelet & Papadopoulou, 2012; 2013; 2014; 2015) led us to further examine the effects of these design assets (background/foreground colors, text and layout) on purchase and revisit intentions as variables in the proposed research model. Hence, the authors developed certain research hypotheses proposing links between these variables, guided by the results of the previously mentioned studies together with the literature. The research hypotheses were then empirically tested with a quantitative study in a real-world conditions experiment, principally with a non-student sample focusing on older consumers in order to reinforce the validity of the results.

LITERATURE REVIEW

The design aesthetics of a website relate to perceived attractiveness, defined as the degree to which a person believes that the website is aesthetically pleasing to the eye (Van der Heijden, Verhagen, & Creemers, 2003). Visual design attracts consumers and encourages shopping (Garett, 2003). Following Cyr, Head and Ivanov (2006), design aesthetics may be expressed through the elements of colors, shapes, language, music or animation. It is important to distinguish legibility from readability. Legibility reflects how comprehensible or identifiable written characters are, based on appearance, whereas readability is a measure of how easy it is to read and understand a piece of text. When viewed from the angle of design, legibility relates to the characteristics inherent to its design, whereas readability is more affected by other factors, such as justification, line spacing and text size, among others (Bix, Lockhart, Cardoso, & Selke, 2003). This is a clear indication that improving legibility enhances readability. They are combined to make information quicker and even more enjoyable to read.

A proper presentation of these elements with an image header, decorative font, and colorful graphical buttons can positively affect a user's positive impression of the site (Schultz, 2005). Nevertheless, a well-designed interface is relatively difficult to achieve in an m-commerce context since the physical limitations related to a tiny and portable mobile handheld device prove to be extremely important and constraining (Tarasewich, 2003).

Research has indicated different intentions to revisit the website. The first one is how usability defines the promptness of how the system could be used (Davis, 1989), and how user friendly, easy to navigate and organized it is (Parasuraman, Zeithaml, & Malhorta, 2005). Then, privacy which is linked to personal information protection (Parasuraman et al., 2005; Suh, Ahn, & Pedersen, 2013) has to be taken into account. Raman, Stephenaus, Alam and Kuppusamy (2008) have paid attention to the reliability which informs of how properly the website service is performed (Raman et al., 2008; Suh et al., 2013). The crucial intention is how the information defines whether the website content is coherent with customers' needs (Li et al., 2012). The important factor is appearance, which covers layout, graphical style and multimedia effects (Suh et al., 2013). Suh et al. (2013) focused attention on the e-service quality and satisfaction as core factors positively related to revisiting the website. By means of the Internet, consumers desire more efficient transfer of information, more interaction with others. And prompt, diverse information is important for customers who intend to revisit the website (Hsieh & Tsao, 2014). Proper choice of font and color contrast creates reliability, e-service quality or perception if the website is user friendly (Lowry, Wilson, & Haig, 2014; Lee & Koubek, 2010). The layout could influence consumer evaluations and behavior (Visinescu, Sidorova, Jones, & Prybutok, 2015).

Prior studies show that website design influences consumer behavior and purchase intention (Cyr, Head, & Ivanov, 2006; Lowry et al., 2014; Kim, Choi, & Lee, 2015; Lim, 2015; Suh et al., 2013; Visinescu et al., 2015). Therefore, the authors can hypothesize that:

H1a: The design of an m-commerce website has an impact on purchase intention.
H1b: The design of an m-commerce website has an impact on intention to revisit.

COLOR CONTRAST, FONT, AND LAYOUT OF WEBSITE

In his recent work, Pelet (2014) demonstrates the importance of the use of the contrast variable as a component of the interface of an e-commerce/m-commerce website. Norman (2003) points out that the contrast of a dark text over a bright background, representing a negative contrast according to Scharff and Ahumada (2002), leads consumers to feel negative emotions and more likely to immediately leave a website, regardless of the content or information. This is mainly due to visual discomfort, displeasure and strain. It should be noted that increasing the colored contrast when designing websites serves to enhance the readability and legibility of the information presented, which directly correlates with the information absorbed and retained by consumers (Gao, Ebert, Chen, & Ding, 2011; Pelet & Papadopoulou, 2012). Therefore, in the experiment the researchers present the background and foreground colors by using both negative and positive contrasts. In this study, the color combinations used are yellow on green for positive contrast, with the negative contrast being green on yellow, following the findings of Hill and Scharff (1997).

Including contrast as an element of the website interface is also significant as color has been found to support the layout and contrast of the actual information the website is trying to convey. For example, some research findings show that a positive contrast of bright text over a dark background setting provides better overall legibility (Greco, Stucchi, Zavagno, & Marino, 2008). Previous research has generally recommended providing spacing between lines and letters, although not too generously, as it would require more effort for elderly people (W3C, 2011; Wang, Sato, Rau, Fujimura, Gao, & Asano, 2008; Wright, 2000).

Lowry et al. (2014) argue that using the right font and color contrast to make the website user-friendly and provide sufficient information is an effective way of encouraging the consumer to want to revisit. Moreover, Akutsu, Legge, Ross and Schuebel (1991) suggested that font size tends to influence and enhance reading for both children and adults. The second main theme with regard to the usage of fonts concerns the readability performance based on using serif or sans serif fonts. The same literature on font sizes usually supports serif fonts to enhance readability compared to sans serif ones (Bernard, Chaparro, Mills, & Halcomb, 2003). However, the literature seems to have a preference for sans serif type faces exclusively in online domains, according to previous research (Morrell & Echt, 1997; Bernard, Lida, Riley, Hackler, & Janzen, 2002) which, in addition to examining the influence of font types on readability, examined in particular the influence of Times and Arial on readability. These fonts were chosen as the most popular fonts in online mediums (Ramsden, 2000). This study found no significant differences in readability, but it suggested that respondents had more difficulty in reading Times at both 10pt and 12pt sizes when compared to Arial. Additionally, a study by the CNIB[1] (2011) examined the font type preferences of people with eyesight problems and identified Arial or Verdana (both sans serif fonts) as preferable font types for people with poor vision. Consequently, drawing on this literature, two fonts are used in this study: Arial (a sans serif font) and Times New Roman (a serif font). In addition, the researchers use two different font sizes (2em corresponding to 24 pts vs 1.5 em corresponding to 18 pts), to ensure that respondents can easily read the information of the fictional m-commerce website used during the experimentation.

With regard to the layout[2] of a website, another important aspect relates to how the navigational aspects are displayed on the screen (Brandtzæg, Lüders, & Skjetne, 2010), and this is particularly important because it influences the successful navigation of the website (Brandtzæg et al., 2010). Furthermore, according to a number of authors (Scalf, Colcombe, McCarley, Erickson, Alvarado, Kim, & Kramer, 2007; Sekuler, 2000), items located on the wide edges of the screen are less likely to be seen by older consumers due to a decline in the peripheral processing and useful field of vision (UFOV). Consequently, in this study the researchers apply two different layouts, a dense one (line-height = 12pt) and an airy one (line-height = 30pt).

According to previous studies the color, font, and layout of the website are important factors that lead to purchase and revisit intentions (e.g. Bauman, 2015; Kirlidog, 2014; Cyr, Head, & Larios, 2010; Tsiaousis & Giaglis, 2014; Pelet & Papadopoulou, 2012). Therefore, the researchers propose the following hypotheses:

H2a: The impact of the design of an m-commerce website on purchase intention is stronger with a positive contrast than with negative contrast.

H2b: The impact of the design of an m-commerce website on intention to revisit is stronger with a positive contrast than with negative contrast.

H3a: The impact of the design of an m-commerce website on purchase intention is stronger with a positive contrast and a sans serif font.

H3b: The impact of the design of an m-commerce website on intention to revisit is stronger with a positive contrast and a sans serif font.

H4a: The impact of the design of an m-commerce website on purchase intention is stronger with a positive contrast and an airy layout.

H4b: The impact of the design of an m-commerce website on intention to revisit is stronger with a positive contrast and an airy layout.

The proposed model follows the general pattern of consumer behavior in an outdoor environment. The model explains how the design of an m-commerce website and specifically its attributes – color contrast, fonts and layout – can moderate the buyer's behavioural intention (Figure 1).

Figure 1. Research model

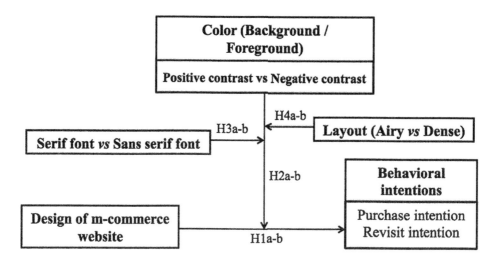

METHODOLOGY

Colors, fonts and layout were manipulated through three experiments in order to examine interaction effects between these elements of website design. As reported by Holmes *et al.* (2013), people appreciate shopping via mobile phones, thanks to the factors of convenience and accessibility. However, very few studies examine intentional behavior of consumers as a result of smartphone marketing (Persaud & Azhar, 2012). For these reasons, in this investigation, the authors applied the proposed framework to the specific context of websites in m-commerce on smartphones.

STIMULI AND MEASURING INSTRUMENTS

A website selling music CDs was especially designed for the experiment. There were 60 CDs available in 10 categories (3 CDs/category). For each CD, detailed information included the album title, the artist's name and seven other items of information (e.g. music category, online store price, condition -new or second hand-, delivery time). In addition, a CD description of 100 characters (the equivalent of 20 words) appeared next to the CD cover. This detailed information was displayed by clicking on the cover image or the title of a CD.

The variables of interest in the study were measured using established scales from the marketing literature and social personality psychology (see Appendix 1). For each scale, items were measured on a 5-point Likert scale ranging from strongly disagree (1) to strongly agree (5). Website design was operationalized using a five-items scale developed by Bressoles (2006). The authors operationalized purchase intention with four items from Limayem and Rowe (2006) and revisit intention with three items from Mukherjee and Nath (2007).

EXPERIMENTAL PROCEDURE AND DATA COLLECTION

Each participant visited the website with a graphic chart which was randomly selected among the schemes prepared for the experiment. Participants could select a category on the left side of the webpage and see the 3 CDs of this category on the right side of the same webpage. Participants had to look into the details of a minimum of two CDs of their choice; after that, an easy-to-see link appeared and respondents were asked to complete a questionnaire. All subjects were then tested for color blindness using the highly reliable Ishihara test[3] (Roullet, Ben Dahmane Mouelhi, & Droulers, 2003; Lanthony, 2005). The test was conducted following the questionnaire in order to prevent respondents from being aware of the importance of color in this experiment and thus avoiding any possible bias in the responses. This guaranteed the validity of the sample's responses, by retaining only people with perfect color vision.

The use of mobile devices means that consumers have the ability to move the device or themselves, closer or further away. Since the ambient lighting and the hygrometry of the environment were found to be important in the context of m-commerce (Tarasewich, 2003), it had to be considered for the empirical study. Therefore, this study conducted measurements in real environmental conditions. An experiment was set up especially to allow for the control of the ambient lighting of the physical setting. The hygrometry was also measured for each obtained response. Since users do not behave in the same way in the street when it is cold and when it is warm, they use their mobile differently: therefore the temperature was also measured during each respondent's visit. These measurements enabled us to neutralize variables and were reported on each questionnaire of each respondent (see Appendix 2.1).

SAMPLE

Most previous studies considered students only (Kirlidog, 2014; Pallud & Straub, 2014; Lee & Koubek, 2010) and they recognized that student samples have often been criticized for their lack of generalizability and their inability to represent the population. For this study, the researchers investigated people over the age of 45 because they represent a growing part of the population of consumers, according to the U.S. Census Bureau, who reports that the older population nationwide will nearly double in the next 20 years (Rios, 2014). This represents a growing audience with more disposable income than other age groups. They are also used to new technologies and handheld devices. Moreover, researchers argue that color emotion responses may change with the advance of age, due to different psychological or social requirements for different life stages (Ou, Luo, Sun, Hu, & Chen, 2012; Terwogt & Hoeksma, 1995). This research was conducted in France and involved 219 non-student participants over the age of 45. The sample comprised 57% females and 43% males. Participants were randomly assigned to the three experiments, approximately 65 respondents per experiment (cell). Individuals who participated in one experimental condition didn't participate in any of the other conditions.

MEASUREMENT VALIDATION

All constructs have high levels of reliability; the Cronbach's alpha (α) and Jöreskog's rhô (ρ) values are greater than .9 for all scales. Discriminant and convergent validity of the scales were additionally examined. Convergent validity is shown when the average variance extracted (AVE) is greater than .5 (Anderson & Gerbing, 1988). The constructs' AVE values are greater than 0.5 for all countries. Discriminant validity is shown when a construct's AVE is greater than the squared correlations of this construct with other constructs (Fornell & Larcker, 1981). Discriminant validity is fulfilled for all constructs. Confirmatory factor analysis was conducted to evaluate the global measurement model and showed a good fit (RMSEA = .07, CFI = .96, $\chi^2/df = 2.77$).

EXPERIMENT 1

Experiment 1 manipulates the levels of color contrast, positive contrast (light text on a dark background) *versus* negative contrast (dark text on light background). The positive contrast being yellow on green and the negative one being green on yellow, following Hill and Scharff's (1997) findings (see Appendix 2.2).

Results of Experiment 1

Composite scores of the variables of model (website design, purchase and revisit intentions) were calculated. Direct and interaction effects were tested by the *bootstrap* method from Hayes (2013) using PROCESS macro in SPSS 20. Results show significant direct effects of website design on purchase ($t=$

5.65, $p<.01$) and revisit intentions ($t=9.36, p<.01$). Results also indicate a significant effect of interaction between website design and the level of contrast on revisit intention ($t = 2.12, p<.05$), but non-significant interaction effect on purchase intention ($t = .94, p>.05$). Nevertheless, the effects of website design on purchase ($t= 10.31, p<.01$) and revisit ($t= 11.52, p<.01$) intentions are stronger with positive contrast than negative contrast (see Tables 1 and 2).

EXPERIMENT 2

In this experiment, contrast and font were manipulated in an experimental plan of 2 (negative *vs* positive contrast) x 2 (Serif font *vs* sans Serif font). This factorial plan varies with 4 CSS (Cascade Style Sheet): 2 charts yellow on green using a Serif font *vs* sans Serif font, and 2 charts green on yellow using a Serif font *vs* sans Serif font (see Appendix 2.3).

Results of Experiment 2

The website design has a positive influence on purchase ($t = 6.94, p<.01$) and revisit intentions ($t = 5.72, p<.01$) when interacting with color contrast and font. Even if interaction effects between design*contrast*font on intentions were not significant ($p>.05$), the results show that t-values are higher for purchase ($t = 7.56, p<.01$) and revisit intentions ($t = 8.94, p<.01$) when website design is conceived with positive contrast and font sans serif (see Tables 3 and 4).

EXPERIMENT 3

In experiment 3, contrast and layout were manipulated with a factorial plan 2 (negative *vs* positive contrast) x 2 (dense *vs* an airy layout). 4 CSS were conceived: 2 charts yellow on green using a dense *vs* an airy layout, and 2 charts green on yellow using a dense *vs* an airy layout (see Appendix 2.4).

Table 1. Conditional effect of website design on purchase intention at levels of color contrast

Contrast	Effect	se	t	p	LLCI	ULCI
Negative	.63	.10	8.12	0.00	.46	.98
Positive	.79	.14	10.31	0.00	.35	1.08

Table 2. Conditional effect of website design on intention to revisit at levels of color contrast

Contrast	Effect	se	t	p	LLCI	ULCI
Negative	.67	.11	9.36	0.00	.42	.77
Positive	.82	.15	11.52	0.00	.62	1.09

Table 3. Conditional effect of website design on purchase intention at levels of color contrast and font

Font	Contrast	Effect	se	t	p	LLCI	ULCI
Serif	Negative	.84	.12	6.94	.00	.60	1.07
Serif	Positive	.86	.12	7.20	.00	.62	1.09
sans serif	Negative	.52	.10	5.34	.00	.33	.71
sans serif	Positive	.74	.10	7.56	.00	.55	.94

Table 4. Conditional effect of website design on intention to revisit at levels of color contrast and font

Font	Contrast	Effect	se	t	p	LLCI	ULCI
Serif	Negative	.81	.11	7.17	.00	.58	1.03
Serif	Positive	.79	.11	7.09	.00	.57	1.01
sans serif	Negative	.60	.09	6.56	.00	.42	.77
sans serif	Positive	.82	.09	8.94	.00	.64	1.00

Results of Experiment 3

The website design still has a positive impact on behavioral intentions ($t = 5.65$ for purchase and $t = 5.72$ for revisit, $p<.01$). Results show a non-significant effect of interaction between design*contrast*layout on intentions ($p>.05$), but the effects of website design on purchase ($t = 7.80$, $p<.01$) and revisit intentions ($t = 9.67$, $p<.01$) are stronger with a positive contrast and an airy layout (see Tables 5 and 6).

Table 5. Conditional effect of website design on purchase intention at levels of color contrast and layout

Layout	Contrast	Effect	se	t	p	LLCI	ULCI
Dense	Negative	.64	.11	5.65	.00	.41	.86
Dense	Positive	.78	.10	6.57	.00	.58	.97
Airy	Negative	.63	.11	5.84	.00	.41	.84
Airy	Positive	.81	.12	7.80	.00	.56	1.05

Table 6. Conditional effect of website design on intention to revisit at levels of color contrast and layout

Layout	Contrast	Effect	se	t	p	LLCI	ULCI
Dense	Negative	.59	.10	5.72	.00	.39	.79
Dense	Positive	.88	.09	6.26	.00	.70	1.06
Airy	Negative	.74	.10	7.47	.00	.54	.93
Airy	Positive	.71	.11	9.67	.00	.48	.93

DISCUSSIONS

As the results indicate, mobile website design influences the intentions to purchase and to revisit, hence, H1a and H1b are statistically supported. The elderly especially, prefer to revisit website designs with positive contrast (yellow-on-green), but the interaction effect between website design and the level of contrast on purchase intention was not significant. H2b receives statistical support, but H2a was not supported. Even if the effects of website design on purchase and revisit intentions are stronger with a positive contrast, font sans serif and an airy layout, the interaction effects were not statistically significant. H3a-b and H4a-b were not supported. The findings reinforce past studies in the electronic and m-commerce literature more specifically, a number of elements has to be present on the website in order to increase the intention to buy and to revisit a website (Büttner & Göritz, 2008; Wang et al., 2008). One of the most important considerations relates to the website appearance (i.e. color, graphics, and layout) which improves the mobile usability and enhances the level of satisfaction and leads the consumer to commit to online buying behavior.

These results are consistent with earlier research that suggests that the preferences for colors differ according to age (Ou et al., 2012; Terwogt & Hoeksma, 1995). So, people aged over 45 in France seem to prefer websites with color contrast yellow-on-green (positive contrast), an airy content and font sans serif. further suggestions include that web design for older people should avoid moving targets, and complex menu hierarchies (Zaphiris, Kurniawan, & Ghiawadwala, 2006) should not demand fast and repetitive movements for interaction such as double clicks with a short interval and fast clicking on touch screens (Zaphiris et al., 2006) and should minimize demands for working memory. A study conducted by applying eye-tracking method has found that older people looked at the central part of the screen more frequently than younger participants and that they took longer to first look at the peripheral top part of the screen (Bergstrom, 2013). The same study suggests that website designers consider repeating important content in the center and the periphery, to use a clean and uncluttered layout, and to reduce unnecessary distractions that may draw older users' eyes to a part of the screen that does not provide essential information. Zhou, Rau and Salvendy (2013) also found that accepting new functions is different from accepting a product: i.e. readability and finding a specific function, which are critical for older adults' acceptance of feature phones, are not determinants of their acceptance of new functions in smartphones.

Previous research suggests that age is an important variable that affects online consumption (Punj, 2011; Coursaris, Swierenga, & Watrall, 2008). Ou et al. (2012) argue that color emotion responses may change with the advance of age and provide evidence that colors are rated as less active, less liked, and cooler for older observers than for young observers. In regard to color preferences, a seminal study by Simon (1999) showed that Western countries preferred bright colors with more images to make the site appear more modern compared to Asian countries. In terms of web design and color and cultural preferences, Tolba (2003) found that users from different cultures were found to have different web design and structure preferences, in addition to different criteria for accepting the website. Cyr et al. (2010) conducted experiments accross cultures (Canada, Germany, Japan) and focused on the impact of differents website colors (yellow, blue, gray) on user trust, satisfaction and e-loyalty. They found differences among cultures.

CONCLUSION

Although mobile phone penetration is reasonably high, the prevalence of m-commerce purchasing is low. The infancy of m-commerce can be attributed to a combination of reasons. Prominent among them are consumer-related reasons, where purchase activity using mobile is not seen as physically convenient due to mobile shop design issues. Active m-commerce users browse and purchase a few items on the mobile web but many brands still do not have the proper interface and many consumers still prefer a bigger screen. More specifically, the Physical properties of the m-commerce environment (brightness, weather, location) should be taken into account in order to achieve a good interface design.

There is general agreement that there are three basic elements that must be considered when developing a highly readable web page: colors (color contrast between the foreground and background colors), typefaces (fonts) and spacing (airy vs dense layout). However, there is a lack of studies that deeply investigate the use of these elements on mobile phone for special targets, like elderly people with sight problems. With an ever-increasing, aging population, the matter of conceiving mobile website design universally accessible to all types of people needs to be investigated more thoroughly in the world of digital marketing. Future research should offer current guidelines on good design for m-commerce, which will be extremely beneficial for designers. The contrast between foreground and background pages, the presence or absence of serifs, layout and interline spacing can all affect intention and behavior, mostly for older groups of people. To better understand the consumer, it is relevant to vary the study's context (different services; different products with different levels of involvement - high involvement, such as with luxury products (Kim *et al.*, 2010), and low involvement products. For example, Parboteeah, Valacich and Wells (2009) encouraged researchers to consider digital music products.

Color preferences also depend on climate, quality of sunshine and clarity of atmosphere. Indeed, the intensity of sunlight affects vision, which can lead to red short-sightedness and hence the preference for warm tones. Thus the use of warm colors in countries with hot climates is favored whereas cold colors are favored in cold climates. Color appears, therefore, to be a contextual factor worthy of a theoretical interest rather than a managerial one. But results for all work seem partly divergent and still insufficiently consensual. Moreover, color seems to comfort the consumer when it is soft and encourages trust during the act of buying. Practitioners have been encouraged to use colors conservatively, emphasizing strong contrast between the foreground and background, especially for text messages. Other context characteristics, such as time of day and temperature, can be used in addition to location. New or modified interaction techniques may be necessary to compensate for the limited visual display of the devices. The environment should thereby be taken into account more deeply. A meteorological report announcing cold weather could initiate warm CSSs on the websites visited at that time, for example. This is totally feasible technically. This research was restricted to understanding the effects of design cues on behavioral intentions towards m-commerce website. Future research is recommended to examine affective reactions (emotion and mood) towards m-commerce purchasing. Specifically, earlier research suggests that enjoyment and emotions are important drivers for the adoption of mobile shopping (Lee & Chang, 2011; Li et al., 2012; Yang & Kim, 2012). The researchers created a website for the requirements of the study (the participants in the experiment experienced real navigation conditions) and thereby increased the reliability of the results. More applied conditions than the use of a fictive website could be pursued. More realistic and detailed e-learning or e-commerce prototypes could be created and examined, or existing sites could be evaluated in an applied context. Finally, a more general examination of the impact of mobile website design cues in other countries would be interesting.

REFERENCES

Akutsu, H., Legge, G. E., Ross, J. A., & Schuebel, K. (1991). Psychophysics of reading: X. Effects of age-related changes in vision. *The Journals of Gerontology. Series B, Psychological Sciences and Social Sciences*, *46*, 325–331. PMID:1940088

Anderson, J. C., & Gerbing, D. W. (1988). Structural equation modeling in practice: A review and preferences. *The Journal of Consumer Research*, *27*(2), 233–248.

Bauman, A. (2015). The use of the repertory grid technique in online trust research. *Qualitative Market Research*, *18*(3), 362–382. doi:10.1108/QMR-08-2014-0080

Bergstrom, J. C. R., Olmsted-Hawala, E. L., & Jans, M. E. (2013). Age-Related Differences in Eye Tracking and Usability Performance: Website Usability for Older Adults. *Journal of Human-Computer Interaction*, *29*(8), 541–548. doi:10.1080/10447318.2012.728493

Bernard, M., Lida, B., Riley, S., Hackler, T., & Janzen, K. (2002). *A Comparison of Popular Online Fonts: Which Size and Type is Best?* Retrieved June 26, 2016, from http://usabilitynews.org/a-comparison-of-popular-online-fonts-which-size-and-type-is-best/html

Bernard, M. L., Chaparro, B. S., Mills, M. M., & Halcomb, C. G. (2003). Comparing the Effects of Text Size and Format on the Readability of Computer-Displayed Times New Roman and Arial Text. *International Journal of Human-Computer Studies*, *59*(6), 823–835. doi:10.1016/S1071-5819(03)00121-6

Bix, L., Lockhart, H., Cardoso, F., & Selke, S. (2003). The Effect of Color Contrast on Message Legibility. *Journal of Design Communication*, 5.

Brandtzæg, P. B., Lüders, M., & Skjetne, J. H. (2010). Too many Facebook friends? Content sharing and sociability versus the need for privacy in social network sites. *International Journal of Human-Computer Interaction*, *26*(11-12), 1006–1030. doi:10.1080/10447318.2010.516719

Bressolles, G. (2006). La qualité de service électronique: NetQu@l. Proposition dune échelle de mesure appliquée aux sites marchands et effets modérateurs. *Recherche et Applications en Marketing*, *21*(3), 19–45. doi:10.1177/076737010602100302

Büttner, O. B., & Göritz, A. S. (2008). Perceived trust worthiness of online shops. *Journal of Consumer Behaviour*, *7*(1), 35–50. doi:10.1002/cb.235

Canadian National Institute for the Blind. (2011). *Clear print accessibility guidelines*. Retrieved July 19, 2016, from http://www.cnib.ca/en/services/resources/clearprint/html

Chadwick-Dias, A., Bergel, M., & Tullis, T. S. (2007). Senior surfers 2.0: A re-examination of the older Web user and the dynamic web. In C. Stephanidis (Ed.), *Universal access in human computer interaction: Coping with diversity* (pp. 868–876). Berlin: Springer. Retrieved March, 2016, from http://www.springerlink.com/content /08q65578604130w8/html

Coursaris, C., Swierenga, S., & Watrall, E. (2008). An empirical investigation of color temperature and gender effects on web aesthetics. *Journal of Usability Studies*, *3*(3), 103–117.

Criteo. (2015). *State of Mobile Commerce*. Retrieved March, 2016, from http://www.criteo.com/resources/mobile-commerce-q1-2015/html

Cyr, D., Head, M., & Ivanov, A. (2006). Design aesthetics leading to m-loyalty in mobile commerce. *Information & Management, 43*(8), 950–963. doi:10.1016/j.im.2006.08.009

Cyr, D., Head, M., & Larios, H. (2010). Colour appeal in website design within and across cultures: A multi-method evaluation. *International Journal of Human-Computer Studies, 68*(1-2), 1–21. doi:10.1016/j.ijhcs.2009.08.005

Davis, F. D. (1989). Perceived Usefulness, Perceived Ease of Use, and User Acceptance of Information Technology. *Management Information Systems Quarterly, 13*(3), 319–340. doi:10.2307/249008

Deeb, S. S., & Motulsky, A. G. (2015). *Red-Green Color Vision Defects*. Retrieved October 12, 2015, from http://www.ncbi.nlm.nih.gov/books/NBK1301/html

Fornell, C., & Larcker, D. F. (1981). Evaluating structural equation models with unobservable variables and measurement error. *JMR, Journal of Marketing Research, 18*(1), 39–50. doi:10.2307/3151312

Gao, Q., Ebert, D., Chen, X., & Ding, Y. (2011). Design of a Mobile Social Community Platform for Older Chinese People in Urban Areas. *Human Factors and Ergonomics in Manufacturing & Service Industries, 25*(1), 66–89.

Gil-Or, O. (2010). The Potential of Facebook in creating commercial value for service companies. *Advances in Management, 3*(2), 20–25.

Greco, M., Stucchi, N., Zavagno, D., & Marino, B. (2008). On the portability of computer-generated presentations: The effect of text-background color combinations on text legibility. *Human Factors: The Journal of the Human Factors and Ergonomics Society, 50*(5), 821–833. doi:10.1518/001872008X354156 PMID:19110842

Hayes, A. F. (2013). *Introduction to Mediation, Moderation, and Conditional Process Analysis: A Regression-Based Approach*. New York: Guilford Publications.

Hill, A., & Scharff, L. V. (1997). Readability of websites with various foreground/background color combinations, font types and word styles. *Proceedings of 11th National Conference in Undergraduate Research, 2*, 742-746.

Holmes, A., Byrne, A., & Rowley, J. (2013). Mobile shopping behaviour: Insights into attitudes, shopping process involvement and location. *International Journal of Retail & Distribution Management, 42*(1), 25–39. doi:10.1108/IJRDM-10-2012-0096

Hsieh, M., & Tsao, W. (2014). Reducing perceived online shopping risk to enhance loyalty: A website quality perspective. *Journal of Risk Research, 17*(2), 241–261. doi:10.1080/13669877.2013.794152

Kim, H., Choi, Y., & Lee, Y. (2015). Web atmospheric qualities in luxury fashion brand websites. *Journal of Fashion Marketing and Management, 19*(4), 384–401. doi:10.1108/JFMM-09-2013-0103

Kim, H., Park, K., & Schwarz, N. (2010). Will This Trip Really Be Exciting? The Role of Incidental Emotions in Product Evaluation. *The Journal of Consumer Research, 36*(6), 983–991. doi:10.1086/644763

Kirlidog, M. (2014). Effect of Colours in Manual Data Typing. *Fourth International conference on Computer Science & Information Technology*. doi:10.5121/csit.2014.4206

Lanthony, P. (2005). *La perception des couleurs sur écran. Intervention dans le cadre d'un séminaire sur la couleur, 3C S*. France: A. Abbaye de Royaumont.

Lee, S., & Koubek, R. J. (2010). The effects of usability and web design attributes on user preference for e-commerce web sites. *Computers in Industry*, *61*(4), 329–341. doi:10.1016/j.compind.2009.12.004

Li, M., Dong, Z. Y., & Chen, X. (2012). Factors influencing consumption experience of mobile commerce: A study from experiential view. *Internet Research*, *22*(2), 120–141. doi:10.1108/10662241211214539

Lim, W. M. (2015). Antecedents and consequences of e-shopping: An integrated model. *Internet Research*, *25*(2), 184–217. doi:10.1108/IntR-11-2013-0247

Limayem, M., & Rowe, F. (2006). Comparaison des facteurs influençant les intentions d'achat à partir du Web à Hong Kong et en France: Influence sociale, risques et aversion pour la perte de contact. *Revue Française du Marketing*, *209*(4/5), 25–48.

Lowry, P. J., Wilson, D. W., & Haig, W. L. (2014). A picture is worth a thousand words: Source Credibility Theory applied to logo and website design for heightened credibility and consumer trust. *International Journal of Human-Computer Interaction*, *30*(1), 63–93. doi:10.1080/10447318.2013.839899

Mobile Institute. (2015). *Raport mShopper, Polacy na zakupach mobilnych*. Retrieved July 19, 2016, from http://www.aptusshop.pl/sklepy_internetowe/przyszlosc-m-commerce-w-polsce.php/html

Mobile Marketing Association. (2013). *Le baromètre trimestriel du Marketing Mobile en France 3ᵉ édition, T2 2013*. Retrieved July 19, 2016, from http://mmaf.fr/html

Morrell, R. W., & Echt, K. V. (1997). Designing written instructions for older adults: Learning to use computers. In A. D. Fisk & W. A. Rogers (Eds.), *Handbook of human factors and the older adult* (pp. 335–361). San Diego, CA: Academic Press.

Mukherjee, A., & Nath, P. (2007). Role of electronic trust in online retailing: A re-examination of the commitment-trust theory. *European Journal of Marketing*, *41*(9/10), 1173–1202. doi:10.1108/03090560710773390

Mutti, D. O., & Zadnik, K. (2000). *Age-related decreases in the prevalence of myopia: Longitudinal change or cohort effect?* Retrieved December 9, 2015, from http://iovs.arvojournals.org/article.aspx?articleid/html

Norman, D. A. (2003). *Emotional Design: Why we love (or hate) everyday things*. New York: Basic Books.

Ou, L. C., Luo, R. M., Sun, P. L., Hu, N. C., & Chen, H. S. (2012). Age Effects on Color Emotion, Preference, and Harmony. *Color Research and Application*, *37*(2), 92–105. doi:10.1002/col.20672

Pallud, J., & Straub, D. W. (2014). Effective website design for experience-influenced environments: The case of high culture museums. *Information & Management*, *51*(3), 359–373. doi:10.1016/j.im.2014.02.010

Parasuraman, A., Zeithaml, V., & Malhorta, A. (2005). E-S-QUAL: A multiple-item scale for assessing electronic service quality. *Journal of Retailing*, *64*(1), 12–40.

Parboteeah, D. V., Valacich, J. S., & Wells, J. D. (2009). The Influence of Website Characteristics on a Consumers Urge to Buy Impulsively. *Information Systems Research*, *20*(1), 60–78. doi:10.1287/isre.1070.0157

Pelet, J. É. (2014). Investigating the Importance of Website Color Contrast in E-Commerce: Website Color Contrast in E-Commerce. In *Encyclopedia of Information Science and Technology* (3rd ed.). Hershey, PA: IGI Global; doi:10.4018/978-1-4666-5888-2

Pelet, J. É., & Papadopoulou, P. (2012). The effect of colors of e-commerce websites on consumer mood, memorization and buying intention. *European Journal of Information Systems*, *21*(4), 438–467. doi:10.1057/ejis.2012.17

Pelet, J. É., & Papadopoulou, P. (2013). *Handbook of Research on User Behavior in Ubiquitous Online Environments*. Retrieved from http://www.igi-global.com/book/user-behavior-ubiquitous-online-environments/76724

Pelet, J. É., & Papadopoulou, P. (2014). Investigating Social Media in M-Commerce. *International Journal of Technology and Human Interaction*, *10*(4). doi:10.4018/ijthi.2014100103

Pelet, J. É., & Papadopoulou, P. (2015). Tablet and social media adoption in m-commerce: An exploratory study. *International Journal of Strategic Innovative Marketing*, *2*, 1. doi:10.15556/IJSIM.02.01.004

Persaud, A., & Azhar, I. (2012). Innovative mobile marketing via smartphones: Are consumers ready? *Marketing Intelligence & Planning*, *30*(4), 408–444. doi:10.1108/02634501211231883

Piqueras-Fiszman, B., Ares, G., Alcaide-Marzal, J., & Diego-Más, J. (2010). Comparing older and younger users perceptions of mobile phones and watches using CATA questions and preference mapping on the design characteristics. *Journal of Sensory Studies*, *26*(1), 1–12. doi:10.1111/j.1745-459X.2010.00315.x

Prałat, E. (2013). *M-commerce – rozwój na świecie i w Polsce*. Retrieved July 19, 2016, from http://www.ur.edu.pl/html

Punj, G. (2011). Effect of consumer beliefs on online purchase behavior: The influence of demographic characteristics and consumption values. *Journal of Interactive Marketing*, *25*(3), 134–144. doi:10.1016/j.intmar.2011.04.004

Raman, M., Stephenaus, R., Alam, N., & Kuppusamy, M. (2008). Information technology in Malaysia: E-service quality and update of Internet banking. *Journal of Internet Banking and Commerce*, *13*(2), 1–18.

Ramsden, A. (2000). *Annabella's HTML help*. Retrieved June 17, 2016, from http://www.geocities.com/annabella.geo/fontface.html

Refuel Agency. (2015). *Millenial Teens, Digital Explorer*. Retrieved July 19, 2016, from http://research.refuelagency.com/wp-content/uploads/2015/07/Millennial-Teen-Digital-Explorer.pdf

Ríos, M. (2014). *Fueled by Aging Baby Boomers, Nation's Older Population to Nearly Double in the Next 20 Years*. Retrieved July 7, 2016, from https://www.census.gov/newsroom/press-releases/2014/cb14-84.html

Roullet, B., Ben Dahmane Mouelhi, N., & Droulers, O. (2003). The Impact of background color on product attitudes among three different cultures. *Proceedings of International Congress of the French Association.*

Scalf, P. E., Colcombe, S. J., McCarley, J. S., Erickson, K. I., Alvarado, M., Kim, J. S., & Kramer, A. F. (2007). The neural correlates of an expanded functional field of view. *Journal of Gerontology, 62B*(1), 32–44. doi:10.1093/geronb/62.special_issue_1.32 PMID:17565163

Scharff, L. F. V., & Ahumada, A. J., Jr. (2002). Predicting the readability of transparent text. *Journal of Vision, 2*(9), 653-666. Retrieved from http://www.journalofvision.org/2/9/7/html

Schultz, L. (2005). *Effects of graphical elements on perceived usefulness of a library.* Retrieved from http://www.tarleton.edu/~schultz/finalprojectinternetsvcs.html

Sekuler, R., & Sekuler, A. B. (2000). Visual perception and cognition. *Oxford Textbook of Geriatric Medicine, 2*, 874–880.

Shankar, V., Venkatesh, A., Hofacker, C., & Naik, P. (2010). Mobile Marketing in the Retailing Environment: Current Insights and Future Research Avenues. *Journal of Interactive Marketing, 24*(2), 111–120. doi:10.1016/j.intmar.2010.02.006

Simon, J. S. (1999). *A cross-cultural analysis of website design: an empirical study of global web users.* Paper Presented at the Seventh Cross-Cultural Consumer Business Studies Research Conference, Cancun, Mexico.

Suh, Y., Ahn, T., & Pedersen, P. M. (2013). Examining the effects of team identification, e-service quality (e-SQ) and satisfaction on intention to revisit sports websites. *International Journal of Sports Marketing & Sponsorship, 14*(4), 2–19. doi:10.1108/IJSMS-14-04-2013-B002

Tarasewich, P. (2003). Designing mobile commerce applications. *Communications of the ACM, 46*(12), 57–60. doi:10.1145/953460.953489

Terwogt, M. M., & Hoeksma, J. B. (1995). Colors and Emotions: Preferences and Combinations. *The Journal of General Psychology, 122*(1), 5–17. doi:10.1080/00221309.1995.9921217 PMID:7714504

Tolba, R. H. (2003). The cultural aspects of design Jordanian websites: An empirical evaluation of university, news, and government website by different user groups. *The International Arab Journal of Information Technology, 1*, 51–59.

Tsiaousis, A. S., & Giaglis, G. M. (2014). Mobile websites: Usability evaluation and design. *International Journal of Mobile Communications, 12*(1), 29–55. doi:10.1504/IJMC.2014.059241

Van der Heijden, H., Verhagen, T., & Creemers, M. (2003). Understanding online purchase intentions: Contributions from technology and trust perspectives. *European Journal of Information Systems, 12*(1), 41–48. doi:10.1057/palgrave.ejis.3000445

Varnali, K., & Toker, A. (2010). Mobile marketing research: The state-of-the-art. *International Journal of Information Management, 30*(2), 144–151. doi:10.1016/j.ijinfomgt.2009.08.009

Visinescu, L. L., Sidorova, A., Jones, M. C., & Prybutok, V. R. (2015). The influence of website dimensionality on customer experiences, perceptions and behavioral intentions: An exploration of 2D vs. 3D web design. *Information & Management*, *52*(1), 1–17. doi:10.1016/j.im.2014.10.005

W3C. (2011). *Web content accessibility guidelines (WCAG) overview.* Retrieved August 22, 2011, from http://www.w3.org/WAI/intro/wcag/html

Wang, L., Sato, H., Rau, P. L. P., Fujimura, K., Gao, Q., & Asano, Y. (2008). Chinese text spacing on mobile phones for senior citizens. *Educational Gerontology*, *35*(1), 77–90. doi:10.1080/03601270802491122

Wright, P. (2000). Supportive documentation for older people. In P. Westendorp, C. H. Jansen, & R. Punselie (Eds.), *Interface design and document design* (pp. 81–100). Amsterdam, The Netherlands: Rodopi.

Wu, J. H., & Wang, S. C. (2005). What drives mobile commerce?: An empirical evaluation of the revised technology acceptance model. *Information & Management*, *42*(5), 719–729. doi:10.1016/j.im.2004.07.001

Yang, K., & Kim, H. Y. (2012). Mobile shopping motivation: An application of multiple discriminant analysis. *International Journal of Retail & Distribution Management*, *40*(10), 778–789. doi:10.1108/09590551211263182

Zaphiris, P., Kurniawan, S., & Ghiawadwala, M. (2006). A systematic approach to the development of research-based web design guidelines for older people. *Universal Access in the Information Society*, *6*(1), 59–75. doi:10.1007/s10209-006-0054-8

Zhou, J., Rau, P. L. P., & Salvendy, G. (2013). Older adults use of smartphones: An investigation of the factors influencing the acceptance of new functions. *Behaviour & Information Technology*, *33*(6), 552–560. doi:10.1080/0144929X.2013.780637

Ziefle, M. (2010). Information presentation in small screen devices: The trade-off between visual density and menu foresight. *Applied Ergonomics*, *41*(6), 719–730. doi:10.1016/j.apergo.2010.03.001 PMID:20382372

ENDNOTES

[1] Canadian National Institute for the Blind.
[2] The page layout refers to the arrangement of text, images, and other objects on a page in order to customize the appearance of the website (http://www.collinsdictionary.com/dictionary/english/layout).
[3] The Ishihara test is the most common colorblindness test used today (Deeb & Motulsky, 2015). It consists of a number of plates, 24 or 38, the Ishihara plates, each containing a circle of dots which appear in random colors and sizes. Within the circle are dots forming a number which should be clearly visible to viewers with normal color vision and hard to see or invisible to viewers with defective color vision. To pass the test participants should recognize the number in every plate.
[4] Luxmeter (Apple: https://itunes.apple.com/fr/app/luxmeter-pro/id408369821?mt=8/Android: https://play.google.com/store/apps/details?id=at.muehlburger.android.lightluxmeter&hl=fr)

5   Thermometer (Apple: https://itunes.apple.com/fr/app/thermo-temperature/id414215658?mt=8/ Android: https://play.google.com/store/apps/details?id=jp.metersfree)

6   Hygrometer (Apple: https://itunes.apple.com/fr/app/hygrometer-check-your-humidity/id515673004?mt=8/Android: https://play.google.com/store/apps/details?id=com.bti.hygroMeter&hl=fr)

7   Ishihara's test (Apple: https://itunes.apple.com/fr/app/eye-test-free-snellen-chart/id473782361?mt=8/ Android: https://play.google.com/store/apps/details?id=com.vaughnweb.colortest&hl=fr)

8   For more explanation on this guideline, please read here: https://www.w3.org/WAI/UA/work/wiki/Guideline_1.4_Text_Customization_Proposal

APPENDIX 1

Table 7. Measures

Construct	Source
Purchase intention Purch1. I intend to make purchases from this website Purch2. It is likely that I will make purchases from this website in the near future Purch3. I expect to make purchases from this website in the near future Purch4. I will definitely buy products from this website in the near future	Limayem and Rowe (2006)
Revisit intention Revis1. I would continue using this website for the next six months Revis2. I plan to continue using this website for the next year Revis3. I would like to continue using this website for the next two years	Mukherjee and Nath (2007)
Website design Desig1. The organization and layout of this commercial website makes information searches easy Desig2. It is easy to search for information on this commercial website Desig3. It is easy to navigate and find what I am looking for on this site Desig4. The colors of this commercial website make information searches easy Desig5. The layout of this commercial website is clear and simple	Bressolles (2006)

APPENDIX 2: APPS USED TO CONTROL ENVIRONMENTAL CONDITIONS, BOTH ON iOS AND ANDROID (SMARTPHONES USED FOR THE EXPERIMENTATION)

Tools installed on the smartphone of the interviewer in his country:

Luxmeter[4]

Description: Luxmeter Pro emulates a Lux meter to measure light intensity. LuxMeter is used to measure the lighting of the environments in reference tables that indicate the value LUX required under the type of environment.

Thermometer[5]

Description: Thermometer is the market leader, it is undoubtedly one of the most useful utility apps. Anywhere around the world, you can obtain the temperature in a second without having to set anything up.

Hygrometer[6]

Description: Check your humidity!

Ishihara's Test[7]

Description: Are you colorblind? Know somebody who is colorblind?

APPENDIX 3: PRESENTATION OF THE GRAPHIC CHARTS OF EXPERIMENT 1 — YELLOW ON GREEN AND GREEN ON YELLOW

Table 8.

Plan	Background				Foreground				Contrast	Explanation of Plan
	Name	H	B	S	Name	H	B	S		
VJ, Foreground: #006500 / Background: #FFFFAC, Positive Contrast										
1	Lemon Chiffon	40	201	240	Bottle Green	80	48	240	7,04* Passed at Level 2	The exact reverse of plan 1
Hex.	#FFFFAC				#006500					
JV, Foreground: #FFFFAC / Background: #006500, Negative Contrast										
2	Bottle Green	80	48	240	Lemon Chiffon	40	201	240	7,04* Passed at Level 2	Hill and Scharff (1997) showed that the sharp contrasts of this chart offered users the fastest reading speed possible.
Hex.	#006500				#FFFFAC					

* Text or diagrams and their background must have a luminosity contrast ratio of at least 5:1 for level 2 conformance to guideline 1.4 of the W3C[8].

** Text or diagrams and their background must have a luminosity contrast ratio of at least 10:1 for level 3 conformance to guideline 1.4 W3C.

APPENDIX 4: PRESENTATION OF THE GRAPHIC CHARTS OF EXPERIMENT 2

Table 9.

Experiment 2	
Color Positive contrast: yellow (#FFFFAC) on green (#006500) *Versus* Negative contrast: green (#006500) on yellow (#FFFFAC)	Font Sans serif font: Arial (2em corresponding to 24 pts) *Versus* Serif font: Times New Roman (2em corresponding to 24 pts)

APPENDIX 5: PRESENTATION OF THE GRAPHIC CHARTS OF EXPERIMENT 3

Table 10.

Experiment 3	
Color Positive contrast: yellow (#FFFFAC) on green (#006500) *Versus* Negative contrast: green (#006500) on yellow (#FFFFAC)	Layout Dense content with a small font and a small interline (line-height = 12pt, font = 1,5 em) *Versus* Airy content with a large font and a large interline (line-height = 30pt, font = 2 em)

Chapter 17
Mobile Customer Relationship Management:
An Overview

Tolga Dursun
Abant İzzet Baysal Üniversity, Turkey

Süleyman Çelik
Abant İzzet Baysal Üniversity, Turkey

ABSTRACT

Electronic platforms provide many advantages both customers and companies due to development of communication technology. Today almost every people have smartphones and tablets. Thus mobile customer relationship management became an significant concept for generating long-term relationships and increasing customer satisfaction, retention and loyalty. In addition companies use mobile CRM to facilitate salespeople for better performance in marketing activities. M-CRM offers interactive relationships between firms and companies. In this study, we define what is customer relationship management and origins of CRM. After that we stated electronic customer relationship management concept and finally we mentioned about mobile CRM especially benefits and characteristics of it.

INTRODUCTION

Today, companies have to develop long-term relationships with their current and potential customers to survive and maintain their lifes in intensive competitive environment. In this context, customer relationship management is considered as an important tool to achieve lasting relationships. Especially advances in internet and technology shifted these relationships to the electronic environenment. Furthermore, developments in mobile broadband connections (3G,4G) and with the advent of smartphones, world is seeing a huge migration to the mobile technologies and mobile CRM which is a type of e-CRM is emerged. Consequently, mobile CRM has become vital to determine customers' needs and requirements properly and boots satisfaction of customer.

DOI: 10.4018/978-1-5225-2469-4.ch017

EVOLUTION OF CUSTOMER RELATIONSHIP MANAGEMENT

The roots of CRM can be traced back to the term of relationship marketing (RM) (Zablah et al, 2004). Since the competition has changed and structural changes in operations have led to the emergence of the relationship concept for generating long-term relationships among customers and suppliers. Due to the globalisation of business, internationalisation, information technology progression, shorter product life cycles, and the evolving recognition of the relationship between customer retention and profitability (Morgan and Hunt, 1994; Zineldin and Jonsson, 2000; Chandra and Kumar, 2000; Sahay, 2003; Stefanou et al., 2003; cited in Osarenkhoe and Bennani, 2007).

According to Parvatiyar and Sheth (1995), "developing customer relationships has historical antecedents going back into the pre-industrial era. Much of it was due to direct interaction between producers of agricultural products and their consumers. Similarly, artisans often developed customized products for each customer. Such direct interaction led to relational bonding between the producer and the consumer. It was only after the advent of mass production in the industrial era and the advent of middlemen that interaction between producers and consumers became less frequent leading to transaction oriented marketing" (Parvatiyar and Sheth, 2001).

After industrialization and mass production, companies have lost their control over their customers and it became very hard to manage and remember informations about on a large number of consumers. But parallelly advances in information technology, computer technology and wireless communication led to start new possibilities to companies for customer relations.

Recently, several elements have contributed to thedevelopment and evolution of customer relationship management rapidly. These elements are the changes in intermediation process in many industries due to the progression of advanced computer and telecommunication technologies that allow producers to directly communicate with customers. For instance, in many industries such as the airline, banking, insurance, computer software, or household appliances industries and even consumables, the de-intermediation process is fast changing the nature of marketing and consequently making relationship marketing more popular. Databases and direct marketing instrumentsprovide these industries the means topersonalize their marketing endeavours. (Parvatiyar and Sheth, 2001).

As a result since many companies compete in industries, customers have many options to buy product or services and customers became the center of competition. So in addition to advances in technology led to development of CRM, changes in customers' behaviours play an active role in the emergence of CRM.

Definition of Customer Relationship Management

In the marketing literature there are many definitions of customer relationship management from different perspectives. Vavra (1992) defines customer relationship management only as providing customer retention through implementing several after marketing tactics that lead to customer bonding or staying in touch with the customer after a sale is made.Another perspective of CRM database marketing underlying the promotional aspects of marketing linked to database efforts (Bickert, 1992).

According to Hamilton (2001) CRM is the process of storing and analyzing the vast amount of data produced by sales calls, customer service centers, actual purchased, supposedly yielding greater insight into customer behavior. As a result CRM can be defined as an interactive process achieving the excellent balance between corporate investments and the satisfaction of customer needs to generate the maximum profit. It involves (Gebert, et al, 2002):

Figure 1. The CRM continuum
(Payne, A. and Frow, P., 2005)

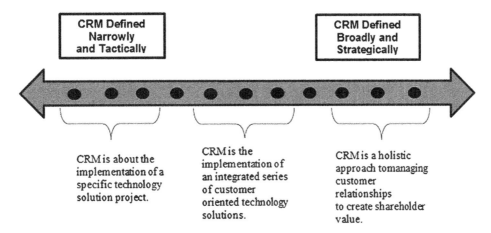

- Measuring both inputs across all functions including marketing sales and service costs and outputs in terms of customer revenue, profit and value.
- Acquiring and continuously updating knowledge about customer needs, motivations and behavior over the lifetime of the relationship.
- Applying customer knowledge to continuously improve performance through a process of learning from successes and failures.
- Integrating the activities of marketing, sales and service to achieve a common goal.
- Implementing appropriate systems to support customer knowledge acquisition, sharing and measuring CRM effectiveness.
- Constantly flexing the balance between marketing, sales and service inputs against changing customer needs to increase profit.

THREE LEVELS OF CUSTOMER RELATIONSHIP MANAGEMENT

As we see, there is no absolute definition of CRM. Some of the complexity comes out since the term is used in a number of different ways. CRM can be seen at three levels: strategic, operational, analytical (Buttle, 2004).

Strategic CRM

Strategic CRM is focused on the development of a customer-centric business culture. This culture is dedicated to winning and keeping customers by creating and delivering value better than competitors (Buttle, 2004).There has been not only customer centric but also it could be found three others orientations such as production, product, sales (Kotler, 2003).

Product-oriented businesses believe that customers always choose best quality products with best design or features. These businesses are often highly innovative and entrepreneurial. Many new business start-ups are product-oriented. In these firms the customer's voice generally to be missing when important marketing decisions are made. There is a little or no customer research is carryed out. Management makes assumptions about what customers want. The outcome is that products are overspecified or overengineered for the requirements of the market, and therefore too costly for the majority of customers (Rogers, 1962; cited in Buttle, 2004).

Sales-oriented businesses believe that if they focus on investing money in sales promotions, advertising, public relations, customers will be keen on buying products of companies. Very often, a sales orientation follows a production orientation. The company produces low-cost products and then they have to promote them heavily to shift inventory (Buttle, 2004).

Operational CRM

Operational CRM contributes and automates customer supporting programs in the business process by using software for marketing, service and selling functions to be integrated. Operational CRM are divided into three forms: Marketing automation, Sales force automation, Service automation (Buttle, 2004).

Marketing automation (MA) uses technology to carry out marketing efforts. A variety of compentences are offered by MA software: customer segmentation, campaign management and event-based marketing. Software enables users to determine their customer data for the purpose of developing targeted communications and offers. In addition MA enables companies to develop budget and execute communication campaigns. MA can also audit and analyse campaign performance, and direct leads from advertising campaigns to the most appropriate sales channel. Sales-force automation (SFA) was the original form of CRM. It applies technology to the management of a company's selling activities. Sales-force automation software enables companies automatically to record leads and track opportunities as they progress through the sales pipeline towards closure. (Buttle, 2004).Service automation facilitate activities of businesses to retain customers by offering best quality of service and building strong relationship. It includes issue management to fix customers' problems, customer call management to handle incoming/outgoing calls, service label management to trace quality of service based on key performance indicators (www.techonestop.com).

Analytical CRM

This is concerned with customer related data such as obtaining, storing, interpreting, integrating, reporting and distributing data to improve not only customer loyalty but also firm's value too. Those data could be internal such as sales data, marketing data, financial data and service data and also external data such as geodemographic data by measuring customer's propensity, response, and value (Greenberg, 2002).

From the customer's point of view, analytical CRM can offers better and more timely, further personally customized solutions to the customer's problems, thereby providing customer satisfaction. From the company's point of view, analytical CRM offers the prospect of more powerful crossselling and up-selling programmes, and more effective customer retention and customer acquisition programmes (Buttle, 2004).

Electronic Customer Relationship Management

Developments ininternet-based technology led to change the way of customers behavior and customers become online consumers. In parallel with companies also changed the way of managing relationships with customers. Thus electronic customer relationship management was emerged. Since traditional

CRM become incapable of meeting requirements of online customers. Companies sustain exercises online platforms for retention and acquisition of customers to survive in competitive environment and to gain financial benefits. Because electronic CRM presents faster and more effective way to interact with customers.

With the rapid advancement in technology especially information and communication technology has helped the scale and scope of customer relationship management. Thus it leads to the increasing use of E-CRM. By integrating and simplifying the customer-related processes through the internet, E-CRM helps to improve customer development, customer acquisition and customer retention (Chang, Liao & Hsiao, 2005).

Although electronic CRM is a new form, there are many different definitions. E-CRM also has no universal specific definition like CRM concept. According to Dyche (2001) E-CRM refers to electronic customer relationship management or, more simply, CRM that is web-based. Dyche (2001) also suggested that there are two main types of e-CRM. These are operational e-CRM and analytical e-CRM. Operational e-CRM is dealt with customer touch points, that is, all methods of customer contact, including Web-based, in person, e-mail, telephone, direct sales, and fax. Analytical e-CRM focuses on technology to process vast amounts of customer data. The purpose is to build new business opportunities via analysing informations such as customer demographics, purchasing patterns, and other factors.

E–CRM is a part of E-business, which describes the use of electronic platforms to conduct a company's business. Electronic business has been heavily influenced by the Internet, which enables firms to serve the customers faster, more accurately, over a wide range of time and space, at a reduced cost, and with the ability to customize and personalize customer offerings (Kotler, 2003).

According to Gilbert, and Mannicom (2003) E-CRM refers to the marketing activities, tools and techniques delivered over the Internet with a specific aim to locate, build and improve long-term customer relationships to enhance their individual potential (Harrigan et al, 2011).

Benefits of E-CRM

E-CRM provides a wide variety of benefits to companies such as improving customer retention, gaining potential customers, determining customers' needs, ensuring satisfied customers and increasing companies' profits etc. All benefits can be achieved by implementing successful E-CRM applications
Adebanjo (2003) determined fundamental benefits of E-CRM as below:

- Reducing the cost of contacting customers by making customer details readily available, customer contact personnel have better opportunities to resolve customer enquiries in less time, thereby freeing them for other productive work.
- Transferring some responsibility to the customer (e.g. product configuration, order tracking, online customer details collection) reduces administrative and operational costs for the organization and therefore, increases the value that an ECRM solution will deliver to the organization.
- Integration of E-CRM applications with back-office systems such as finance supply chains and production can enhance work flow and consequently, the efficiency of the organization, thereby delivering cost savings. For example, field salespeople could use hand-held devices to initiate orders, check stock, track orders, check production status, request invoices and with minimum cost and effort.

Table 1. Tangible and intangible benefits of E-CRM

Tangible Benefits of E-CRM	Intangible Benefits of E-CRM
Decrease internal cost	Improve customer service
Increase revenues and profitability	Streamlined business process
Higher employee productivity	Increased dept and effectiveness of customer Segmentation
Higher customer retention rates	Acute targeting and portfolio of customers
Preserved marketing investments	Increase customer satisfaction
Decrease marketing cost	Better understanding of customer requirements
Maximized returns	Closer contact management

(Chen and Chen, 2004

- E-CRM applications also can improve sales by customer profiling, automated campaign management, e-mail marketing, etc., thereby improving the bottom line for the organization.
- Improving the overall interaction with customers would lead to better service and improve customer satisfaction, loyalty and ultimately customer life-time value.

In addition Chen and Chen (2004) considered E-CRM benefits with two dimensions as tangible and intangible:

INTRODUCTION OF MOBILE CUSTOMER RELATIONSHIP MANAGEMENT (M-CRM)

The mass migration of internet users from desktop PCs to mobile devices is global and universal. The proportion of internet traffic coming from mobile devices grew from 1% in 2009 to 13% in 2013, and it's still growing (http://www.salesforce.com/uk/crm/mobile-crm/). Mobile is at the forefont of the new digital age. Mobile is driving the development of new services in areas such as social networking, digital content and electronic commerce. Mobile is delivering a new and vibrant ecosystem which is based on mobile broadband networks, advanced tablets and smartphones and a growing range of other connected technological devices and objects (GSMA, 2015).

According to GSMA Global Mobile Economy Report (2015) The world is seeing a rapid technology migration to both higher speed mobile broadband networks and the increased adoption of smartphones and other connected devices. Mobile broadband connections will account for almost 70% of the global base by 2020, up from just under 40% at the end of 2014. Smartphone adoption is already reaching critical mass in developed markets, with the devices now accounting for 60% of connections. It is the developing world—driven by the increased affordability of devices—that will produce most of the future growth, adding a further 2.9 billion smartphone connections by 2020.

The mobile industry contiunes to scale rapidly, with a total of 3.6 billion unique mobile subscribers at the end of 2014. Half of the word's population now has mobile subscription. An additional one billion subscirbers are predicted by 2020 taking the global penetration rate to approximately %60. There were 7.1 billion global sim connections at the end of 2014. In addition, At the end of 2014, the number of people using the mobile internet reached 2.4 billion. This is expected to rise to 3.8 billion by 2020 (GSMA, 2015).

There is an accelerating technology shift to mobile broadband networks across the world. Mobile broadband connections (i.e. 3G and 4G technologies) accounted for just under 40% of total connections at the end of 2014, but by 2020 will increase to almost 70% of the total. This migration is being driven by greater availability and affordability of smartphones, more extensive and deeper network coverage, and in some cases by operator. The increasing proportion higher speed connections largely reflects the accelerating rate of smartphone adoption. Adoption rates have already reached 60% of the connection base in the developed word. Over the next four years, smartphone adoption in the developed word is expected to reach the 70-80% ceiling, the level at which growth tends to slow (GSMA, 2015). 2012, about a quarter of all mobile users were smartphone users. By 2018, this number is expected to double, reaching 50 percent. The number of smartphone users worldwide is expected to grow by one billion in a time span of five years, which means the number of smartphone users in the world is expected to reach 2.6 billion by 2019 (http://www.statista.com/statistics/274774/forecast-of-mobile-phone-users-worldwide/).

All these statistics show that the most people use smartphones every hour of the day all over the world to fulfill their daily transactions thanks to technological development in mobile industry. Since mobile devices mostly are used nowadays, Mobile CRM is the critical factor for companies to develop personalized long-term relationships with their potential and current customers. What companies have to do is to set up proper mobile CRM systems to survive in the competitive environment.

Mobile CRM

Mobile CRM can be defined as the communication, either one way or mutual, that is related to marketing, sales and customer service activities through mobile technologies in attemp to build and maintain relationships between the costumer and the company (Kim et al, 2015). Mobile CRM allows companies' employees use mobile devices such as smartphones and tablets to access, update and interact with customer data wherever and whenever they are (http://www.tendigits.com/about-mobile-crm.html).

According to Nguyen and Waring (2013) m-CRM can be defined from two perspectives. From the perspective of technology, m-CRM is seen as a technological tool applied to marketing for the purpose of reducing costs and increasing the efficiency of the processing information between buyer and seller. On the other side, from strategic perspective, m-CRM is seen as a long-term management approach that companies fulfill through mobile channels in order to get benefits (San- Martin et al, 2015).

Recently, Software Advice (2014) carried out a survey including severeal questions and 1,940 responses are acquired from sales professionals in the U.S. who currently access their company's CRM system through a mobile device. Their aim was to find out how it benefits their companies, which features they commonly use and what requirements they feel are most important for a mobile CRM system to meet.

Accordingly their research nearly half of 1,940 sales professionals who are CRM users Access their system via smartphones (48 percent) and/or tables (45 percent). Furthermore, a significant majority (81 percent) say they access their system on multiple types of devices and 20 percent use the combination of three of the above listed devices and 29 percent of participants use four devices to Access CRM system (Software Advice, 2014).

Another important output of research shows that most common used mobile crm applications and features are consecutively sales content management, review/input customer data, sales reporting and analytics, business card scanning and geolocation (Software Advice, 2014).

Figure 2. Devices used to access CRM system
(Software Advice, 2014)

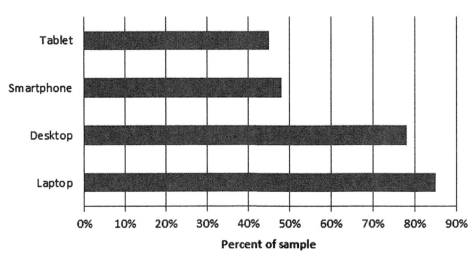

37 percent reported regularly using their mobile CRM system to manage sales content, such as slide presentations and reports. This reflects the increasing importance of tablets, in particular, as a vector for delivering sales content during presentations. In addition 31 percent said that they regularly use mobile CRM to arrange customer informations. 26 percent routinely used business-card scanning—a relatively niche mobile feature—as used such core functionality as reporting or database access (Software Advice, 2014).

Chracteristics and Benefits of Mobile CRM

According to Deans (2004), the convergence of mobile internet and wireless communication technology has promised users the concept "anytime anywhere", which implies access to information for work and

Figure 3. Most used mobile CRM applications and features
(Software Advice, 2014)

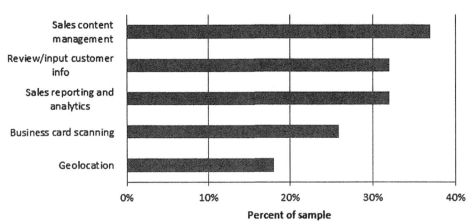

Mobile Customer Relationship Management

personal communication. The mobile medium and wireless technology enable companies' four reasons to build relationships with its customers, which are:

- Personalize content and services.
- Track customers or users across media and over time.
- Contribute content and service at the point of need.
- Contribute content with highly engaging characteristics.

According to Durlacher Research (1999) m-CRM overcomes existing traditional CRM limitations such as obtaining customer information through face-to-face interactions and wired networks by enabling the ability to easily obtain customer information anytime, anywhere. In addition, employees can benefit from rapid and continuous information updates and engage in real-time marketing (Kim et al., 2015).

Verma and Verma (2013) mentioned that m-CRM provides personalized and two- way communication with customers, thereby improving customers' intelligence by making employees easier to gather data on each customer. This allows employees to figure out customer needs better and develop suitable responses as well as to improve interactions with customers by retaining a record of their inquiries, transactions, complaints, and problems solved (Kim et al., 2015).

Turban et al. (2004) proposed mobility and accessibility as the most important characteristics of mobile computing and business and suggested that mobility enables employees to access systems through wireless networks and devices to execute real-time business as well as to search for and process information (Kim et all., 2015).

According to Li and Mao (2012) m-CRM can provide optimal information and services by synthetically considering information on customers, including their location, personal identification, personal background, individual preference, and purchase history as well as other types of information extracted from the CRM database (Kim et al., 2015).

According to the authors Sinisalo et. al. (2006), consider mobile medium of being a powerful opportunity to reach customers, by offering different ways for companies to plan and implement more advanced ways to communicate with their customers. One particular way is SMS, which is seen to be immediate, automated, reliable, personal, and customized channel, which allows an effective way to reach customer directly. Other benefits of mCRM are that mobile medium allows high speed message delivery, relatively low cost and high retention rates. Mobile CRM also provides interactive communication in real-time between companies and their customers (Belachew et al., 2007).

According to Software Advice Research (2014), There are important benefits using mobile CRM such as increased efficiency, better decision making, better follow ups, higher end user adoption. Research results revealed that participants who use smartphones and tablets get higher benefits than only one device users. Important outputs of research as follows:

50 percent of the respondents say that mobile CRM increased their efficiency, and 42 percent say it facilitated "faster, more informed decision-making." In both cases this was over twice the number of single-device users who said the same and furthermore 23 percent of multiple-device users and 15 percent of single-device users said mobile CRM access resulted in higher end-user adoption.

A research made by Nucleus (2012) found that mobile access to CRM increases productivity of sales people by 14.6 percent. This significant increase in productivity is driven by the development of custom, device-specific applications that take advantage of the form factors of individual device. Vendors and

Figure 4. Top benefits realized through mobile CRM
(Software Advice, 2014)

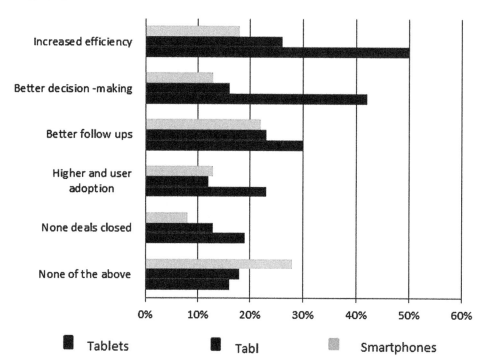

consultants are increasingly delivering task-specific, role- and vertical-based views of mobile CRM data that make it easier for salespeople to go beyond updating their pipeline through their smartphones.

Furthermore, research by Innoppl Technologies showed that 65% of sales representatives that adopted mobile CRM achieved their sales quotas while only 22% of representatives using non-mobile CRM reached the same targets (www.hso.com).

Figure 5. Effect of having mobile CRM on sales quotas
(www.hso.com)

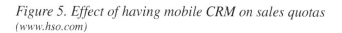
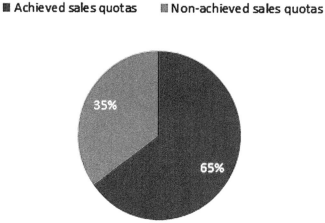

Figure 6. Effect of not having mobile CRM on sales quotas
(www.hso.com)

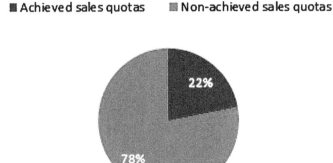

Kim et al. (2015) mentioned that characteristics of mobile CRM are information quality, system quality and service quality. They examined the role of characteristics of m-CRM on employees' personal performance through the full mediating effects of user satisfaction and system use. They found that m-CRM is crucial not only for firms' growth but also for the significant improvement of employees' performance.

San-Martin et al. (2015) examined the perception of companies on the benefits of implementing a m-CRM strategy from relationship marketing and TOE model and found that companies perceived better technologial compentence is more willingness to innovative, the more employee support and the better management of customer information which results in improving customer loyalty and increasing firm's global profits.

Mobile CRM activities also provides trustworthiness which leads to customer loyalty. Sohn et al. (2011) analysed the effect of mobile CRM activities on trust-based commitment and found that trust plays the role of mediator between commitment and mobile CRM activities. Customers who experience mobile CRM activities trust the company and finally commit to the company.

CONCLUSION

Developments in mobile technologies led to smartphones is essential tool for people in the world. Thus mobilizing customer relationship management is necessity for companies to improve and to maintain its relationships with customers. Mobile CRM provides interactive communication with customers on the contrary traditional CRM. Therefore m-CRM enables companies to increase sales people performance, to improve customer service and to facilitate other marketing activities. Thanks to mobile CRM, sales people can use it anywhere and anytime which impacts performance of users positively. If companies want to be successful, they have to improve relationships with their customers. In this context, mobile CRM is a neccessity for companies to have satisfied and loyal customers.

REFERENCES

Adebanjo, D. (2003). Classifying and Selecting E-CRM Applications: An Analysis Based Proposal. *Management Decision, 41*(6), 570–577. doi:10.1108/00251740310491517

Belachew, Y., Hoang, A., & Kourieh, J. (2007). *Mobile Customer Relationship Management: A Study of mCRM adoption in the Swedish Market* (Unpublished Master's Thesis). Jönköping University, Sweden.

Bickert, J. (1992). *The Database Revolution.* Target Marketing.

Buttle, F. (2004). *Customer relationship management: concepts and tools*. Oxford, UK: Elsevier Butterworth-Heinemann.

Chang, T., Liao, L., & Hsiao, W. (2005). An Empirical Study on the e-CRM Performance Influence Model of Service Sectors in Taiwan. *Proceedings of the 2005 IEEE International Conference on e-Technology, e-Commerce and e-Service (EEE'05) on e-Technology, e-Commerce and e-Service*, 240-245. doi:10.1109/EEE.2005.33

Changsu, K., In-Seok, L., Tao, W., & Mirsobit, M. (2015). Evaluating effects of mobile CRM on employees performance. *Industrial Management & Data Systems, 115*(4), 740–764. doi:10.1108/IMDS-08-2014-0245

Chen, Q., & Chen, H. M. (2004). Exploring the Success Factors of E-CRM Strategies in Practice. *Journal of Database Marketing & Customer Strategy Marketing, 11*(4), 333–343. doi:10.1057/palgrave.dbm.3240232

Deans, P. C. (2004). *E-commerce and M-commerce technologies*. IRM Press. doi:10.4018/978-1-59140-566-5

Dyche, J. (2001). *The CRM handbook: A business guide to customer relationship management*. Boston: Addison-Wesley.

Gebert, H., Geib, M., Kolbe, L., & Riempp, G. (2002). Towards Customer Knowledge Management. *Proceedings of the 2nd International Conference on Electronic Business (ICEB 2002)*.

Greenberg, P. (2002). *CRM at the speed of light: capturing and keeping customers in internet real time*. Berkeley, CA: McGraw-Hill.

Hamilton, D. P. (2001, May 21). Making Sense of It All. *The Asia Wall Street Journal*, p. T4.

Harrigan, P., Ramsey, E., & Ibbotson, P. (2011). Critical factors underpinning the e-CRM activities of SMEs. *Journal of Marketing Management, 27*(5-6), 503–529. doi:10.1080/0267257X.2010.495284

Kotler, P. (2003). *Marketing management* (11th ed.). Prentice Hall.

Nucleus Research. (2012). *Market focus report: The value of mobile and social CRM* (Report No. M13). Author.

Osarenkhoe, A., & Bennani, A. E. (2007). An exploratory study of implementation of customer relationship management strategy. *Business Process Management Journal, 13*(1), 139–164. doi:10.1108/14637150710721177

Parvatiyar, A., & Sheth, J. N. (2001). Customer relationship management: Emerging practice, process and discipline. *Journal of Economic & Social Research, 3*(2), 1–34.

Payne, A., & Frow, P. (2005, October). A Strategic Framework for Customer Relationship Management. *Journal of Marketing, 69*(4), 167–176. doi:10.1509/jmkg.2005.69.4.167

San-Martín, S. Jimenez N.H. Lopez-Catalan, B. (2015). The firms benefits of mobile CRM from the relationship marketing approach and the TOE model. *Spanish Journal of Marketing – ESIC, 26,* 18-29.

Sohn, C., Lee, D. I., & Lee, H. (2011). The effects of mobile CRM activities on trust-based commitment. *International Journal of Electronic Customer Relationship Management., 5*(2), 130–152. doi:10.1504/IJECRM.2011.041262

Vavra, T. G. (1992). *After marketing: how to keep customer life through relationship marketing.* Homewood, IL: Business One – Irwin.

Verma, D., & Verma, D. S. (2013). Managing customer relationships through mobile CRM In organized retail outlets. *International Journal of Engineering Trends and Technology, 4*(5), 1696–1701.

Zablah, A. R., Bellenger, D. N., & Johnston, W. J. (2004). An evaluation of divergent perspectives on customer relationship management: Towards a common understanding of an emerging phenomenon. *Industrial Marketing Management, 33*(6), 475–489. doi:10.1016/j.indmarman.2004.01.006

ADDITIONAL READING

GSMA. (2016) GSMA mobile economy 2016. *GSMA.* Retrieved from: http://www.gsmamobileeconomy.com/GSMA_Global_Mobile_Economy_Report_2015.pdf

HSO. (n.d.) The rise of CRM and how it's transforming business interactions. *HSO.* Retrieved from: https://www.hso.com/fileadmin/user_upload/CRM_manufacturing_Whitepaper.pdf

Sales Force. (n.d.) Mobile CRM – Responding to a connected world. *Sales Force.* Retrieved from: http://www.salesforce.com/uk/crm/mobile-crm/

Software Advice. (2014) Mobile CRM Software UserView. *Software Advice.* Retrieved from: http://www.softwareadvice.com/crm/userview/mobile-report-2014/

Statista (n.d.) Number of mobile phone users worldwide from 2013-2019. *Statista.* Retrieved from: http://www.statista.com/statistics/274774/forecast-of-mobile-phone-users-worldwide/

Taylor, M. (2016) 18 CRM Statistics You need to Know for 2017. *Super Office.* Retrieved from: http://www.superoffice.com/blog/crm-software-statistics/

TechOneStop. (n.d.) Types of CRM- operational, analytical, collaborative. *TechOneStop.* Retrieved from: http://techonestop.com/types-of-crm-operational-analytical-collaborative

Ten Digits. (n.d.) About Mobile CRM. *Ten Digits.* Retrieved from; http://www.tendigits.com/about-mobile-crm.html

Chapter 18
Benefits of Using Social Media Commerce Applications for Businesses

Ardian Hyseni
South East European University, Kosovo

ABSTRACT

Social media commerce has changed the way of commerce globally; customers are affected more and more by social media, in decision making for buying a product or a service. While in the past people were affected by traditional marketing ways like newspapers, televisions and radios for buying a product, nowadays, through social media customers can find feedbacks and reviews on social media and can see thousands of photos of a single product with less a minute of searching in a social networking sites like. With the growth of social media's impact on businesses, social commerce has become a trending way of making commerce. In this paper it demonstrated a platform for businesses to make commerce through Facebook which is called Facebook commerce.

INTRODUCTION

Social media ecommerce is a trending way of making business over the internet, despite ecommerce that is made by websites, nowadays social media has made it available to buy and sell products over social networking sites like Facebook, twitter etc. Social media isn't just about posting a feed on Facebook or twitter, or putting like button or comment on your website, it is about connecting customer directly with your website and making visitors loyal customers. Ecommerce dates since 1994 when Jeff Bezos founded Amazon.com and in 1995 when Pierre Omidyar started P2P marketplace eBay (Shaefer, 2014). Both these sites were social networks seen as marketplaces for products with discount price. Customers could leave feedback, post reviews and comment for the products they bought. This was new era of commerce through internet was born. After eBay and Amazon, in 2004, Facebook was founded by a group of Harvard students. Facebook now is a leading social networking site based on the number of users and fan / business pages (Collier, 2012).

DOI: 10.4018/978-1-5225-2469-4.ch018

Businesses want to connect with people and customers which they do business, they want their customers opinion and reviews (Safko, 2009). By using social media, companies now can easily create interaction between company product and the customer. To understand and hear the voice of their customers, businesses need to keep up to date with the technology. Social media marketing is constantly evolving; sites like Facebook, Twitter, LinkedIn are leaders in the online networking which are the current communication trends (Corson-Knowles, 2013). Businesses need to combine new technologies with traditional marketing to increase the sales revenue (Varela, 2015). Social media is not just another way of marketing; it has to become a integrative part of a company. It is understandable that businesses should take more seriously the involvement and planning of social media for commercial gain.

Most of the ecommerce sites have started implementing social media networking site services in order to improve interaction and increase active participation of user in order to increase revenues (Kwahk & Ge, 2012). In a social networking sites users can get answers and interact with each other in order to get more information for a specific product or service. When user wants to order a product online, he or she can ask and find out more information on the social networking site. The aim is to examine how much people do use social media commerce and what is the impact of social media in people's decision making during the buying process in ecommerce.

Social networking sites consist of large number of users who are potential content generators and massive source of information (Underhill, 2008). Users generate new ideas, advertise and add a value for a little cost while increasing the effectiveness by understanding customer needs, identifying potential customers, and building customer loyalty (Kwahk & Ge, 2012). The increased number of users in social networks has led to a new shopping trend where people leverage social networks to make the right purchase. While businesses spend thousands of money in marketing, and it is considered as a temporary investment, in a TV or Newspaper, in social networking sites people who engage at your page they become a lifelong loyal customers (Kwahk & Ge, 2012). Businesses do not need to pay for advertising in social shopping, they can post products in their business page and all customers who follow that page are able to see it (Chaffey, 2011).

Problem Statement

Before discussing social media commerce, we need to analyze a bit of traditional marketplaces that were formally created in Europe around 1000 AD (Schaefer, 2014). Towns and villages were competing with each other towns and villages for commerce legal or illegal until few rules were applied usually by church or mosque leaders. Traditional marketplaces were highly personal and interactive between seller and customer. People usually were standing face to face with the seller, looked them in the eye saw also not only as a seller but also as a personality, and bought with a firm handshake (Schaefer, 2014). People purchased goods from people they knew and trusted or were recommended by a friend or relative. People visited or passed through workshop or farm on the way to market and expected transparency and loyalty from seller where they could buy products right in front of them (Mikalef, Giannokos & PAteli, 2012). If people felt cheated they knew where and whom they bought from and they could be knocking at seller's door.

In traditional marketplaces success of sellers depended on word of mouth or people's recommendation, feedback on quality, service and pricing was constant and immediate [6]. That time, there was no advertising so you needed to treat people right, it was sellers themselves who created their own authority of their shop or workshop. So if seller cheated a buyer, the word would spread throughout the marketplace like a plague, so that, the authority of the seller would be depended on buyer's feedback (Schaefer, 2014). In traditional marketplaces it wasn't just about buying and selling; it was also the social aspect of marketplaces where people talked about news, gossips and themselves. Than around 1400s things started to change, the invention of printing press and newspapers magazines flyers soon followed which was era of advertising, it was a new step away from traditional successes of sellers where advertising was word of mouth, people's recommendation and interactions between buyer and seller (Collier, 2012). Till 1920s still commerce was done through neighbors and small family-owned businesses on every town, until the first commercial broadcast radio station went on the air, and era of communication had begun. After radio came out Television, the internet and websites which made commerce easier and increased number of customers, but all of these traditional ways are evaporating. Newspapers are losing print circulation all around the world, people read online through web or social sites, television programs now are watched more likely through Netflix or Hulu, or buying movies and series for their iPads (Schaefer, 2014).

Traditional media channels are fading away, so where are people going? (Mikalef, Giannakos & Pateli, 2012) The social networking sites like Facebook Twitter LinkedIn Pinterest Google+ are new town squares for people to share news, photos, videos, and their personal life events. Latest trend is social media commerce, from where people through these social media networking sites are looking for products and finding answers to problems (Farooq & Jan, 2012). Research field is social media commerce, which is going to be new way, and most common way of making commerce through social media sites like Facebook and other social networking sites.

Motivation

Motivation of this thesis is based on social shopping trend that is taking a lead for commerce over the internet. Social shopping applications are new trending ways of commerce over the social media networking sites. At the moment there are only a few types of these applications that are circulating over the internet. The weakness of these applications is expensive membership per month. Small businesses want a cheap and flexible app for their own store.

There is a fact that 85% of orders from social media sites come from Facebook? Average orders 55$. According to US Social Commerce – Statistics & Trends it is expected that in 2015 social media commerce will represent five percent of all retail revenue in 2015 (Bennett, 2014). This five percent revenue is more than 15 billion, but according to Booz & Company (2011) it is predicted that social media commerce will revenue more than 30 billion dollar in 2015. These statistics tell us that social media commerce is rapidly raising and is a trend of commerce. Let's take a look some statistics for social media commerce:

- 90% of all orders are influenced according social media (Cheshire & Rowan, 2011).
- 90 percent of customers trust recommendations from people they know (Nielsen, 2009).
- 85% of orders from social media sites come from Facebook (Zephoria, n.d.).
- 9.5% of bump up price comes from positive product reviews (Booz & Company, 2011).
- 11.5% of negative reviews change person's intent of purchase (Zephoria, n.d.).
- 33% of costumers act on a promotion of brands from social media page (Statista, 2014).
- 30 million small business pages are active in Facebook and more than 1 billion active users (Bazaarvoice, n.d.).

Based on above statistics about impact of social media in commerce, it is concluded that in 2015 it is expected a huge increase of commerce through social media comparing to previous years. Social media is leading us to new way and trending way of commerce, just like in the past when Amazon and eBay started doing commerce. Firstly, it began just in USA, but few years later there was as a dramatic increase of sales over the internet all around the world (hou, Zhang & Zimmerman, 2013) . The trend of commerce in the past is going to be the same with social media commerce. People are looking for new trends of commerce, which is social media commerce (Yang, Kim & Dhalwani, 2008). It began with just a billion in 2010 and now it is increasing significantly.

LITERATURE REVIEW

Before social media networking sites emerged, lots of business activities existed in the online electronic world, which was called e-commerce. E-commerce websites existed in the past and people were able to buy and sell products and services over the internet through e-commerce websites (Yang, Kim & Dhalwani, 2008). Social media commerce is a combination of social media and e-commerce which together make social media commerce, which stands for buying and selling products over social media networking sites. Social media has changed and revolutionized the way that people communicate and share information. Globally, internet users spend more than four and a half hours per week on social networking sites, which make social networking sites huge potential market for customers and retailers (Booz & Company, 2011).

Related Work

Social commerce is not new, it began since 2005 when firstly Yahoo presented this word (Ysearchblog.com, 2005). But with the growing of e-commerce, social commerce started to enhance in different ways. Firstly were recommends, reviews and comments, but only in e-commerce sites. Afterwards there were needed to develop new ways for selling products through the social networking sites. Like and comment buttons connected to websites were implemented in e-commerce sites, that many people think it is social commerce, it is but not exactly (Guo, Wang, & Leskovic, 2011). Social media commerce is buying and selling directly from social networking sites like Facebook or twitter or any other social site (Marsden, n.d.). Social media commerce isn't just a trending way of business, it's the future and low cost commerce with not much advertising needed.

In Figure 1 it is shown the trend on how social media commerce has increased since 2010. And are based on this chart, one can see that it has at least 100-200% increase each year. In upcoming years, one can imagine how much of commerce will be done through social media sites.

Why Social Media?

There are many reasons why people should invest on social media, as there are thousands different purposes of using social media. In this thesis there are claimed some of the reasons why should be investing on social media, but not only for fun, mainly for business purposes. While social media has become a place for grouping people in one site for chatting, sharing photos and experiences, there are a few more reasons why social media can be good for business purposes.

Figure 1. Social commerce 2010-2015
(Booz & Company, 2011)

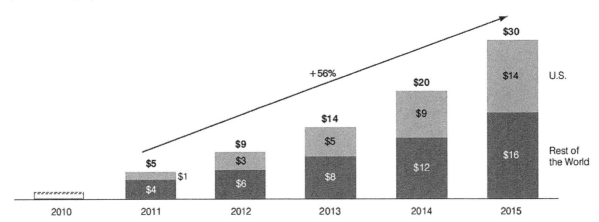

List of reasons why to invest in social media commerce:

1. Worldwide, there are over 1.39 billion monthly active Facebook users (MAUs) which is a 13 percent increase year over year (Zephoria, n.d.). This information helps our research to be concentrated on Facebook social networking site.
2. 4.5 billion likes generated daily as of May 2013 (Zephoria, n.d.). If there are 4.5 billion likes generated daily, even if there is only 1% for a product promoted this would be enough.
3. 890 million people log onto Facebook daily (DAU) for December 2014 (Zephoria, n.d.), If we provide a good app for business pages imagine having 890 million people daily active, it's like having 890 million customers available.
4. There are 1.9 billion mobile active users (MAU) (Zephoria, n.d.) .The use of Facebook advertising methods, business can get at any of these 1.9 billion mobile active users via Facebook.
5. Age 25 to 34, at 29.7% of users, is the most common age demographic (E-Marketer, 2014). Why it is important for a business the age? It is because most of online shoppers are between 18-35 years old.
6. Five new profiles are created every second. (Zephoria, n.d.) .The idea is to invest in any app for Facebook, imagine having 5 new potential customers every second in your store or app.
7. Photo uploads total 300 million per day (Chan, 2012). Imagine, from 300 million only 10% of these photos, to be photo of products, it is a huge marketing for free by social media for businesses.
8. Average time spent per Facebook visit is 20 minutes (Socialskinny, 2012). It's like having a potential customer that has 20 minutes to visit your store on Facebook.
9. Every minute on Facebook: 510 comments are posted, 293,000 statuses are updated, and 136,000 photos are uploaded (Hubspot, n.d.). Nowadays anyone who buys a product or a wear takes a picture and posts on Facebook so the key of marketing is to get to peoples eye and ea,r in the past was marketing of word of mouth no is word of social media.
10. 42% of marketers report that Facebook is critical or important to their business (Hubspot, n.d.). If in Facebook, there are around 1billion daily active users and your business doesn't have a page it's like being out of a huge potential market.

11. 16 Million Local business pages have been created as of May 2013 which is a 100 percent increase from 8 million in June 2012 (Zephoria, n.d.). If we have the latest statistics this number would be a lot bigger than this. 16 million local business pages in one website like Facebook make things much easier. Facebook seems to take a lead over Google Ads in aspect of marketing.
12. 7.5 million Promoted posts have been made from June 2012 to May 2013 (Zephoria, n.d.). Marketing way is changing; social media has become a powerful way of marketing. Newspapers, TVs and radios are going to fade away in aaspect of marketing.

After analyzing this statistics and explaining the business benefit from Facebook it is decided that for the application development to be concentrated on Facebook application (Corson-Knowles, 2013).

RESEARCH METHODOLOGY

In this chapter, the research methodology, research questions, hypothesis and survey questions will be presented in order to explore the way which will be taken for the purpose to test and examine the hypothesis and results of the research. Finally, how the data collected and results will be analyzed.

Research Methodology

In order to examine the questions for our research, in this thesis we raised few hypothesis. For the purpose of evaluating the hypotheses, some techniques will be used as quantitative approach through the questionnaire. To test the developed model and hypothesis, an effective research and methodology is required. It should be considered the conditions such as research questions, purpose of questions, and amount of time (Zhou, Zhang & Zimmerman, 2013). Quantitative and qualitative researches are the best way to explain and understand any phenomena. The research can be achieved by analyzing experiences and thoughts of individuals and groups. Quantitative research questions are the best strategy for gathering data, in this case survey/questionnaire. For this purpose, survey is the best strategy suggested for gathering data and finding the answers. Through survey, in short time can get bigger number of participants and data collected.

Research Questions

Research questions will be prepared according to the hypothesis which will be raised based on experiences and expectations of participants in this research. In this research questionnaire, there are 23 questions, which all of these have its own importance, its own purpose on the research. Through these questions that are prepared, results will be examined and find answers to the research findings. Results of the answers provided in the survey from participants will lead the thesis in which social networking site to be focused on preparing the application. Questions will be classified through a process, from which will lead to the adequate answers and steps will be taken to proceed with the questionnaire.

Survey contains of three parts of questions. First part contains three questions about participant's education level, department and gender they belong which are needed for profiling. Second part contains questions about social media commerce what knowledge do participants have and what do they think about the future of social media commerce. Third part contains questions about measuring how much impact does social media has in participant's decision making, and the last question is for estimating how much do participants do believe that social media commerce will be a leading way of commerce in the future.

Survey

In this research questionnaire, there are 10 questions, which all of these have its own importance, its own purpose on the research. Through these questions that are prepared, results will be examined and find answers to the research findings.

RESEARCH FINDINGS

In this chapter there are presented the research findings from survey taken by 100 participants. Participants are from Kosovo and Macedonia mostly, students from South East European University and others from different Universities around Europe who are friends and colleagues from different departments. From research results, one can get the demographics, such as the department that the participants belongs to, the gender of participants, education level, their shopping experience and expectations for social media commerce in the future.

Results

In this section are shown the results of the findings from the questionnaire taken by seven companies who have implemented social commerce application on their Facebook business page.

What is the Industry of the Company?

In Figure 2 are shown the sectors of the companies who participated on the research.

Table 1. Research questions

	Questions
1	What is the industry of the company?
2	Do you have e-commerce website?
3	How would you rate the benefits of Using Social Media Commerce apps on your business?
4	What is the number of sales that your company has made from social commerce apps since implementation?
5	What are the lacks of social commerce application?
6	Which are advantages of using Social Media Commerce?
7	Social Media Commerce will be a leading way of commerce in the future?

Figure 2. Sectors of the companies who participated on the research

Answer Choices	Responses	
Jewelry	14.29%	1
Electrical	0.00%	0
Lighting	28.57%	2
Clothing	0.00%	0
Sanitary	14.29%	1
Furniture	0.00%	0
Bookstore	0.00%	0
Flower Store	14.29%	1
Industrial Power tools	28.57%	2
Total		7

Figure 3. Results whether companies who participated have e-commerce website

Answer Choices	Responses	
Yes	57.14%	4
No	42.86%	3
Total		7

Do You Have E-Commerce Website?

In Figure 3 are shown the results whether companies who participated have e-commerce website in order to see how familiar companies are with online trading, Out of 7 companies who participated 4 companies have e-commerce website.

How Would You Rate the Benefits of Using Social Media Commerce Apps on Your Business?

In Figure 4, it clearly shown that social media commerce apps increase the effectiveness of the company on social media, but increase the sales revenue remains low in percentage.

What Is the Number of Sales That Your Company Has Made From Social Commerce Apps Since Implementation?

In Figure 5, it is clear that number of sales remains too low, in next question it will seen which are the reasons of low number of sales through social media.

Figure 4. Benefits of Using Social Media Commerce apps for a business

	1	2	3	4	5	N/A	Total	Weighted Average
Customer satisfaction on responding for prices	14.29% 1	14.29% 1	14.29% 1	14.29% 1	42.86% 3	0.00% 0	7	3.57
Increase the sales revenue	28.57% 2	0.00% 0	0.00% 0	14.29% 1	28.57% 2	28.57% 2	7	3.20
Increase the effectiveness of the company on social media	28.57% 2	0.00% 0	0.00% 0	0.00% 0	71.43% 5	0.00% 0	7	3.86
Catalog product and price-list on the social networking page	14.29% 1	14.29% 1	14.29% 1	14.29% 1	42.86% 3	0.00% 0	7	3.57

Figure 5. Number of sales that the company has made from social commerce apps since implementation

Answer Choices	Responses	
0-9	85.71%	6
10-49	14.29%	1
50-99	0.00%	0
100-999	0.00%	0
1000+	0.00%	0
Total		7

What Are the Lacks of Social Commerce Application?

In Figure 6, it is see that one of the main reasons that business think that customers do not buy through social media: is the lack of experience to do shopping via social media.

Which Are Advantages of Using Social Media Commerce?

In Figure 7, it is clear that Social Media Commerce has many advantages and benefits to businesses. All of above can be seen in Figure 7.

Figure 6. Lacks of social commerce applications

Answer Choices	Responses	
Price of platform for social commerce is expensive	14.29%	1
Customers aren't used to buy via social networking sites	100.00%	7
Social Media Commerce still remains new trend and needs time.	71.43%	5
Non of Above	0.00%	0
Total Respondents: 7		

Figure 7. Advantages of using social media commerce

Answer Choices	Responses	
Social Media Commerce is replacing E-Commerce sites?	71.43%	5
You have your own store on Facebook business page no need for a e-commerce website?	85.71%	6
Social media ads have a targeting people option which is more useful than traditional media like (TV, Radio, Newspaper) and it is much cheaper	71.43%	5
Your clients become potential customers once they become fan of your social media page and each product or promotion you make people are able to see on their news feed	100.00%	7

Total Respondents: 7

Social Media Commerce Will Be a Leading Way of Commerce in the Future?

In Figure 8, It clear that businesses are skeptic that social media commerce will be a leading way of commerce in the future, so most of the companies are neutral.

DESIGN OF NEW SOCIAL MEDIA COMMERCE APP

The whole idea behind this application is to create an e-commerce website that will be integrated to Facebook for the sole purpose of selling. When one mentions e-commerce website, it should be clear that the idea is to allow for every user to create an account and to be able to have its own store. Each user will have its own admin area for managing the store and to have the possibility of publishing its store on Facebook business page. To do this there is needed for a good plan on developing this application.

The Idea

The idea for creating a social media commerce application is based on the new trending way of commerce which is a new topic for most of companies. While Facebook has 1.4 billion users and 300 million active users daily it makes us think what all those people are doing in a single web (Booz & Company, 2011). By using social media and e-commerce, a combination of these technologies, leads us to social media commerce. There are a few applications for buying and selling products through social media sites, but it hasn't emerged totally and people do not know much about this trending way of commerce. While looking at these few social commerce applications over the Internet, it is decided to come up with a new application and new idea. The application is going to be called SABOF.

Figure 8. Will social commerce be a leading way of commerce in the future?

Answer Choices	Responses	
Strongly Disagree	14.29%	1
Disagree	0.00%	0
Neutral	42.86%	3
Agree	28.57%	2
Strongly Agree	14.29%	1
Total		7

What Is SABOF?

SABOF is acronym of Selling And Buying Over Facebook – Sabof, is a new trend of shopping application (app) where you can create your page and post your products, from where you are able to sell and accept money via PayPal. The purpose of this app is to sell products through Facebook. It is simple to use and enough completed to set up a store and start selling. It is absolutely free! You don't need to pay for setting up a store and you can start selling from your Facebook business page for free.

How Does It Work?

The idea is simple, just pointing the browser to SABOF.com and click to a button create online store with Facebook, User will have to be signed up with Facebook through the login button, after which the user will be directed to his/her admin are. In admin area the user will have to give the name to the store and fill in the store with categories and products. After creating categories and products, the user will be able to write information about his/her company, payments, privacy policy, shipping and contact form.

Adding SABOF to Facebook

After filling the user store with the products and categories, in admin area there is the Facebook Tab, which when user clicks on it he/she will be redirected to the Facebook business page and the user will have to choose which Facebook business page wants to install the store in and start selling products online.

Before setting up the store, the user has PayPal link in admin area where he/she is supposed to enter his/her PayPal account details, in order payment's money to be transferred to user's PayPal account.

Design of UI Prototype

The design proposed for SABOF will look like Figure 9. Each user will have the store published in his/her Facebook business page within the shop tab. This is the prototype proposed, where in the next part it will be shown how it really looks like.

The Admin area prototype that is proposed looks like Figure 9. Here user will be able to manage its own store. In the next part of this chapter, there will be an example on how the user will create the store.

Proposed Technology

This application uses the PHP scripting language and the MySQL database application as the foundation of the websites (Ullman, 2013). Of course HTML is involved, as it is also CSS for any website. For design it will be used also Bootstrap and some free templates form Bootstrap technology.

Java script and jQuery framework are used to enhance the application, to add some extra functionalities. As long as our sign up and log-in in the system, it will use Facebook, therefore it is required to use Facebook SDK for PHP or JavaScript.

Benefits of Using Social Media Commerce Applications for Businesses

Figure 9. Design proposed for the store page

Figure 10. Proposed admin prototype

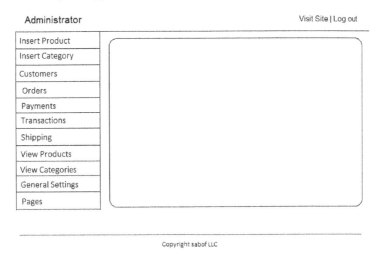

Developing the System and Integration with Facebook

After the User Interface design is proposed and final idea is finalized, next step is to write down HTML and SQL commands, or creating the application. But before that, few steps should be analyzed. Each step follows the other: From the beginning with planning till improving, as shown in the Figure 11.

After the programming part is finished, let's take a look at the server organization. In Figure 12 and Figure 13 are shown the files organized in the server side.

The Few Steps to Use SABOF

After everything is prepared, SABOF app is going to be tried. To see if everything is programmed well and the app is functional, a simple shop is going to be created. Firstly it is needed to sign up or login with Facebook credentials. After the user logs in, he is redirected to admin area where he/she should fill the

Figure 11. The development process

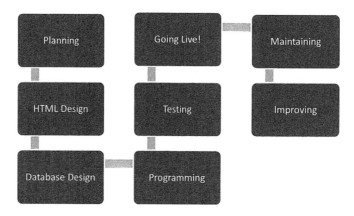

Figure 12. System organization: Files in the server side

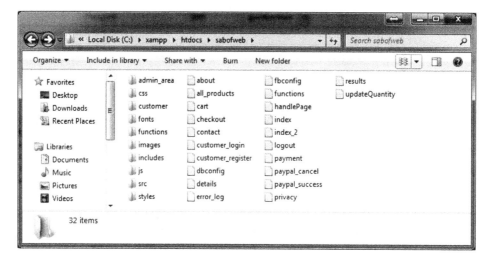

Figure 13. System organization: Files of admin area in the server side

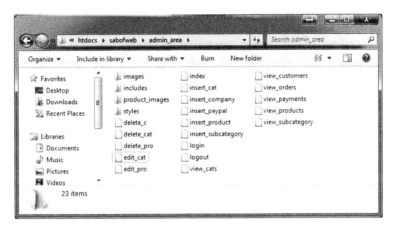

store with the info and products. After filling the web data, the user should go to PayPal section to write his/hers PayPal account details that are required for receiving the payments. After the PayPal account is registered, the user must go to the Facebook tab, where he/she will be directed to his/her Facebook business page, and user can see the store on Facebook (user must have Facebook business page in order to publish its store to Facebook business page),

For creating a SABOF store and to post it to Facebook business page, user must go to www.sabof.com. In the front page there is a button "Create Store Now" as shown in Figure 14, which will direct user to Log In or Sign Up page, as shown on Figure 15.

After being redirected to Sign Up page, user must click to Login with Facebook button because the credentials and authentication are handled by Facebook (Figure 15). User must have Facebook account in order to sign up.

After the user is logged in, he/she will be redirected to Admin Area (Figure 16), where the user will be able to write company details, insert categories, insert products, insert PayPal account and insert the store into Facebook business page.

In Figure 16 is shown the company details page, where the user will be able to give a name to the store, write email address, put a logo of the company, and write about company and terms and conditions of payments of the company.

Where as in Figure 17 is shown the page that is used for inserting the categories for the products in the store.

In Figure 18 there is a screen of the categories that are created, and that can be edited or deleted by user in the future.

Figure 14. Create online store

Figure 15. Sign up or log in area

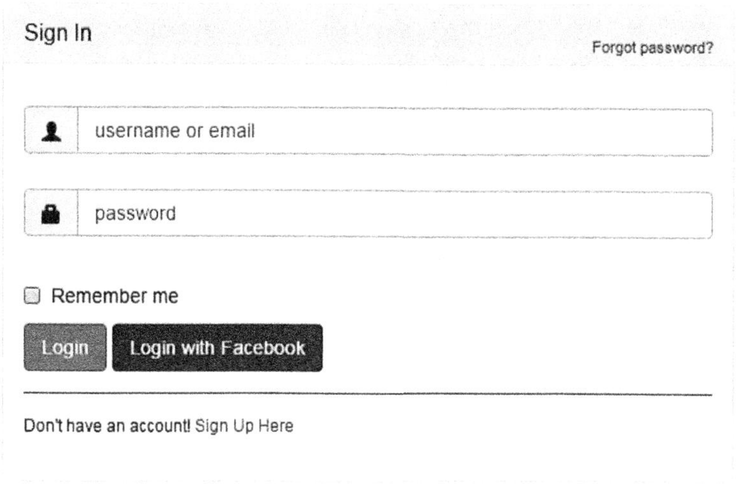

Figure 16. Admin area: Company details

Figure 17. Inserting category

Benefits of Using Social Media Commerce Applications for Businesses

Figure 18. Categories

The Figure 19 shows the page for inserting products to the store. In the product page, the user can write the name of the product, insert the category, insert subcategory, and insert product picture, product price, product details and keywords.

In Figure 20 is shown the page of the products which are created for our test store. In the view products page, are the number of products, product title, product image, product price, product edit button and delete button.

In Figure 21 is shown the page for inserting the PayPal email account, in order payment money to be delivered to users PayPal account. PayPal account should be written before the store is published in the Facebook business page.

In Figure 23 is shown the Facebook shop tab page. In this page user will be able to post the store to his/her Facebook business page. There is also 'Add Another Page Tab' link in case user has more than one Facebook business page.

Figure 19. Insert product

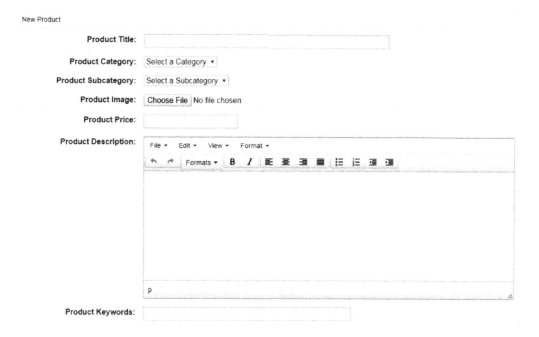

Figure 20. View products

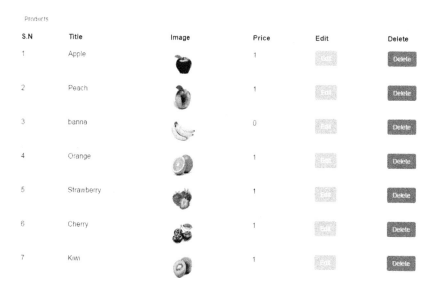

Figure 21. Insert PayPal account

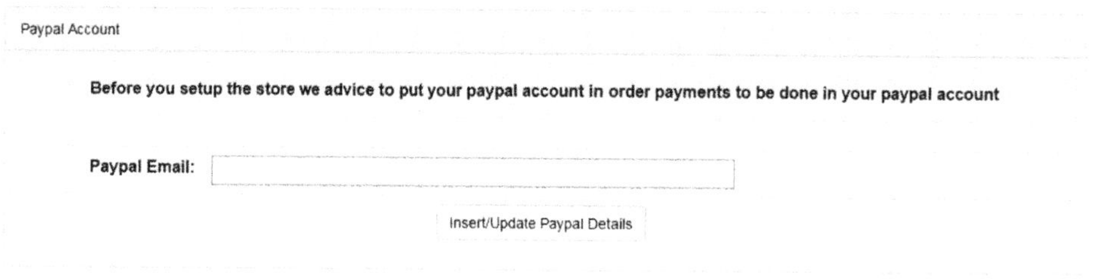

Figure 22. Facebook shop tab

In Figure 24 is shown the actual store, which is created for the test purpose for this thesis. As it is seen the design, is good and simple to use. All the categories and products which were created are shown in this test page.

In Figure 25 is shown the cart page. In the cart, quantity of the products can be updated, and also user can continue the shopping. After final order is decided, user can proceed with check out.

Figure 23. Store on Facebook business page

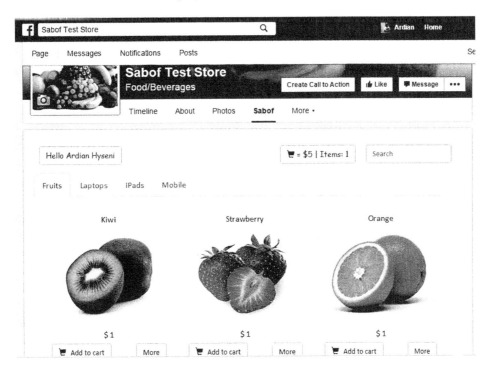

Figure 24. Cart page

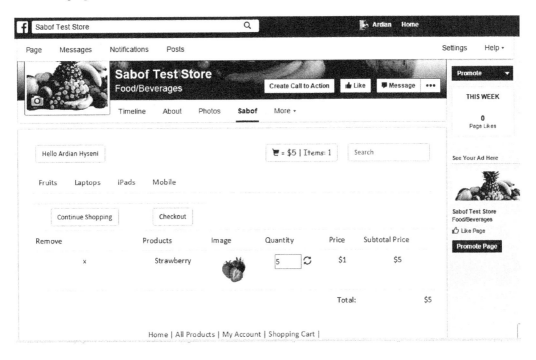

Figure 25. Payments by PayPal

In Figure 26 is shown the check-out page, where is a PayPal button, which is for directing the user to the PayPal payment page.

In Figure 27 is shown the order summary and on the right is PayPal Login account.

In Figure 28 is shown the page, when user Logs In and continues with payment of the order.

Figure 26. PayPal log in

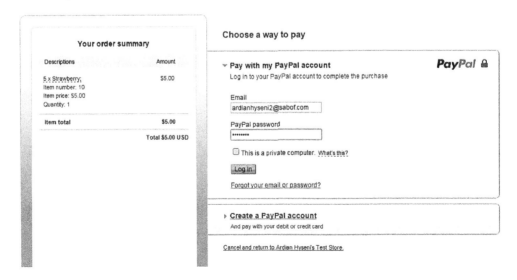

Benefits of Using Social Media Commerce Applications for Businesses

Figure 27. PayPal account logged in

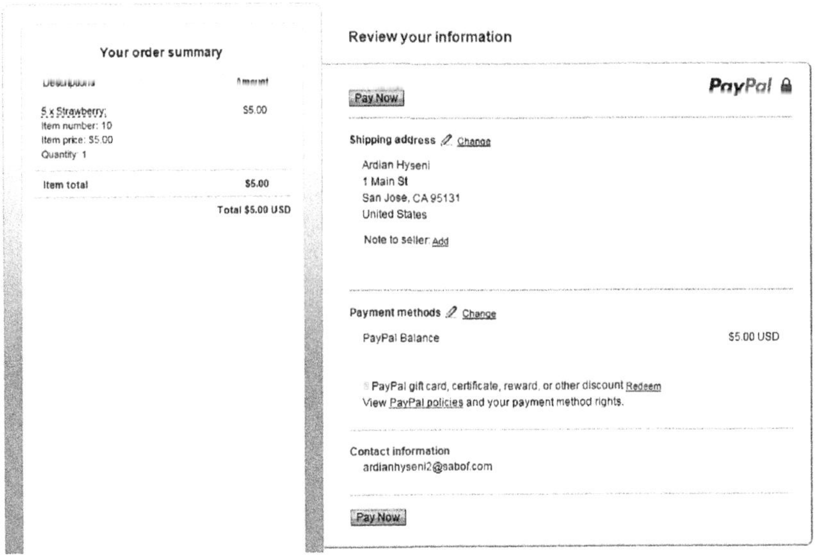

Figure 28. Payment completed

In Figure 29 is shown the message that thanks for the order, and tells that the payment was finished successfully. Finally user will be redirected to the store.

In Figure 30 is shown the message that tells that payment was successful. And user can be directed to his account.

In Figure 31 is shown the invoice that is delivered to customer's account. User can see the invoice and has a link, which can direct user to his account to see other order details.

Figure 29. Payment approval

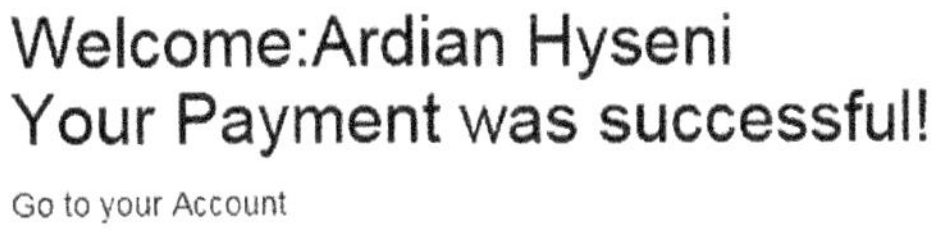

Figure 30. Invoice delivered to customers account

Future Development

The app provides the basic things for buying and selling products from a personal store with a PayPal payment method and shipping. For the future it will be worked on Currencies, Languages, Shipping, Membership, Payment methods and product gallery.

Currencies (Figure 31) will be put, to make possible for stores from different countries to be able to make payments with their own currencies and put the currencies in their personal store for their payment process.

Another future development that is proposed is going to be a membership (Figure 32), for the number of products per store. The reason is that servers need capacity for saving a huge data for many stores. The membership, it will not be too expensive and it will be calculated according to the number of products per store.

Figure 31. Currency

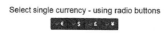

Figure 32. Membership feature

Benefits of Using Social Media Commerce Applications for Businesses

Another future that is proposed is integration of other payment methods like Credit Card. It is proposed for customers to be able to choose which payment method they want, PayPal or Credit Card (Figure 33).

Another important feature that is proposed for the application is integration of other shipping methods to the application (Figure 34). As it is known, most useful shipping methods around the world are DHL, TNT, UPS, FedEx, EMS.

Another feature that is being under construction is the bazaar or shop.sabof.com, where all products from all shops created in SABOF application will be added in one page with its own category (Figure 35). And for the future it is being planned how to implement the payment system for all products from different stores in one store.

Figure 33. Credit card payment method

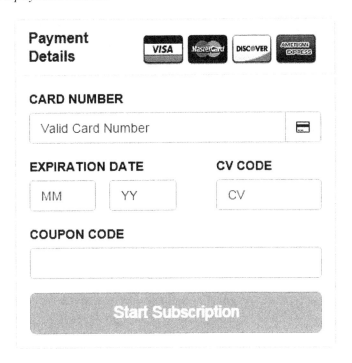

Figure 34. Integration of other shipping methods

Figure 35. Bazaar for all shops

CONCLUSION

Generally social media commerce is an emerging shopping trend that is increasing significantly. So many apps for selling products through social networking sites are developed for a very short time. Based on the previous years the rise of social commerce it tends to be the leading way of commerce in the coming years.

Finally, based on the social trend and the app that is developed. It can be concluded that social commerce applications will have positive impact in businesses in all aspects of shopping experiences.

REFERENCES

Bazaarvoice. (n.d.). *Social commerce statistics*. Retrieved from: http://www.bazaarvoice.com/research-and-insight/social-commerce-statistics/

Bennett, S. (2014). U.S. social commerce – statistics & trends (infographic). *AdWeek*. Retrieved from: http://www.adweek.com/socialtimes/social-commerce-stats-trends/500895

Booz & Company. (2011). *Turning "Like" to "Buy" Social Media Emerges as a Commerce Channel*. Booz & Company Retrieved from: http://www.strategyand.pwc.com/reports/turning-like-social-media-emerges

Chaffey, D. (2011). *E-business and e-commerce management: strategy, implementation, and practice* (4th ed.). Upper Saddle River, NJ: Prentice Hall.

Chan, C. (2012). What Facebook Deals with Everyday: 2.7 Billion Likes, 300 Million Photos Uploaded and 500 Terabytes of Data. *Gizmodo*. Retrieved from: http://gizmodo.com/5937143/what-Facebook-deals-with-everyday-27-billion-likes-300-million-photos-uploaded-and-500-terabytes-of-data

Cheshire, T., & Rowan, D. (2011). Commerce gets social: How social networks are driving what you buy. *Wired Magazine UK*. Retrieved from: http://www.wired.co.uk/magazine/archive/2011/02/features/social-networks-drive-commerce

Corson-Knowles, T. (2013). *Facebook for Business Owners: Facebook Marketing For Fan Page Owners and Small Businesses (Social Media Marketing)* (vol. 2). TCKPublishing.com.

E-Marketer. (2014). Younger Users Spend More Daily Time on Social Networks. *E-Marketer*. Retrieved from: http://www.emarketer.com/Article/Younger-Users-Spend-More-Daily-Time-on-Social-Networks/1011592

Farooq, F., & Jan, Z. (2012). The Impact of Social Networking to Influence Marketing through Product Reviews. *International Journal of Information and Communication Technology Research*, 2(8).

Guo, S., Wang, M., & Leskovec, J. (2011). The Role of Social Networks in Online Shopping: Information Passing, Price of Trust, and Consumer Choice. *EC '11 Proceedings of the 12th ACM Conference on Electronic Commerce*, 157-166.

Hubspot. (n.d.). *State of Inbound Marketing 2012: The ultimate list of marketing statistics*. Retrieved from: http://www.hubspot.com/marketing-statistics

Kwahk, K. Y., & Ge, X. (2012). The Effects of Social Media on E-commerce: A Perspective of Social Impact Theory. *2012 45th Hawaii International Conference on System Sciences*. IEEE.

Marsden, P. (n.d.). *Social Commerce: Monetizing Social Media*. Syzygy Group. Retrieved from: http://digitalintelligencetoday.com/downloads/White_Paper_Social_Commerce_EN.pdf

Mikalef, P., Giannakos, M., & Pateli, A. (2012). Shopping and Word-of-Mouth Intentions on Social Media. *Journal of Theoretical and Applied Electronic Commerce Research*, 8(1), 17–34.

Nielsen. (2009). Global advertising consumers trust real friends and virtual strangers the most. *Nielsen*. Retrieved from: http://www.nielsen.com/us/en/insights/news/2009/global-advertising-consumers-trust-real-friends-and-virtual-strangers-the-most.html

Safko, L. (2009). *The Social Media Bible: Tactics, Tools, and Strategies for Business Success*. Hoboken, NJ: Wiley.

Schaefer, M.W. (2014). Social media explained: Untangling the world's most understood business trend. In *Social Media Commerce For Dummies*. Wiley.

Socialskinny. (2012). 100 social media statistics for 2012. *Social Skinny*. Retrieved from: http://thesocialskinny.com/100-social-media-statistics-for-2012/

Statista. (2014). *Social media and user-generated content tools used by omnichannel retailers in the United States as of 1st quarter 2014*. Retrieved from: http://www.statista.com/statistics/308539/us-omnichannel-retailer-social-media-tools/

Ullman, L. (2013). Effortless E-Commerce with PHP and MySQL (2nd ed.). New Riders.

Underhill, P. (2008). Why We Buy: The Science of Shopping--Updated and Revised for the Internet, the Global Consumer, and Beyond. Simon & Schuster.

Varela, C. (2015). *Facebook Marketing for Business: Learn to create a successful campaign, Advertise your Business, Strategies to generate traffic and Boost sales (Social Media)*. Amazon.

Yang, T. A., Kim, D. J., & Dhalwani, V. (2008). *Social Networking as a New Trend in e-Marketing. In Research and Practical Issues of Enterprise Information Systems II* (pp. 847–856). Boston: Springer.

Ysearchblog.com (2005). *Social Commerce via the Shoposphere & Pick Lists*. Retrieved from: Ysearchblog.com

Zephoria. (n.d.). The top 20 valuable Facebook statistics. *Zephoria*. Retrieved from: https://zephoria.com/top-15-valuable-Facebook-statistics/

Zhou, L., Zhang, P., & Zimmerman, H. D. (2013). Social Commerce Research: An integrated view. *Electronic Commerce Research and Applications*, *12*(2), 61–68. doi:10.1016/j.elerap.2013.02.003

Compilation of References

Abdi, L., Abdallah, F. B., & Meddeb, A. (2015). In-vehicle augmented reality traffic information system: A new type of communication between driver and vehicle. *Procedia Computer Science*, *73*, 242–249. doi:10.1016/j.procs.2015.12.024

Abt, E., Bader, J. D., & Bonetti, D. (2012). A practitioners guide to developing critical appraisal skills: Translating research into clinical practice. *The Journal of the American Dental Association*, *143*(4), 386–390. doi:10.14219/jada.archive.2012.0181 PMID:22467699

Ackermann, T. (2012). *IT Security Risk Management: Perceived IT Security Risks in the Context of Cloud Computing*. Berlin, Germany: Springer-Gabler.

Adamopoulos, P. (2013). *What makes a great MOOC? An interdisciplinary analysis of student retention in online courses*. Paper presented at the 34th International Conference on Information Systems, Milan, Italy.

Adebanjo, D. (2003). Classifying and Selecting E-CRM Applications: An Analysis Based Proposal. *Management Decision*, *41*(6), 570–577. doi:10.1108/00251740310491517

Adlakha, M. (2016). Mobile commerce security and its prevention. In S. Madan & J. Arora (Eds.), *Securing transactions and payment systems for m-commerce* (pp. 141–157). Hershey, PA: IGI Global. doi:10.4018/978-1-5225-0236-4.ch007

Agarwal, R., & Prasad, J. (1998). A conceptual and operational definition of personal innovativeness in the domain of information technology. *Information Systems Research*, *9*(2), 204–215. doi:10.1287/isre.9.2.204

Aghaee, N. M. (2010). *Social media use in academia; campus students perceptions of how using social media supports educational learning* (Unpublished master's thesis). Uppsala University.

Agnihotri, R., Dingus, R., Hu, M. Y., & Krush, M. T. (2016). Social media: Influencing customer satisfaction in B2B sales. *Industrial Marketing Management*, *53*, 172–180. doi:10.1016/j.indmarman.2015.09.003

Aitken, M., & Gauntlett, C. (2013). *Patient apps for improved healthcare: from novelty to mainstream*. Parsippany, NJ: IMS Institute for Healthcare Informatics.

Ajami, R., Qirim, N. A., & Ramadan, N. (2012). Privacy Issues in Mobile Social Networks. *Procedia Computer Science*, *10*, 672–679. doi:10.1016/j.procs.2012.06.086

Ajzen, I. (1985). From Intention to Actions: A Theory of Planned Behavior. In J. Kuhl & J. Beckman (Eds.), *Action Control: From Cognition to Behavior* (pp. 11–39). New York: Springer-Verlag. doi:10.1007/978-3-642-69746-3_2

Ajzen, I. (1991). The theory of planned behavior. *Organizational Behavior and Human Decision Processes*, *50*(2), 179–211. doi:10.1016/0749-5978(91)90020-T

Aker, C. J., & Mbiti, I. M. (2010). *Mobile Phones and Economic Development in Africa*. Retrieved in May 13, 2015, from http://www.cgdev.org/files/1424175_file_Aker_Mobile_wp211_FINAL.pdf

Akter, S., Dambra, J., Ray, P., & Hani, U. (2013). Modelling the impact of mHealth service quality on satisfaction, continuance and quality of life. *Behaviour & Information Technology*, *32*(12), 1225–1241. doi:10.1080/0144929X.2012.745606

Akter, S., & Ray, P. (2010). mHealth-an ultimate platform to serve the unserved. *Yearbook of Medical Informatics*, *2010*, 94–100. http://doi.org/me10010094 PMID:20938579

Akter, S., Ray, P., & DAmbra, J. (2013). Continuance of mHealth services at the bottom of the pyramid: The roles of service quality and trust. *Electronic Markets*, *23*(1), 29–47. doi:10.1007/s12525-012-0091-5

Akturan, U., & Tezcan, N. (2012). Mobile Banking Adoption of the Youth Market: Perceptions and Intentions. *Marketing Intelligence & Planning*, *30*(4), 4–4. doi:10.1108/02634501211231928

Akutsu, H., Legge, G. E., Ross, J. A., & Schuebel, K. (1991). Psychophysics of reading: X. Effects of age-related changes in vision. *The Journals of Gerontology. Series B, Psychological Sciences and Social Sciences*, *46*, 325–331. PMID:1940088

Aldas-Manzano, J., Ruiz-Mafe, C., & Sanz-Blas, S. (2009). Mobile commerce adoption in Spain: The influence of consumer attitudes and ICT usage behaviour. In B. Unhelkar (Ed.), Handbook of research in mobile business, second edition: Technical, methodological and social perspectives (pp. 282–292). Hershey, PA: IGI Global.

Aldas-Manzano, J., Ruiz-Mafé, C., & Sanz-Blas, S. (2009). Exploring individual personality factors as drivers of M-shopping acceptance. *Industrial Management & Data Systems*, *109*(6), 739–757. doi:10.1108/02635570910968018

Alfahl, H., Sanzogni, L., Houghton, L., & Sandhu, K. (2014). Mobile commerce adoption in organizations: A literature review and preliminary findings. In I. Lee (Ed.), *Trends in e-business, e-services, and e-commerce: Impact of technology on goods, services, and business transactions* (pp. 47–68). Hershey, PA: IGI Global. doi:10.4018/978-1-4666-4510-3.ch003

Al-khamayseh, S., Lawrence, E., & Zmijewska, A. (2006). *Towards Understanding Success Factors in Interactive Mobile Government* Paper presented at the Consortium International.

Al-Razgan, M. S., Al-Khalifa, H. S., & Al-Shahrani, M. D. (2014). Heuristics for evaluating the usability of mobile launchers for elderly people. LNCS, 8517, 415-424.

Alsumait, A., & Al-Osaimi, A. (2010). Usability heuristics evaluation for child E-learning applications. *J. Softw.*, *5*(6), 654–661. doi:10.4304/jsw.5.6.654-661

Amin, E. M. (2005). *Social Science Research: Conception, Methodology and Analysis*. Makerere University.

Amin, H., Hamid, M. R. A., Lada, S., & Anis, Z. (2008). The adoption of mobile banking in Malaysia: The case of Bank Islam Malaysia Berhad. *International Journal of Business and Society*, *9*(2), 43–53.

Anastasopoulou, K., Tryfonas, T., & Kokolakis, S. (2013). Strategic Interaction Analysis of Privacy-Sensitive End-Users of Cloud-Based Mobile Apps. In L. Marinos & I. Askoxylakis (Eds.), *Human Aspects of Information Security, Privacy, and Trust: First International Conference, HAS 2013, Held as Part of HCI International 2013, Las Vegas, NV, USA, July 21-26, 2013. Proceedings* (pp. 209-216). Berlin: Springer Berlin Heidelberg. doi:10.1007/978-3-642-39345-7_22

Andersen, D., Popescu, V., Cabrera, M. E., Shanghavi, A., Gomez, G., Marley, S., & Wachs, J. P. et al. (2016). Medical telementoring using an augmented reality transparent display. *Surgery*, *159*(6), 1646–1653. doi:10.1016/j.surg.2015.12.016 PMID:26804823

Anderson, J. C., & Gerbing, D. W. (1988). Structural equation modeling in practice: A review and preferences. *The Journal of Consumer Research*, *27*(2), 233–248.

Anderson, J. C., & Gerbing, D. W. (1988). Structural equation modeling inpractice: A review and recommended two-step approach. *Psychological Bulletin*, *103*(3), 411–423. doi:10.1037/0033-2909.103.3.411

Compilation of References

Angell, J. M. (2015). *Psychology: Chapter 8: Imagination.* Retrieved in May 13, 2015, from https://brocku.ca/Mead-Project/Angell/Angell_1906/Angell_1906_h.html

AppBrain. (2016). *Top categories.* Retrieved September 18, 2016, from http://www.appbrain.com/stats/android-market-app categories

Ardito, C., De Marsico, M., Lanzilotti, R., Levialdi, S., Roselli, T., Rossano, V., & Tersigni, M. (2004, May). Usability of e-learning tools. In *Proceedings of the working conference on Advanced visual interfaces.* Gallipoli, Italy: ACM. doi:10.1145/989863.989873

Argade, D., & Chavan, H. (2015). Improve accuracy of prediction of users future m-commerce behaviour. *Procedia Computer Science, 49,* 111–117. doi:10.1016/j.procs.2015.04.234

Arhippainen, L. (2009). *Studying User Experience: Issues And Problems Of Mobile Services– Case Adamos: User Experience (Im)Possible To Catch?* Doctoral Dissertation.

Armstrong, G., & Kotler, P. (2005). *Marketing: An Introduction* (7th ed.). Upper Saddle River, NJ: Prentice Hall.

Arora, J. B. (2016). Regulatory framework of mobile commerce. In S. Madan & J. Arora (Eds.), *Securing transactions and payment systems for m-commerce* (pp. 176–192). Hershey, PA: IGI Global. doi:10.4018/978-1-5225-0236-4.ch009

Ashida, K., Takamoto, J., Ibuki, M., Yamamoto, S., Matsui, H., Kumazaki, D., & Suga, A. (2008). *U.S. Patent No. D559254.* Washington, DC: US Patent Office.

Atienza, A. A., & Patrick, K. (2011). Mobile health: The killer app for cyberinfrastructure and consumer health. *American Journal of Preventive Medicine, 40*(5), S151–S153. doi:10.1016/j.amepre.2011.01.008 PMID:21521588

Atkinson, L. (2013). Smart shoppers? Using QR codes and green smartphone apps to mobilize sustainable consumption in the retail environment. *International Journal of Consumer Studies, 37*(4), 387–393. doi:10.1111/ijcs.12025

Aung, Y. M., & Al-Jumaily, A. (2011). Development of augmented reality rehabilitation games integrated with biofeedback for upper limb. *Proceedings of the 5th International Conference on Rehabilitation Engineering & Assistive Technology,* 1-4.

Au, Y., & Kauffman, R. (2008). The economics of mobile payments: Understanding stakeholder issues for an emerging financial technology application. *Electronic Commerce Research and Applications, 7*(2), 141–164. doi:10.1016/j.elerap.2006.12.004

Avgerou, A. D., Nastou, P. E., Nastouli, D., Pardalos, P., & Stamatiou, Y. C. (2016). On the Deployment of Citizens Privacy Preserving Collective Intelligent eBusiness Models in Smart Cities. *International Journal of Security and Its Applications, 10*(2), 171–184. doi:10.14257/ijsia.2016.10.2.16

Awad, M. A. (2005). *A Comparison between Agile and Traditional Software Development Methodologies* (Unpublished master dissertation). The University of Western Australia, Perth, Australia.

Aydin, S., Özer, G., & Arasil, Ö. (2005). Customer loyalty and the effect of switching costs as a moderator variable: A case in the Turkish mobile phone market. *Marketing Intelligence & Planning, 23*(1), 89–103. doi:10.1108/02634500510577492

Bady, A. (2013). The MOOC moment and the end of reform. *Liberal Education, 99*(4), 6–15.

Balagué, C., & Fayon, D. (2010). *Facebook, Twitter and the others: integrate social networks into a business strategy.* Pearson Editions.

Baldwin, R. (2012). *Path Uploads Your Entire Address Book to Its Servers Without Your Explicit Permission.* Retrieved 18 June, 2016, from http://gizmodo.com/5883118/path-uploads-your-entire-address-book-to-find-your-friends

Bardin, L. (1996). Content analysis. Paris: SAGE Publications Ltd.

Baron, R. M., & Kenny, D. A. (1986). The moderator-mediator variable distinction in social psychological research: Conceptual, strategic, and statistical considerations. *Journal of Personality and Social Psychology, 51*(6), 1173–1182. doi:10.1037/0022-3514.51.6.1173 PMID:3806354

Battarbee, K., & Mattelmäki, T. (2002) Meaningful Product Relationships. *Proceedings of Design and Emotion Conference.*

Bauman, A. (2015). The use of the repertory grid technique in online trust research. *Qualitative Market Research, 18*(3), 362–382. doi:10.1108/QMR-08-2014-0080

Baumeister, R. F., Vohs, K. D., DeWall, C. N., & Zhang, L. (2007). *How emotion shapes behavior*: Feedback, anticipation, and reflection, rather than direct causation. *Personality and Social Psychology Review, 11*(2), 167–203. doi:10.1177/1088868307301033 PMID:18453461

Baumgartner, J. (2012). The Psychology of Dress. *The Psychology of Fashion.* Retrieved in May 13 2015, from https://www.psychologytoday.com/blog/the-psychology-dress/201202/the-psychology-fashion

Bay, H., Ess, A., Tuytelaars, T., & van Gool, L. (2008). SURF: Speed up robust features. *Computer Vision and Image Understanding, 110*(3), 346–359. doi:10.1016/j.cviu.2007.09.014

Bazaarvoice. (n.d.). *Social commerce statistics.* Retrieved from: http://www.bazaarvoice.com/research-and-insight/social-commerce-statistics/

BBC. (2012, March 14). *India census: Half of homes have phones but no toilets.* Retrieved from http://www.bbc.com/news/world-asia-india-17362837

Beck, K. (2001). *Manifesto for Agile Software Development.* Retrieved February 5, 2016, from http://agilemanifesto.org/

Beckhard, R., & Reuben, T. H. (1987). *Organizational Transition: Managing Complex Change.* Reading, MA: Addison-Wesley.

Belachew, Y., Hoang, A., & Kourieh, J. (2007). *Mobile Customer Relationship Management: A Study of mCRM adoption in the Swedish Market* (Unpublished Master's Thesis). Jönköping University, Sweden.

Belk, W. (2013). Extended Self in a Digital World. *The Journal of Consumer Research, 40*(3), 477–500. doi:10.1086/671052

Belk, W. R., & Wallendorf, M. (1990). The sacred meanings of money. *Journal of Economic Psychology, 11*(1), 35–67. doi:10.1016/0167-4870(90)90046-C

Beltran, M., & Belle, S. (2013, December). The use of internet-based social media as a tool in enhancing student's learning experiences in biological sciences. *Higher Learning Research Communications, 3*(4).

Ben Dahmane Mouelhi, N., Hassen, S., & Souissi, N. (2009). An exploratory approach to sources of irritation felt by the French customer in hotels in Tunisia. *Revue Marocaine de Recherche en Management et Marketing,* (2-3), 137 – 156.

Benferdia, Y., & Zakaria, H. (n.d.). A Systematic Literature Review of Content-Based Mobile Health. *Journal of Information Systems Research and Innovation.* Retrieved from http://seminar.utmspace.edu.my/jisri/

Bennett, S. (2014). U.S. social commerce – statistics & trends (infographic). *AdWeek.* Retrieved from: http://www.adweek.com/socialtimes/social-commerce-stats-trends/500895

Berger, J. (2011). Arousal Increases Social Transmission of Information. *Psychological Science.* doi:10.1177/0956797611413294 PMID:21690315

Berget, J. P. (2009). *Why Legacy is Important in Marketing*. Retrieved from http://slymarketing.com/why-legacy-important-marketing/

Bergstrom, J. C. R., Olmsted-Hawala, E. L., & Jans, M. E. (2013). Age-Related Differences in Eye Tracking and Usability Performance: Website Usability for Older Adults. *Journal of Human-Computer Interaction*, *29*(8), 541–548. doi:10.1080/10447318.2012.728493

Bernard, M., Lida, B., Riley, S., Hackler, T., & Janzen, K. (2002). *A Comparison of Popular Online Fonts: Which Size and Type is Best?* Retrieved June 26, 2016, from http://usabilitynews.org/a-comparison-of-popular-online-fonts-which-size-and-type-is-best/html

Bernard, M. L., Chaparro, B. S., Mills, M. M., & Halcomb, C. G. (2003). Comparing the Effects of Text Size and Format on the Readability of Computer-Displayed Times New Roman and Arial Text. *International Journal of Human-Computer Studies*, *59*(6), 823–835. doi:10.1016/S1071-5819(03)00121-6

Bertini, E., Gabrielli, S., & Kimani, S. (2006). Appropriating and assessing heuristics for mobile computing. In *Proceedings of the Working Conference on Advanced Visual Interfaces*. Venezia, Italy: ACM. doi:10.1145/1133265.1133291

Betts, G. H. (1916). *Imagination: The mind and its education*. D. Appleton and Company.

Bevan, N. (2010). *What is the difference between the purpose of usability and user experience evaluation methods?* Retrieved in April 10 2016, from http://www.nigelbevan.com/papers/What_is_the_difference_between_usability_and_user_experience_evaluation_methods.pdf

Bickert, J. (1992). *The Database Revolution*. Target Marketing.

Bidmon, S., Terlutter, R., & Röttl, J. (2014). What explains usage of mobile physician-rating apps? Results from a web-based questionnaire. *Journal of Medical Internet Research*, *16*(6), e148. doi:10.2196/jmir.3122 PMID:24918859

Billinghurst, M., Kato, H., & Poupyrev, I. (2008). Article. *ACM SIGGRAPH Asia*, *7*, 1–11.

Billington, P. J., & Fronmueller, M. P. (2013). MOOCs and the future higher education. *Journal of Higher Education Theory and Practice*, *13*(4), 37–43.

BinDhim, N. F., Freeman, B., & Trevena, L. (2014). Pro-smoking apps for smartphones: The latest vehicle for the tobacco industry? *Tobacco Control*, *23*(1), e4–e4. doi:10.1136/tobaccocontrol-2012-050598 PMID:23091161

Binsaleh, M., & Hassan, S. (2013). Systems development methodology for mobile commerce applications: Agile vs. traditional. In H. El-Gohary (Ed.), *Transdisciplinary marketing concepts and emergent methods for virtual environments* (pp. 264–278). Hershey, PA: IGI Global. doi:10.4018/978-1-4666-1861-9.ch018

Bix, L., Lockhart, H., Cardoso, F., & Selke, S. (2003). The Effect of Color Contrast on Message Legibility. *Journal of Design Communication*, *5*.

Blum, T., Kleeberger, V., Bichlmeier, C., & Navab, N. (2012). Mirracle: An augmented reality magic mirror system for anatomy education. *2012 IEEE Virtual Reality Workshops*, 115-116. doi:10.1109/VR.2012.6180909

Boden, R. (2016). *Apple pay launches in New Zealand*. Retrieved September 13, 2016, from http://www.nfcworld.com/2016/10/13/347807/apple-pay-launches-new-zealand/

Bogdan, R., & Biklen, S. K. (2006). *Qualitative research for education: An introduction to theories and methods* (5th ed.). Pearson.

Bolt, R. A. (1980). Put-that-there: Voice and gesture at the graphics interface. *Computer Graphics*, *14*(3), 262–270. doi:10.1145/965105.807503

Booz & Company. (2011). *Turning "Like" to "Buy" Social Media Emerges as a Commerce Channel.* Booz & Company Retrieved from: http://www.strategyand.pwc.com/reports/turning-like-social-media-emerges

Bosch, T. E. (2009). Using online social networking for teaching and learning: Facebook use at the University of Cape Town. *Communication, 35*(2), 185–200.

Boudreaux, E. D., Waring, M. E., Hayes, R. B., Sadasivam, R. S., Mullen, S., & Pagoto, S. (2014). Evaluating and selecting mobile health apps: Strategies for healthcare providers and healthcare organizations. *Translational Behavioral Medicine, 4*(4), 363–371. doi:10.1007/s13142-014-0293-9 PMID:25584085

Bourcieu, S., & Léon, O. (2013). Les MOOC, alliés ou concurrents des business schools? *LExpansion Management Review, 149*(2), 14–24. doi:10.3917/emr.149.0014

Bowcock, J., & Pope, S. (2008). iPhone App Store Downloads Top 10 Million in First Weekend. *Apple Press Info.* Retrieved 10 June, 2016, from http://www.apple.com/pr/library/2008/07/14iPhone-App-Store-Downloads-Top-10-Million-in-First-Weekend.html

Bowen, W. G. (2012). The "cost disease" in higher education: Is technology the answer. *The Tanner Lectures.* Retrieved 02/02/15, from http://edf.stanford.edu/sites/default/files/Bowen%20lectures%20SU%20102.pdf

Bowers, B. (2001). *Sir Charles Wheatstone FRS 1802-1875.* London: The Institution of Electrical Engineers in Association with the Science Museum.

Boyatt, R., Joy, M., Rocks, C., & Sinclair, J. (2014). *What (use) is a MOOC?* Paper presented at the The 2nd International Workshop on Learning Technology for Education in Cloud. doi:10.1007/978-94-007-7308-0_15

Boyce, B. (2014). Nutrition apps: Opportunities to guide patients and grow your career. *Journal of the Academy of Nutrition and Dietetics, 114*(1), 13–15. doi:10.1016/j.jand.2013.10.016 PMID:24342602

boyd, d., & Ellison, N. B. (2007). Social Network Sites: Definition, History, and Scholarship. *Journal of Computer-Mediated Communication, 13*(2), 210–230.

Brady, M., Fellenz, M. R., & Brookes, R. (2008). Researching the role of information and communications technology (ICT) in contemporary marketing practices. *Journal of Business and Industrial Marketing, 23*(2), 108–114. doi:10.1108/08858620810850227

Brands, S. (2000). *Rethinking Public Key Infrastructures and Digital Certificates: Building in Privacy.* MIT Press.

Brandtzæg, P. B., Lüders, M., & Skjetne, J. H. (2010). Too many Facebook friends? Content sharing and sociability versus the need for privacy in social network sites. *International Journal of Human-Computer Interaction, 26*(11-12), 1006–1030. doi:10.1080/10447318.2010.516719

Brenner, A. (2003). *Fantasy.* Academic Press.

Bressolles, G. (2006). La qualité de service électronique: NetQu@l. Proposition dune échelle de mesure appliquée aux sites marchands et effets modérateurs. *Recherche et Applications en Marketing, 21*(3), 19–45. doi:10.1177/076737010602100302

Brooks, L. (2009, May-June). Social learning by design: The role of social media. *Knowledge Quest, 37*(5), 58–60.

Bruner, G. C. II, & Kumar, A. (2005). Explaining consumer acceptance of handheld Internet Devices. *Journal of Business Research, 58*(5), 553–558. doi:10.1016/j.jbusres.2003.08.002

Budiu, R. (2014). *The Reciprocity Principle: Give Before You Take in Web Design.* Retrieved in April 10 2016, from https://www.nngroup.com/articles/reciprocity-principle/

Bull, S., & Ezeanochie, N. (2015). From Foucault to Freire Through Facebook: Toward an Integrated Theory of mHealth. *Health Education & Behavior*, *43*(4), 399–411. doi:10.1177/1090198115605310 PMID:26384499

Burke, J. W., McNeill, M. D. J., Charles, D. K., Morrow, P. J., Crosbie, J. H., & McDonough, M. S. (2010). Augmented reality games for upper-limb stroke rehabilitation. *2010 Second International Conference on Games and Virtual Worlds for Serious Applications*, 75-78. doi:10.1109/VS-GAMES.2012.21

Business Dictionaries. (2015). *Socio-cultural environment*. Retrieved in April 10 2016, from http://www.businessdictionary.com/definition/socio-cultural-environment.html

Bussone, A. (2016). *Disclose-It-Yourself: Security and Privacy for People Living with HIV*. Academic Press.

Buttle, F. (2004). *Customer relationship management: concepts and tools*. Oxford, UK: Elsevier Butterworth-Heinemann.

Büttner, O. B., & Göritz, A. S. (2008). Perceived trust worthiness of online shops. *Journal of Consumer Behaviour*, *7*(1), 35–50. doi:10.1002/cb.235

Büyüközkan, G. (2009). Determining the mobile commerce user requirements using an analytic approach. *Computer Standards & Interfaces*, *31*(1), 144–152. doi:10.1016/j.csi.2007.11.006

Buyyaa, R., Yeoa, C. S., Venugopala, S., Broberga, J., & Brandicc, I. (2009, June). Cloud computing and emerging IT platforms: Vision, hype, and reality for delivering computing as the 5th utility. *Future Generation Computer Systems*, *25*(6), 599–616. doi:10.1016/j.future.2008.12.001

Buzzetto-More, N. A. (2014). An examination of undergraduate student's perceptions and predilections of the use of youtube in the teaching and learning process. *Interdisciplinary Journal of E-Learning and Learning Objects*, *10*, 17–32.

Byrne, Z. S., Dvorak, K. J., Peters, J. M., Ray, I., Howe, A., & Sanchez, D. (2016). From the users perspective: Perceptions of risk relative to benefit associated with using the internet. *Computers in Human Behavior*, *59*, 456–468. doi:10.1016/j.chb.2016.02.024

Cafaro, M., & Aloisio, G. (2010). *Grids, Clouds and Virtualization*. Springer-Verlag New York, Inc.

Camenisch, J., & Lysyanskaya, A. (2001). An Efficient non-transferable anonymous multi-show credential system with optional anonymity revocation. (2001). In *Proceedings of the 30th Annual International Conference on the Theory and Applications of Cryptographic Techniques (EUROCRYPT 2001)* (vol. 2045, pp. 93-118). LNCS, Springer.

Campbell-Kelly, M., Garcia-Swartz, D., Lam, R., & Yang, Y. (2014). Economic and business perspectives on smartphones as multi-sided platforms. *Telecommunications Policy*, *39*(8), 717–734. doi:10.1016/j.telpol.2014.11.001

Campbirdge Dictionary. (2015). Retrieved in April 10 2016, from reminder.http://dictionary.cambridge.org/dictionary/english/reminder

Canadian National Institute for the Blind. (2011). *Clear print accessibility guidelines*. Retrieved July 19, 2016, from http://www.cnib.ca/en/services/resources/clearprint/html

Carlile, P.R., & Christensen, C.M. (2004). *The Cycles of Theory Building in Management Research*. Academic Press.

Carter, E. (2008). Mobile marketing and generation Y African–American mobile consumers: The issues and opportunities. *International Journal of Mobile Marketing*, *3*(1), 62–66.

Cartwright, P., & Noone, L. (2006). *Critical imagination*: A pedagogy for engaging pre-service teachers in the university classroom. *College Quarterly*, *9*(4), 1.

Cauwenberghe, P. V. (2016). *Agile Fixed Price Projects part 1: "The Price Is Right"*. Retrieved February 5, 2016, from http://www.nayima.be/html/fixedpriceprojects.pdf

Cawsey, T., & Rowley, J. (2016). Social media brand building strategies in B2B companies. *Marketing Intelligence & Planning, 34*(Iss: 6), 754–776. doi:10.1108/MIP-04-2015-0079

Cernezel, A., Karakatic, S., & Brumen, B. (2014). *Predicting grades based on students' online course activities*. Paper presented at the Ninth International Knowledge Management in Organizations Conference, Santiago, Chile. doi:10.1007/978-3-319-08618-7_11

Chadwick-Dias, A., Bergel, M., & Tullis, T. S. (2007). Senior surfers 2.0: A re-examination of the older Web user and the dynamic web. In C. Stephanidis (Ed.), *Universal access in human computer interaction: Coping with diversity* (pp. 868–876). Berlin: Springer. Retrieved March, 2016, from http://www.springerlink.com/content/08q65578604130w8/html

Chae, M. H., & Kim, J. W. (2003). Whats so different about the mobile Internet? *Communications of the ACM, 46*(12), 240–247. doi:10.1145/953460.953506

Chae, M., Kim, J., Kim, H., & Ryu, H. (2002). Information Quality for Mobile Internet Services: A Theoretical Model with Empirical Validation. *Electronic Markets, 12*(1), 38–46. doi:10.1080/101967802753433254

Chaffey, D. (2011). *E-business and e-commerce management: strategy, implementation, and practice* (4th ed.). Upper Saddle River, NJ: Prentice Hall.

Chan, M. (2015). Examining the Influences of news use patterns, motivations, and age cohort on mobile news use: The case of Hong Kong. *Mobile Media & Communication, 3*(2), 179–195. doi:10.1177/2050157914550663

Chan, C. (2012). What Facebook Deals with Everyday: 2.7 Billion Likes, 300 Million Photos Uploaded and 500 Terabytes of Data. *Gizmodo*. Retrieved from: http://gizmodo.com/5937143/what-Facebook-deals-with-everyday-27-billion-likes-300-million-photos-uploaded-and-500-terabytes-of-data

Chan, F. T. S., & Chong, A. Y. L. (2013). Analysis of the determinants of consumers m-commerce usage activities. *Online Information Review, 37*(3), 443–461. doi:10.1108/OIR-01-2012-0012

Chang, J. E., Simpson, T. W., Rangaswamy, A., & Tekchadaney, J. R. (2002). *A good website can convey the wrong brand image! A preliminary report*. Working paper. E-Business Research Center (EBRC), University of Pennsylvania.

Chang, C. C., & Lin, I. C. (2006). A new solution for assigning cryptographic keys to control access in mobile agent environments. *Wireless Communications and Mobile Computing, 6*(1), 137–146. doi:10.1002/wcm.276

ChangingMinds.org. (2016). *How we change what others think, feel, believe & do*. Retrieved in Feb 19 2016, from http://changingminds.org/

Changsu, K., In-Seok, L., Tao, W., & Mirsobit, M. (2015). Evaluating effects of mobile CRM on employees performance. *Industrial Management & Data Systems, 115*(4), 740–764. doi:10.1108/IMDS-08-2014-0245

Chang, T., Liao, L., & Hsiao, W. (2005). An Empirical Study on the e-CRM Performance Influence Model of Service Sectors in Taiwan. *Proceedings of the 2005 IEEE International Conference on e-Technology, e-Commerce and e-Service (EEE'05) on e-Technology, e-Commerce and e-Service*, 240-245. doi:10.1109/EEE.2005.33

Chan-Olmsted, S., Rim, H., & Zerba, A. (2013). Mobile news adoption among young adults: Examining the roles of perceptions, news consumption, and media usage. *Journalism & Mass Communication Quarterly, 90*(1), 126–147. doi:10.1177/1077699012468742

Chan, S. S., & Fang, X. (2003). Mobile commerce and usability. In E. Lim & K. Siau (Eds.), *Advances in mobile commerce technologies* (pp. 235–257). Hershey, PA: Idea Group Publishing. doi:10.4018/978-1-59140-052-3.ch011

Chan, S. S., & Fang, X. (2009). Interface design issues for mobile commerce. In D. Taniar (Ed.), *Mobile computing: Concepts, methodologies, tools, and applications* (pp. 526–533). Hershey, PA: IGI Global. doi:10.4018/978-1-60566-054-7.ch045

Charbaji, R., Rebeiz, K., & Sidani, Y. (2010). Antecedents and consequences of the risk taking behavior of mobile commerce adoption in Lebanon. In H. Rahman (Ed.), *Handbook of research on e-government readiness for information and service exchange: Utilizing progressive information communication technologies* (pp. 354–380). Hershey, PA: IGI Global. doi:10.4018/978-1-60566-671-6.ch018

Charette, R. (2005). Why software fails. *IEEE Spectrum*, *42*(9), 42–49. doi:10.1109/MSPEC.2005.1502528

Chatti, M. A., Jarke, M., & Frosch-Wilke, D. (2007). The future of e-learning: A shift to knowledge networking and social software. *International Journal of Knowledge and Learning*, *3*(4), 404–420. doi:10.1504/IJKL.2007.016702

Chattopadhyay, S. (2015). Mobile news culture: news apps, journalistic practices and the 2014 Indian General Election. In E. Thorsen & C. Sreedharan (Eds.), *India Election 2014: First Reflections* (pp. 143–157). The Centre for the Study of Journalism, Culture and Community: Bournemouth University.

Chattratichart, J., & Jordan, P. W. (2003). Simulating 'lived' user experience - virtual immersion and inclusive design. In M. Rauterberg, M. Menozzi, & J. Wesson (Eds.), *Proceedings of Human-Computer Interaction - INTERACT'03* (pp. 721–724). Amsterdam: IOS Press.

Chen, L. D. (2008). A model of consumer acceptance of mobile payment. *International Journal of Mobile Communications*, *6*(1), 32–52. doi:10.1504/IJMC.2008.015997

Chen, Q., & Chen, H. M. (2004). Exploring the Success Factors of E-CRM Strategies in Practice. *Journal of Database Marketing & Customer Strategy Marketing*, *11*(4), 333–343. doi:10.1057/palgrave.dbm.3240232

Chen, S. C. (2012). To use or not to use: Understanding the factors affecting continuance intention of mobile banking. *International Journal of Mobile Communications*, *10*(5), 490–507. doi:10.1504/IJMC.2012.048883

Chen, Y., & Lan, Y. (2014). An empirical study of the factors affecting mobile shopping in Taiwan. *International Journal of Technology and Human Interaction*, *10*(1), 19–30. doi:10.4018/ijthi.2014010102

Chen, Z. S., Li, R., Chen, X., & Xu, H. (2011). A survey study on consumer perception of mobile-commerce applications. *Procedia Environmental Sciences*, *11*, 118–124. doi:10.1016/j.proenv.2011.12.019

Cheok, A. D. (2010). Human Pacman: A mobile augmented reality entertainment system based on physical, social, and ubiquitous computing. In A. D. Cheok (Ed.), *Art and Technology of Entertainment Computing and Communication* (pp. 19–57). Springer London. doi:10.1007/978-1-84996-137-0_2

Cherry, K. (2014). *Social Learning Theory: How People Learn By Observation*. Retrieved in Feb 19 2016, from http://psychology.about.com/od/developmentalpsychology/a/sociallearning.htm

Cheshire, T., & Rowan, D. (2011). Commerce gets social: How social networks are driving what you buy. *Wired Magazine UK*. Retrieved from: http://www.wired.co.uk/magazine/archive/2011/02/features/social-networks-drive-commerce

Chien, C., Chen, C., & Jeng, T. (2010). An interactive augmented reality system for learning anatomy structure. *Proceedings of the International Multiconference of Engineers and Computer Scientists*, *2010*, 370–375.

Childers, T. L., Carr, C. L., Peck, J., & Carson, S. (2001). Hedonic and Utilitarian Motivations for Online Retail Shopping Behavior. *Journal of Retailing*, *77*(4), 511–535. doi:10.1016/S0022-4359(01)00056-2

Choi, J., Seol, H., Lee, S., Cho, H., & Park, Y. (2008). Customer satisfaction factors of mobile commerce in Korea. *Internet Research*, *183*(3), 313–335. doi:10.1108/10662240810883335

Choi, W., & Stvilia, B. (2013). Use of mobile wellness applications and perception of quality. *Proceedings of the American Society for Information Science and Technology*, *50*(1), 1–4.

Chong, A. Y. L. (2013). A two-staged SEM-neural network approach for understanding and predicting the determinants of m-commerce adoption. *Expert Systems with Applications*, *40*(4), 1240–1247. doi:10.1016/j.eswa.2012.08.067

Chong, A. Y. L. (2013). Mobile commerce usage activities: The roles of demographic and motivation variables. *Technological Forecasting and Social Change*, *80*(7), 1350–1359. doi:10.1016/j.techfore.2012.12.011

Chong, A. Y. L., Chan, F. T. S., & Ooi, K. B. (2011). Predicting consumer decisions to adopt m-commerce: Cross country empirical examination between China and Malaysia. *Decision Support Systems*, *53*(1), 34–43. doi:10.1016/j.dss.2011.12.001

Chong, A. Y. L., Darmawan, N., Ooi, K. B., & Lin, B. (2010). Adoption of 3G services among Malaysian consumers: An empirical analysis. *International Journal of Mobile Communications*, *8*(2), 129–149. doi:10.1504/IJMC.2010.031444

Chong, J. L., Chong, A. Y. L., Ooi, K. B., & Lin, B. (2011). An empirical analysis of the adoption of m-learning in Malaysia. *International Journal of Mobile Communications*, *9*(1), 1–18. doi:10.1504/IJMC.2011.037952

Chorppath, A. K., & Alpcan, T. (2013). Trading privacy with incentives in mobile commerce: A game theoretic approach. *Pervasive and Mobile Computing*, *9*(4), 598–612. doi:10.1016/j.pmcj.2012.07.011

Chou, T. H., & Seng, J. L. (2012). Telecommunication e-services orchestration enabling business process management. *Transactions on Emerging Telecommunications Technologies*, *23*(7), 646–659. doi:10.1002/ett.2520

Christin, D., Engelmann, F., & Hollick, M. (2014). Usable Privacy for Mobile Sensing Applications. In D. Naccache & D. Sauveron (Eds.), *Information Security Theory and Practice. Securing the Internet of Things: 8th IFIP WG 11.2 International Workshop, WISTP 2014, Heraklion, Crete, Greece, June 30 – July 2, 2014. Proceedings* (pp. 92-107). Berlin: Springer Berlin Heidelberg. doi:10.1007/978-3-662-43826-8_7

Christin, D., Reinhardt, A., Kanhere, S. S., & Hollick, M. (2011). A survey on privacy in mobile participatory sensing applications. *Journal of Systems and Software*, *84*(11), 1928–1946. doi:10.1016/j.jss.2011.06.073

Christin, D., Roßkopf, C., & Hollick, M. (2013). uSafe: A privacy-aware and participative mobile application for citizen safety in urban environments. *Pervasive and Mobile Computing*, *9*(5), 695–707. doi:10.1016/j.pmcj.2012.08.005

Chung, K. C., & Holdsworth, D. (2012). Culture and behavioural intent to adopt mobile commerce among the Y Generation: Comparative analyses between Kazakhstan, Morocco and Singapore. *Young Consumers: Insight and Ideas for Responsible Marketers*, *13*(3), 3. doi:10.1108/17473611211261629

Chun, S. G., Chung, D., & Shin, Y. B. (2013). Are Students Satisfied with the Use of Smartphone Apps? *Issues in Information Systems*, *14*(2), 23–33.

Cialdini, R. B. (2007). *Influence: The Psychology of Persuasion*. New York, NY: HarperCollins Publishers Inc.

Cleff, E. (2007). Privacy issues in mobile advertising. *International Review of Law Computers & Technology*, *21*(3), 225–236. doi:10.1080/13600860701701421

Clow, D. (2013). *Moocs and the Funnel of Participation.* Paper presented at the The Third International Conference on Learning Analytics and Knowledge, Leuven, Belgium. doi:10.1145/2460296.2460332

Cockton, G., & Woolrych, A. (2001). Understanding inspection methods: lessons from an assessment of heuristic evaluation. In People and Computers XV. Interaction without Frontiers, 171-191. doi:10.1007/978-1-4471-0353-0_11

Coetzee, D., Fox, A., Hearst, M. A., & Hartmann, B. (2014). *Should your MOOC forum use a reputation system?* Paper presented at the Computer on supported cooperative work & social computing, Baltimore, MD. doi:10.1145/2531602.2531657

Cole-Lewis, H., & Kershaw, T. (2010). Text messaging as a tool for behavior change in disease prevention and management. *Epidemiologic Reviews, 32*(1), 56–69. doi:10.1093/epirev/mxq004 PMID:20354039

Committee, A. S. (2000). *The Link flight trainer: A historic mechanical engineering landmark. Roberson Museum and Science Center.* Binghamton, NY: ASME International, History and Heritage Committee.

Comscore. (2016). *The 2016 U.S. Mobile App Report.* Comscore.

Corbeil, J. R., & Valdes-Corbeil, M. E. (2007). Are You Ready for Mobile Learning. *Education Quarterly, 30*(2), 51–58.

Cormier, D., & Siemens, G. (2010). The Open Course: Through the Open Door – Open Courses as Research, Learning and Engagement. *EDUCAUSE Review, 45*(4), 30–32.

Cormode, G., Krishnamurthy, B., & Willinger, W. (2010). A manifesto for modeling and measurement in social media. *First Monday, 15*(9). doi:10.5210/fm.v15i9.3072

Cornelius, D. (2013). *SMAC and transforming innovation.* Paper presented at the meeting of the 2013 PMI Global Congress, New Orleans, LA.

Cornwall, J., Fette, I., Hsieh, G., Prabaker, M., Rao, J., Tang, K., . . . Hong, J. (2007). *User-controllable security and privacy for pervasive computing.* Paper presented at the 8th IEEE workshop on mobile computing systems and applications (HotMobile 2007). doi:10.1109/HotMobile.2007.9

Corson-Knowles, T. (2013). *Facebook for Business Owners: Facebook Marketing For Fan Page Owners and Small Businesses (Social Media Marketing)* (vol. 2). TCKPublishing.com.

Cosseboom, L. (2014). *Asia is dominating the mCommerce market, puts US and Europe to shame.* Retrieved September 13, 2016, from https://www.techinasia.com/digi-capital-reports-asia-as-dominant-player-mcommerce-industry-insights

Costabile, M. F., De Marsico, M., Lanzilotti, R., Plantamura, V. L., & Roselli, R. (2005). On the usability evaluation of E-learning application. In *Proceedings of the 38th Hawaii International Conference on System Sciences.* IEEE. doi:10.1109/HICSS.2005.468

Coursaris, C., & Hassanein, K. (2002). Understanding m-Commerce: A Consumer-Centric Model. *Quarterly Journal of Electronic Commerce, 3*(3), 247–271.

Coursaris, C., Swierenga, S., & Watrall, E. (2008). An empirical investigation of color temperature and gender effects on web aesthetics. *Journal of Usability Studies, 3*(3), 103–117.

Criteo. (2015). *State of mobile commerce apps and cross-device lead mobile business.* Retrieved September 13, 2016, from http://www.criteo.com/media/2501/criteo-state-of-mobile-commerce-report-q2-2015-ppt.pdf

Criteo. (2015). *State of Mobile Commerce.* Retrieved March, 2016, from http://www.criteo.com/resources/mobile-commerce-q1-2015/html

Critéo. (2016). *State of Mobile Commerce 2016.* Author.

Criteo. (2016). *The State of Mobile Commerce (2016)*. Retrieved from http://www.criteo.com/resources/mobile-commerce-report

Curty, R. G., & Zhang, P. (2011). Social commerce: Looking back and forward. *Proceedings of the American Society for Information Science and Technology, 48*(1), 1–10. doi:10.1002/meet.2011.14504801096

Cyr, D., Head, M., & Ivanov, A. (2006). Design aesthetics leading to m-loyalty in mobile commerce. *Information & Management, 43*(8), 950–963. doi:10.1016/j.im.2006.08.009

Cyr, D., Head, M., & Larios, H. (2010). Colour appeal in website design within and across cultures: A multi-method evaluation. *International Journal of Human-Computer Studies, 68*(1-2), 1–21. doi:10.1016/j.ijhcs.2009.08.005

D'Ambrosio, S., De Pasquale, S., Iannone, G., Malandrino, D., Negro, A., Patimo, G., . . . Spinelli, R. (2016). Energy consumption and privacy in mobile Web browsing: Individual issues and connected solutions. *Sustainable Computing: Informatics and Systems*. doi: 10.1016/j.suscom.2016.02.003

Dahlberg, T., Mallat, N., Ondrus, J., & Zmijewska, A. (2006). *M-payment market and research-past, present and future*. Helsinki Mobility Roundtable, Helsinki, Finland.

Dahnil, M. I., Marzuki, K. M., Langgat, J., & Fabeil, N. F. (2014). Factors influencing SMEs adoption of social media marketing. *Procedia: Social and Behavioral Sciences, 148*, 119–126. doi:10.1016/j.sbspro.2014.07.025

Dai, H., & Palvia, P. (2008). Factors affecting mobile commerce adoption: A cross-cultural study in China and the United States. *The Data Base for Advances in Information Systems, 40*(4), 43–61. doi:10.1145/1644953.1644958

Damani, V. (2016, May 15). *What's Holding Back Mobile Data Uptake in India, and How Will It Change?* Retrieved from http://gadgets.ndtv.com/telecom/features/whats-holding-back-mobile-data-uptake-in-india-and-how-will-it-change-818783

Daniel, J. (2012). Making sense of MOOCs: Musings in a maze of myth, paradox and possibility. *Journal of Interactive Media in Education, 3*, 1–20.

Daniels, D. (2015). *Pragmatic Marketing: Secrets of a Winning Product Launch*. Retrieved in Feb 19 2016, from http://pragmaticmarketing.com//resources/6-secrets-of-a-winning-product-launch?p=2

Daştan, İ., & Gürler, C. (2016). Factors Affecting the Adoption of Mobile Payment Systems: An Empirical Analysis. *Emerging Markets Journal, 6*, 1. Retrieved from http://emaj.pitt.edu

DAstous, A. (2000). Irritating aspects of the shopping environment. *Journal of Business Research, 49*(2), 149–156. doi:10.1016/S0148-2963(99)00002-8

Davis, F. D. (1989). Perceived usefulness, perceived ease of use, and user acceptance of information technology. *Management Information Systems Quarterly, 13*(3), 319–339. doi:10.2307/249008

Davis, R., Sajtos, L., & Chaudhri, A. A. (2011). Do Consumers Trust Mobile Service Advertising? *Contemporary Management Research, 7*(4), 245–270. doi:10.7903/cmr.9696

De Haes, S., & Grembergen, W. (2009). *Enterprise governance of information technology: Achieving strategic alignment and value*. New York: Springer. doi:10.1007/978-0-387-84882-2

De Leon, N. I., Bhatt, S. K., & Al-Jumaily, A. (2014). Augmented reality game based multi-usage rehabilitation therapist for stroke patients. *International Journal on Smart Sensing and Intelligent Systems, 7*(3), 1044–1058.

Compilation of References

De Santo, A., & Gaspoz, C. (2015). Influence of Users' Privacy Risks Literacy on the Intention to Install a Mobile Application. In A. Rocha, M. A. Correia, S. Costanzo, & P. L. Reis (Eds.), *New Contributions in Information Systems and Technologies* (Vol. 1, pp. 329–341). Cham: Springer International Publishing. doi:10.1007/978-3-319-16486-1_33

De Waard, I. (2011). *Explore a New Learning Frontier: MOOCs.* Learning Solutions Magazine.

Deans, P. C. (2004). *E-commerce and M-commerce technologies.* IRM Press. doi:10.4018/978-1-59140-566-5

Deci, E. L., & Ryan, R. M. (1985). *Intrinsic motivation and self-determination in human behaviour.* New York: Plenum. doi:10.1007/978-1-4899-2271-7

Deeb, S. S., & Motulsky, A. G. (2015). *Red-Green Color Vision Defects.* Retrieved October 12, 2015, from http://www.ncbi.nlm.nih.gov/books/NBK1301/html

DeFanti, T. A., & Sandin, D. J. (1977). *Final Report to the National Endowment of the Arts* (U.S. NEA R60-34-163). Chicago: University of Illinois.

Degirmenci, K., Guhr, N., & Breitner, M. H. (2013). *Mobile Applications and Access to Personal Information: A Discussion of Users' Privacy Concerns.* Paper presented at the International Conference on Information Systems, Milan, Italy.

DeMers, J. (2015) *The Importance Of Social Validation In Online Marketing.* Forbes. Retrieved in Feb 19 2016, from http://www.forbes.com/sites/jaysondemers/2015/02/19/the-importance-of-social-validation-in-online-marketing/

Deng, Z., Lu, Y., Wei, K. K., & Zhang, J. (2010). Understanding customer satisfaction and loyalty: An empirical study of mobile instant messages in China. *International Journal of Information Management, 30*(4), 289–300. doi:10.1016/j.ijinfomgt.2009.10.001

Dennis, C., Merrilees, B., Jayawardhena, C., & Wright, L. (2009). E-consumer behavior. *European Journal of Marketing, 43*(9/10), 1121–1139. doi:10.1108/03090560910976393

Desmet, P. M. A., & Hekkert, P. (2007). Framework of product experience. *International Journal of Design, 1*(1), 57–66.

DeVries, W. A. (1988). *Hegel's theory of mental activity: An introduction to theoretical spirit.* Ithaca, NY: Cornell University Press. Retrieved August 18, 2011, ttp://pubpages.unh.edu/~wad/HTMA/HTMAfrontpage

Dholakiya, P. (2013). *Want Conversions? Start with User-Friendly, Useful Landing Pages.* Retrieved in Feb 19 2016, from https://blog.kissmetrics.com/want-conversions/

Dimmick, J., Feaster, J. C., & Hoplamazian, G. J. (2011). News in the interstices: The niches of mobile media in space and time. *New Media & Society, 13*(1), 23–39. doi:10.1177/1461444810363452

Dimmick, J., Powers, A., Mwangi, S., & Stoycheff, E. (2011). The fragmenting mass media marketplace. In W. Lowrey & P. J. Gade (Eds.), *Changing the news. The forces shaping journalism in uncertain times* (pp. 177–192). London, UK: Routledge.

Doron, A., & Jeffery, R. (2013). *The Great Indian Phone Book: How the Cheap Cell Phone Changes Business, Politics, and Daily Life.* Cambridge, MA: Harvard University Press. doi:10.4159/harvard.9780674074248

Doueihi, M. (2011). L'esthète du numérique. *Le Monde.* Retrieved from www.lemonde.fr/idees/article/2011/10/07/l-esthete-du-numerique_1583940_3232.html

Downes, J., & Palmer, R. A. (2005). The Influences on Contemporary Marketing Practices In High-Technology Companies: Research Programme and Preliminary Findings. In *IMP Conference, 2005, Rotterdam, Netherlands.* (Unpublished draft proceeding)

Dringus, L. P., & Cohen, M. S. (2005). An adaptable usability heuristic checklist for online courses. In *Proceedings of 35th Annual Conference Frontiers in Education*. IEEE. doi:10.1109/FIE.2005.1611918

Dunlap, J. C., & Lowenthal, P. R. (2009). Tweeting the night away: Using Twitter to enhance social presence. *Journal of Information Systems Education, 20*(2), 129–136.

Dunne, E. (2014). *Study: Good Design Is Good For Business*. Retrieved in August 21 2015, from http://www.fastcodesign.com/3026287/study-good-design-really-is-good-for-business

Dyche, J. (2001). *The CRM handbook: A business guide to customer relationship management*. Boston: Addison-Wesley.

Eastin, M. S., Brinson, N. H., Doorey, A., & Wilcox, G. (2016). Living in a big data world: Predicting mobile commerce activity through privacy concerns. *Computers in Human Behavior, 58*, 214–220. doi:10.1016/j.chb.2015.12.050

Eckfeldt, B., Madden, R., Horowitz, J., & Grotta, E. (2005). Selling Agile: Target-Cost Contract. In *Proceedings of 2012 Agile Conference* (pp. 160-166). Denver, CO: IEEE Computer Society.

Edosomwan, S., Prakasan, S. K., Kouame, D., Watson, J., & Seymour, T. (2011). The history of social media and its impact on business. *Journal of Applied Management and Entrepreneurship, 16*(3), 79–91.

Eilu, E., Baguma, R., & Pettersen, J. S. (2014). *Persuasion and Acceptance of Mobile Phones as a Voting Tool in Developing Countries*. Accepted for Presentation at the 4th International Conference on M4D Mobile Communication for Development, Dakar, Senegal.

Eltoweissy, M., Jajodia, S., & Mukkamala, R. (2008). Secure multicast for mobile commerce applications: Issues and challenges. In A. Becker (Ed.), *Electronic commerce: Concepts, methodologies, tools, and applications* (pp. 930–951). Hershey, PA: IGI Global. doi:10.4018/978-1-59904-943-4.ch077

E-Marketer. (2014). Younger Users Spend More Daily Time on Social Networks. *E-Marketer*. Retrieved from: http://www.emarketer.com/Article/Younger-Users-Spend-More-Daily-Time-on-Social-Networks/1011592

Emmanouilidis, C., Koutsiamanis, R.-A., & Tasidou, A. (2013). Mobile guides: Taxonomy of architectures, context awareness, technologies and applications. *Journal of Network and Computer Applications, 36*(1), 103–125. doi:10.1016/j.jnca.2012.04.007

Enriquez, M. A. S. (2014). Students' perceptions on the effectiveness of the use of edmodo as a supplementary tool for learning. *The DLSU Research Congress*.

Erev, I., Rodensky, D., Levi, M. A., Hershler, M.E., Admi, H., Donchin, Y. (2010). *The value of 'gentle reminder' on safe medical behaviour: Quality improvement report*. Academic Press.

Eroglu, S. A., Machleit, K. A., & Davis, L. M. (2001). Atmospheric Qualities of Online Retailing: A Conceptual Model and Implications. *Journal of Business Research, 54*(2), 177–184. doi:10.1016/S0148-2963(99)00087-9

Eurostat. (2014). *Internet use statistics - individuals - Statistics Explained*. Retrieved from http://epp.eurostat.ec.europa.eu/statistics_explained/index.php/Internet_use_statistics_-_individuals#

Eze, U. C., & Poong, Y. S. (2016). Investigating the moderating roles of age and ethnicity in mobile commerce acceptance. In J. Prescott (Ed.), *Handbook of research on race, gender, and the fight for equality* (pp. 90–112). Hershey, PA: IGI Global. doi:10.4018/978-1-5225-0047-6.ch005

Fadzilah, F. M., & Arshad, N. I. (2015). Evaluating the Impact of Non-Clinical M-Health Application : Towards Development of a Framework Akademia Baru. *Journal of Advanced Research Design, 14*(1), 28–38.

Compilation of References

Falk, A., & Fischbacher, U. (2000). *A Theory of Reciprocity*. Retrieved in August 21 2015, from http://e-collection.library.ethz.ch/eserv/eth:25511/eth-25511-01.pdf

Falk, M., & Hagsten, E. (2015). E-commerce trends and impacts across Europe. *International Journal of Production Economics, 170*, 357–369. doi:10.1016/j.ijpe.2015.10.003

Faqih, K. M. S., & Jaradat, M. R. M. (2015). Assessing the moderating effect of gender differences and individualism-collectivism at individual-level on the adoption of mobile commerce technology: TAM3 perspective. *Journal of Retailing and Consumer Services, 22*, 37–52. doi:10.1016/j.jretconser.2014.09.006

Farooq, F., & Jan, Z. (2012). The Impact of Social Networking to Influence Marketing through Product Reviews. *International Journal of Information and Communication Technology Research, 2*(8).

Faudree, B., & Ford, M. (2013). Security and Privacy in Mobile Health. *CIO Journal*. Retrieved May 20, 2016, from http://deloitte.wsj.com/cio/2013/08/06/security-and-privacy-in-mobile-health/

Fayaz-Bakhsh, A., & Geravandi, S. (2015). Medical Students Perceptions Regarding the Impact of Mobile Medical Applications on their Clinical Practice. *Journal of Mobile Technology in Medicine, 4*(2), 51–52. doi:10.7309/jmtm.4.2.8

Feaster, J. C. (2009). The repertoire niches of interpersonal media: Competition and coexistence at the level of the individual. *New Media & Society, 11*(6), 965–984. doi:10.1177/1461444809336549

Feinberg, R. A. (1986). Credit cards as spending facilitating stimuli: A conditioning interpretation. *The Journal of Consumer Research, 13*(12), 348–356. doi:10.1086/209074

Feldman, B. (2014). *Social Proof: Your Key to More Magnetic Marketing*. Retrieved in August 21 2015, from https://blog.kissmetrics.com/social-proof/

Felix, R., Rauschnabel, P. A., & Hinsch, C. (2016). Elements of Strategic Social Media Marketing: A Holistic Framework. *Journal of Business Research*.

Felt, A. P., Ha, E., Egelman, S., Haney, A., Chin, E., & Wagner, D. (2012). *Android permissions: user attention, comprehension, and behavior*. Paper presented at the Eighth Symposium on Usable Privacy and Security, Washington, DC. doi:10.1145/2335356.2335360

Feng, H., Hoegler, T., & Stucky, W. (2006). *Exploring the Critical Success Factors for Mobile Commerce*. Paper presented at the International Conference on Mobile Business, Copenhagen, Denmark. doi:10.1109/ICMB.2006.15

Ferrier, M. (Ed.). (2015). *Leading in the SMAC age*. Bangalore, India: Wipro.

Fiordelli, M., Diviani, N., & Schulz, P. J. (2013). Mapping mhealth research: A decade of evolution. *Journal of Medical Internet Research, 15*(5), e95. doi:10.2196/jmir.2430 PMID:23697600

Fischer, J., Neff, M., Freudenstein, D., & Bartz, D. (2004, June). Medical augmented reality based on commercial image guided surgery. In S. Coquillart (Chair), *10th Eurographics Symposium on Virtual Environments*. Symposium conducted at the meeting of the Eurographics Association, Grenoble, France.

Fishbein, M., & Ajzen, I. (1975). *Belief, Attitude, Intentions and Behavior: An Introduction to Theory and Research*. Boston, MA: Addison-Wesley.

Flatraaker, D. (2008). Mobile, Internet and electronic payments: The key to unlocking the full potential of the internal payments market. *Journal of Payments Strategy &Systems, 3*(1), 60–70.

Fogg, B. J. (2002). *Persuasive Technology: Using Computers to Change What We Think and Do*. Morgan Kaufmann.

Fogg, B. J. (2008). Mass interpersonal persuasion: An early view of a new phenomenon. In *Proc. Third International Conference on Persuasive Technology, Persuasive 2008*. Berlin: Springer. doi:10.1007/978-3-540-68504-3_3

Fogg, B.J. (2004). *Captology Understanding How Computers Manipulate People*. Persuasive Technology Lab, Stanford University.

Forlizzi, J., & Ford, S. (2000). *The Building Blocks of Experience: An Early Framework for Interaction Designers*. DIS '00, Brooklyn, NY.

Forlizzi, J., & Battarbee, K. (2004). Understanding experience in interactive systems. In *Proceedings of the 5th Conference on Designing Interactive Systems: Processes, Practices, Methods,Techniques - DIS'04* (pp. 261-268). New York: ACM Press.

Fornell, C. (1992). A national customer satisfaction barometer: The Swedish experience. *Journal of Marketing, 56*(1), 1, 6–21. doi:10.2307/1252129

Fornell, C., & Larcker, D. F. (1981). Evaluating structural equation models with unobservable variables and measurement error. *JMR, Journal of Marketing Research, 18*(1), 39–50. doi:10.2307/3151312

Forsé, M. (2012). Today's Social Networks. *OFCE Revue, 7*, 155–169.

Franklin, T. (2008). Adventures in Agile Contracting: Evolving from Time and Materials to Fixed Price, Fixed Scope Contracts. In *Proceedings of IEEE Agile 2008 Conference* (pp. 269-273). Los Alamitos, CA: IEEE Computer Society doi:10.1109/Agile.2008.88

Frenkel, K. A. (2013). Mobile Apps Need Better Security. *CIO Insight*, 1.

Fritz, J., U-Thainual, P., Ungi, T., Flammang, A. J., Fichtinger, G., Iordachita, I. I., & Carrino, J. A. (2012). Augmented reality visualization with use of image overlay technology for MR imaging-guided interventions: Assessment of performance in cadaveric shoulder and hip arthrography at 1.5 T. *Radiology, 265*(1), 254–263. doi:10.1148/radiol.12112640 PMID:22843764

Gaebel, M. (2013). *MOOCs: Massive Open Online Courses*. Retrieved from http://www.eua.be/Libraries/Publication/EUA_Occasional_papers_MOOCs.sflb.ashx

Ganes, A., & Nævdal, S. (2008). *Software Contracting and Agile Development in the Norwegian ICT Industry: A Qualitative Survey* (Unpublished Master dissertation). Norwegian University of Science and Technology (NTNU), Trondheim, Norway.

Gao, Q., Ebert, D., Chen, X., & Ding, Y. (2011). Design of a Mobile Social Community Platform for Older Chinese People in Urban Areas. *Human Factors and Ergonomics in Manufacturing & Service Industries, 25*(1), 66–89.

Garland, R. (1991). The Mid-Point on a Rating Scale: Is it Desirable? *Marketing Bulletin, 2*(3), 66–70.

Gebert, H., Geib, M., Kolbe, L., & Riempp, G. (2002). Towards Customer Knowledge Management. *Proceedings of the 2nd International Conference on Electronic Business (ICEB 2002)*.

George, P., Thouvenin, I., Frémont, V., & Cherfaoui, V. (2012). DARIA: Driver assistance by augmented reality for intelligent automotive. *Intelligent Vehicles Symposium (IV)*, 1043-1048.

Ghosh, A. K., & Swaminatha, T. M. (2001). Software security and privacy risks in mobile e-commerce. *Communications of the ACM, 44*(2), 51–57. doi:10.1145/359205.359227

Compilation of References

Gilbert, P., Chun, B.-G., Cox, L. P., & Jung, J. (2011). *Vision: automated security validation of mobile apps at app markets*. Paper presented at the second international workshop on Mobile cloud computing and services, Bethesda, MD. doi:10.1145/1999732.1999740

Gil Or, O. (2010). The Potential of Facebook in creating commercial value for service companies. *Advances in Management, 3*(2), 20–25.

Gitau, L., & Nzuki, D. (2014). Analysis of Determinants of M-Commerce Adoption by Online Consumers. *International Journal of Business Human Technology, 4*, 3.

Gleason, A. W. (2015). mHealth — Opportunities for Transforming Global Health Care and Barriers to Adoption. *Journal of Electronic Resources in Medical Libraries, 12*(2), 114–125. doi:10.1080/15424065.2015.1035565

Godbole, N. (2006). Relating mobile computing to mobile commerce. In B. Unhelkar (Ed.), *Handbook of research in mobile business: Technical, methodological, and social perspectives* (pp. 463–486). Hershey, PA: IGI Global. doi:10.4018/978-1-59140-817-8.ch033

Goldsmith, R. E., & Flynn, L. R. (2004). Psychological and behavioural drivers of online clothing purchase. *Journal of Fashion Marketing and Management, 8*(1), 84–95. doi:10.1108/13612020410518718

Gorder, L. M. (2008). A study of teacher perceptions of instructional technology integration in the classroom. *Delta Pi Epsilon Journal, 50*(2).

Gowda, A. D., & A S, M. (2016). Smart shopping using augmented reality on android OS. *International Journal of Engineering Research and General Science, 4*(3), 211–217.

Greco, M., Stucchi, N., Zavagno, D., & Marino, B. (2008). On the portability of computer-generated presentations: The effect of text-background color combinations on text legibility. *Human Factors: The Journal of the Human Factors and Ergonomics Society, 50*(5), 821–833. doi:10.1518/001872008X354156 PMID:19110842

Greenberg, P. (2002). *CRM at the speed of light: capturing and keeping customers in internet real time*. Berkeley, CA: McGraw-Hill.

Greenwald, M. (2014). Of the Best Product Demos Ever and why they're so effective. *Forbes*. Retrieved in August 21 2015, from http://www.forbes.com/sites/michellegreenwald/2014/06/04/8-of-the-best-product-demos-ever-and-why-theyre-so-effective/

Guerra-Casanova, J., Sánchez-Ávila, C., de Santos Sierra, A., & del Pozo, G. B. (2011). Score optimization and template updating in a biometric technique for authentication in mobiles based on gestures. *Journal of Systems and Software, 84*(11), 2013–2021. doi:10.1016/j.jss.2011.05.059

Guillén, A., Herrera, L. J., Pomares, H., Rojas, I., & Liébana-Cabanillas, F. (2016). Decision support system to determine intention to use mobile payment systems on social networks: A methodological analysis. *International Journal of Intelligent Systems, 31*(2), 153–172. doi:10.1002/int.21749

Guo, S., Wang, M., & Leskovec, J. (2011). The Role of Social Networks in Online Shopping: Information Passing, Price of Trust, and Consumer Choice. *EC '11 Proceedings of the 12th ACM Conference on Electronic Commerce*, 157-166.

Gupta, P. (2016). Exploring barriers affecting the acceptance of mobile commerce. In S. Madan & J. Arora (Eds.), *Securing transactions and payment systems for m-commerce* (pp. 234–250). Hershey, PA: IGI Global. doi:10.4018/978-1-5225-0236-4.ch012

Gupta, R., Muttoo, S. K., & Pal, S. K. (2016). Understanding fraudulent activities through m-commerce transactions. In S. Madan & J. Arora (Eds.), *Securing transactions and payment systems for m-commerce* (pp. 68–93). Hershey, PA: IGI Global. doi:10.4018/978-1-5225-0236-4.ch004

Gütl, C., Hernández Rizzardini, R., Chang, V., & Morales, M. (2014). Attrition in MOOC: Lessons learned from drop-out students. In L. Uden, J. Sinclair, Y.-H. Tao, & D. Liberona (Eds.), *Learning Technology for Education in Cloud. MOOC and Big Data* (Vol. 446, pp. 37–48). Springer International Publishing. doi:10.1007/978-3-319-10671-7_4

Hair, J. F., Black, W. C., Babin, B. J., Anderson, R. E., & Tatham, R. L. (2009). *Multivariate Data Analysis* (7th ed.). Upper Saddle River, NJ: Pearson Education Inc.

Ha, K. H., Canedoli, A., Baur, A. W., & Bick, M. (2012). Mobile banking-insights on its increasing relevance and most common drivers of adoption. *Electronic Markets*, 1–11.

Hakansson, H., Ford, D., Gadde, L., & Snehota, I. (2011). *Managing Business Relationships*. Hoboken, NJ: John Wiley & Sons.

Halbach, M., & Gong, T. (2013). What predicts commercial bank leaders' intention to use mobile commerce?: The roles of leadership behaviors, resistance to change, and technology acceptance model. In M. Khosrow-Pour (Ed.), *E-commerce for organizational development and competitive advantage* (pp. 151–170). Hershey, PA: IGI Global. doi:10.4018/978-1-4666-3622-4.ch008

Hallikainen, H., & Laukkanen, T. (2016). Consumer Trust towards an Online Vendor in High-vs. Low-Context Cultures. In *2016 49th Hawaii International Conference on System Sciences (HICSS)* (pp. 3536-3545). IEEE.

Hall, R. H., & Hanna, P. (2003). The impact of web page text-background colour combinations on readability, retention, aesthetics and behavioural intention. *Behaviour & Information Technology*, *23*(3), 183–195. doi:10.1080/01449290410001669932

Hamilton, D. P. (2001, May 21). Making Sense of It All. *The Asia Wall Street Journal*, p. T4.

Hamza-Lup, F. G., Santhanam, A. P., Imielinska, C., Meeks, S. L., & Rolland, J. P. (2007). Distributed augmented reality with 3-D lung dynamics – a planning tool concept. *IEEE Transactions on Information Technology in Biomedicine*, *11*(1), 40–46. doi:10.1109/TITB.2006.880552 PMID:17249402

Handel, M. J. (2011). mHealth (mobile health)-using Apps for health and wellness. *Explore (New York, N.Y.)*, *7*(4), 256–261. doi:10.1016/j.explore.2011.04.011 PMID:21724160

Han, J., Yang, Y., Huang, X., Yuen, T. H., Li, J., & Cao, L. (2016). Accountable mobile e-commerce scheme via identity-based plaintext-checkable encryption. *Information Sciences*, *345*, 143–155. doi:10.1016/j.ins.2016.01.045

Harrigan, P., Ramsey, E., & Ibbotson, P. (2011). Critical factors underpinning the e-CRM activities of SMEs. *Journal of Marketing Management*, *27*(5-6), 503–529. doi:10.1080/0267257X.2010.495284

Harris, M. A., Brookshire, R., & Chin, A. G. (2016). Identifying factors influencing consumers intent to install mobile applications. *International Journal of Information Management*, *36*(3), 441–450. doi:10.1016/j.ijinfomgt.2016.02.004

Harrison, R., Flood, D., & Duce, D. (2013). Usability of mobile applications: Literature review and rationale for a new usability model. *Journal of Interaction Science*, *1*(1), 1–16. doi:10.1186/2194-0827-1-1

Hartmann, M. (2013). From domestication to mediated mobilism. *Mobile Media & Communication*, *1*(1), 42–49. doi:10.1177/2050157912464487

Hartnett, M., St. George, A., & Dron, J. (2011). Examining Motivation in Online Distance Learning Environments: Complex, Multifaceted, and Situation-Dependent. *International Review of Research in Open and Distance Learning*, *12*(6), 20–37. doi:10.19173/irrodl.v12i6.1030

Hashemi, M., Azizinezhad, M., Najafi, V., & Nesari, A. J. (2011). What is Mobile Learning? Challenges and Capabilities. *Procedia: Social and Behavioral Sciences*, *30*, 2477–2481. doi:10.1016/j.sbspro.2011.10.483

Hassenzahl, M., & Tractinsky, N. (2006, March-April). User experience – a research agenda. *Behaviour & Information Technology*, *25*(2), 91–97. doi:10.1080/01449290500330331

Hayes, A. F. (2013). *Introduction to Mediation, Moderation, and Conditional Process Analysis: A Regression-Based Approach*. New York: Guilford Publications.

Hebly, P. (2012). *Willingness to pay for mobile apps*. Erasmus Universiteit Rotterdam. Retrieved from http://thesis.eur.nl/pub/13913/Heblij-P.G.-375422ph-.docx

Heikkinen, J., Olsson, T., & Väänänen-Vainio-Mattila, K. (2009). Expectations for user experience in haptic communication with mobile devices. *Proceedings of the 11th International Conference on Human-Computer Interaction with Mobile Devices and Services, MobileHCI'09*. Retrieved in August 21 2015, from http://womeninbusiness.about.com/od/marketingpsychology/a/prineciprocity.htm

Heilig, M. (1962). *U.S. Patent No. 3050870*.

Heinrichs, J. H., Lim, J. S., & Lim, K. S. (2011). Influence of social networking site and user access method on social media evaluation. *Journal of Consumer Behaviour*, *10*(6), 347–355. doi:10.1002/cb.377

Helme-Guizon, A. (2002). *Sources and consequences of the irritation felt during navigation on a retail website. An exploratory study*. 18th AFM International Symposium, Lille, France.

Henderson, S. J., & Feiner, S. (2008). Opportunistic controls: leveraging natural affordances as tangible user interfaces for augmented reality. *Proceedings of the 2008 ACM symposium on Virtual reality software and technology* (pp. 211-218). New York: ACM. doi:10.1145/1450579.1450625

Hermawati, S., & Lawson, G. (2016). Establishing usability heuristics for heuristics evaluation in a specific domain: Is there a consensus? *Applied Ergonomics*, *56*, 34-51. doi:10.1016/j.apergo.2015.11.016

Herzberg, A. (2003). Payments and banking with mobile personal devices. *Communications of the ACM*, *46*(5), 53–58. doi:10.1145/769800.769801

Heutte, J., Kaplan, J., Fenouillet, F., Caron, P., Rosselle, M., & Uden, L. (2014). MOOC User Persistence. In L. Uden, J. Sinclair, Y.-H. Tao, & D. Liberona (Eds.), *Learning Technology for Education in Cloud. MOOC and Big Data* (Vol. 446, pp. 13–24). Springer International Publishing.

Hew, J. J. (2017). Hall of fame for mobile commerce and its applications: A bibliometric evaluation of a decade and a half 2000–2015. *Telematics and Informatics*, *34*(1), 43–66. doi:10.1016/j.tele.2016.04.003

Hew, J. J., Lee, V. H., Ooi, K. B., & Lin, B. (2016). Mobile social commerce: The booster for brand loyalty? *Computers in Human Behavior*, *59*, 142–154. doi:10.1016/j.chb.2016.01.027

Hill, A., & Scharff, L. V. (1997). Readability of websites with various foreground/background color combinations, font types and word styles. *Proceedings of 11th National Conference in Undergraduate Research*, *2*, 742-746.

Hill, S. R., & Troshani, I. (2010). Factors influencing the adoption of personalisation mobile services: Empirical evidence from young Australians. *International Journal of Mobile Communications*, *8*(2), 150–168. doi:10.1504/IJMC.2010.031445

Hirschman, E. C. (1979). Differences in Consumer Purchase Behavior by Credit Card Payment System. *The Journal of Consumer Research*, *6*(6), 58–66. doi:10.1086/208748

Hoda, R., Noble, J., & Marshall, S. (2009). Negotiating Contracts for Agile Projects: A Practical Perspective. In *Proceedings of 10th International Conference on Agile Processes in Software Engineering and Extreme Programming* (pp. 186-191). Berlin: Springer. doi:10.1007/978-3-642-01853-4_25

Hoe, J. M. C. (2012). Facebook and learning: Students' perspective on a course. *The Journal of the NUS Teaching Academy*, *2*(3), 131–143.

Hofstede, G. (2011). Dimensionalising Cultures: The Hofstede Model in Context. *Online Readings in Psychology and Culture*, *2*(1). doi:10.9707/2307-0919.1014

Hogan, M., & Sokol, A. (Eds.). (2013). NIST Cloud Computing Standards Roadmap, NIST Special Publication, 500-291. Gaithersburg, MD: National Institute of Standards and Technology (NIST).

Holmes, A., Byrne, A., & Rowley, J. (2013). Mobile shopping behaviour: Insights into attitudes, shopping process involvement and location. *International Journal of Retail & Distribution Management*, *42*(1), 25–39. doi:10.1108/IJRDM-10-2012-0096

Hsiao, C.-H., Chang, J.-J., & Tang, K.-Y. (2016). Exploring the influential factors in continuance usage of mobile social Apps: Satisfaction, habit, and customer value perspectives. *Telematics and Informatics*, *33*(2), 342–355. doi:10.1016/j.tele.2015.08.014

Hsieh, M., & Tsao, W. (2014). Reducing perceived online shopping risk to enhance loyalty: A website quality perspective. *Journal of Risk Research*, *17*(2), 241–261. doi:10.1080/13669877.2013.794152

Hsu, C.-L., & Lin, J. C.-C. (2016). An empirical examination of consumer adoption of Internet of Things services: Network externalities and concern for information privacy perspectives. *Computers in Human Behavior*, *62*, 516–527. doi:10.1016/j.chb.2016.04.023

Huang, H. M., & Liaw, S. S. (2005). Exploring users attitudes and intentions toward the web as a survey tool. *Computers in Human Behavior*, *21*(5), 729–743. doi:10.1016/j.chb.2004.02.020

Hubspot. (n.d.). *State of Inbound Marketing 2012: The ultimate list of marketing statistics*. Retrieved from: http://www.hubspot.com/marketing-statistics

Hughes, N., & Lonie, S. (2010). M-PESA: Mobile money for the "unbanked". *Innovations: Technology, Governance, Globalization*, *2*(1–2), 63–81.

Huitt, W. (2003). The affective system: Educational Psychology Interactive. Valdosta, GA: Valdosta State University. Retrieved from http://www.edpsycinteractive.org/topics/affect/affsys.html

Hu, J., Hoang, X. D., & Khalil, I. (2011). An embedded DSP hardware encryption module for secure e-commerce transactions. *Security and Communication Networks*, *4*(8), 902–909. doi:10.1002/sec.221

Hung, M. C., Yang, S. T., & Hsieh, T. C. (2012). An examination of the determinants of mobile shopping continuance. *International Journal of Electronic Business Management*, *10*(1), 29–37.

Huron, D. (2006). Sweet anticipation: Music and the psychology of expectation. Cambridge, MA: A Bradford Book, The MIT Press.

Hurwitz, J. (2010). *Cloud computing for dummies*. Wiley Publications.

Hussain, M., Al-Haiqi, A., Zaidan, A., Zaidan, B., Kiah, M., Anuar, N. B., & Abdulnabi, M. (2015). The landscape of research on smartphone medical apps: Coherent taxonomy, motivations, open challenges and recommendations. *Computer Methods and Programs in Biomedicine, 122*(3), 393–408. doi:10.1016/j.cmpb.2015.08.015 PMID:26412009

Husson, T. (2016). *The Future of Messaging Apps*. Comscore.

Hu, W. (2009). Fundamentals of mobile commerce systems. In W. Hu (Ed.), *Internet-enabled handheld devices, computing, and programming: Mobile commerce and personal data applications* (pp. 1–25). Hershey, PA: IGI Global. doi:10.4018/978-1-59140-769-0.ch001

Hu, W., Yeh, J., Yang, H., & Lee, C. (2008). Mobile handheld devices for mobile commerce. In A. Becker (Ed.), *Electronic commerce: Concepts, methodologies, tools, and applications* (pp. 152–162). Hershey, PA: IGI Global. doi:10.4018/978-1-59904-943-4.ch015

Hu, W., Zuo, Y., Kaabouch, N., & Chen, L. (2010). A technological perspective of mobile and electronic commerce systems. In M. Khosrow-Pour (Ed.), *E-commerce trends for organizational advancement: New applications and methods* (pp. 16–35). Hershey, PA: IGI Global. doi:10.4018/978-1-60566-964-9.ch002

Hyppölä, J., Martínez, H., & Laukkanen, S. (2014, March). Experiential learning theory and virtual and augmented reality applications. In R. Fisher (Chair), *4th Global Conference on Experiential Learning in Virtual Worlds*. Symposium conducted at the meeting of Inter-Disciplinary.Net, Prague, Czech Republic.

IE The Indian Express. (2015, July 21). *IAMAI says India will have 500 Million Internet users by2017*. Retrieved from http://indianexpress.com/article/technology/tech-news-technology/iamai-says-india-to-have-236-million-mobile-internet-users-by-2016/

Ilie, V., van Slyke, C., Green, G., & Lou, H. (2005). Gender differences in perceptions and use communication technologies: A diffusion of innovation approach. *Information Resources Management Journal, 18*(3), 13–31. doi:10.4018/irmj.2005070102

Inostroza, R., & Rusu, C. (2014). Mapping usability heuristics and design principles for touchscreen-based mobile devices. In *Proceedings of the 7th Euro American Conference on Telematics and Information Systems*. Valparaiso, Chile: ACM. doi:10.1145/2590651.2590677

Inostroza, R., Rusu, C., Roncagliolo, S., Jimenez, C., & Rusu, V. (2012a). Usability heuristics for touchscreen-based mobile devices. In *Proceedings of the 9th International Conference on Information Technology*. IEEE.

Inostroza, R., Rusu, C., Roncagliolo, S., Jimenez, C., & Rusu, V. (2012b). Usability heuristics validation through empirical evidences: a touchscreen-based mobile devices proposal. In *Proceedings of the 31th International Conference of the Chilean Computer Science Society*. IEEE. doi:10.1109/SCCC.2012.15

Inostroza, R., Rusu, C., Roncagliolo, S., Jimenez, C., & Rusu, V. (2013). Usability heuristics for touchscreen-based mobile devices: update. In *Proceedings of the 1st Chilean Conference of Computer-Human Interaction*. Temuco, Chile: ACM. doi:10.1145/2535597.2535602

iOS app approvals. (2016). Retrieved 20 June, 2016, from https://en.wikipedia.org/wiki/IOS_app_approvals

Irwin, L. D. B., Ball, L., Desbrow, B., & Leveritt, M. (2012). Students perceptions of using Facebook as an interactive learning resource at university. *Australasian Journal of Educational Technology, 28*(7), 1221–1232. doi:10.14742/ajet.798

Islam, M. A., Khan, M. A., Ramayah, T., & Hossain, M. M. (2011). The Adoption of Mobile Commerce Service among Employed Mobile Phone Users in Bangladesh: Self-efficacy as A Moderator. *International Business Research, 4*(2), 80–89. doi:10.5539/ibr.v4n2p80

ISO DIS 9241-210. (2008). *Ergonomics of Human Systems Interaction-Part 210: Human centered design for interactive systems*. ISO Switzerland.

Ito, M., Horst, H., Bittanti, M., Boyd, D., Herr-Stephenson, B., Lange, P. G., & Robinson, L. (2008). *Living and Learning with New Media: Summary of Findings from the Digital Youth Project. The John D. and Catherine T. MacArthur Foundation Reports on Digital Media and Learning*. The MIT Press.

Jaafar. (2012). Malaysian Health System Review. *Health Systems in Transition, 3*(1), 1–103. Retrieved from http://www.wpro.who.int/asia_pacific_observatory/hits/series/Malaysia_Health_Systems_Review2013.pdf

Jacob, C. (2005). *The influence of the music of a business website on consumer responses* (PhD Thesis). Rennes1 University.

Jahns, R.-G. (2013). *The market for mHealth app services will reach $26 billion by 2017*. Retrieved from http://research-2guidance.com/the-market-for-mhealth-app-services-will-reach-26-billion-by-2017/

Jain, A. K., & Shanbhag, D. (2012). Addressing Security and Privacy Risks in Mobile Applications. *IT Professional, 14*(5), 28–33. doi:10.1109/MITP.2012.72

Jayaswal, K., Kallakurchi, K., Houde, D., & Shah, D. (2014). *Cloud Computing Black Book*. New Delhi, India: Dreamtech Press.

Jeong, S.-H., Kim, H., Yum, J.-Y., & Hwang, Y. (2016). What type of content are smartphone users addicted to?: SNS vs. games. *Computers in Human Behavior, 54*, 10–17. doi:10.1016/j.chb.2015.07.035

Jeremiah, D. (2014). *Frost and Sullivan: New Zealand's mobile commerce market to reach nearly $10 billion by 2019*. Retrieved September 13, 2016, from http://ww2.frost.com/news/press-releases/frost-sullivan-new-zealands-mobile-commerce-market-reach-nearly-10-billion-2019/

Jones, N. (2002). *Gatner research: Citizens to Vote With Mobile Phone Messages in U.K. Elections*. Publication Date: 8 February 2002 ID Number: FT-15-4730.

Jordan, K., & King, M. (2011). Augmented reality assisted upper limb rehabilitation following stroke. In A. Y. C. Nee (Ed.), *Augmented reality: Some emerging application areas* (pp. 155–174). Rijeka, Croatia: InTech. doi:10.5772/25954

Joshua, A. J., & Koshy, M. P. (2011). Usage patterns of electronic banking services by urban educated customers: Glimpses from India. *Journal of Internet Banking and Commerce, 16*(1), 1–12.

Joyce, G., & Lilley, M. (2014). Towards the development of usability heuristics for native smartphone mobile applications. LNCS, 8517, 465-474. doi:10.1007/978-3-319-07668-3_45

Juan, S. (2006). *Which comes first: imagination or fantasy?* Retrieved in June 25 2015, from http://www.theregister.co.uk/2006/05/19/the_odd_body_imagination_fantasy/

Jung, Y., Perez-Mira, B., & Wiley-Patton, S. (2009). Consumer adoption of mobile TV: Examining psychological flow and media content. *Computers in Human Behavior, 25*(1), 123–129. doi:10.1016/j.chb.2008.07.011

Kaikkonen, A. (2009). *Internet On Mobiles: Evolution Of Usability And User Experience*. Doctoral Dissertation. Retrieved in June 25 2015, from URL: http://lib.tkk.fi/Diss/2009/isbn9789522481900/

Kalliny, M., & Minor, M. (2006). The antecedents of m-commerce adoption. *Journal of Strategic E-Commerce, 4*(1/2), 81–98.

Kamphuis, C., Barsom, E., Schijven, M., & Christoph, N. (2014). Augmented reality in medical education? *Perspectives on Medical Education, 3*(4), 300–311. doi:10.1007/s40037-013-0107-7 PMID:24464832

Kang, J., Kim, H., Cheong, Y. G., & Huh, J. H. (2015). Visualizing Privacy Risks of Mobile Applications through a Privacy Meter. In J. Lopez & Y. Wu (Eds.), *Information Security Practice and Experience: 11th International Conference, ISPEC 2015, Beijing, China, May 5-8, 2015, Proceedings* (pp. 548-558). Cham: Springer International Publishing. doi:10.1007/978-3-319-17533-1_37

Kang, S. (2014). Factors influencing intention of mobile application use. *International Journal of Mobile Communications*, *12*(4), 360–379. doi:10.1504/IJMC.2014.063653

Kanwalvir, K., & Aggarwal, H. (2013). Critical Factors in Consumers Perception towards Mobile Commerce in E-Governance Implementation: An Indian Perspective. *International Journal of Engineering and Advanced Technology*, *2*, 3.

Kaplan, J. (2014). Co-regulation in technology enhanced learning environments. In L. Uden, J. Sinclair, Y.-H. Tao, & D. Liberona (Eds.), *Learning Technology for Education in Cloud. MOOC and Big Data* (Vol. 446, pp. 72–81). Springer International Publishing.

Kapoor, N. (2016). Consumer perception to mobile commerce. In S. Madan & J. Arora (Eds.), *Securing transactions and payment systems for m-commerce* (pp. 217–233). Hershey, PA: IGI Global. doi:10.4018/978-1-5225-0236-4.ch011

Karsenti, T. (2013). The MOOC: What the research says. *International Journal of Technologies in Higher Education*, *10*(2), 23–37.

Kasemsap, K. (2016c). Mastering big data in the digital age. In M. Singh & D. G. (Eds.), *Effective big data management and opportunities for implementation* (pp. 104–129). Hershey, PA: IGI Global. doi:10.4018/978-1-5225-0182-4.ch008

Kasemsap, K. (2016a). Advocating electronic business and electronic commerce in the global marketplace. In S. Dixit & A. Sinha (Eds.), *E-retailing challenges and opportunities in the global marketplace* (pp. 1–24). Hershey, PA: IGI Global. doi:10.4018/978-1-4666-9921-2.ch001

Kasemsap, K. (2016b). Implementing electronic commerce in global marketing. In I. Lee (Ed.), *Encyclopedia of e-commerce development, implementation, and management* (pp. 591–602). Hershey, PA: IGI Global. doi:10.4018/978-1-4666-9787-4.ch043

Kasemsap, K. (2016d). Role of social media in brand promotion: An international marketing perspective. In A. Singh & P. Duhan (Eds.), *Managing public relations and brand image through social media* (pp. 62–88). Hershey, PA: IGI Global. doi:10.4018/978-1-5225-0332-3.ch005

Kasemsap, K. (2016e). Investigating the roles of mobile commerce and mobile payment in global business. In S. Madan & J. Arora (Eds.), *Securing transactions and payment systems for m-commerce* (pp. 1–23). Hershey, PA: IGI Global. doi:10.4018/978-1-5225-0236-4.ch001

Kasemsap, K. (2016f). Promoting service quality and customer satisfaction in global business. In U. Panwar, R. Kumar, & N. Ray (Eds.), *Handbook of research on promotional strategies and consumer influence in the service sector* (pp. 247–276). Hershey, PA: IGI Global. doi:10.4018/978-1-5225-0143-5.ch015

Kasemsap, K. (2016g). Encouraging supply chain networks and customer loyalty in global supply chain. In N. Kamath & S. Saurav (Eds.), *Handbook of research on strategic supply chain management in the retail industry* (pp. 87–112). Hershey, PA: IGI Global. doi:10.4018/978-1-4666-9894-9.ch006

Kasemsap, K. (2017a). Software as a service, Semantic Web, and big data: Theories and applications. In A. Turuk, B. Sahoo, & S. Addya (Eds.), *Resource management and efficiency in cloud computing environments* (pp. 264–285). Hershey, PA: IGI Global. doi:10.4018/978-1-5225-1721-4.ch011

Kasemsap, K. (2017b). Mastering business process management and business intelligence in global business. In M. Tavana, K. Szabat, & K. Puranam (Eds.), *Organizational productivity and performance measurements using predictive modeling and analytics* (pp. 192–212). Hershey, PA: IGI Global. doi:10.4018/978-1-5225-0654-6.ch010

Kasemsap, K. (2017c). Mastering consumer attitude and sustainable consumption in the digital age. In N. Suki (Ed.), *Handbook of research on leveraging consumer psychology for effective customer engagement* (pp. 16–41). Hershey, PA: IGI Global. doi:10.4018/978-1-5225-0746-8.ch002

Kasemsap, K. (2017d). Professional and business applications of social media platforms. In V. Benson, R. Tuninga, & G. Saridakis (Eds.), *Analyzing the strategic role of social networking in firm growth and productivity* (pp. 427–450). Hershey, PA: IGI Global. doi:10.4018/978-1-5225-0559-4.ch021

Kasemsap, K. (2017e). Mastering social media in the modern business world. In N. Rao (Ed.), *Social media listening and monitoring for business applications* (pp. 18–44). Hershey, PA: IGI Global. doi:10.4018/978-1-5225-0846-5.ch002

Kasemsap, K. (2017f). Mastering web mining and information retrieval in the digital age. In A. Kumar (Ed.), *Web usage mining techniques and applications across industries* (pp. 1–28). Hershey, PA: IGI Global. doi:10.4018/978-1-5225-0613-3.ch001

Katell, M. A., Mishra, S. R., & Scaff, L. (2016). *A Fair Exchange: Exploring How Online Privacy is Valued.* Paper presented at the 2016 49th Hawaii International Conference on System Sciences (HICSS).

Kaufmann, H., & Meyer, B. (2008). Simulating educational physical experiments in augmented reality. *Proceedings of ACM SIGGRAPH ASIA 2008 Educators Programme*, 1-8. doi:10.1145/1507713.1507717

Kaur, R., & Malhotra, H. (2016). SWOT analysis of m-commerce. In S. Madan & J. Arora (Eds.), *Securing transactions and payment systems for m-commerce* (pp. 48–67). Hershey, PA: IGI Global. doi:10.4018/978-1-5225-0236-4.ch003

Keengwe, J. (2014). *Promoting Active Learning through the Integration of Mobile and Ubiquitous Technologies.* IGI Global.

Keith, M. J., Babb, J. S., & Lowry, P. B. (2014). *A Longitudinal Study of Information Privacy on Mobile Devices.* Paper presented at the 2014 47th Hawaii International Conference on System Sciences.

Keith, M. J., Babb, J. S., Lowry, P. B., Furner, C. P., & Abdullat, A. (2015). The role of mobile-computing self-efficacy in consumer information disclosure. *Information Systems Journal*, *25*(6), 637–667. doi:10.1111/isj.12082

Keith, M. J., Thompson, S. C., Hale, J., Lowry, P. B., & Greer, C. (2013). Information disclosure on mobile devices: Re-examining privacy calculus with actual user behavior. *International Journal of Human-Computer Studies*, *71*(12), 1163–1173. doi:10.1016/j.ijhcs.2013.08.016

Kelley, P. G., Consolvo, S., Cranor, L. F., Jung, J., Sadeh, N., & Wetherall, D. (2012). *A conundrum of permissions: installing applications on an android smartphone.* Paper presented at the 16th international conference on Financial Cryptography and Data Security, Bonaire. doi:10.1007/978-3-642-34638-5_6

Kemp, E. A., Thompson, A. J., & Johnson, R. (2008). Interface evaluation for invisibility and ubiquity e an example from E-learning. In *Proceedings of International Conference on Human-Computer Interaction.* Wellington, New Zealand: ACM.

Kennedy-Eden, H., & Gretzel, U. (2012). A Taxonomy of Mobile Applications in Tourism. *e-Review of Tourism Research*, *10*(2).

Khaled, R. (2008). *Culturally-Relevant Persuasive Technology.* PhD Thesis.

Khalifa, M., & Shen, K. N. (2008a). Drivers for Transactional B2C M-Commerce Adoption: Extended Theory of Planned Behavior. *Journal of Computer Information Systems*, *48*(3), 111–117.

Compilation of References

Khalifa, M., & Shen, K. N. (2008b). Explaining the adoption of transactional B2C mobile commerce. *Journal of Enterprise Information Management, 21*(2), 110–124. doi:10.1108/17410390810851372

Khalil, H., & Ebner, M. (2013). *How satisfied are you with your MOOC? - A Research Study on Interaction in Huge Online Courses.* Paper presented at the World Conference on Educational Media and Technology.

Khambete, P., & Athavankar, U. (2010). *Grounded Theory: An Effective Method for User Experience Design Research*. Academic Press.

Khan, J., Belk, W. R., & Craig-Lees, M. (2012). *Cash and Cards: Perceptions of Payment Mode*. Paper presented at Australian & New Zealand Marketing Academy Conference (ANZMAC) the Ehrenberg-Bass Institute, University of South Australia.

Khan, A. N., Mat Kiah, M. L., Khan, S. U., & Madani, S. A. (2013). Towards secure mobile cloud computing: A survey. *Future Generation Computer Systems, 29*(5), 1278–1299. doi:10.1016/j.future.2012.08.003

Khan, J., Abbas, H., & Al-Muhtadi, J. (2015). Survey on Mobile Users Data Privacy Threats and Defense Mechanisms. *Procedia Computer Science, 56*, 376–383. doi:10.1016/j.procs.2015.07.223

Khan, J., Belk, R. W., & Craig-Lees, M. (2015). Measuring consumer perceptions of payment mode. *Journal of Economic Psychology, 47*, 34–49. doi:10.1016/j.joep.2015.01.006

Khan, J., & Craig-Lees, M. (2009). "Cashless" transactions: Perceptions of money in mobile payment. *International Business and Economics Review, 1*(1), 23–32.

Khansa, L., Zobel, C. W., & Goicochea, G. (2012). Creating a taxonomy for mobile commerce innovations using social network and cluster analyses. *International Journal of Electronic Commerce, 16*(4), 19–52. doi:10.2753/JEC1086-4415160402

Kim, C., Mirusmonov, M., & Lee, I. (2010a). An empirical examination of factors influencing the intention to use mobile payment. *Computers in Human Behavior, 26*(3), 310–322. doi:10.1016/j.chb.2009.10.013

Kim, C., Tao, W., Shin, N., & Kim, K. (2010b). An empirical study of customers perceptions of security and trust in e-payment systems. *Electronic Commerce Research and Applications, 9*(1), 84–95. doi:10.1016/j.elerap.2009.04.014

Kim, D., Chun, H., & Lee, H. (2014). Determining the factors that influence college students adoption of smartphones. *Journal of the Association for Information Science and Technology, 65*(3), 578–588. doi:10.1002/asi.22987

Kim, G., Shin, B., & Lee, H. G. (2009). Understanding dynamics between initial trust and usage intentions of mobile banking. *Information Systems Journal, 19*(3), 283–311. doi:10.1111/j.1365-2575.2007.00269.x

Kim, H., Choi, Y., & Lee, Y. (2015). Web atmospheric qualities in luxury fashion brand websites. *Journal of Fashion Marketing and Management, 19*(4), 384–401. doi:10.1108/JFMM-09-2013-0103

Kim, H., Park, K., & Schwarz, N. (2010). Will This Trip Really Be Exciting? The Role of Incidental Emotions in Product Evaluation. *The Journal of Consumer Research, 36*(6), 983–991. doi:10.1086/644763

Kim, K. K., Shin, H. K., & Kim, B. (2011). The role of psychological traits and social factors in using new mobile communication services. *Electronic Commerce Research and Applications, 10*(4), 408–417. doi:10.1016/j.elerap.2010.11.004

Kim, K.-J., & Frick, T. W. (2011). Changes in Student Motivation During Online Learning. *Journal of Educational Computing Research, 44*(1), 1–24. doi:10.2190/EC.44.1.a

Kim, M. J., Chung, N., Lee, C. K., & Preis, M. W. (2015). Motivations and use context in mobile tourism shopping: Applying contingency and task–technology fit theories. *International Journal of Tourism Research*, *17*(1), 13–24. doi:10.1002/jtr.1957

Kim, M. K., Park, M. C., & Jeong, D. H. (2004). The effects of customer satisfaction and switching barrier on customer loyalty in Korean mobile telecommunication services. *Telecommunications Policy*, *28*(2), 145–159. doi:10.1016/j.telpol.2003.12.003

Kim, S., & Dey, A. K. (2009). Simulated augmented reality windshield display as a cognitive mapping aid for elder driver navigation. *Proceedings of ACM CHI 2009 Conference on Human Factors in Computing Systems*, 133-142. doi:10.1145/1518701.1518724

Kiritani, Y., & Shirai, S. (2003). Effects of background colours on user's experience in reading website. *Journal of the Asian Design International Conference*, *1*(64).

Kirlidog, M. (2014). Effect of Colours in Manual Data Typing. *Fourth International conference on Computer Science & Information Technology*. doi:10.5121/csit.2014.4206

Kivijärvi, M., Laukkanen, T., & Cruz, P. (2007). Consumer trust in electronic service consumption: A cross-cultural comparison between Finland and Portugal. *Journal of Euromarketing*, *16*(3), 51–65. doi:10.1300/J037v16n03_05

Kivity, A. (2007). KVM: the linux virtual machine monitor. *Proceedings of the Linux Symposium*, 1, 225–230.

Kjaerside, K., Kortbek, K. J., Hedegaard, H., & Gronbaek, K. (2005, June). Ardresscode: Augmented dressing room with tag-based motion tracking and real-time clothes simulation. In J. Zara (Chair), *Central European Multimedia and Virtual Reality Conference 2005*. Symposium conducted at the meeting of The Eurographics Association, Prague, Czech Republic.

Kjeldskov, J. (2002). "Just-in-Place" information for mobile device interfaces. In *International Conference on Mobile Human-Computer Interaction*. Pisa, Italy: Springer Berlin Heidelberg.

Kleijnen, M., Ruyter, K. D., & Wetzels, M. (2007). An assessment of value creation in mobile service delivery and the moderating role of time consciousness. *Journal of Retailing*, *83*(1), 33–46. doi:10.1016/j.jretai.2006.10.004

Knowledge@Wharton (Producer). (2013). MOOCs on the Move: How Coursera Is Disrupting the Traditional Classroom. *Innovation and Entrepreneurship*. Retrieved from http://knowledge.wharton.upenn.edu/article.cfm?articleid=3109

Ko, E., Kim, E. Y., & Lee, E. K. (2009). Modeling consumer adoption of mobile shopping for fashion products in Korea. *Psychology and Marketing*, *26*(7), 669–687. doi:10.1002/mar.20294

Koening-Lewis, N., Palmer, A., & Moll, A. (2010). Predicting young consumers take up of mobile banking services. *International Journal of Bank Marketing*, *28*(5), 410–432. doi:10.1108/02652321011064917

Koh, J., Kim, Y. G., Butler, B., & Bock, G. W. (2007). Encouraging participation in virtual communities. *Communications of the ACM*, *50*(2), 68–73. doi:10.1145/1216016.1216023

Koh, , Wan, J. K., Selvanathan, S., Vivekananda, C., Lee, G. Y., & Ng, C. T. (2014). Medical students perceptions regarding the impact of mobile medical applications on their clinical practice. *Journal of Mobile Technology in Medicine*, *3*(1), 46–53. doi:10.7309/jmtm.3.1.7

Kolowich, S. (2013). The professors who make the MOOCs. *The Chronicle of Higher Education*. Retrieved from http://chronicle.com/article/The-Professors-Behind-the-MOOC/137905/?cid=at&utm_source=at&utm_medium=en#id=overview

Konchada, V., Shen, Y., Burke, D., Argun, O. B., Weinhaus, A., Erdman, A. G., & Sweet, R. M. (2011). The Minnesota pelvic trainer: A hybrid vr/physical pelvis for providing virtual mentorship. *Studies in Health Technology and Informatics*, *163*, 280–282. PMID:21335805

Kop, R. (2011). The challenges to connectivist learning on open online networks: Learning experiences during a massive open online course. *International Review of Research in Open and Distance Learning*, *12*(3), 19–38. doi:10.19173/irrodl.v12i3.882

Korhonen, H., & Koivisto, E. M. I. (2006). Playability heuristics for mobile games. In *Proceedings of Mobile Human-Computer Interactions*. Helsinki, Finland: ACM. doi:10.1145/1152215.1152218

Köster, A., Matt, C., & Hess, T. (2016). Carefully choose your (payment) partner: How payment provider reputation influences m-commerce transactions. *Electronic Commerce Research and Applications*, *15*, 26–37. doi:10.1016/j.elerap.2015.11.002

Kotler, P. (2003). *Marketing management* (11th ed.). Prentice Hall.

Kotler, P., & Keller, K. L. (2015). *Marketing Management* (15th ed.). Pearson Education Limited.

Koufaris, M., & Hampton-Sosa, W. (2004). The development of initial trust in an online company by new customers. *Information & Management*, *41*(3), 377–397. doi:10.1016/j.im.2003.08.004

Kozma, R. B. (1994, June). Will media influence learning? Reframing the debate. *Educational Technology Research and Development*, *42*(2), 7–19. doi:10.1007/BF02299087

Krebs, P., & Duncan, D. T. (2015). Health App Use Among US Mobile Phone Owners: A National Survey. *JMIR mHealth and uHealth*, *3*(4), 1–12. doi:10.2196/mhealth.4924

Krejcie, R. V., & Morgan, D. W. (1970). Determining sample size for research activities. *Educational and Psychological Measurement*, *30*(3), 607–610. doi:10.1177/001316447003000308

Krotov, V., Junglas, I., & Steel, D. (2015). The mobile agility framework: an exploratory study of mobile technology enhancing organizational agility. *Journal of Theory and Applications in Electronic Commerce Research*, *10*.

Kumar, K. J., & Amos, T. (2006). Telecommunications and development: The cellular mobile revolution in India and China. *Journal of Creative Communications*, *1*(3), 297–309. doi:10.1177/097325860600100306

Kumar, K., & Lu, Y. H. (2010). Cloud Computing for Mobile Users: Can Offloading Computation Save Energy? *Computer*, *43*(4), 51–56. doi:10.1109/MC.2010.98

Kumar, , Nilsen, W. J., Abernethy, A., Atienza, A., Patrick, K., Pavel, M., & Swendeman, D. et al. (2013). Mobile health technology evaluation: The mHealth evidence workshop. *American Journal of Preventive Medicine*, *45*(2), 228–236. doi:10.1016/j.amepre.2013.03.017 PMID:23867031

Kuo, Y. F., Wu, C. M., & Deng, W. J. (2009). The relationships among service quality, perceived value, customer satisfaction, and post-purchase intention in mobile value-added services. *Computers in Human Behavior*, *25*(4), 887–896. doi:10.1016/j.chb.2009.03.003

Kurnia, S., Smith, S. P., & Lee, H. (2006). Consumers' perception of mobile internet in Australia. *e-. Business Review (Federal Reserve Bank of Philadelphia)*, *5*(1), 19–32.

Kwahk, K. Y., & Ge, X. (2012). The Effects of Social Media on E-commerce: A Perspective of Social Impact Theory. *2012 45th Hawaii International Conference on System Sciences*. IEEE.

Laakkonen, K. (2014). *Contracts in Agile Software Development* (Unpublished Master Dissertation). Aalto University, Helsenki, Finland.

Lamberton, C., & Stephen, A. T. (2016). A thematic exploration of digital, social media, and mobile marketing researches evolution from 2000 to 2015 and an agenda for future research. *Journal of Marketing, 27*(6), 146–172. doi:10.1509/jm.15.0415

Lander, K. (2014). *Contracted to deliver outcomes.* Retrieved February 5, 2016, from http://www.energizedwork.com/weblog/2014/09/agile-beach-2014-contracted-deliver-outcomes

Lane, J., & Kinser, K. (2015). *MOOC's and the McDonaldization of Global Higher Education.* Retrieved 11/12, 2015, from http://chronicle.com/blogs/worldwise/moocs-mass-education-and-the-mcdonaldization-of-higher-education/30536

Lanthony, P. (2005). *La perception des couleurs sur écran. Intervention dans le cadre d'un séminaire sur la couleur, 3C S.* France: A. Abbaye de Royaumont.

Larman, C., & Vodde, B. (2010). *Practices for Scaling Lean & Agile Development.* Boston, MA: Addison-Wesley & Pearson Education Inc.

Laukkanen, T. (2015). How uncertainty avoidance affects innovation resistance in mobile banking: The moderating role of age and gender. In *System Sciences (HICSS), 2015 48th Hawaii International Conference on* (pp. 3601-3610). IEEE.

Laukkanen, T. (2016). Consumer adoption versus rejection decisions in seemingly similar service innovations: The case of the Internet and mobile banking. *Journal of Business Research, 69*(7), 2432–2439. doi:10.1016/j.jbusres.2016.01.013

Laukkonen, T., & Passanen, M. (2008). Mobile banking innovations and early adopters: How they differ from another online user? *Journal of Financial Services Marketing, 23*(2), 86–94. doi:10.1057/palgrave.fsm.4760077

Law, E. L.C., Roto, V., Hassenzahl, M., Vermeeren, A. P.O.S., Kort, J. (2009). *Understanding, Scoping and Defining User Xperience: A Survey Approach.* CHI 2009 User Experience.

Lee, C. C., Cheng, H. K., & Cheng, H. H. (2007). An empirical study of mobile commerce in insurance industry: Task–technology fit and individual differences. *Decision Support Systems, 43*(1), 95–110. doi:10.1016/j.dss.2005.05.008

Lee, C., Hu, W., & Yeh, J. (2009). Mobile commerce technology. In M. Khosrow-Pour (Ed.), *Encyclopedia of information science and technology* (2nd ed., pp. 2584–2589). Hershey, PA: IGI Global. doi:10.4018/978-1-60566-026-4.ch412

Lee, C., Kou, W., & Hu, W. (2008). Mobile commerce security and payment methods. In A. Becker (Ed.), *Electronic commerce: Concepts, methodologies, tools, and applications* (pp. 292–306). Hershey, PA: IGI Global. doi:10.4018/978-1-59904-943-4.ch027

Lee, S., & Koubek, R. J. (2010). The effects of usability and web design attributes on user preference for e-commerce web sites. *Computers in Industry, 61*(4), 329–341. doi:10.1016/j.compind.2009.12.004

Lee, T. M. (2005). The impact of perceptions of interactivity on customer trust and transaction intentions in mobile commerce. *Journal of Electronic Commerce Research, 6*(3), 165–180.

Lee, V., Shineider, H., & Schell, R. (2005). *Aplicações móveis: arquitetura, projeto e desenvolvimento.* São Paulo: Pearson.

Lee, Y. E., & Benbasat, I. (2003). Interface design for mobile commerce. *Communications of the ACM, 46*(12), 48–52. doi:10.1145/953460.953487

Lee, Y. E., & Benbasat, I. (2004). A Framework for the Study of Customer Interface Design for Mobile Commerce. *International Journal of Electronic Commerce, 8*(3), 79–102.

Lenhart, A., & Madden, M. (2007). Social networking websites and teens: An overview. Pew Research Center/Internet.

Leonard, L. N. (2010). C2C mobile commerce: Acceptance factors. In I. Lee (Ed.), *Encyclopedia of e-business development and management in the global economy* (pp. 759–767). Hershey, PA: IGI Global. doi:10.4018/978-1-61520-611-7.ch076

Leong, L. Y., Ooi, K. B., Chong, A. Y. L., & Lin, B. (2011). Influence of individual characteristics, perceived usefulness and perceived ease of use on mobile entertainment adoption in Malaysia – an SEM approach. *International Journal of Mobile Communications*, *9*(44), 359–382. doi:10.1504/IJMC.2011.041141

Leu, F. Y., Huang, Y. L., & Wang, S. M. (2015). A secure m-commerce system based on credit card transaction. *Electronic Commerce Research and Applications*, *14*(5), 351–360. doi:10.1016/j.elerap.2015.05.001

Levin, T., & Wadmany, R. (2006). Listening to students voices on learning with information technologies in a rich technology-based classroom. *Journal of Educational Computing Research*, *34*(3), 281–317. doi:10.2190/CT6Q-0WDG-CDDP-U6TJ

Lewin, T. (2012, March 4). Instruction for masses knocks down campus walls. *New York Times*.

Lewis, S., Pea, R., & Rosen, J. (2010). Beyond participation to co-creation of meaning: Mobile social media in generative learning communities. *Social Sciences Information. Information Sur les Sciences Sociales*, *49*(3), 351–369. doi:10.1177/0539018410370726

Li, X. (2013). Innovativeness, personal initiative, news affinity and news utility as predictors of the use of mobile phones as news devices. *Chinese Journal of Communication*, *6*(3), 350–373. doi:10.1080/17544750.2013.789429

Li, H., Liu, Y., & Heikkilä, J. (2014). Understanding the factors driving nfc-enabled mobile payment adoption: an empirical investigation. *PACIS Proceedings*. Retrieved from http://aisel.aisnet.org/pacis2014/231

Li, X., Ren, S., Cheng, W., Xiang, L., & Liu, X. (2014). Smartphone: Security and Privacy Protection. *Pervasive Computing and the Networked World*, 289–302. doi:10.1007/978-3-319-09265-2_30

Liang, C. (2013). *The predictive model of imagination stimulation*. Academic Press.

Liang, C., Hsu, Y., Chang, C. C., & Lin, L. J. (2012). In search of an index of imagination for virtual experience designers. *International Journal of Technology and Design Education*. doi:10.1007/s10798-012-9224

Li, C., Li, D. Y., Miklau, G., & Suciu, D. (2013). A theory of pricing private data. In *Proceeding of the 16th International Conference on Database Theory (ICDT '13)* (pp. 33-44). New York: ACM Press.

Lickerman, A. (2010). *The Value of A Good Reputation; Why we should care about how others perceive us*. Retrieved in May 12 2015, from https://www.psychologytoday.com/blog/happiness-in-world/201004/the-value-good-reputation

Li, F., Pieńkowski, D., van Moorsel, A., & Smith, C. (2012). A holistic framework for trust in online transactions. *International Journal of Management Reviews*, *14*(1), 85–103. doi:10.1111/j.1468-2370.2011.00311.x

Li, M., Dong, Z. Y., & Chen, X. (2012). Factors influencing consumption experience of mobile commerce: A study from experiential view. *Internet Research*, *22*(2), 120–141. doi:10.1108/10662241211214539

Limayem, M., & Rowe, F. (2006). Comparaison des facteurs influençant les intentions d'achat à partir du Web à Hong Kong et en France: Influence sociale, risques et aversion pour la perte de contact. *Revue Française du Marketing*, *209*(4/5), 25–48.

Lim, W. M. (2015). Antecedents and consequences of e-shopping: An integrated model. *Internet Research*, *25*(2), 184–217. doi:10.1108/IntR-11-2013-0247

Lin Lu, Z., & Dosher, B. A. (2007). Cognitive psychology. *Scholarpedia*, *2*(8), 2769. doi:10.4249/scholarpedia.2769

Lin, H. F. (2010). An empirical investigation of mobile banking adoption: The effect of innovation attributes and knowledge-based trust. *International Journal of Information Management*, *30*(6), 33–45.

Lin, H. H., & Wang, Y. S. (2006). An examination of the determinants of customer loyalty in mobile commerce contexts. *Information & Management*, *43*(3), 271–282. doi:10.1016/j.im.2005.08.001

Lin, J., Lu, Y., Wang, B., & Wei, K. K. (2011). The role of inter-channel trust transfer in establishing mobile commerce trust. *Electronic Commerce Research and Applications*, *10*(6), 615–625. doi:10.1016/j.elerap.2011.07.008

Lin, J., Wang, B., Wang, N., & Lu, Y. (2014). Understanding the evolution of consumer trust in mobile commerce: A longitudinal study. *Information Technology Management*, *15*(1), 37–49. doi:10.1007/s10799-013-0172-y

Lin, K. Y., & Lu, H. P. (2015). Predicting mobile social network acceptance based on mobile value and social influence. *Internet Research*, *25*(1), 107–130. doi:10.1108/IntR-01-2014-0018

Liou, C. H., & Liu, D. R. (2012). Hybrid recommendations for mobile commerce based on mobile phone features. *Expert Systems: International Journal of Knowledge Engineering and Neural Networks*, *29*(2), 108–123.

List of most downloaded Android applications. (2016). Retrieved 2016, 20 June, from https://en.wikipedia.org/wiki/List_of_most_downloaded_Android_applications

Liu, C., Zhu, Q., Holroyd, K. A., & Seng, E. K. (2011). Status and trends of mobile-health applications for iOS devices: A developers perspective. *Journal of Systems and Software*, *84*(11), 2022–2033. doi:10.1016/j.jss.2011.06.049

Liu, D. R., & Liou, C. H. (2011). Mobile commerce product recommendations based on hybrid multiple channels. *Electronic Commerce Research and Applications*, *10*(1), 94–104. doi:10.1016/j.elerap.2010.08.004

Liu, M., Kang, J., Cao, M., Lim, M., Ko, Y., Myers, R., & Weiss, A. S. (2014). Understanding MOOCs as an emerging online tool: Perspectives from the students. *American Journal of Distance Education*, *28*(3), 147–159. doi:10.1080/08923647.2014.926145

Liu, Y. (2014). User control of personal information concerning mobile-app: Notice and consent? *Computer Law & Security Report*, *30*(5), 521–529. doi:10.1016/j.clsr.2014.07.008

Liu, Z., Bonazzi, R., & Pigneur, Y. (2016). Privacy-based adaptive context-aware authentication system for personal mobile devices. *Journal of Mobile Multimedia*, *12*(1&2), 159–180.

Li, Y. M., & Yeh, Y. S. (2010). Increasing trust in mobile commerce through design aesthetics. *Computers in Human Behavior*, *26*(4), 673–684. doi:10.1016/j.chb.2010.01.004

Lovell, S. (2015). 'Try before you buy' – why brands should use product demonstration?. Retrieved from http://gottabemarketing.co.uk/try-before-you-buy-product-deminstration/

Lowry, P. J., Wilson, D. W., & Haig, W. L. (2014). A picture is worth a thousand words: Source Credibility Theory applied to logo and website design for heightened credibility and consumer trust. *International Journal of Human-Computer Interaction*, *30*(1), 63–93. doi:10.1080/10447318.2013.839899

Luarn, P., & Lin, H. H. (2005). Toward an understanding of the behavioral intention to use mobile banking. *Computers in Human Behavior*, *21*(6), 873–891. doi:10.1016/j.chb.2004.03.003

Luca, J., & Mcloughlin, C. (2004). *Using Online Forums to Support a Community of Learning*. Paper presented at the EdMedia: World Conference on Educational Media and Technology 2004, Lugano, Switzerland. Retrieved from http://www.editlib.org/p/12668

Compilation of References

Lu, H. P., & Su, Y. J. (2009). Factors affecting purchase intention on mobile shopping web sites. *Internet Research, 19*(4), 442–458. doi:10.1108/10662240910981399

Lu, J. (2014). Are personal innovativeness and social influence critical to continue with mobile commerce? *Internet Research, 24*(2), 134–159. doi:10.1108/IntR 05 2012-0100

Lu, J., Yao, J. E., & Yu, C. S. (2005). Personal innovativeness, social influences and adoption of wireless internet services via mobile technology. *The Journal of Strategic Information Systems, 14*(3), 245–268. doi:10.1016/j.jsis.2005.07.003

Lupton, D., & Jutel, A. (2015). Its like having a physician in your pocket! A critical analysis of self-diagnosis smartphone apps. *Social Science & Medicine, 133*, 128–135. doi:10.1016/j.socscimed.2015.04.004 PMID:25864149

Lu, Y. B., Yang, S. Q., Chau, P. Y. K., & Cao, Y. Z. (2011). Dynamics between the trust transfer progress and intention to use mobile payment services: Across-environment perspective. *Information & Management, 48*(8), 393–403. doi:10.1016/j.im.2011.09.006

Lu, Y., Deng, Z., & Wang, B. (2010). Exploring factors affecting Chinese consumers usage of short message service for personal communication. *Information Systems Journal, 20*(2), 183–208. doi:10.1111/j.1365-2575.2008.00312.x

Lu, Y., Wang, X., & Ma, Y. (2013). Comparing user experience in a news website across three devices: iPhone, iPad, and desktop. *Proceedings of the American Society for Information Science and Technology, 50*(1), 1–4. doi:10.1002/meet.14505001133

Lv, Z., Halawani, A., Feng, S., Réhman, S., & Li, H. (2015). Touch-less interactive augmented reality game on vision based wearable device. *Personal and Ubiquitous Computing, 19*(3), 551–567. doi:10.1007/s00779-015-0844-1

Lytle, R. (2012, October). Twitter improves student learning in college classrooms (Study). *U.S. News and World Report and World Report. Education.*

MacIntyre, B., Gandy, M., Dow, S., & Bolter, J. D. (2004). DART: A toolkit for rapid design exploration of augmented reality experiences. In S. K. Feiner (Ed.), *Proceedings of the 17th Annual ACM Symposium on User Interface Software and Technology* (pp. 197-206). Santa Fe, NM: ACM. doi:10.1145/1029632.1029669

Magura, B. (2003). What hooks m-commerce customers? *MIT Sloan Management Review, 44*(3), 9–16.

Mahatanankoon, P., & OSullivan, P. (2008). Attitude toward mobile text messaging: An expectancy-based perspective. *Journal of Computer-Mediated Communication, 13*(4), 973–992. doi:10.1111/j.1083-6101.2008.00427.x

Mahatanankoon, P., Wen, H. J., & Lim, B. (2005). Consumer-based m-commerce: Exploring consumer perception of mobile applications. *Computer Standards & Interfaces, 27*(4), 347–357. doi:10.1016/j.csi.2004.10.003

Maher, M. S. J. (1915). *Psychology; Empirical Rational: Imagination*. Retrieved in May 12 2015, from https://www3.nd.edu/~maritain/jmc/etext/psych008.htm

Mäkelä, A., & Fulton Suri, J. (2001). Supporting Users' Creativity: Design to Induce Pleasurable Experiences. *Proceedings of the International Conference on Affective Human Factors Design*, 387-394.

Malaysian Communications And Multimedia Commission. (2015). Hand phone users survey 2014. *Malaysian Communication and Multimedia Commission*, 48.

Malgaonkar, Koul, Thorat, & Zawar. (2011). Mapping of Virtual Machines in Private Cloud. *International Journal of Computer Trends and Technology, 2*(2), 54-57.

Malik, A., Kumra, R., & Srivastava, V. (2013). Determinants of Consumer Acceptance of M-Commerce. *South Asian Journal of Management, 20*(2), 102–126.

Mallya, H. (2015, Nov 26). *Is regional language content the next frontier to reach India's 1.2 billion people?* Retrieved from http://yourstory.com/2015/11/news-aggregators-vernacular/

MAMPU. (2015). *GAMMA : Galeri Mudah Alih Kerajaan.* Retrieved from http://www.moh.gov.my/mgpwa2015/Mesyuarat Teknikal ICT 2015/GAMMA.pdf

MAMPU. (n.d.). *Gallery of Malaysian Government Mobile Application (GAMMA).* Retrieved October 1, 2016, from http://gamma.malaysia.gov.my/#/home

Mao, E., Srite, M., Thatcher, J. B., & Yaprak, O. (2005). A Research Model for Mobile Phone Service Behaviors: Empirical Validation in the U.S. and Turkey. *Journal of Global Information Technology Management, 8*(4), 4, 7–28. doi:10.1080/1097198X.2005.10856406

Margalit, L. (2014). *The Psychology of Online Customization.* Retrieved in May 12 2015, from https://www.psychologytoday.com/blog/behind-online-behavior/201503/the-psychology-online-customization

Marks, N. (2015). *The myth of IT risk.* Retrieved September 09, 2015, from https://normanmarks.wordpress.com/2015/08/28/the-myth-of-it-risk/

Marsden, P. (n.d.). *Social Commerce: Monetizing Social Media.* Syzygy Group. Retrieved from: http://digitalintelligencetoday.com/downloads/White_Paper_Social_Commerce_EN.pdf

Marshall, C., Lewis, D., & Whittaker, M. (2013). *mHealth technologies in developing countries: a feasibility assessment and a proposed framework.* Working paper series (Vol. 25). Retrieved from http://www.uq.edu.au/hishub/docs/WP25/WP25 mHealth_web.pdf

Martínez-Pérez, B., de la Torre-Díez, I., & López-Coronado, M. (2014). Privacy and Security in Mobile Health Apps: A Review and Recommendations. *Journal of Medical Systems, 39*(1), 1–8. doi:10.1007/s10916-014-0181-3 PMID:25486895

Masters, K. (2011). A Brief Guide to Understanding Moocs. *The Internet Journal of Medical Education, 1*(2).

Mastorakis, G., Mavromoustakis, C., & Pallis, E. (2015). *Resource Management of Mobile Cloud Computing Networks and Environments.* Hershey, PA: IGI Global. doi:10.4018/978-1-4666-8225-2

Mayol, S. (2011). *Marketing 3.0.* Dunod Editions.

McAuley, A., Stewart, B., Siemens, G., & Cormier, D. (2010). *The MOOC model for digital practice.* Retrieved from http://www.elearnspace.org/Articles/MOOC_Final.pdf

McIlroy, S., Ali, N., & Hassan, A. E. (2015). Fresh apps: An empirical study of frequently-updated mobile apps in the Google play store. *Empirical Software Engineering, 21*(3), 1346–1370. doi:10.1007/s10664-015-9388-2

McKnight, D. H., & Chervany, N. L. (2001). *Conceptualizing trust: A typology and e-commerce customer relationships model.* Paper presented at the 34th Annual Hawaii International Conference on System Sciences (HICSS 2001), Maui, HI. doi:10.1109/HICSS.2001.927053

McLeod, S. (2015). *Cognitive Psychology.* Retrieved in September 05 2015, from http://www.simplypsychology.org/cognitive.html

McLuhan, M. (1994). *Understanding Media: The extensions of man: Critical Edition.* London: The MIT Press.

McMullin, J. (2004). *The Experience Cycle model.* Academic Press.

Mechael, P. N. (2009). The Case for mHealth in Developing Countries. *Innovations: Technology, Governance, Globalization, 4*(1), 103–118. doi:10.1162/itgg.2009.4.1.103

Compilation of References

Mehlenbacher, B., Bennett, L., Bird, T., Ivey, M., Lucas, J., Morton, J., & Whitman, L. (2005). Usable e-learning: A conceptual model for evaluation and design. *Proceedings of HCI International 2005: 11th International Conference on Human-Computer Interaction, 4*.

Mellinger, D., Bonna, K., Sharon, M., & SantRamet, M. (2004). *Socialight: A Mobile Social Networking System*. Paper presented at the 6th International Conference on Ubiquitous Computing, Nottingham, UK.

Merino, C., Pino, S., Meyer, E., Garrido, J. M., & Gallardo, F. (2015). Realidad aumentada para el diseño de secuencias de enseñanza-aprendizaje en química. *Educación en la Química, 26*(2), 94–99.

Merriam-Webster Dictionary. (2015). *Full Definition of Reciprocity*. Retrieved in September 05 2015, from http://www.merriam-webster.com/dictionary/reciprocity

Metheny, M. (2013). *Federal Cloud Computing: The Definitive Guide for Cloud Service Providers*. Waltham, MA: Elsevier.

Meulendijk, M. C., Meulendijks, E. A., Jansen, P. A. F., Numans, M. E., & Spruit, M. R. (2014). *What concerns users of medical apps? Exploring non-functional requirements of medical mobile applications*. Paper presented at the ECIS 2014 22nd European Conference on Information Systems. http://aisel.aisnet.org/cgi/viewcontent.cgi?article=1004&context=ecis2014

Miao, M., & Jayakar, K. (2016). Mobile payments in Japan, South Korea and China: Cross-border convergence or divergence of business models? *Telecommunications Policy, 40*(2/3), 182–196. doi:10.1016/j.telpol.2015.11.011

Mikalef, P., Giannakos, M., & Pateli, A. (2012). Shopping and Word-of-Mouth Intentions on Social Media. *Journal of Theoretical and Applied Electronic Commerce Research, 8*(1), 17–34.

Milgram, S. (1969). Interdisciplinary thinking and the small world problem. *Interdisciplinary Relationships in the Social Sciences*, 103-20.

Ming, . (2016). *Use of Medical Mobile Applications Among Hospital Pharmacists in Malaysia*. Therapeutic Innovation & Regulatory Science. doi:10.1177/2168479015624732

Ministry of Health Malaysia. (2015). *Malaysia National Health Accounts Health Expenditure Report 1997-2013*. Retrieved from http://www.moh.gov.my/index.php/file_manager/dl_item/554756755a584a6961585268626939515a57356c636d4a706447467549656305957316

Miserski, R. (1982). An attribution explanation of the disproportionate influence of unfavorable information. *The Journal of Consumer Research, 9*(3), 301–310. doi:10.1086/208925

Mishra, H., Mishra, A., & Nayakankuppam, D. (2006). Money: A bias for the whole. *The Journal of Consumer Research, 32*(4), 541–549. doi:10.1086/500484

Mishra, S. (2014). Adoption of M-commerce in India: Applying Theory of Planned Behaviour Model. *Journal of Internet Banking and Commerce, 19*, 1. Retrieved from http://www.arraydev.com/commerce/jibc/

Mobile Institute. (2015). *Raport mShopper, Polacy na zakupach mobilnych*. Retrieved July 19, 2016, from http://www.aptusshop.pl/sklepy_internetowe/przyszlosc-m-commerce-w-polsce.php/html

Mobile Marketing Association. (2013). *Le baromètre trimestriel du Marketing Mobile en France 3ᵉ édition, T2 2013*. Retrieved July 19, 2016, from http://mmaf.fr/html

Mobile/Tablet Operating System Market Share. (2016). Retrieved 2016, 8 June, from https://www.netmarketshare.com/operating-system-market-share.aspx?qprid=8&qpcustomd=1

MOH. (2016). *My Blue Book*. Retrieved from https://play.google.com/store/apps/details?id=air.mypharmacisthouse.mybluebook

Mohd-Tahir, N. A., Paraidathathu, T., & Li, S. C. (2015). Quality use of medicine in a developing economy: Measures to overcome challenges in the Malaysian healthcare system. *SAGE Open Medicine*, *3*. doi:10.1177/2050312115596864 PMID:26770795

Möhlenbruch, D., Dölling, S., & Ritschel, F. (2010). Interactive customer retention management for mobile commerce. In K. Pousttchi & D. Wiedemann (Eds.), *Handbook of research on mobile marketing management* (pp. 437–456). Hershey, PA: IGI Global. doi:10.4018/978-1-60566-074-5.ch023

Molina-Castillo, F., & Meroño-Cerdan, A. (2014). Drivers of mobile application acceptance by consumers: A meta analytical review. *International Journal of E-Services and Mobile Applications*, *6*(3), 34–47. doi:10.4018/ijesma.2014070103

Molm, L. D., Schaefer, D. R., & Collett, J. L. (2007). The Value of Reciprocity. *The Value of Reciprocity Social Psychology Quarterly*, *70*(2), 199–217. doi:10.1177/019027250707000208

Moqbel, A., Yani-De-Soriano, M., & Yousafzai, S. (2012). Mobile commerce use among UK mobile users: An experimental approach based on a proposed mobile network utilization framework. In A. Zolait (Ed.), *Knowledge and technology adoption, diffusion, and transfer: International perspectives* (pp. 78–111). Hershey, PA: IGI Global. doi:10.4018/978-1-4666-1752-0.ch007

Morosan, C., & DeFranco, A. (2015). Disclosing personal information via hotel apps: A privacy calculus perspective. *International Journal of Hospitality Management*, *47*, 120–130. doi:10.1016/j.ijhm.2015.03.008

Morrell, R. W., & Echt, K. V. (1997). Designing written instructions for older adults: Learning to use computers. In A. D. Fisk & W. A. Rogers (Eds.), *Handbook of human factors and the older adult* (pp. 335–361). San Diego, CA: Academic Press.

Mort, G. S., & Drennan, J. (2007). Mobile communications: A study of factors influencing consumer use of m-services. *Journal of Advertising Research*, *47*(3), 302–312. doi:10.2501/S0021849907070328

Motiwalla, L. F. (2007). Mobile learning: A framework and evaluation. *Computers & Education*, *49*(3), 581–596. doi:10.1016/j.compedu.2005.10.011

Mousavi Hondori, H., Khademi, M., Dodakian, L., Cramer, S. C., & Lopes, C. V. (2013). A spatial augmented reality rehab system for post-stroke hand rehabilitation. *Medicine Meets Virtual Reality*, *20*, 279–285. PMID:23400171

Mukherjee, A., & Nath, P. (2007). Role of electronic trust in online retailing: A re-examination of the commitment-trust theory. *European Journal of Marketing*, *41*(9/10), 1173–1202. doi:10.1108/03090560710773390

Mulligan, C. E. A., & Olsson, M. (2013). Architectural Implications of Smart City Business Models: An Evolutionary Perspective. *IEEE Communications Magazine, Vol*, *51*, 80–85. doi:10.1109/MCOM.2013.6525599

Muñoz, A., & Maña, A. (2011). TPM-based protection for mobile agents. *Security and Communication Networks*, *4*(1), 45–60. doi:10.1002/sec.158

Murthy, D. N. P. (2007). *Product reliability &warranty: an overview &future research*. Retrieved in September 05 2015, from http://www.scielo.br/scielo.php?script=sci_arttext&pid=S0103-65132007000300003

Mutti, D. O., & Zadnik, K. (2000). *Age-related decreases in the prevalence of myopia: Longitudinal change or cohort effect?* Retrieved December 9, 2015, from http://iovs.arvojournals.org/article.aspx?articleid/html

Compilation of References

Mylonas, A., Kastania, A., & Gritzalis, D. (2013). Delegate the smartphone user? Security awareness in smartphone platforms. *Computers & Security, 34*, 47–66. doi:10.1016/j.cose.2012.11.004

Narang, B., & Arora, J. B. (2016). Present and future of mobile commerce: Introduction, comparative analysis of m commerce and e commerce, advantages, present and future. In S. Madan & J. Arora (Eds.), *Securing transactions and payment systems for m-commerce* (pp. 293–308). Hershey, PA: IGI Global. doi:10.4018/978-1-5225-0236-4.ch015

National Audit Office. (2014). *A snapshot of the use of Agile delivery in central government*. London: Author.

Nel, F., & Westlund, O. (2012). The 4 Cs of mobile news. Channels, conversation, content and commerce. *Journalism Practice, 6*(5–6), 744–753. doi:10.1080/17512786.2012.667278

Nemec Zlatolas, L., Welzer, T., Heričko, M., & Hölbl, M. (2015). Privacy antecedents for SNS self-disclosure: The case of Facebook. *Computers in Human Behavior, 45*(0), 158–167. doi:10.1016/j.chb.2014.12.012

Neto, O. M., & Pimentel, M. D. G. (2013). Heuristics for the assessment of interfaces of mobile devices. In *Proceedings of the 19th Brazilian symposium on Multimedia and the web*. Salvador, Brazil: ACM. doi:10.1145/2526188.2526237

Neuman, W. L. (2011). Social Research Methods: Qualitative and Quantitative Approaches (D. Musslewhite Ed.; 7th ed.). Pearson.

Newman, N., & Levy, A. L. D. (Eds.). (2013). Reuters Institute digital news report 2013: Tracking the future of news. Oxford, UK: Reuters Institute for the Study of Journalism.

Newman, N. (2014). The Costs of Lost Privacy: Consumer Harm and Rising Economic Inequality in the Age of Google. *William Mitchell Law Review, 40*(2), 849–889.

Ngai, E. W. T., & Gunasekaran, A. (2007). A review for mobile commerce research and applications. *Decision Support Systems, 43*(1), 3–15. doi:10.1016/j.dss.2005.05.003

Niantic. (2016). *Pokémon Go*. Retrieved from the Niantic, Inc. website: http://pokemongo.nianticlabs.com/es/

Nicolau, S., Soler, L., Mutter, D., & Marescaux, J. (2011). Augmented reality in laparoscopic surgical oncology. *Surgical Oncology, 20*(3), 189–201. doi:10.1016/j.suronc.2011.07.002 PMID:21802281

Nielsen, J. (1994). Heuristic evaluation. In Usability Inspection Methods (pp. 155-163). New York: John Wiley & Sons.

Nielsen. (2009). Global advertising consumers trust real friends and virtual strangers the most. *Nielsen*. Retrieved from: http://www.nielsen.com/us/en/insights/news/2009/global-advertising-consumers-trust-real-friends-and-virtual-strangers-the-most.html

Nielsen. (2013). *The Mobile Consumer: A Global Snapshot*. Retrieved from http://www.nielsen.com

Nielsen, J. (1992). Finding usability problems through heuristic evaluation. In *Proceedings of the SIGCHI conference on Human factors in computing systems*. Monterey, CA: ACM.

Nielsen, J., & Budiu, E. (2015). *Usabilidade Móvel*. Rio de Janeiro: Elsevier.

Nielsen, J., & Molich, R. (1990). Heuristic evaluation of user interfaces. In *Proceedings of ACM CHI'90 Conference*. Seattle, WA: ACM. doi:10.1145/97243.97281

Nilashi, , Ibrahim, O., Reza Mirabi, V., Ebrahimi, L., & Zare, M. (2015). The role of Security, Design and Content factors on customer trust in mobile commerce. *Journal of Retailing and Consumer Services, 26*, 57–69. doi:10.1016/j.jretconser.2015.05.002

Niranjanamurthy, M., Kavyashree, N., Jagannath, S., & Chahar, D. (2013). Analysis of e-commerce and m-commerce: Advantages, limitations and security issues. *International Journal of Advanced Research in Computer and Communication Engineering, 2*(6).

Norman, D. A. (2003). *Emotional Design: Why we love (or hate) everyday things*. New York: Basic Books.

Norris, A. C., Stockdale, R. S., & Sharma, S. (2009). A strategic approach to m-health. *Health Informatics Journal, 15*(3), 244–253. doi:10.1177/1460458209337445 PMID:19713398

Nucleus Research. (2012). *Market focus report: The value of mobile and social CRM* (Report No. M13). Author.

Nutrition Division, Ministry of Health, Malaysia. (n.d.). Retrieved from http://nutrition.moh.gov.my/

Ofori, K. S., Larbi-Siaw, O., Fianu, E., Gladjah, R. E., & Boateng, E. O. Y. (2016). Factors Influencing the Continuance Use of Mobile Social Media: The effect of Privacy Concerns. *Journal of Cyber Security and Mobility, 4*(2), 105–124. doi:10.13052/jcsm2245-1439.426

Okazaki, S., Li, H., & Hirose, M. (2009). Consumer privacy concerns and preference for degree of regulatory control. *Journal of Advertising, 38*(4), 63–77. doi:10.2753/JOA0091-3367380405

Okazaki, S., & Mendez, F. (2013). Exploring convenience in mobile commerce: Moderating effects of gender. *Computers in Human Behavior, 29*(3), 1234–1242. doi:10.1016/j.chb.2012.10.019

Oliver, R. L., & Bearden, W. O. (1983). The role of involvement in satisfaction processes. In R. P. Bagozzi & A. M. Tybout (Eds.), *Advances in Consumer Research* (pp. 250–255). Ann Arbor, MI: Association for Consumer Research.

Olson, D. & Peters, S. (2011). Managing Software Intellectual Assets in Cloud Computing, Part 1. *Journal of Licensing Executives Society International, H*(3), 160-165.

Omar, H. M., Yusof, Y. H. M., & Sabri, N. M. (2010). Development and potential analysis of heuristic evaluation for courseware. In *Proceedings of the 2nd International Congress on Engineering Education*. IEEE.

Opelt, A., Gloger, B., Pfarl, W., & Mittermayr, R. (2013). *Agile Contracts: Creating and Managing Successful Projects with Scrum*. Hoboken, NJ: John Wiley & Sons. doi:10.1002/9781118640067

Osarenkhoe, A., & Bennani, A. E. (2007). An exploratory study of implementation of customer relationship management strategy. *Business Process Management Journal, 13*(1), 139–164. doi:10.1108/14637150710721177

Osman. (2011). An exploratory study on the trend of smartphone usage in a developing country. Communications in Computer and Information Science, 387–396. http://doi.org/ doi:10.1007/978-3-642-22603-8_35

Ou, L. C., Luo, R. M., Sun, P. L., Hu, N. C., & Chen, H. S. (2012). Age Effects on Color Emotion, Preference, and Harmony. *Color Research and Application, 37*(2), 92–105. doi:10.1002/col.20672

Oxford Dictionaries. (2015). *Aesthetic*. Retrieved in September 05 2015, from http://www.oxforddictionaries.com/definition/english/aesthetic

Ozdalga, E., Ozdalga, A., & Ahuja, N. (2012). The Smartphone in Medicine: A Review of Current and Potential Use Among Physicians and Students. *Journal of Medical Internet Research, 14*(5), e128. doi:10.2196/jmir.1994 PMID:23017375

Ozuem, W., & Mulloo, B. N. (2016). Manifested consumption: Mobile storefront. In A. Diab (Ed.), *Self-organized mobile communication technologies and techniques for network optimization* (pp. 356–373). Hershey, PA: IGI Global. doi:10.4018/978-1-5225-0239-5.ch013

Paavilainen, J. (2002). *Mobile business strategies: understanding the technologies and opportunities*. Addison-Wesley.

Compilation of References

Pahl, C., & Donnellan, D. (2002). *Data minning technology for the evaluation of web-based teaching and learning systems.* Paper presented at the World conference on e-learning in Corporate, Government, Healthcare and Higher Education.

Pai, H. T., & Wu, F. (2011). Prevention of wormhole attacks in mobile commerce based on non-infrastructure wireless networks. *Electronic Commerce Research and Applications, 10*(4), 384–397. doi:10.1016/j.elerap.2010.12.004

Pal & Pattnaik. (2012). Efficient architectural Framework of Cloud Computing. *International Journal of Cloud Computing and Services Science, 1*(2), 66-73.

Pal, Mohanty, Pattnaik, & Mund. (2012). A Virtualization Model for Cloud Computing. *Proceedings of International Conference on Advances in Computer Science*, 10-16.

Pal, S., & Mohanty, S. (n.d.). *An Approach to Cross-Cloud Live Migration of Virtual Machines in Cloud Computing Environment.* Unpublished.

Pallud, J., & Straub, D. W. (2014). Effective website design for experience-influenced environments: The case of high culture museums. *Information & Management, 51*(3), 359–373. doi:10.1016/j.im.2014.02.010

Parasuraman, A., Zeithaml, V., & Malhorta, A. (2005). E-S-QUAL: A multiple-item scale for assessing electronic service quality. *Journal of Retailing, 64*(1), 12–40.

Parboteeah, D. V., Valacich, J. S., & Wells, J. D. (2009). The Influence of Website Characteristics on a Consumers Urge to Buy Impulsively. *Information Systems Research, 20*(1), 60–78. doi:10.1287/isre.1070.0157

Park, J. K., Yang, S. J., & Lehto, X. (2007). Adoption of mobile technologies for Chinese consumers. *Journal of Electronic Commerce Research, 8*(3), 196–206.

Park, S. Y. (2009). An Analysis of the Technology Acceptance Model in Understanding University Students' Behavioral Intention to Use e-Learning. *Journal of Educational Technology & Society, 12*(3), 150–162.

Parvatiyar, A., & Sheth, J. N. (2001). Customer relationship management: Emerging practice, process and discipline. *Journal of Economic & Social Research, 3*(2), 1–34.

Patel, N., & Puri, R. (n.d.). *A complete guide to understanding consumer psychology.* Retrieved in September 05 2015, from https://www.quicksprout.com/the-complete-guide-to-understand-customer-psychology-chapter-4/

Patel, A. (2006). Mobile commerce in emerging economics. In B. Unhelkar (Ed.), *Handbook of research in mobile business: Technical, methodological, and social perspectives* (pp. 429–434). Hershey, PA: IGI Global. doi:10.4018/978-1-59140-817-8.ch030

Pattan, N., & Madamanchi, D. (2010). Study of Usability of Security and Privacy in Context Aware Mobile Applications. In T. Phan, R. Montanari & P. Zerfos (Eds.), *Mobile Computing, Applications, and Services: First International ICST Conference, MobiCASE 2009, San Diego, CA, USA, October 26-29, 2009, Revised Selected Papers* (pp. 326-330). Berlin: Springer Berlin Heidelberg. doi:10.1007/978-3-642-12607-9_21

Payal, V. N. (2014). GSM: Improvement of Authentication and Encryption Algorithms. *International Journal of Computer Science and Mobile Computing, 3*(7), 393-408.

Payne, A., & Frow, P. (2005, October). A Strategic Framework for Customer Relationship Management. *Journal of Marketing, 69*(4), 167–176. doi:10.1509/jmkg.2005.69.4.167

Pedersen, P. E., Methlie, L. B., & Thorbjornsen, H. (2002). *Understanding mobile commerce end-user adoption: A triangulation perspective and suggestions for an exploratory service evaluation framework.* Paper presented at the 35th Annual Hawaii International Conference on System Sciences (HICSS 2002), Maui, HI. doi:10.1109/HICSS.2002.994011

Pelet, J. É., & Papadopoulou, P. (2013). *Handbook of Research on User Behavior in Ubiquitous Online Environments*. Retrieved from http://www.igi-global.com/book/user-behavior-ubiquitous-online-environments/76724

Pelet, J.-É., & Uden, L. (2014). *Mobile learning platforms to assist individual knowledge management*. Paper presented at the 9th International Conference on Knowledge Management in Organizations, Santiago, Chile. doi:10.1007/978-3-319-08618-7_26

Pelet, J.-É., Khan, J., Papadopoulou, P., & Bernardin, E. (2014). m-learning: exploring the use of mobile devices and social media. In N. Baporikar (Ed.), Handbook of Research on Higher Education in the MENA Region: Policy and Practice (pp. 261-296). Hershey, PA: IGI Global.

Pelet, J. É. (2014). Investigating the Importance of Website Color Contrast in E-Commerce: Website Color Contrast in E-Commerce. In *Encyclopedia of Information Science and Technology* (3rd ed.). Hershey, PA: IGI Global; doi:10.4018/978-1-4666-5888-2

Pelet, J. É., & Papadopoulou, P. (2012). The effect of colors of e-commerce websites on consumer mood, memorization and buying intention. *European Journal of Information Systems*, *21*(4), 438–467. doi:10.1057/ejis.2012.17

Pelet, J. É., & Papadopoulou, P. (2014). Consumer Behavior in the Mobile Environment: An Exploratory Study of M-Commerce and Social Media. *International Journal of Technology and Human Interaction*, *10*(4), 36–48. doi:10.4018/ijthi.2014100103

Pelet, J. É., & Papadopoulou, P. (2015). Tablet and social media adoption in m-commerce: An exploratory study. *International Journal of Strategic Innovative Marketing*, *2*, 1. doi:10.15556/IJSIM.02.01.004

Pelet, J. E., Taieb, B., & Ben Dahmane Mouelhi, N. (2016). From m-commerce website's design to behavioral intentions. *Information and Management Association Symposium*.

Pelet, J.-É. (2010). Effets de la couleur des sites web marchands sur la mémorisation et sur lintention dachat. *Systèmes dInformation et Management*, *15*(1), 97–131. doi:10.3917/sim.101.0097

Pelet, J.-É. (Ed.). (2013). *E-learning 2.0 technologies and web applications in higher education*. Hershey, PA: Information Science Reference.

Pelet, J.-É., & Papadopoulou, P. (Eds.). (2013). *User Behavior in Ubiquitous Online Environments*. Hershey, PA: IGI Global.

Peng, J., Quan, J., & Zhang, S. (2013). Mobile phone customer retention strategies and Chinese e-commerce. *Electronic Commerce Research and Applications*, *12*(5), 321–327. doi:10.1016/j.elerap.2013.05.002

Perdue, K. (2003). *Imagination. The Chicago school of media theory*. Retrieved October 16, 2012, from http://lucian.uchicago.edu/blogs/mediatheory/keywords/imagination/

Persaud, A., & Azhar, I. (2012). Innovative mobile marketing via smartphones: Are consumers ready? *Marketing Intelligence & Planning*, *30*(4), 408–444. doi:10.1108/02634501211231883

Peters, C. (2012). Journalism to go. *Journalism Studies*, *13*(5-6), 695–705. doi:10.1080/1461670X.2012.662405

Petrova, K. (2008). Mobile commerce applications and adoption. In A. Becker (Ed.), *Electronic commerce: Concepts, methodologies, tools, and applications* (pp. 889–897). Hershey, PA: IGI Global. doi:10.4018/978-1-59904-943-4.ch072

Pew Research Centre. (2010, March 15). *The state of the news media 2010: Audience behavior*. Retrieved from http://stateofthemedia.org/2010/online-summary-essay/audience-behavior/

Picone, I., Courtois, C., & Paulussen, S. (2014). When news is everywhere. Understanding participation, cross-mediality and mobility in journalism from a radical user perspective. *Journalism Practice*, *9*(1), 35–49. doi:10.1080/17512786.2014.928464

Piech, C., Huang, J., Chen, Z., Do, C., Ng, A., & Koller, D. (2013). *Tuned Models of Peer Assessment in MOOCs*. Retrieved from http://www.stanford.edu/~cpiech/bio/papers/tuningPeerGrading.pdf

Pierantonelli, M., Perna, A., & Gregori, G. L. (2015). Interaction between Firms in New Product Development. *International Conference on Marketing and Business Development Journal*, *1*(1), 144-152.

Pierre, S. (2009). Security issues concerning mobile commerce. In D. Taniar (Ed.), *Mobile computing: Concepts, methodologies, tools, and applications* (pp. 2653–2659). Hershey, PA: IGI Global. doi:10.4018/978-1-60566-054-7.ch201

Pihlström, M., & Brush, G. J. (2008). Comparing the perceived value of information and entertainment mobile services. *Psychology and Marketing*, *25*(8), 732–755. doi:10.1002/mar.20236

Piqueras-Fiszman, B., Ares, G., Alcaide-Marzal, J., & Diego-Más, J. (2010). Comparing older and younger users perceptions of mobile phones and watches using CATA questions and preference mapping on the design characteristics. *Journal of Sensory Studies*, *26*(1), 1–12. doi:10.1111/j.1745-459X.2010.00315.x

Podsakoff, P. M., MacKenzie, S. B., Lee, J. Y., & Podsakoff, N. P. (2003). Common method biases in behavioral research: A critical review of the literature and recommended remedies. *The Journal of Applied Psychology*, *88*(5), 879–903. doi:10.1037/0021-9010.88.5.879 PMID:14516251

Polzin, R. S., Kipman, A. A., Finocchio, M. J., Geiss, R. M., Stone Perez, K., Tsunoda, K., & Bennett, D. A. (2014). *U.S. Patent No. 8744121*. Washington, DC: US Patent Office.

Poore, J., & Tevfik, B. (2011). *Augmented reality games for neurological rehabilitation*. Retrieved from Colorado State University, Department of Electrical and Computer Engineering website: http://projects-web.engr.colostate.edu/ece-sr-design/AY11/rehabilitation/documents/first_report.pdf

Poppendieck, M., & Poppendieck, T. (2013). *The Lean Mindset: Ask the Right Questions*. Westford, MA: Addison-Wesley.

Poushter, J. (2016, Feb 22). Smartphone Ownership and Internet Usage Continues to Climb in Emerging Economies. *Pew Research Center*. Retrieved from http://www.pewglobal.org/2016/02/22/smartphone-ownership-and-internet-usage-continues-to-climb-in-emerging-economies/

Pousttchi, K., Schiessler, M., & Wiedemann, D. (2009). Proposing a comprehensive framework for analysis and engineering of mobile payment business models. *Information Systems and E-Business Management*, *7*(3), 363–393. doi:10.1007/s10257-008-0098-9

Powell, A. C., Landman, A. B., & Bates, D. W. (2014). In search of a few good apps. *Journal of the American Medical Association*, *311*(18), 1851–1852. doi:10.1001/jama.2014.2564 PMID:24664278

Prałat, E. (2013). *M-commerce – rozwój na świecie i w Polsce*. Retrieved July 19, 2016, from http://www.ur.edu.pl/html

Preacher, J., Derek, K., Rucker, D., & Hayes, A. F. (2007). Addressing moderated mediation hypotheses: Theory, methods, and prescriptions. *Multivariate Behavioral Research*, *42*(1), 185–227. doi:10.1080/00273170701341316 PMID:26821081

Preece, J., Rogers, Y., & Sharp, H. (2002). *Interaction Design: beyond human-computer interaction*. New York: John Wiley & Sons Inc.

Prelec & Loewenstein. (1998). The red and the black: mental accounting of savings and debt. *Marketing Science*, *17*, 4–28.

Prelec, D., & Simester, D. (2001). Always leave home without it: A further investigation of the credit-card Effect on willingness to pay. *Marketing Letters*, *12*(1), 5–12. doi:10.1023/A:1008196717017

Prensky, M. (2008). *The role of technology in teaching and the classroom*. Retrieved from http://www.marcprensky.com/writing/Prensky-The_Role_of_Technology-ET-11-12-08.pdf

Prensky, M. (2001). Digital natives, digital immigrants part 1. *On the horizon*, *9*(5), 1–6. doi:10.1108/10748120110424816

Pritchard, D., & Warnakulasooriya, R. (2005). *Data from a Web-based Homework Tutor can predict Student's Final Exam Score*. Paper presented at the World Conference on Educational Media and Technology.

Psychologist World. (2016). *Cognitive Approach (Psychology) Introduction to the cognitive approach in psychology*. Author.

Punj, G. (2011). Effect of consumer beliefs on online purchase behavior: The influence of demographic characteristics and consumption values. *Journal of Interactive Marketing*, *25*(3), 134–144. doi:10.1016/j.intmar.2011.04.004

Purcell, K., Project, P. I., Entner, R., President, S. V., Practice, T., Company, T. N., & Henderson, N. (2010). *The Rise of Apps Culture. Group*. Retrieved from http://www.pewinternet.org/~/media//Files/Reports/2010/PIP_Nielsen Apps Report.pdf

Puschel, J., Mazzon, J. A., & Hernandez, J. M. C. (2010). Mobile banking: Proposition of an integrated adoption intention framework. *International Journal of Bank Marketing*, *28*(5), 389–409. doi:10.1108/02652321011064908

Puttaswamy, K. P. N., & Zhao, B. Y. (2010). *Preserving privacy in location-based mobile social applications*. Paper presented at the Eleventh Workshop on Mobile Computing Systems and Applications, Annapolis, MD. doi:10.1145/1734583.1734585

PWC. (2014). *2014/2015 Malaysian Tax and Business Booklet*. Retrieved from https://www.pwc.com/my/en/assets/publications/2015-malaysian-tax-business-booklet.pdf

Pyke, R. L. (1926). *Report on the legibility of print* (Vol. 110). London: HM Stationery Office.

Pymnts. (2016). *Consumers in France aren't using mobile devices for payment*. Retrieved September 8, 2016, from http://www.pymnts.com/news/mobile-payments/2016/france-consumers-mobile-devices-payment/

Qi, J., Li, L., Li, Y., & Shu, H. (2009). An extension of technology acceptance model: Analysis of the adoption of mobile data services in China. *Systems Research and Behavioral Science*, *26*(3), 391–407. doi:10.1002/sres.964

Raghubir, P., & Srivastava, J. (2008). Monopoly money: The effect of payment coupling and form on spending behavior. *Journal of Experimental Psychology. Applied*, *14*(3), 213–225. doi:10.1037/1076-898X.14.3.213 PMID:18808275

Raman, M., Stephenaus, R., Alam, N., & Kuppusamy, M. (2008). Information technology in Malaysia: E-service quality and update of Internet banking. *Journal of Internet Banking and Commerce*, *13*(2), 1–18.

Ramsden, A. (2000). *Annabella's HTML help*. Retrieved June 17, 2016, from http://www.geocities.com/annabella.geo/fontface.html

Rannenberg, K., Camenisch, J., & Sabouri, A. (Eds.). (2015). *Attribute-based Credentials for Trust: Identity in the Information Society*. Springer International Publishing. doi:10.1007/978-3-319-14439-9

Rassia, S. Th., & Pardalos, P. M. (Eds.). (2014). *Cities for Smart Environmental and Energy Futures*. Springer-Verlag Berlin Heidelberg. doi:10.1007/978-3-642-37661-0

Redecker, C. K. A.-M., & Punie, Y. (2010). Learning 2.0 - the impact of social media on learning in Europe. *European Communities*, 1-13.

Redecker, C., Ala-Mutka, K., & Punie, Y. (2010). *Learning 2.0 - The impact of social media on learning in Europe.* Retrieved from http://ftp.jrc.es/EURdoc/JRC56958.pdf

Reeves, T. C., Benson, L., Elliott, D., Grant, M., Holschuh, D., Kim, B., & Loh, S. et al. (2002). Usability and Instructional Design Heuristics for E-Learning Evaluation. In *Proceedings of World Conference on Educational Multimedia, Hypermedia and Telecommunications.* Norfolk, VA: AACE.

Refuel Agency. (2015). *Millenial Teens, Digital Explorer.* Retrieved July 19, 2016, from http://research.refuelagency.com/wp-content/uploads/2015/07/Millennial-Teen-Digital-Explorer.pdf

Reinfelder, L., Benenson, Z., & Gassmann, F. (2014). Differences between Android and iPhone Users in Their Security and Privacy Awareness. In C. Eckert, S. K. Katsikas & G. Pernul (Eds.), *Trust, Privacy, and Security in Digital Business: 11th International Conference, TrustBus 2014, Munich, Germany, September 2-3, 2014. Proceedings* (pp. 156-167). Cham: Springer International Publishing. doi:10.1007/978-3-319-09770-1_14

Reinfelder, L., Benenson, Z., & Gassmann, F. (2014). Differences between Android and iPhone users in their security and privacy awareness. Lecture Notes in Computer Science, 8647, 156–167. doi:10.1007/978-3-319-09770-1_14

Research2Guidance The App Market Specialist. (2015). *mHealth App Developer Economics 2015.* Retrieved from www.mHealthEconomics.com

Research2Guidance. (2015). *mHealth App Development Economic 2015.* Retrieved from http://research2guidance.com/r2g/r2g-mHealth-App-Developer-Economics-2015.pdf

Resnick, P., Kuwabara, K., Zeckhauser, R., & Friedman, E. (2000). Reputation systems. *Communications of the ACM, 43*(12), 45–48. doi:10.1145/355112.355122

Review, H. B. (2011). *Harvard Business Review on Aligning Technology with Strategy.* Boston, MA: Harvard Business School Publishing.

Ribot, T. (1906). *Essay on the creative imagination.* Chicago, IL: Open Court. doi:10.1037/13773-000

Rieck, D. (1997). Design, legibility and unnatural acts. *Direct Marketing, 60*(6), 23–25.

Riedel, J.C., & Fransoo, V.C.S. (2006). *Modelling dynamics in decision support systems.* Academic Press.

Riegler, B.R., & Riegler, G.L. (2012). *Cognitive Psychology: Applying The Science of the Mind.* Pearson.

Ríos, M. (2014). *Fueled by Aging Baby Boomers, Nation's Older Population to Nearly Double in the Next 20 Years.* Retrieved July 7, 2016, from https://www.census.gov/newsroom/press-releases/2014/cb14-84.html

Riquelme, H., & Rios, R. E. (2010). The moderating effect of gender in the adoption of mobile banking. *International Journal of Bank Marketing, 28*(5), 328–341. doi:10.1108/02652321011064872

Rivard, R. (2013). Coursera's contractual Elitism. *Inside Higher Education.* Retrieved from the Internet December 11, 2015 at https://www.insidehighered.com/news/2013/03/22/coursera-commits-admitting-only-elite-universities

Robbins, S. P., & Judge, T. A. (2009). *Organisational decisions* (13th ed.). Prince Hall.

Robert, J., Sternberg, C. J., & Sternberg, K. (2012). Cognitive Psychology (6th ed.). Academic Press.

Rocha, L. C., Andrade, R. M., & Sampaio, A. L. (2014). Heurísticas para avaliar a usabilidade de aplicações móveis: estudo de caso para aulas de campo em Geologia. *Proceedings of XIX Conferência Internacional sobre Informática na Educação.*

Rochwerger, B., Breitgand, D., Levy, E., Galis, A., Nagin, K., Llorente, I. M., & Galan, F. et al. (2009). The RESERVOIR model and architecture for open federated cloud computing. *IBM Journal of Research and Development*, *53*(4), 1–11. doi:10.1147/JRD.2009.5429058

Rodriguez, O. (2013). The concept of openness behind c and x-MOOCs. *Open Praxis*, *5*(1), 67–73. doi:10.5944/openpraxis.5.1.42

Rogers, E. M. (1995). *Diffusion of Innovation*. New York: Free Press.

Roto, V., Lee, M., Pihkala, K., Castro, B., Vermeeren, A. P. O. S., & Law, E. (2010). *All about UX: Information for user experience professionals*. Retrieved in September 22 2015, from http://www.allaboutux.org/

Roullet, B., Ben Dahmane Mouelhi, N., & Droulers, O. (2003). The Impact of background color on product attitudes among three different cultures. *Proceedings of International Congress of the French Association*.

Round, C. (2013). *The Best MOOC Provider: A Review of Coursera, Udacity and edX. Skilledup.com*. Skilledup.

Rowe. (2011). The Impact of Cloud on Mid-size Businesses. In *Implementing and Developing Cloud Computing Applications*. Taylor and Francis Group, LLC.

Roy, S. (2011). *The Psychology of Fantasy*. Retrieved in September 22 2015, from http://www.futurehealth.org/articles/The-Psychology-of-Fantasy-by-Saberi-Roy-100901-178.html

Rubin, J., & Chisnell, D. (2008). *Handbook of usability testing: how to plan, design and conduct effective tests*. John Wiley & Sons.

Ruddra, A. (2015, Feb 12). *Why Do We Make Apps in English in India?* Retrieved from http://anshumaniruddra.com/2015/02/11/make-apps-english-india/

Ryan, V. (2004). *Colours and Cultures*. Retrieved in September 22 2015, from http://www.technologystudent.com/despro2/colcul1.htm

Ryan, R. M., & Deci, E. L. (2000). Self-determination theory and the facilitation of intrinsic motivation, social development, and well-being. *The American Psychologist*, *55*(1), 68–78. doi:10.1037/0003-066X.55.1.68 PMID:11392867

Ryan, R. M., & Deci, E. L. (2006). Self-regulation and the problem of human autonomy: Does psychology need choice, self-Determination, and will? *Journal of Personality*, *74*(6), 1557–1585. doi:10.1111/j.1467-6494.2006.00420.x PMID:17083658

Sadeh, N., Hong, J., Cranor, L., Fette, I., Kelley, P., Prabaker, M., & Rao, J. (2009). Understanding and capturing peoples privacy policies in a mobile social networking application. *Personal and Ubiquitous Computing*, *13*(6), 401–412. doi:10.1007/s00779-008-0214-3

Sadia, S. (2011). User acceptance decision towards mobile commerce technology: A study of user decision about acceptance of mobile commerce technology. *Interdisciplinary Journal of Contemporary Research in Business*, *2*(12), 535–547.

Safko, L. (2009). *The Social Media Bible: Tactics, Tools, and Strategies for Business Success*. Hoboken, NJ: Wiley.

Sakas, D. P., Dimitrios, N. K., & Kavoura, A. (2015). The Development of Facebooks Competitive Advantage for Brand Awareness. *Procedia Economics and Finance*, *24*, 589–597. doi:10.1016/S2212-5671(15)00642-5

Salles Junior, F. M., Pinho, A. L., Santa Rosa, G. J., & Ramos, M. A. (2016). Pedagogical usability: a theoretical essay for e-learning. *Holos*, *32*(1), 3-15. doi:10.15628/holos.2016.2593

San Martín-Gutiérrez, S., López-Catalán, B., & Ramón-Jerónimo, M. (2012). Determinants of involvement in mobile commerce. The moderating role of gender. *EsicMarket, 141,* 69–101.

San-Martín, S. Jimenez N.H. Lopez-Catalan, B. (2015). The firms benefits of mobile CRM from the relationship marketing approach and the TOE model. *Spanish Journal of Marketing – ESIC, 26,* 18-29.

Sarathy, Narayan, & Mikkilineni. (2010). Next generation cloud computing architecture -enabling real-time dynamism for shared distributed physical infrastructure. *19th IEEE International Workshops on Enabling Technologies: Infrastructures for Collaborative Enterprises (WETICE'10),* 48-53.

Sargut, G., & McGrath, R. (2011). Learning to Live with Complexity. *Harvard Business Review, 89*(9), 68–76. PMID:21939129

Sari, P. K., Candiwan, & Trianasari, N. (2014). *Information security awareness measurement with confirmatory factor analysis.* Paper presented at the Technology Management and Emerging Technologies (ISTMET), 2014 International Symposium on.

Sari, A., & Bayram, P. (2015). Challenges of Internal and External Variables of Consumer Behaviour towards Mobile Commerce. *International Journal of Communications Network and System Sciences, 8*(13), 578–596. doi:10.4236/ijcns.2015.813052

Sarrina Li, S. C. (2001). New media and market competition: A niche analysis of television news, electronic news, and newspaper news in Taiwan. *Journal of Broadcasting & Electronic Media, 45*(2), 259–276. doi:10.1207/s15506878jobem4502_4

Saylor, M. (2012). The Mobile Wave: How Mobile Intelligence Will Change Everything. Perseus Books.

Scalf, P. E., Colcombe, S. J., McCarley, J. S., Erickson, K. I., Alvarado, M., Kim, J. S., & Kramer, A. F. (2007). The neural correlates of an expanded functional field of view. *Journal of Gerontology, 62B*(1), 32–44. doi:10.1093/geronb/62.special_issue_1.32 PMID:17565163

Schaefer, M.W. (2014). Social media explained: Untangling the world's most understood business trend. In *Social Media Commerce For Dummies.* Wiley.

Scharff, L. F. V., & Ahumada, A. J., Jr. (2002). Predicting the readability of transparent text. *Journal of Vision, 2*(9), 653-666. Retrieved from http://www.journalofvision.org/2/9/7/html

Schierz, P. G., Schilke, O., & Wirtz, B. W. (2010). Understanding consumer acceptance of mobile payment services: An empirical analysis. *Electronic Commerce Research and Applications, 9*(3), 209–216. doi:10.1016/j.elerap.2009.07.005

Schilit, B., Adams, N., & Want, R. (1994). *Context-aware computing applications.* Paper presented at the Mobile Computing Systems and Applications.

Schnall, R., Higgins, T., Brown, W., Carballo-Dieguez, A., & Bakken, S. (2015). Trust, Perceived Risk, Perceived Ease of Use and Perceived Usefulness as Factors Related to mHealth Technology Use. *MEDINFO 2015: eHealth-enabled Health, 216.* doi: 10.3233/978-1-61499-564-7-467

Schrøder, K. C., & Christian, K. (2010). Towards a typology of cross-media news consumption: A qualitative-quantitative synthesis. *Northern Lights, 8*(1), 115–137. doi:10.1386/nl.8.115_1

Schrøder, K. C. (2014). News media old and new. *Journalism Studies, 16*(1), 60–78. doi:10.1080/1461670X.2014.890332

Schultz, L. (2005). *Effects of graphical elements on perceived usefulness of a library.* Retrieved from http://www.tarleton.edu/~schultz/finalprojectinternetsvcs.html

Schweitzer, J., & Synowiec, C. (2012). The Economics of eHealth and mHealth. *Journal of Health Communication, 17*(October), 73–81. doi:10.1080/10810730.2011.649158 PMID:22548602

Sefton, P. (1980). *Privacy and data control in the era of cloud computing. OECD Guidelines on the Protection of Privacy and Transborder Flows of Personal Data*. Organisation for Economic Cooperation and Development.

Segerståhl, K., & Oinas-Kukkonen, H. (2008). Distributed User Experience in Persuasive Technology Environments. Linnanmaa.

Sekuler, R., & Sekuler, A. B. (2000). Visual perception and cognition. *Oxford Textbook of Geriatric Medicine, 2*, 874–880.

Selwyn, N. (2012). Social media in higher education. *The Europa World of Learning*. Retrieved from http://sites.jmu.edu/flippEDout/files/2013/04/sample-essay-selwyn.pdf

Sensor Tower. (2016). *Nomura research*. Author.

Setaro, J. L. (2001). *If you build it, will they come? Distance learning through wireless devices*. Retrieved from http://www.unisysworld.com/monthly/2001/07/wireless.shtml

Shalan, M. A. (2010). Managing IT Risks in Virtual Enterprise Networks: A Proposed Governance Framework. In S. Panios (Ed.), *Managing Risk in Virtual Enterprise Networks: Implementing Supply Chain Principles* (pp. 115–136). Hershey, PA: IGI Global. doi:10.4018/978-1-61520-607-0.ch006

Shalan, M. A. (2017). Ethics and Risk Governance for the Middle Circle in Mobile Cloud Computing: Outsourcing, Contracting and Service Providers Involvement. In K. Munir (Ed.), *Security Management in Mobile Cloud Computing* (pp. 43–72). Hershey, PA: IGI Global. doi:10.4018/978-1-5225-0602-7.ch003

Shankar, V., Venkatesh, A., Hofacker, C., & Naik, P. (2010). Mobile Marketing in the Retailing Environment: Current Insights and Future Research Avenues. *Journal of Interactive Marketing, 24*(2), 111–120. doi:10.1016/j.intmar.2010.02.006

Shapiro, L. (2011). *Embodied Cognition*. New York: Routledge.

Sharma, M. (2016). Services of mobile commerce. In S. Madan & J. Arora (Eds.), *Securing transactions and payment systems for m-commerce* (pp. 251–274). Hershey, PA: IGI Global. doi:10.4018/978-1-5225-0236-4.ch013

Shaukat, M., & Zafar, J. (2010). Impact of Sociological and Organizational Factors on Information Technology Adoption: An Analysis of Selected Pakistani Companies. *European Journal of Soil Science, 13*(2), 305.

Shaver, P., Schwarz, J., Kirson, D., & OConnor, C. (1987). Emotion Knowledge: Further exploration of a prototype approach. *Journal of Personality and Social Psychology, 52*(2), 1061–1086. doi:10.1037/0022-3514.52.6.1061 PMID:3598857

Sheila, M., Faizal, M. A., & Shahrin, S. (2015). Dimension of mobile security model: Mobile user security threats and awareness. *International Journal of Mobile Learning and Organisation, 9*(1), 66. doi:10.1504/IJMLO.2015.069718

Shelton, T. (2013). *Business Models for the Social Mobile Cloud: Transform Your Business Using Social Media, Mobile Internet, and Cloud Computing*. Indianapolis, IN: John Wiley & Sons. doi:10.1002/9781118555910

Shibli, M. A., Masood, R., Ghazi, Y., & Muftic, S. (2015). MagicNET: Mobile agents data protection system. *Transactions on Emerging Telecommunications Technologies, 26*(5), 813–835. doi:10.1002/ett.2742

Shih, H. P. (2012). Cognitive lock-in effects on consumer purchase intentions in the context of B2C web sites. *Psychology and Marketing, 29*(10), 738–751. doi:10.1002/mar.20560

Shilton, K. (2009). Four billion little brothers? Privacy, mobile phones, and ubiquitous data collection. *Communications of the ACM, 52*(11), 48–53. doi:10.1145/1592761.1592778

Compilation of References

Shin, D. H., & Kim, W. Y. (2008). Forecasting customer switching intention in mobile service: An exploratory study of predictive factors in mobile number portability. *Technological Forecasting and Social Change*, *75*(6), 854–874. doi:10.1016/j.techfore.2007.05.001

Shin, Y. M., Lee, S. C., Shin, B., & Lee, H. G. (2010). Examining influencing factors of post-adoption usage of mobile internet: Focus on the user perception of supplier-side attributes. *Information Systems Frontiers*, *12*(5), 595–606. doi:10.1007/s10796-009-9184-x

Siau, K., Lim, E. P., & Shen, Z. (2001). Mobile commerce: Promises, challenges, and research agenda. *Journal of Database Management*, *12*(3), 3, 4–13. doi:10.4018/jdm.2001070101

Siau, K., Lim, E., & Shen, Z. (2003). Mobile commerce: Current states and future trends. In E. Lim & K. Siau (Eds.), *Advances in mobile commerce technologies* (pp. 1–17). Hershey, PA: Idea Group Publishing. doi:10.4018/978-1-59140-052-3.ch001

Siau, K., & Shen, Z. (2003). Building customer trust in mobile commerce. *Communications of the ACM*, *46*(4), 91–94. doi:10.1145/641205.641211

Sidney, K., Antony, J., Rodrigues, R., Arumugam, K., Krishnamurthy, S., Dsouza, G., & Shet, A. et al. (2012). Supporting patient adherence to antiretrovirals using mobile phone reminders: Patient responses from South India. *AIDS Care*, *24*(5), 612–617. doi:10.1080/09540121.2011.630357 PMID:22150088

Sielhorst, T., Obst, T., Burgkart, R., Riener, R., & Navab, N. (2004). An augmented reality delivery simulator for medical training. In M. Berger, & N. Navab (Chairs), *Workshop AMI-ARCS 2004*. Symposium conducted at the meeting of IRISA, Rennes, France.

Siepmann, F. (2014). *Managing Risk and Security in Outsourcing IT Services: Onshore, Offshore and the cloud*. Boca Raton, FL: Taylor and Francis Group.

Silva, B. M. C., Rodrigues, J. J. P. C., de la Torre Díez, I., López-Coronado, M., & Saleem, K. (2015). Mobile-health: A review of current state in 2015. *Journal of Biomedical Informatics*, *56*, 265–272. doi:10.1016/j.jbi.2015.06.003 PMID:26071682

Simon, J. S. (1999). *A cross-cultural analysis of website design: an empirical study of global web users*. Paper Presented at the Seventh Cross-Cultural Consumer Business Studies Research Conference, Cancun, Mexico.

Sinclair, J., Boyatt, R., Foss, J., Rocks, C., & Uden, L. (2014). A tale of two modes: Initial reflections on an innovative MOOC. In L. Uden, J. Sinclair, Y.-H. Tao, & D. Liberona (Eds.), *Learning Technology for Education in Cloud. MOOC and Big Data* (Vol. 446, pp. 49–60). Springer International Publishing. doi:10.1007/978-3-319-10671-7_5

Singh, P., & Pant, S. (2014, Jan 24). *Unstoppable! Smartphone surge in India continues*. Retrieved from http://www.nielsen.com

Singh, V. R. (2014). An Overview of Mobile Commerce in India. *International Journal of Management Research and Review*, *4*(3), 354–366.

Skulmowsky, A., Pradel, S., Kühnert, T., Brunnett, G., & Daniel, R. G. (2016). Embodied learning using a tangible user interface: The effects of haptic perception and selective pointing on a spatial learning task. *Computers & Education*, *92-93*, 64–75. doi:10.1016/j.compedu.2015.10.011

Slatten, A. L. D. (2010). An Application and Extension of the Technology Acceptance Model to Nonprofit Certification. Journal for Non-Profit Management, 14.

Smutz, W. (2013). MOOCs Are No Education Panacea, but Here's What can Make Them Work. *Forbes*. Retrieved from http://www.forbes.com/sites/forbesleadershipforum/2013/04/08/moocs-are-no-education-panacea-but-heres-what-can-make-them-work/?cid=dlvr.it

Socialskinny. (2012). 100 social media statistics for 2012. *Social Skinny*. Retrieved from: http://thesocialskinny.com/100-social-media-statistics-for-2012/

Sohn, C., Lee, D. I., & Lee, H. (2011). The effects of mobile CRM activities on trust-based commitment. *International Journal of Electronic Customer Relationship Management.*, 5(2), 130–152. doi:10.1504/IJECRM.2011.041262

Soman, D. (2001). Effects of payment mechanism on spending behavior: The role of rehearsal and immediacy of payments. *The Journal of Consumer Research*, 27(March), 460–474. doi:10.1086/319621

Soman, D. (2003). The effect of payment transparency on consumption: Quasi-experiments from the Field. *Marketing Letters*, 14(3), 5–12. doi:10.1023/A:1027444717586

Somro, S., Ahmad, W. F. W., & Sulaiman, S. (2012). A preliminary study on heuristics for mobile games. In *Proceedings of International Conference on Computer & Information Science*. IEEE. doi:10.1109/ICCISci.2012.6297177

Sonia, S. M.-G., & Blanca, L.-C. (2013). How can a mobile vendor get satisfied customers? *Industrial Management & Data Systems*, 113(2), 156–170. doi:10.1108/02635571311303514

Sosinsky, B. (2011). *Cloud Computing Bible*. Hoboken, NJ: Wiley Publishing, Inc.

Spaulding, E., & Perry, C. (2013). *Making it personal: Rules for success in product customization*. Retrieved from http://www.bain.com/publications/articles/making-it-personal-rules-for-success-in-product-customization.aspx

Sripalawat, J., Thongmak, M., & Ngramyarn, A. (2011). M-banking in metropolitan Bangkok and a comparison with other countries. *Journal of Computer Information Systems*, 51(3), 67–76.

Stahl, G., Koschmann, T., & Suthers, D. (2006). Computer-supported collaborative learning: an historical perspective. In R. K. Sawyer (Ed.), *Cambridge Handbook of the Learning Sciences* (pp. 409–426). Cambridge, UK: Cambridge University Press.

Starkey, G. (2013). Trust in the diverging, convergent multi-platform media environment. *Communication Management Quarterly*, 26(26), 73–98. doi:10.5937/comman1326073S

Statista. (2014). *Number of smartphone users worldwide from 2014 to 2019 (in millions)*. Retrieved July 10, 2016, from http://www.statista.com/statistics/330695/number-of-smartphone-users-worldwide/

Statista. (2014). *Social media and user-generated content tools used by omnichannel retailers in the United States as of 1st quarter 2014*. Retrieved from: http://www.statista.com/statistics/308539/us-omnichannel-retailer-social-media-tools/

Statista. (2016). *Global mobile retail commerce revenue from 2012 to 2018 in billion U.S. dollars*. Retrieved from https://www.statista.com/statistics/324636/mobile-retail-commerce-revenue-worldwide/

Statista. (2016). *Number of apps available in leading app stores as of June 2016*. Retrieved June 1, 2016, from http://www.statista.com/statistics/276623/number-of-apps-available-in-leading-app-stores/

Statista. (2016). *Number of mobile app downloads worldwide from 2009 to 2017 (in millions)*. Retrieved 26 February, 2016, from http://www.statista.com/statistics/266488/forecast-of-mobile-app-downloads/

Stone, A., Briggs, J., & Smith, C. (2002). *SMS and interactivity-some results from the field, and its implications on effective uses of mobile technologies in education*. Paper presented at the Wireless and Mobile Technologies in Education. doi:10.1109/WMTE.2002.1039238

Compilation of References

Stone, D., Jarret, C., Woodroffe, M., & Minocha, S. (2005). *User interface design and evaluation*. San Francisco: Elsevier.

Subramaniam, V., & Hunt, A. (2006). *Practices of an Agile Developer*. Frisco, TX: Pragmatic Bookshelf.

Suh, Y., Ahn, T., & Pedersen, P. M. (2013). Examining the effects of team identification, e-service quality (e-SQ) and satisfaction on intention to revisit sports websites. *International Journal of Sports Marketing & Sponsorship, 14*(4), 2–19. doi:10.1108/IJSMS-14-04-2013-B002

Sumita, U., & Yoshii, J. (2010). Enhancement of e-commerce via mobile accesses to the Internet. *Electronic Commerce Research and Applications, 9*(3), 217–227. doi:10.1016/j.elerap.2009.11.006

Sung, Y. T., Chang, K. E., & Liu, T. C. (2016). The effects of integrating mobile devices with teaching and learning on students learning performance: A meta-analysis and research synthesis. *Computers & Education, 94*, 252–275. doi:10.1016/j.compedu.2015.11.008

Suoranta, M., & Mattila, M. (2004). Mobile banking and consumer behavior: New insights into the diffusion pattern. *Journal of Financial Services Marketing, 8*(4), 354–366. doi:10.1057/palgrave.fsm.4770132

Su, Q., & Adams, C. (2009). Mobile commerce adoption: A novel buyer-user-service payer metric. *Journal of Electronic Commerce in Organizations, 7*(4), 59–72. doi:10.4018/jeco.2009070106

Sutherland, I. E. (1965). The ultimate display. In W. A. Kalenich (Ed.), *Proceedings of IFIP Congress* (pp. 506-508). New York: Spartan Books, Macmillan.

Sutherland, I. E. (1968). A head-mounted three dimensional display. *AFIPS Conference Proceedings, 33, I* (pp. 757-764). San Francisco, CA: The Thompson Book Company.

Sweller, J. (1988). Cognitive load during problem solving: Effects on learning. *Cognitive Science, 12*(2), 257–285. doi:10.1207/s15516709cog1202_4

Sweller, J., van Merrienboer, J. J., & Paas, F. G. (1998). Cognitive architecture and instructional design. *Educational Psychology Review, 10*(3), 251–296. doi:10.1023/A:1022193728205

Tanakinjal, G. H., Deans, K. R., & Gray, B. J. (2010). Third screen communication and the adoption of mobile marketing: A Malaysia perspective. *International Journal of Marketing Studies, 2*(1), 36–47.

Taneja, H., Webster, J. G., Malthouse, E. C., & Ksiazek, T. B. (2012). Media consumption across platforms: Identifying user-defined repertoires. *New Media & Society, 14*(6), 951–968. doi:10.1177/1461444811436146

Tarasewich, P. (2003). Designing mobile commerce applications. *Communications of the ACM, 46*(12), 57–60. doi:10.1145/953460.953489

Teo, T. S. H., & Pok, S. H. (2003). Adoption of WAP-enabled mobile phones among Internet users. *Omega: The International Journal of Management Science, 31*(6), 483–498. doi:10.1016/j.omega.2003.08.005

Terwogt, M. M., & Hoeksma, J. B. (1995). Colors and Emotions: Preferences and Combinations. *The Journal of General Psychology, 122*(1), 5–17. doi:10.1080/00221309.1995.9921217 PMID:7714504

Thaler, R. (1980). Towards a positive theory of consumer choice. *Journal of Economic Behavior & Organization, 1*(1), 39–60. doi:10.1016/0167-2681(80)90051-7

The Free Dictionary. (2015). *Tailoring*. Retrieved in September 22 2015, from http://www.thefreedictionary.com/tailoring

The Hindu. (2016, Feb. 3). *With 220mn users, India is now world's second-biggest smartphone market*. Retrieved from http://www.thehindu.com/news/cities/mumbai/business/article8186543.ece

The Times of India. (2010, March 14). *Indiaspeak: English is our 2nd language.* Retrieved from http://timesofindia.indiatimes.com/india/Indiaspeak-English-is-our-2nd-language/articleshow/5680962.cms

Thomas, B., Close, B., Donoghue, J., & Squires, J. (2000). ARQuake: An outdoor/indoor augmented reality first person application. *Proceedings of the Fourth International Symposium on Wearable Computers* (pp. 139-146). Atlanta, GA: IEEE. doi:10.1109/ISWC.2000.888480

Thomas, M., Desai, K. K., & Seenavasin, S. (2011). How credit card payments increase unhealthy food purchases: Visceral regulation of vices. *The Journal of Consumer Research, 38*(June), 126–139. doi:10.1086/657331

Thorson, E., Shoenberger, H., Karaliova, T., Kim, E., & Fidler, R. (2015). News use of mobile media: A contingency model. *Mobile Media & Communication, 3*(2), 160–178. doi:10.1177/2050157914557692

Tianfield, H. (2011). Cloud Computing Architectures. *Proceedings of Systems, Man, and Cybernetics (SMC), 2011 IEEE International Conference,* 1394 – 1399.

Tiwana, A., & Bush, A. A. (2007). A Comparison of Transaction Cost, Agency, and Knowledge-Based Predictors of IT Outsourcing Decisions: A U.S.–Japan Cross-Cultural Field Study. *Journal of Management Information Systems, 24*(1), 259–300. doi:10.2753/MIS0742-1222240108

Tolba, R. H. (2003). The cultural aspects of design Jordanian websites: An empirical evaluation of university, news, and government website by different user groups. *The International Arab Journal of Information Technology, 1,* 51–59.

Touzani, M., Khedri, M., & Ben Dahmane Mouelhi, N. (2007), An exploratory approach to sources of irritation felt during a shopping activity: case of food predominantly malls. *6th Marketing Trends International Symposium.*

TRAI. (2016, July 31). *Highlights of Telecom Subscription Data.* Retrieved from http://trai.gov.in/WriteReadData/PressRealease/Document/PR-TSD-Nov-15.pdf

Trant, M., & Ravi, R. (2014). *Cloud bound: Advice from organizations in outsourcing relationships.* Retrieved February 5, 2016, from http://www.ibm.com/smarterplanet/us/en/centerforappliedinsights/article/cloudbound.html

Traxler, J. (2007). Defining, Discussing and Evaluating Mobile Learning: The moving finger writes and having writ.... *International Review of Research in Open and Distance Learning, 8*(2). doi:10.19173/irrodl.v8i2.346

Tsiaousis, A. S., & Giaglis, G. M. (2014). Mobile websites: Usability evaluation and design. *International Journal of Mobile Communications, 12*(1), 29–55. doi:10.1504/IJMC.2014.059241

Turley, L. W., & Milliman, R. E. (2000). Atmospheric effects on shopping behavior: A review of the experimental evidence. *Journal of Business Research, 49*(2), 193–211. doi:10.1016/S0148-2963(99)00010-7

Ullman, L. (2013). Effortless E-Commerce with PHP and MySQL (2nd ed.). New Riders.

Ullmer, B., Ishii, H., & Jacob, R. J. K. (2005). Token+constraint systems for tangible interaction with digital information. *ACM Transactions on Computer-Human Interaction, 12*(1), 81–118. doi:10.1145/1057237.1057242

Underhill, P. (2008). Why We Buy: The Science of Shopping--Updated and Revised for the Internet, the Global Consumer, and Beyond. Simon & Schuster.

Underkoffler, J. (1999). *The I/O bulb and the luminous room* (Doctoral dissertation). Retrieved from http://tmg-trackr.media.mit.edu/publishedmedia/Papers/300-The%20IO%20Bulb/Published/PDF

Usability Net. (2006). *The business case for usability.* Retrieved in September 22 2015, from http://www.usabilitynet.org/management/c_business.htm

Väänänen, K.V.M., Roto, V., & Hassenzahl, M. (2009). *Towards Practical User Experience Evaluation Methods.* Academic Press.

Väänänen-Vainio-Mattila, K., & Wäljas, M. (2009). Development of evaluation heuristics for web service user experience. In *Extended Abstracts of the 27th International Conference on Human Factors in Computing Systems - CHI'09* (pp. 3679-3684). New York: ACM Press. doi:10.1145/1520340.1520554

Väänänen-Vainio-Mattila, K., & Wäljas, M. (2009). Developing an expert evaluation method for user experience of cross-platform web services. In *Proceedings of MindTrek*. Tampere, Finland: ACM. doi:10.1145/1621841.1621871

Van Damme, K., Courtois, C., Verbrugge, K., & Marez, L. D. (2015). Whats APPening to news? A mixed-method audience-centered study on mobile news consumption. *Mobile Media & Communication, 3*(2), 196–213. doi:10.1177/2050157914557691

van der Heijden, H., Verhagen, T., & Creemers, M. (2003). Understanding online purchase intentions: Contributions from technology and trust perspectives. *European Journal of Information Systems, 12*(1), 41–48. doi:10.1057/palgrave.ejis.3000445

Varela, C. (2015). *Facebook Marketing for Business: Learn to create a successful campaign, Advertise your Business, Strategies to generate traffic and Boost sales (Social Media)*. Amazon.

Varnali, K., & Toker, A. (2010). Mobile marketing research: The-state-of-the-art. *International Journal of Information Management, 30*(2), 144–151. doi:10.1016/j.ijinfomgt.2009.08.009

Varshney, U. (2007). Supporting dependable group-oriented mobile transactions: Redundancy-based architecture and performance. *International Journal of Network Management, 17*(3), 219–229. doi:10.1002/nem.619

Varshney, U. (2008). A middleware framework for managing transactions in group-oriented mobile commerce services. *Decision Support Systems, 46*(1), 356–365. doi:10.1016/j.dss.2008.07.005

Varshney, U., & Vetter, R. (2002). Mobile Commerce: Framework, Applications and Networking Support. *Mobile Networks and Applications, 7*(3), 185–198. doi:10.1023/A:1014570512129

Vaufrey, C. (2015). *MOOC "Economie du Web"*. Thot Cursus.

Vavra, T. G. (1992). *After marketing: how to keep customer life through relationship marketing*. Homewood, IL: Business One – Irwin.

Veijalainen, J., & Weske, M. (2003). Modeling static aspects of mobile electronic commerce environments. In E. Lim & K. Siau (Eds.), *Advances in mobile commerce technologies* (pp. 137–170). Hershey, PA: Idea Group Publishing. doi:10.4018/978-1-59140-052-3.ch007

Venkatesh, V., Ramesh, V., & Massey, A. P. (2003). Understanding usability in mobile commerce. *Communications of the ACM, 46*(12), 53–56. doi:10.1145/953460.953488

VentureBeat. (2015). *How marketers are tailoring their content strategy for a multi-channel, multi-device world*. Retrieved in June 12 2015, from http://venturebeat.com/2015/08/31/how-marketers-are-tailoring-their-content-strategy-for-a-multi-channel-multi-device-world/

Verma, D., & Verma, D. S. (2013). Managing customer relationships through mobile CRM In organized retail outlets. *International Journal of Engineering Trends and Technology, 4*(5), 1696–1701.

Verschuren, P., & Hartog, R. (2005). *Evaluation in Design-Oriented Research*. Springer.

Vesanen, J. (2007). What is personalization? A conceptual framework. *European Journal of Marketing, 41*(5/6), 409–418. doi:10.1108/03090560710737534

Vieira, J., Sousa, M., Arsénio, A., & Jorge, J. (2015). Augmented reality for rehabilitation using multimodal feedback. *Proceedings of the 3rd 2015 Workshop on ICTs for improving Patients Rehabilitation Research Techniques*, 38-41. doi:10.1145/2838944.2838954

Viola, P., & Jones, M. J. (2004). Robust real-time face detection. *International Journal of Computer Vision, 57*(2), 137–154. doi:10.1023/B:VISI.0000013087.49260.fb

Visinescu, L. L., Sidorova, A., Jones, M. C., & Prybutok, V. R. (2015). The influence of website dimensionality on customer experiences, perceptions and behavioral intentions: An exploration of 2D vs. 3D web design. *Information & Management, 52*(1), 1–17. doi:10.1016/j.im.2014.10.005

Vital Wave Consulting. (2009). mHealth for Development: The Opportunity of Mobile Technology for Healthcare in the Developing World. *Technology, 46*. http://doi.org/10.1145/602421.602423

Volle, P. (2000). From marketing of points of sale to the merchant websites: Specificities, opportunities and research questions. *Revue Française du Marketing, 177/178*(2/3), 83–101.

Vošner, H. B., Bobek, S., Kokol, P., & Krečič, M. J. (2016). Attitudes of active older internet users towards online social networking. *Computers in Human Behavior, 55*, 230–241. doi:10.1016/j.chb.2015.09.014

Vygotsky, S. L. (2004). Imagination and Creativity in Childhood. *Journal of Russian & East European Psychology, 42*(1).

W3C. (2011). *Web content accessibility guidelines (WCAG) overview*. Retrieved August 22, 2011, from http://www.w3.org/WAI/intro/wcag/html

Wakefield, R. L., Stocks, M. H., & Wilder, W. M. (2004). The role of web site characteristics in initial trust formation. *Journal of Computer Information Systems, 45*(1), 94–103.

Wang, C. C., Lo, S. K., & Fang, W. (2008). Extending the technology acceptance model to mobile telecommunication innovation: The existence of network externalities. *Journal of Consumer Behaviour, 7*(2), 101–110. doi:10.1002/cb.240

Wang, C., Chiang, Y., & Wang, M. (2015). Evaluation of an augmented reality embedded on-line shopping system. *Procedia Manufacturing, 3*, 5624–5630. doi:10.1016/j.promfg.2015.07.766

Wang, H., Hsu, C., Chiu, C., & Tsai, S. (2011). The design and implementation of augmented reality gaming system in hand rehabilitation. *Communications in Information Science and Management Engineering, 1*(8), 37–40. doi:10.5963/CISME0108009

Wang, J. (2007). Mobile commerce. In D. Taniar (Ed.), *Encyclopedia of mobile computing and commerce* (pp. 455–460). Hershey, PA: IGI Global. doi:10.4018/978-1-59904-002-8.ch075

Wang, L., Sato, H., Rau, P. L. P., Fujimura, K., Gao, Q., & Asano, Y. (2008). Chinese text spacing on mobile phones for senior citizens. *Educational Gerontology, 35*(1), 77–90. doi:10.1080/03601270802491122

Wang, T., Duong, T. D., & Chen, C. C. (2016). Intention to disclose personal information via mobile applications: A privacy calculus perspective. *International Journal of Information Management, 36*(4), 531–542. doi:10.1016/j.ijinfomgt.2016.03.003

Wang, X., Hong, Z., Xu, Y., Zhang, C., & Ling, H. (2014). Relevance judgments of mobile commercial information. *Journal of the Association for Information Science and Technology, 65*(7), 1335–1348. doi:10.1002/asi.23060

Wang, Y. S., & Liao, Y. W. (2007). The conceptualization and measurement of m-commerce user satisfaction. *Computers in Human Behavior, 23*(1), 381–398. doi:10.1016/j.chb.2004.10.017

Wang, Y. S., Lin, H. H., & Luarn, P. (2006). Predicting consumer intention to use mobile service. *Information Systems Journal, 16*(2), 157–179. doi:10.1111/j.1365-2575.2006.00213.x

Weber, H. (2014). *Path has just 5 million daily active users globally*. Retrieved 17 June, 2016, from http://venturebeat.com/2014/09/10/path-has-just-5-million-daily-active-users-globally/

Wei, R., Lo, V., Xu, X., Chen, K., & Zhang, G. (2013). Predicting mobile news use among college students: The role of press freedom in four Asian cities. *New Media & Society, 16*(4), 637–654. doi:10.1177/1461444813487963

Weintraub, S. (2012). *Apple's iOS problem: Contacts uploading is just the tip of the iceberg. Apps can upload all your photos, calendars or record conversations*. Retrieved 15 June, 2016, from http://9to5mac.com/2012/02/15/apples-ios-problem-contacts-uploading-is-just-the-tip-of-the-iceberg-apps-can-upload-all-your-photos-calendars-or-record-conversations/

Wei, T. T., Marthandan, G., Chong, A. Y. L., Ooi, K. B., & Arumugam, S. (2009). What drives Malaysian m-commerce adoption? An empirical analysis. *Industrial Management & Data Systems, 109*(3), 370–388. doi:10.1108/02635570910939399

Wells, J. (2013). *Complexity and Sustainability*. New York, NY: Routledge.

Wells, J. D., Parboteeah, V., & Valacich, J. S. (2011). Online Impulse Buying: Understanding the Interplay between Consumer Impulsiveness and Website Quality. *Journal of the Association for Information Systems, 12*(1), 32–56.

Wessels, B. (2012). Identification and the practices of identity and privacy in everyday digital communication. *New Media & Society, 14*(8), 1251–1268. doi:10.1177/1461444812450679

Westlund, O. (2013). Mobile news: A review and model of journalism in an age of mobile media. *Digital Journalism, 1*(1), 6–26. doi:10.1080/21670811.2012.740273

Whittaker, Z. (2012). *Twitter uploads contact list data without consent; retains for 18 months*. Retrieved 19 June, 2016, from http://www.zdnet.com/article/twitter-uploads-contact-list-data-without-consent-retains-for-18-months/

Wickens, C. D., Hollands, J. G., Simon, B., & Parasuraman, R. (2013). *Engineering psychology and human performance*. Pearson.

Willging, P. A., & Johnson, S. D. (2004). Factors That Influence Students' Decision to Drop out of Online Courses. *Journal of Asynchronous Learning Networks, 8*(4), 105–118.

Wilska, T. A. (2003). Mobile phone use as part of young peoples consumption styles. *Journal of Consumer Policy, 26*(4), 441–463. doi:10.1023/A:1026331016172

Wolfe, L (2015). *The Principle of Reciprocity and how it applies to business*. Academic Press.

Worcester, R. M., & Burns, T. R. (1975). A statistical examination of the relative precision of verbal scales. *Journal of the Market Research Society, 17*(3), 181–197.

Wright, P. (2000). Supportive documentation for older people. In P. Westendorp, C. H. Jansen, & R. Punselie (Eds.), *Interface design and document design* (pp. 81–100). Amsterdam, The Netherlands: Rodopi.

Wu, J. H., & Wang, S. C. (2005). What drives mobile commerce? An empirical evaluation of the revised technology acceptance model. *Information & Management, 42*(5), 719–729. doi:10.1016/j.im.2004.07.001

Wu, L., Kang, M., & Yang, S. B. (2015). What makes users buy paid smartphone applications? Examining app, personal, and social influences. *Journal of Internet Banking and Commerce, 20*(1), 2–22. doi:10.1007/978-3-531-92534-9_12

Wyse, J. E. (2009). Location-aware query resolution for location-based mobile commerce: Performance evaluation and optimization. In D. Taniar (Ed.), *Mobile computing: Concepts, methodologies, tools, and applications* (pp. 3040–3067). Hershey, PA: IGI Global. doi:10.4018/978-1-60566-054-7.ch229

Xiao, S., & Dong, M. (2015). Hidden semi-Markov model-based reputation management system for online to offline (O2O) e-commerce markets. *Decision Support Systems*, 77, 87–99. doi:10.1016/j.dss.2015.05.013

Xu, H., & Teo, H.-H. (2005). Privacy Considerations in Location-Based Advertising. In C. Sørensen, Y. Yoo, K. Lyytinen & J. I. DeGross (Eds.), *Designing Ubiquitous Information Environments: Socio-Technical Issues and Challenges: IFIP TC8 WG 8.2 International Working Conference, August 1–3, 2005, Cleveland, Ohio, U.S.A.* (pp. 71-90). Boston, MA: Springer US. doi:10.1007/0-387-28918-6_8

Xu, H., Gupta, S., & Pan, S. (2009). *Balancing User Privacy Concerns in the Adoption of Location-Based Services: An Empirical Analysis across Pull-Based and Push-Based Applications.* Paper presented at the iConference (iSociety: Research, Education, and Engagement), University of North Carolina, Chapel Hill, NC. http://hdl.handle.net/2142/15224

Xu, H., Gupta, S., Rosson, M. B., & Carroll, J. M. (2012). *Measuring Mobile Users' Concerns for Information Privacy.* Paper presented at the International Conference on Information Systems.

Xu, H., Rosson, M. B., & Carroll, J. M. (2008). *Mobile User's Privacy Decision Making: Integrating Economic Exchange and Social Justice Perspectives.* Paper presented at the AMCIS.

Xu, D., & Jaggars, S. S. (2014). Performance Gaps Between Online and Face-to-Face Courses: Differences Across Types of Students and Academic Subject Areas. *The Journal of Higher Education*, 85(5), 633–659. doi:10.1353/jhe.2014.0028

Yang, D., Sinha, T., Adamson, D., & Rose, C. P. (2013). *Turn on, tune in, drop out: Anticipating student dropouts in massive open online courses.* Paper presented at the NIPS Data-Driven Education Workshop.

Yang, D., Wen, M., & Rose, C. (2014). Peer influence on attrition in massively open online courses. *Educational Data Mining*. Retrieved from http://www.cs.cmu.edu/~mwen/papers/edm2014-39.pdf

Yang, J. H., & Chang, C. C. (2009). An efficient three-party authenticated key exchange protocol using elliptic curve cryptography for mobile-commerce environments. *Journal of Systems and Software*, 82(9), 1497–1502. doi:10.1016/j.jss.2009.03.075

Yang, K. (2010). Determinants of US consumer mobile shopping services adoption: Implications for designing mobile shopping services. *Journal of Consumer Marketing*, 27(3), 262–270. doi:10.1108/07363761011038338

Yang, K. C. C. (2005). Exploring factors affecting the adoption of mobile commerce in Singapore. *Telematics and Informatics*, 22(3), 257–277. doi:10.1016/j.tele.2004.11.003

Yang, K., & Kim, H. Y. (2012). Mobile shopping motivation: An application of multiple discriminant analysis. *International Journal of Retail & Distribution Management*, 40(10), 778–789. doi:10.1108/09590551211263182

Yang, K., & Lee, H. J. (2010). Gender differences in using mobile data services: Utilitarian and hedonic value approaches. *Journal of Research in Interactive Marketing*, 4(2), 142–156. doi:10.1108/17505931011051678

Yang, T. A., Kim, D. J., & Dhalwani, V. (2008). *Social Networking as a New Trend in e-Marketing. In Research and Practical Issues of Enterprise Information Systems II* (pp. 847–856). Boston: Springer.

Yan, Z., Dong, Y., Niemi, V., & Yu, G. (2013). Exploring trust of mobile applications based on user behaviors: An empirical study. *Journal of Applied Social Psychology*, 43(3), 638–659. doi:10.1111/j.1559-1816.2013.01044.x

Compilation of References

Yeh, Y. S., & Li, Y. M. (2009). Building trust in m-commerce: Contributions from quality and satisfaction. *Online Information Review*, *33*(6), 1066–1086. doi:10.1108/14684520911011016

Yogasara, T. (2014). *Anticipated User Experience in the Early Stages of Product Development*. PhD Thesis.

Yogasara, T., Popovic, V., Kraal, B., & Chamorro-Koc, M. (2011). *General Characteristics of Anticipated User Experience (Aux) with Interactive Products*. Academic Press.

Yogasara, T., Popovic, V., Kraal, B., & Chamorro-Koc, M. (2012). *Anticipating user experience with a desired product*. The AUX Framework.

Yoon, S. (2001). Tangled octopus. *Far Eastern Economic Review*, *164*(6), 40.

Young, H. P. (2003). The Diffusion of Innovations in Social Networks. In *Proc. The Economy as a Complex Evolving System* (Vol. 3, pp. 267–282). Oxford University Press.

Yow, K. C., & Mittal, N. (2008). Mobile commerce multimedia messaging peer. In A. Becker (Ed.), *Electronic commerce: Concepts, methodologies, tools, and applications* (pp. 514–523). Hershey, PA: IGI Global. doi:10.4018/978-1-59904-943-4.ch042

Ysearchblog.com (2005). *Social Commerce via the Shoposphere & Pick Lists*. Retrieved from: Ysearchblog.com

Yuan, E. (2011). News consumption across multiple media platforms: A repertoire approach. *Information Communication and Society*, *14*(7), 998–1016. doi:10.1080/1369118X.2010.549235

Yu, C. S. (2012). Factors affecting individuals to adopt mobile banking: Empirical evidence from the utaut model. *Journal of Electronic Commerce Research*, *13*, 2.

Yu, F. R., Wong, V. W. S., Song, J. H., Leung, V. C. M., & Chan, H. C. B. (2011). Next generation mobility management: An introduction. *Wireless Communications and Mobile Computing*, *11*(4), 446–458. doi:10.1002/wcm.904

Zablah, A. R., Bellenger, D. N., & Johnston, W. J. (2004). An evaluation of divergent perspectives on customer relationship management: Towards a common understanding of an emerging phenomenon. *Industrial Marketing Management*, *33*(6), 475–489. doi:10.1016/j.indmarman.2004.01.006

Zaharias, P., & Koutsabasis, P. (2012). Heuristic evaluation of e-learning courses: A comparative analysis of two e-learning heuristic sets. *Campus-Wide Information Systems*, *29*(1), 45–60. doi:10.1108/10650741211192046

Zaichkowsky, J. L. (1985). Measuring the Involvement Construct. *The Journal of Consumer Research*, *12*(3), 341–352. doi:10.1086/208520

Zaina, L. A. M., Ameida, T. A., & Torres, G. M. (2014). Can the Online Social Networks Be Used as a Learning Tool? A Case Study in Twitter. In L. Uden, J. Sinclair, Y.-H. Tao, & D. Liberona (Eds.), *Learning Technology for Education in Cloud. MOOC and Big Data* (Vol. 446, pp. 114–123). Springer International Publishing. doi:10.1007/978-3-319-10671-7_11

Zaphiris, P., Kurniawan, S., & Ghiawadwala, M. (2006). A systematic approach to the development of research-based web design guidelines for older people. *Universal Access in the Information Society*, *6*(1), 59–75. doi:10.1007/s10209-006-0054-8

Zeithaml, V. A. (1988). Consumer perceptions of price, quality and value: A means-end model and synthesis of evidence. *Journal of Marketing*, *52*(3), 2–22. doi:10.2307/1251446

Zellermayer, O. (1996). *The pain of paying* (Unpublished Ph.D. Dissertation). Carnegie Mellon University, Pittsburgh, PA.

Zephoria. (n.d.). The top 20 valuable Facebook statistics. *Zephoria*. Retrieved from: https://zephoria.com/top-15-valuable-Facebook-statistics/

Zhang, B., & Xu, H. (2016). *Privacy Nudges for Mobile Applications: Effects on the Creepiness Emotion and Privacy Attitudes*. Paper presented at the 19th ACM Conference on Computer-Supported Cooperative Work & Social Computing, San Francisco, CA. doi:10.1145/2818048.2820073

Zhang, D., & Adipat, B. (2005). Challenges, methodologies, and issues in the usability testing of mobile application. *International Journal of Human-Computer Interaction*, *18*(3), 293–308. doi:10.1207/s15327590ijhc1803_3

Zhang, J., & Mao, E. (2008). Understanding the acceptance of mobile SMS advertising among young Chinese consumers. *Psychology and Marketing*, *25*(8), 787–805. doi:10.1002/mar.20239

Zhang, L., Zhu, J., & Liu, Q. (2012). A meta-analysis of mobile commerce adoption and the moderating effect of culture. *Computers in Human Behavior*, *28*(5), 1902–1911. doi:10.1016/j.chb.2012.05.008

Zhang, X., Prybutok, V. R., & Koh, C. E. (2006). The role of impulsiveness in a TAM based online purchasing behavior model. *Information Resources Management Journal*, *19*(2), 54–68. doi:10.4018/irmj.2006040104

Zhao, Y., Ye, J., & Henderson, T. (2016). *The Effect of Privacy Concerns on Privacy Recommenders*. Paper presented at the 21st International Conference on Intelligent User Interfaces, Sonoma, CA. doi:10.1145/2856767.2856771

Zheng, L., Junfeng, Y., Wei, C., & Ronghuai, H. (2014). Emerging approaches for supporting easy, engaged and effective collaborative learning. *Journal of King Saud University - Computer and Information Sciences*, *26*, 11-16.

Zhou, J., Rau, P. L. P., & Salvendy, G. (2013). Older adults use of smartphones: An investigation of the factors influencing the acceptance of new functions. *Behaviour & Information Technology*, *33*(6), 552–560. doi:10.1080/0144929X.2013.780637

Zhou, L., Zhang, P., & Zimmerman, H. D. (2013). Social Commerce Research: An integrated view. *Electronic Commerce Research and Applications*, *12*(2), 61–68. doi:10.1016/j.elerap.2013.02.003

Zhu, W., Owen, C. B., Li, H., & Lee, J. (2004). Personalized in-store e-commerce with the PromoPad: An augmented reality shopping assistant. *Electronic Journal for E-commerce Tools and Applications*, *1*(3), 1–19.

Ziefle, M. (2010). Information presentation in small screen devices: The trade-off between visual density and menu foresight. *Applied Ergonomics*, *41*(6), 719–730. doi:10.1016/j.apergo.2010.03.001 PMID:20382372

Zimmerman, L. (2011). Critical Importance of Social Interaction in Online Courses. *ETC Journal*. Retrieved from http://etcjournal.com/2011/01/02/7050/

Zou, X., & Huang, K. W. (2015). Leveraging location-based services for couponing and infomediation. *Decision Support Systems*, *78*, 93–103. doi:10.1016/j.dss.2015.05.007

About the Contributors

Jean-Éric Pelet holds a PhD in Marketing, an MBA in Information Systems and a BA (Hns) in Advertising. As an assistant professor in management, he works on problems concerning consumer behaviour when using a website or other information system (e-learning, knowledge management, e-commerce platforms), and how the interface can change that behavior. His main interest lies in the variables that enhance navigation in order to help people to be more efficient with these systems. He works as a visiting professor both in France and abroad (England, Switzerland, Thailand, Finland...) teaching e-marketing, ergonomics, usability, and consumer behaviour at Design Schools, Business Schools, and Universities. Dr. Pelet has also actively participated in a number of European Community and National research projects. His current research interests focus on social networks, interface design, and usability. His work has been published in international journals and conferences such as EJIS, SIM, INFMAN, ICIS, EMAC, ANZMAC, AIM, AWBR, VDQS, KMO, LTEC.

* * *

Sharidatul Akma Abu Seman was born on 23 June 1983. She obtained her first degree in Universiti Utara Malaysia in Multimedia and Master's degree from UiTM. She later joined Faculty of Business Management, UiTM Puncak Alam since 2009. Currently, she is a Ph.D. student at Universiti Sains Malaysia. Her research focus is on ICT, healthcare informatics, mHealth apps and mobile computing. She has published in several international conferences and journals.

Mohammed Alfadil is an English Teacher with strong classroom management skills demonstrated through 11 years classroom experience - Ministry of Education - Saudi Arabia a Ph.D. Candidate at the Department of Educational Technology, University of Northern Colorado, Greeley, CO Emphasis Area: Teacher Education and Game Based-Learning Alhababi, H., Alfadil, M., Alzamanan, M. & Williams, M.K. (2015). Students' Perception on the Use of Social Media on Their Academic Learning. In Proceedings of E-Learn: World Conference on E-Learning in Corporate, Government, Healthcare, and Higher Education 2015 (pp. 1211-1217). Chesapeake, VA: Association for the Advancement of Computing in Education (AACE). Under Publishing: Connecting to the Digital Age: Using Technology to Enhance Student Learning.

Nebal Anaim is a professional business adviser with back ground in strategy, organizational culture, standardized operating policies and procedures (SOPs), Quality (e.g. ISO), business software solutions consulting (e.g. systems like ERP, CRM, HRMS, BI etc.), user experience (UX) and Human Resource

Recruitment (HRR). He is currently a doctorate study candidate after his successful completion of several qualifications including: MBA IT from Coventry University, UK, BSc. in Economics from Ottawa University, Certificate in Human Resource Management (HRM) from McGill University, Canada He is known for empirical alignment initiatives between company departments, theory (Academia) and practice (real world) and other scenarios. Moreover, he has extensive experience in management analysis, requirement gathering, workflow documentation, business automation, shifting (change/transition/transformation management), portfolio management, etc. He has been a renowned guest speaker in several events on the above topics in addition to voluntary community subjects such as Pedagogy/Andragogy, parenting & children upbringing, religion, ethics, conflict resolution, etc. He coined the "10Ps Framework", Nebal's Ten Pillars of Success business strategy model for the successful implementation of Management Information Systems (e.g. ERP) projects from the first go.

Artemis Avgerou holds a BSc degree from the Department of Business Administration, University of Patras and she is an MSc student at Imperial College, Business School, at the MSc Strategic Marketing programme. Her research interests include Entrepreneurship, eCommerce and eMarketing, Innovation and Business Modelling in the Digital Economy. Avgerou is especially interested in studying how users' social network interactions influence their attitude towards adopting new products or services and, thus, how social networking can be employed in successful e-marketing strategies.

Norchene Ben Dahmene Mouelhi holds a Ph.D. in Marketing from IAE de Caen (France) and ISG Tunis (Tunisia). Dr. Ben Dahmane Mouelhi is currently full-time faculty at IHEC Carthage. She was also the Co-Director of the Master "Grandes Ecoles" (Co- Diploma with Em Normandie) at IHEC Carthage. Now she is the Co-Director of the Master Marketing. Previous positions include full-time Assistant Professor at ISG Tunis and University of Nabeul and part ime Professor at ISC Paris. She wrote, in 2014, a book "Le marketing du point de vente" about retailing for students, researchers and professional with Kaouther Ghozzi and Ilhem Tekaya. Her work has been published in international journals such as IJRM, DM, RFM, RTM, REMAREM, Revue des sciences de gestion, and conferences (ACR, AMS, AFM, ATM, IBIMA, Marketing trends). She is member of scientific and organization committees of several international conferences and a reviewer in Internationals Journals. Her research interests cover many areas including digital marketing, Social Media, experiential and sensorial marketing, store atmospherics, retailing and consumer behavior.

Saayan Chattopadhyay is an Assistant Professor and head in the Department of Journalism and Mass Communication at Baruipur College, affiliated to Calcutta University, Kolkata. After a stint as a journalist, he is currently engaged in research involving neoliberal techno-culture in India. He has published articles in South Asia Research, Studies in South Asian Film and Media, Journalism Practice, Sarai Reader. He has also contributed chapters in books on media and communication published from Palgrave Macmillan, Springer, Routledge, Sussex Academic Press, IGI Global, among others. His research interests include journalism studies, masculinity studies, and techno-culture in developing nations.

Emmanuel Eilu is a final year Information Technology PhD Student at the School of Computing and Informatics Technology –Makerere University. His areas of research and interest are in HCI (Usability and User Experience), Mobile User Experience, ICT for Development (Adoption of E-Government Systems;

About the Contributors

E-governance; E-voting; Biometric Voter Registration Systems; Electronic Civil Registers). His current PhD Topic is entitled "Bridging the User Experience Gap in Mobile Phone Voting in Developing Countries". He holds a master degree in Information Technology, specializing in E-government. His Master Dissertation was entitled "A systematic Approach to Designing and Implementing Electronic- Government Systems in the Developing World". He has published over 10 research papers in ACM, Springer and IEEE. Since 2007, Ellu has been teaching both post graduate and undergraduate programmes at the School of Computing and IT – Makerere University. He also lectures at the International University of East Africa-Uganda. His major teaching area includes; Usability and User experience, Egovernment, E-governance, SAD, Information Systems Security and Data Communication Networks.

Sana El Mouldi is a second year PhD student IAE Bordeaux / ISG Tunis Member of IRGO Doctorate School.

Saïd Ettis is a lecturer at the Gulf College, Muscat, Sultanate of Oman. He holds the Ph.D. from Higher Institute of Economics and Management – IAE (IEMN-IAE), University of Nantes, France. He teaches Marketing, E-marketing, Market Research, and Management. He published a number of papers in international journals such as Information & Management, International Journal of Higher Education Management, and La Revue Gestion et Organisation. He wrote also some books chapters and presented papers in international conferences. His research interests are related to e-marketing and consumer behavior.

Stéphane Fauvy holds a Ph.D. in Management Science. He is Associate Professor in ESSCA School of Management and he is in charge of Human Resource Management teachings. His research topics are related to the study of HRM processes in different organizational contexts (cluster, SME, MNC, NPO) and their contribution to organizational change situations. He is the author of numerous conferences and scientific papers in national and international level. His current research interests deal with the impact of digital technologies on human resource management.

Marjan Heričko is a professor of Informatics at the Institute of Informatics at the University of Maribor, Faculty of Electrical Engineering and Computer Science where he is the Head of the Institute of Informatics. He was born in 1966 in Maribor, where he graduated in 1989. He received his MSc (1993) and PhD (1998) in Computer Science and Informatics from the University of Maribor. His main research interests include all aspects of IS development with emphasis on software and service engineering, software process improvement, information security and data privacy. He has published more than 80 original scientific papers in journals. He participated/coordinated many basic, applied and industrial research projects. He was awarded for his achievements in knowledge transfer to business/industrial environment by Slovenian Society Informatika and Faculty of Electrical Engineering and Computer Science.

Marko Hölbl is an assistant professor in Computer Science at the Institute of Informatics at the University of Maribor, Faculty of Electrical Engineering and Computer Science. He was born in 1980 in Maribor, where he graduated in 2004. He has earned his PhD in Computer Science in 2009. His research interests include authentication and key agreement, securing data and communication in specific domains (e.g. Iternet of Things (IoT)) and cryptography. He is also working on user aspects of security, particularly privacy issues of social networks, cybersecurity education and protections of data. He was recently involved in a TEMPUS project on the topic of cybersecurity education. He is actively participating in the CEPIS LSI (Council of European Professional Informatics Societies – Legal and Security Issues Special Interest Network) and in the NIS WG 3 (Network and Information Security – Working Group 3 on secure ICT research and innovation).

Ardian Hyseni is a PhD student at Faculty of Economics University of Ljubljana, Department of Information Management. His is currently working in his research topic focused on Business Intelligence and Self Service Business Intelligence. He holds Master in Business Informatics and Bachelor of Computer Engineering at South East European University in Tetovo, Macedonia. Ardian works in private company and he owns his private company. He has a publication and currently working in his PhD research topic.

Despina A. Karayanni (Ph.D. in Marketing, M.B.A., B.Sc. in Business Management) is Associate Professor of Marketing, at the Department of Business Administration, in the University of Patras, Greece. Her research interests include Marketing Strategy, Pharmaceutical Marketing, Relationship and Networking Marketing, Business-to-Business Marketing, Salesforce Management, marketing research, Electronic Commerce and Social Media Marketing and Hi-Tech Marketing. She has published in a number of referred scientific journals and international referred marketing conference proceedings and her work is cited in a few scientific publications. She has also participated in a number of European Union research projects.

Kijpokin Kasemsap received his BEng degree in Mechanical Engineering from King Mongkut's University of Technology, Thonburi, his MBA degree from Ramkhamhaeng University, and his DBA degree in Human Resource Management from Suan Sunandha Rajabhat University. Dr. Kasemsap is a Special Lecturer in the Faculty of Management Sciences, Suan Sunandha Rajabhat University, based in Bangkok, Thailand. Dr. Kasemsap is a Member of the International Economics Development and Research Center (IEDRC), the International Foundation for Research and Development (IFRD), and the International Innovative Scientific and Research Organization (IISRO). Dr. Kasemsap also serves on the International Advisory Committee (IAC) for the International Association of Academicians and Researchers (INAAR). Dr. Kasemsap is the sole author of over 250 peer-reviewed international publications and book chapters on business, education, and information technology. Dr. Kasemsap is included in the TOP 100 Professionals–2016 and in the 10th edition of 2000 Outstanding Intellectuals of the 21st Century by the International Biographical Centre, Cambridge, England.

Jashim Khan, a Lecturer in Marketing with University of Surrey, UK, holds a doctorate in business marketing from AUT University, New Zealand. His current research looks at experimental consumer behaviour in new digital environment with particular interest in how mobile devices influences consumption and informs online-mobile education. He is also an honorary lecturer in marketing with University of Liverpool/ Laureate Education and holds adjunct lecturer role with Dongbei University of Finance and Economics, China. Previously he was a Lecturer in Marketing at AUT University and a visiting professor of marketing with Ramkhamhaeng University, Thailand. Dr Khan's research interest include the use of contactless smart cards, mobile Internet, location based advertising, and social media that are transcending consumption landscape and redefining how we shop for goods and services. His work appears in Journal of Business Research, Journal of Economic Psychology, Association for Consumer Research; Journal of Business and Economics, and International Business and Economic Review.

About the Contributors

Raghvendra Kumar is working as Assistant Professor in Computer Science and Engineering Department at L.N.C.T Group of College Jabalpur, M.P. India. He received B. Tech. in Computer Science and Engineering from SRM University Chennai (Tamil Nadu), India, in 2011, M. Tech. in Computer Science and Engineering from KIIT University, Bhubaneswar, (Odisha) India in 2013, and pursuing Ph.D. in Computer Science and Engineering from Jodhpur National University, Jodhpur (Rajasthan), India. He has published many research papers in international journal including IEEE and ACM. He attends many national and international conferences and also He Received best paper award in IEEE 2013 for his research work in the field of distributed database in Tamil Nadu. His researches areas are Computer Networks, Data Mining, cloud computing and Secure Multiparty Computations, Theory of Computer Science and Design of Algorithms. He authored many computer science books in field of Data Mining, Robotics, Graph Theory, Turing Machine, Cryptography, Security Solutions in cloud computing and Privacy Preservation.

José A. Márquez-Domínguez is co-founder of the Laboratorio de Ciencias Computacionales e Inteligencia Artificial, director of the Computer Science Department, and collaborator of the Biotecnología Agroalimentaria research group; all them at the Universidad de la Cañada. Márquez-Domínguez holds a bachelor's degree in computer science from the Benemérita Universidad Autónoma de Puebla. He earned a MS in computer science from the Benemérita Universidad Autónoma de Puebla. Collaborator of projects related to digital image processing, 3D graphics, databases, web design, and artificial intelligence, among others. Experience in information recovery, pattern recognition, bioinformatics, and system programming. Currently he is involved in the project "motor rehabilitation for patients with Parkinson based on virtual reality".

Lili Nemec Zlatolas was a junior researcher at the Institute of Informatics at the University of Maribor, Faculty of Electrical Engineering and Computer Science from 2010 until 2015. She is a teaching assistant in Computer Science at the Institute of Informatics at the University of Maribor, Faculty of Electrical Engineering and Computer Science from 2015. She was born in 1984 in Slovenia, where she graduated in 2008. She has earned her PhD in Computer Science in 2015. Her research interests include user aspects on security and privacy.

Priyanka Pandey is working as Assistant Professor in Computer Science and Engineering Department at L.N.C.T Group of College Jabalpur, M.P. India. She received B.E. in Information Technology from TIE Tech (RGPV University), Jabalpur, MP, India, in 2013, M. Tech. in Computer Science and Engineering from TIE Tech (RGPV University), Jabalpur, MP, India. She published many research papers in international journal and conferences including IEEE. She attends many national and international conferences, her researches areas are Computer Networks, Data Mining, wireless network and Design of Algorithms.

Prasant Kumar Pattnaik, Ph.D. (Computer Science), Fellow IETE, Senior Member IEEE, is Professor at the School of Computer Engineering, KIIT University, Bhubaneswar. He has more than a decade of teaching research experience. Dr. Pattnaik has published numbers of Research papers in peer reviewed international journals and conferences. His researches areas are Computer Networks, Data

Mining, cloud computing, Mobile Computing. He authored many computer science books in field of Data Mining, Robotics, Graph Theory, Turing Machine, Cryptography, Security Solutions in Cloud Computing, Mobile Computing and Privacy Preservation.

Marlene A. Pratt, Ph.D, is a lecturer within the Griffith Business School at Griffith University on the Gold Coast, Australia. Marlene completed her PhD in 2011 in consumer behaviour in the wine industry. Marlene's research interests include consumer behavior, experiential learning, and wine tourism. Her current research project is a government funded project titled 'Enhancing Student Learning Outcomes with Simulation-based Pedagogies'. She has published conference papers, journal articles, and industry reports. Marlene is also a reviewer for several journals and publishers, such as Journal of Retailing and Consumer Services, Journal of Hospitality and Tourism Research, Journal of Wine Business Research, Cambridge University Press, and Routledge.

Gary Rivers is an Associate Dean for the University of Surrey based in China at the Surrey International Institute. With over 20 involvement with Transnational Education (TNE) programmes he has gained extensive knowledge in the management of TNE. Previous roles have included Deputy Head of International Programmes and Director of China projects with Curtin University of Technology, Australia and Head of Department at the Australian College of Kuwait. Dr Rivers has taught in the areas of Human Resource Management (HRM), Organisational Behaviour (OB) and Industrial Relations (IR) in several locations including Australia, Singapore, Malaysia and and China - Hong Kong and mainland. He has a PhD from the University of Western Australia, a Go8 member, and is a member of the Academy of Management (USA) and a fellow of the Higher Education Association (UK). Dr Rivers has research interests in HRM, OB, buyer behaviour and TNE.

Beatriz A. Sabino-Moxo is co-founder of the Laboratorio de Ciencias Computacionales e Inteligencia Artificial, director of the Information Technology Academy, and collaborator of the Biotecnología Agroalimentaria research group; all them at the Universidad de la Cañada. Sabino-Moxo holds a bachelor's degree in computer science from the Benemérita Universidad Autónoma de Puebla. He earned a MS in computer science from the Benemérita Universidad Autónoma de Puebla. Collaborator of projects related to virtual rehabilitation, image processing, and databases, among others. Experience in information recovery, pattern recognition, bioinformatics, and system programming. Currently he is involved in the project "motor rehabilitation for patients with Parkinson based on virtual reality".

Miguel A. Sánchez-Acevedo is co-founder and director of the Laboratorio de Ciencias Computacionales e Inteligencia Artificial, director of the Roboclub, and collaborator of the Biotecnología Agroalimentaria research group; all them at the Universidad de la Cañada. Sánchez-Acevedo holds a bachelor's degree in computer systems engineering from the Instituto Tecnológico de Tehuacán. He earned a MS in electrical engineering and a truncated PhD in computer science from the CINVESTAV Unidad Guadalajara. Collaborator of projects related to radar simulation, virtual worlds, self-organization of mobile robots. He is particularly interested in swarm intelligence, 3d modelling, sensor networks, augmented reality, and distributed computing.

About the Contributors

Flávio Santos is an Industrial Designer from ESDI/UERJ (1992), Master's at Production Engineering from COPPE/UFSC (1998) and Doctorate at Production Engineering from UFSC (2005). Has experience in Industrial Design, acting on the following subjects: human factors and ergonomics, design methods and design management. Associate professor at UDESC, Design Department, and Design Consultant for private and government organizations.

Mohammad Shalan is a Professional Engineer, with international working experience since 1995 in telecommunications, cloud computing, SMAC contracting, enterprise architecture, project management, risk analysis, audit and governance. He graduated with a Master degree in Telecommunication Engineering, 2005 and a B.Sc. in Electrical Engineering in 1995 from the University of Jordan. He is an author of several articles and published chapters in edited books. He is a holder of several active memberships in many professional organizations with unique certifications including: Project Management Professional, PMP® and Risk Management Professional, PMI-RMP® from Project Management Institute (PMI); Certified in the Governance of Enterprise Information Technology, CGEIT®, Certified Information System Auditor, CISA® and Certified in Risk and Information Systems Control, CRISC® from Information Systems Audit and Control Association (ISACA); Certification in Risk Management Assurance, CRMA® from The Institute of Internal Auditors (The IIA); Information Technology Infrastructure Library, ITIL foundations certification, with high level experience; Jordan Engineers Association (JEA) and the Institute of Electrical and Electronic Engineers (IEEE).

Christofer Ramos holds a bachelor's degree in Graphic Design from ESAMC (2011) and a master's degree in Design from UDESC (2017), and is also a specialist in Higher Education Teaching. He has contributed in studies and projects comprising communicational interfaces, user experience, usability methods, usability heuristic evaluation, m-learning systems, and mobile platforms.

Yannis Stamatiou graduated from the Computer Engineering and Informatics Department, University of Patras, Greece, in 1990. He now serves as Associate Professor at the Department of Business Administration of the University of Patras, Greece and Consultant on Cryptography and Security for the Security Sector of the Computer Technology Institute & Press ("Diophantus") in Patras, Greece. His interests lie in cryptography, modelling of computer viruses/worms in computer networks, cryptanalysis and ICT security with a focus in eVoting and eGovernement related security protocols and systems. He has extensive experience in theoretical and applied computer science with a focus on cryptography and ICT security.

Basma Taieb holds a PhD in Marketing, from Graduate School of Management (IAE Aix-en-Provence, France). She is currently a teacher at the University of Cergy Pontoise (France) and Business Schools. Her current research focuses on design of website, consumer behavior and consumer culture. Her work has been published in international journals and conferences as JBR, M&A, RBIRS and EMAC.

Ramayah Thurasamy is currently a Professor of Technology Management at the School of Management, Universiti Sains Malaysia, Visiting Professor King Saud University, Kingdom of Saudi Arabia and Adjunct Professor at Multimedia University and Universiti Tenaga Nasional, Malaysia. His publications have appeared in Information & Management, Journal of Environmental Management, Technovation, Journal of Business Ethics, Journal of Business Economics and Management, Computers in Human

Behavior, Resources, Conservation and Recycling, International Journal of Information Management, Evaluation Review, Information Research, Asian Journal of Technology Innovation, Social Indicators Research, Quantity & Quality, Service Business, Knowledge Management Research & Practice, Journal of Medical System, International Journal of Production Economics and Telematics and Informatics among others. He also serves on the editorial boards and program committee of several international journals and conferences of repute. His full profile can be accessed from http://www.ramayah.com.

Monique Vandresen, journalist and university professor, has been working since 1988 with Cultural Journalism, and edited the newspaper culture book O Estado and collaborated with publications such as Revista Nova Cosmopolitan, Revista Veja, Revista Empreendedor, Jornal O Catarina, Jornal da Indústria e Comércio, Jornal AN Capital. Worked on TV Cultura and produced videos and roadmaps for UDESC, UFSC and Ministry of Education. PhD in Communication Sciences from University of São Paulo (2005) and develops research in communication, with emphasis on the role of women's magazine in the construction of concepts of fashion and elegance. Associate professor at Universidade Federal de Santa Catarina, held post doctoral stage at University of California-Riverside. Has experience in Fashion, Art and Design, acting on the following topics: Transmedia, New Media, Fashion, Behavior, History, Communication and consumption. Graduated in Social Communication from Universidade Federal de Santa Catarina (1990), with a Masters in Development from Institute Of Social Studies (1993) financed by Dutch Government.

Tatjana Welzer is the head of the Database Technologies Laboratory of Institute of Informatics. She was born in 1961 in Maribor, where she graduated in 1984. She has earned her PhD in Computer Science in 1995. Dr. Welzer is fluent in English and German and has a fairly solid knowledge of Serbian, Spanish and French. Dr. Tatjana Welzer is a full Professor at University of Maribor, Faculty of Electrical Engineering and Computer Science. Her research work covers many research areas, ranging from database technology to cross-cultural communication and problems in media communications. She holds a professorship for the following classes: Databases, Data Modelling and Databases, as well as Information Security classes, all for Computer Science students and International and Cross-cultural Communication, Radio and Radio Programing, Research Methods in Media Communication, Databases in Media, all at Media Communication study programs.

Na Zuo is a business and technology consulting professional with digital economy programme management experience. 19 years of business experience within the retail, fashion, construction, alcohol beverage, financial services, and digital technology. Specialising in sales & marketing, category management, supply chain management, and procurement across various industries in New Zealand, Australia and China.

Index

A

Ad Hoc Network 163, 173
age 8, 30, 34-35, 37, 60, 63-65, 139, 150, 157, 198, 205, 221, 231, 233, 243-244, 255, 259, 266, 270, 276, 289, 294, 297, 314
Agile 115-117, 119-120, 123-127, 129-130, 133-134
Agile Methodology 120, 133
Analytics 2, 19, 115, 117, 134, 226, 236, 243, 315
Anticipated User Experience 175, 177, 190, 196
AR Card 173
Augmented Reality 2, 153-168, 229
Automotive Industry 153-154, 156, 158-159, 168

C

Cash 248-259, 262
Client Enterprise (CE) 115, 133-134
Cloud 106-113, 115, 117, 120-121, 134, 221, 224, 226, 229, 235, 250
Cognitive Psychology 175-176, 178-180, 189-190, 196
Collectivism 251
Commerce 77, 153-154, 156, 159-160, 168, 197-204, 218, 221, 248-250, 258, 262, 264-267, 272, 277, 288, 314, 322-326, 328-331, 344
Constructivist Learning 2, 18-19, 161
Credentials 230-236, 238-239, 244, 334-335
Credit Card 79, 250, 252, 254, 256, 259, 262, 343
CRM 309-311, 313-319
Crowdsourcing 219-221, 231, 237, 241, 244-245

D

Debit Card 250, 252-254, 259, 262
design 1, 3-4, 7-8, 15, 19, 27, 30, 39-42, 64, 85, 106-107, 118, 124, 127, 130, 148-149, 156, 159, 161, 168, 176-177, 183-184, 188, 198-199, 201, 203-204, 218, 231, 243, 256, 273, 277, 279, 288-290, 292-298, 311, 331-334, 338

Developing Country 56-60, 63-65

E

E-Commerce 159, 197-198, 264-266, 270, 273-275, 289, 291, 298, 325, 329, 331
Education 1-3, 6, 8-9, 18, 20, 27-29, 31, 33-34, 39, 87, 136, 153-154, 156, 161-162, 164-165, 168, 197, 221, 244, 255, 270, 327-328
Educational Learning 31, 34
eIdentity 233
E-Learning 1-2, 7-9, 15, 18-19, 21, 28-29, 38-39, 41-42, 45, 51-52, 298
electronic 2, 8, 30, 86, 155, 197, 200, 204-205, 218, 228, 231, 233, 239, 248, 250-251, 257, 262, 265, 288, 297, 309, 312-314, 325
Electronic Commerce 197, 218, 248, 262, 314
Emotional Association with Payment Mode 262
energy consumption 106-107
E-Textile 155, 173
Experience and Engagement 37

F

Facebook 9, 27-31, 34, 60-61, 98, 135-137, 139, 143, 146-149, 198, 322-325, 327-328, 331-332, 334-335, 337-339
Facebook Commerce 322
France 137, 251-254, 257, 259, 289, 294, 297

H

Haptic Device 173
Heuristic Evaluation 38, 40-41, 45-47, 51
Hofstede Cultural Dimension 262

I

IBM 74, 188, 231

India 56-61, 63-65
Individualism 251-253
Indulgence 251-253, 262
Infrastructure Network 163, 174
Innovation 21, 117-119, 129, 134, 199, 242-243, 267, 269
Internet 1-2, 7, 9, 15-16, 19, 27-29, 35, 37, 59, 61, 63, 72, 86, 98, 107, 134-139, 141, 146-150, 153, 159, 197-200, 204-205, 218, 220-222, 224, 229, 236, 242, 244-245, 248, 252-254, 265, 274, 289-290, 309, 313-316, 322, 324-325, 331
Irritating Factors 135-136, 146, 149-150

K

Knowledge Management 2, 14-15, 18-19, 21

L

Literature Review 34-35, 70, 85, 89-91, 98-100, 178, 198, 264, 267, 290, 325
Long-termism 251-253, 262

M

Malaysia Healthcare 72, 74
Masculinity 251, 253
M-Commerce 39, 197-205, 248-249, 258-259, 262, 264-279, 288-293, 297-298
M-Consumer Behavior 278
Medicine 153-154, 164-165, 168
Mental Account 262
mHealth Application 70-71, 73, 78
Microsoft 155, 231
Middle Circle Contractor (MCC) 134
M-Learning 7-8, 18, 21, 38-42, 45-46, 48, 51-52
Mobile Applications 8, 39, 42, 45, 47-48, 51-52, 56-62, 64-65, 69-73, 77, 84-92, 94, 96-100, 108, 111-112, 136, 198-199
mobile cloud networking 106-107
Mobile Commerce 77, 197, 199-203, 218, 248-250, 262, 264-267, 272, 288
Mobile Computing 70, 107-108, 153, 176, 184, 186, 203, 317
Mobile Computing Products 176, 184
Mobile Devices 8, 10, 16, 38-39, 41, 51, 56-57, 59, 69, 73, 84, 86-89, 96, 98, 106-109, 111, 136-137, 146, 149, 153-154, 159, 162-164, 173, 176, 181-182, 186-187, 190, 198-201, 203-204, 219-221, 226, 228, 236-237, 242, 248-250, 262, 264-267, 269, 271-274, 289, 293, 314-315

Mobile Learning 7-8, 39
Mobile Payment 199, 203, 218, 248, 250-252, 254-257, 262, 266, 270
Mobile Telephony 227, 250
Mobile-Payment (M-Payment) 251-255, 257, 259, 262, 267-268, 270-271
MOOC 1-5, 7-8, 10, 15-16, 18, 20-21, 46
M-Payment 251-255, 257, 259, 262, 267-268, 270-271

N

New Zealand 249-255, 257-259
News Apps 56-57, 59-63, 65

O

Opportunity Contract (OC) 115-116, 122, 125, 134
Outsourcing 116-117, 121-122, 134

P

Payment 117, 123, 128, 186, 198-199, 203, 218, 248, 250-259, 262, 266, 270, 274, 332, 337, 340-343
Payment Mode 198, 248, 250-259, 262
Payment mode choice 257
Privacy 74, 79, 84-92, 94, 96-100, 128-129, 147, 156, 168, 197, 199, 201-202, 205, 219, 221, 224, 226, 228-231, 237-238, 241, 244-246, 267, 277, 290, 332
Privacy-ABCs 231-233, 235, 238-240, 245
Product 15-16, 40-41, 78-79, 119-120, 125-127, 154, 160, 175-178, 181-190, 196, 198, 200-201, 203-204, 233, 239, 242-245, 253, 262, 266, 275, 289, 297, 310-311, 322-323, 337, 342
purchase intention 290, 293, 295, 297

R

revisit intention 293, 295
Risk 19, 31, 96-97, 115-119, 121-122, 124-125, 127, 134, 156, 159, 198, 201, 231, 252, 256, 266, 272, 276
Risk Management 120, 134

S

Security 39, 74, 79, 84-92, 94, 96-100, 107, 109-110, 113, 116, 128-129, 147-148, 154, 156, 159, 168, 197, 199, 201-202, 204-205, 221, 226, 228-231, 235, 244, 250, 255, 266

Index

Shopping 74, 139, 159, 168, 197-198, 201-203, 205, 240, 248, 255, 259, 262, 266-268, 274-276, 289-290, 293, 298, 323-324, 328, 330, 332, 338, 344
SMAC 115-130, 133-134
SMAC Service Provider (SSP) 134
Smart Glass 174
Smartphone 8, 15, 46, 58-60, 69-71, 74, 78, 85, 88, 136, 143, 146, 148-149, 199, 218, 249, 275, 288-289, 293, 306, 314-315
SOCIAL Commerce 198, 322, 324-326, 328-331, 344
Social Commerce Application 328, 330
Social Media 1-2, 9-10, 15-16, 18-19, 21, 27-31, 34-35, 60-61, 63, 71, 97, 136-137, 198, 205, 266, 289, 322-326, 328-331, 344
Social Network 27, 29, 31, 33, 137, 220
Strategic Planning 134

T

Technology Acceptance Model 197, 202, 218, 267, 273
Technology Organization (TO) 115, 134
Traditional Contract 116, 119

Transaction 197, 199-202, 205, 218, 222, 250-251, 253, 262, 265, 275, 278, 310
Trust 61, 85-86, 97, 99, 117-118, 121, 124, 186, 197, 199-202, 204-205, 228, 231, 237, 242, 245, 252-253, 276-277, 279, 289, 297-298, 319

U

Uncertainty Avoidance 251-253
User Experience 45, 47, 51, 106, 124, 149, 153, 156, 168, 175-178, 187, 190, 196

V

Video Games 154, 163, 168

W

Wearable Computing 174
Wearable Devices 163, 168
Web Design 15, 297
web services 136

Purchase Print + Free E-Book or E-Book Only

Purchase a print book through the IGI Global Online Bookstore and receive the e-book for free or purchase the e-book only! Shipping fees apply.

www.igi-global.com

Recommended Reference Books

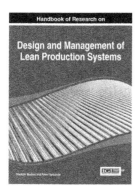

Handbook of Research on Design and Management of Lean Production Systems

ISBN: 978-1-4666-5039-8
© 2014; 487 pp.
List Price: $260

Handbook of Research on Business Ethics and Corporate Responsibilities

ISBN: 978-1-4666-7476-9
© 2015; 508 pp.
List Price: $212

Integrating Social Media into Business Practice, Applications, Management, and Models

ISBN: 978-1-4666-6182-0
© 2014; 325 pp.
List Price: $180

Handbook of Research on Strategic Performance Management and Measurement Using Data Envelopment Analysis

ISBN: 978-1-4666-4474-8
© 2014; 735 pp.
List Price: $276

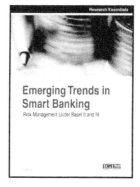

Emerging Trends in Smart Banking: Risk Management Under Basel II and III

ISBN: 978-1-4666-5950-6
© 2014; 290 pp.
List Price: $156

Computer-Mediated Marketing Strategies: Social Media and Online Brand Communities

ISBN: 978-1-4666-6595-8
© 2015; 406 pp.
List Price: $156

*IGI Global now offers the exclusive opportunity to receive a free e-book with the purchase of the publication in print, or purchase any e-book publication only. You choose the format that best suits your needs. This offer is only valid on purchases made directly through IGI Global's Online Bookstore and not intended for use by book distributors or wholesalers. Shipping fees will be applied for hardcover purchases during checkout if this option is selected.

Should a new edition of any given publication become available, access will not be extended on the new edition and will only be available for the purchased publication. If a new edition becomes available, you will not lose access, but you would no longer receive new content for that publication (i.e. updates). The free e-book is only available to single institutions that purchase printed publications through IGI Global. Sharing the free e-book is prohibited and will result in the termination of e-access.

Publishing Information Science and Technology Research Since 1988

www.igi-global.com | Sign up at www.igi-global.com/newsletters | facebook.com/igiglobal | twitter.com/igiglobal

Stay Current on the Latest Emerging Research Developments

Become an IGI Global Reviewer for Authored Book Projects

The overall success of an authored book project is dependent on quality and timely reviews.

In this competitive age of scholarly publishing, constructive and timely feedback significantly decreases the turnaround time of manuscripts from submission to acceptance, allowing the publication and discovery of progressive research at a much more expeditious rate. Several IGI Global authored book projects are currently seeking highly qualified experts in the field to fill vacancies on their respective editorial review boards:

Applications may be sent to:
development@igi-global.com

Applicants must have a doctorate (or an equivalent degree) as well as publishing and reviewing experience. Reviewers are asked to write reviews in a timely, collegial, and constructive manner. All reviewers will begin their role on an ad-hoc basis for a period of one year, and upon successful completion of this term can be considered for full editorial review board status, with the potential for a subsequent promotion to Associate Editor.

If you have a colleague that may be interested in this opportunity,
we encourage you to share this information with them.

InfoSci®-Books
A Database for Progressive Information Science and Technology Research

Maximize Your Library's Book Collection!

Invest in IGI Global's InfoSci®-Books database and gain access to hundreds of reference books at a fraction of their individual list price.

The InfoSci®-Books database offers unlimited simultaneous users the ability to precisely return search results through more than 68,000 full-text chapters from nearly 3,000 reference books in the following academic research areas:

Business & Management Information Science & Technology • Computer Science & Information Technology
Educational Science & Technology • Engineering Science & Technology • Environmental Science & Technology
Government Science & Technology • Library Information Science & Technology • Media & Communication Science & Technology
Medical, Healthcare & Life Science & Technology • Security & Forensic Science & Technology • Social Sciences & Online Behavior

Peer-Reviewed Content:
- Cutting-edge research
- No embargoes
- Scholarly and professional
- Interdisciplinary

Award-Winning Platform:
- Unlimited simultaneous users
- Full-text in XML and PDF
- Advanced search engine
- No DRM

Librarian-Friendly:
- Free MARC records
- Discovery services
- COUNTER4/SUSHI compliant
- Training available

To find out more or request a free trial, visit:
www.igi-global.com/eresources

IGI Global
Proudly Partners with

eContent Pro specializes in the following areas:

Academic Copy Editing
Our expert copy editors will conduct a full copy editing procedure on your manuscript and will also address your preferred reference style to make sure your paper meets the standards of the style of your choice.

Expert Translation
Our expert translators will work to ensure a clear cut and accurate translation of your document, ensuring that your research is flawlessly communicated to your audience.

Professional Proofreading
Our editors will conduct a comprehensive assessment of your content and address all shortcomings of the paper in terms of grammar, language structures, spelling, and formatting.

IGI Global Authors, Save 10% on eContent Pro's Services!

Scan the QR Code to Receive Your 10% Discount

The 10% discount is applied directly to your eContent Pro shopping cart when placing an order through IGI Global's referral link. Use the QR code to access this referral link. eContent Pro has the right to end or modify any promotion at any time.

Email: customerservice@econtentpro.com

econtentpro.com

Become an IRMA Member

Members of the **Information Resources Management Association (IRMA)** understand the importance of community within their field of study. The Information Resources Management Association is an ideal venue through which professionals, students, and academicians can convene and share the latest industry innovations and scholarly research that is changing the field of information science and technology. Become a member today and enjoy the benefits of membership as well as the opportunity to collaborate and network with fellow experts in the field.

IRMA Membership Benefits:

- **One FREE Journal Subscription**
- **30% Off Additional Journal Subscriptions**
- **20% Off Book Purchases**
- Updates on the latest events and research on Information Resources Management through the IRMA-L listserv.
- Updates on new open access and downloadable content added to Research IRM.
- A copy of the Information Technology Management Newsletter twice a year.
- A certificate of membership.

IRMA Membership $195

Scan code or visit **irma-international.org** and begin by selecting your free journal subscription.

Membership is good for one full year.

www.irma-international.org

Encyclopedia of Information Science and Technology, Third Edition (10 Vols.)

Mehdi Khosrow-Pour, D.B.A. (Information Resources Management Association, USA)
ISBN: 978-1-4666-5888-2; EISBN: 978-1-4666-5889-9; © 2015; 10,384 pages.

The **Encyclopedia of Information Science and Technology, Third Edition** is a 10-volume compilation of authoritative, previously unpublished research-based articles contributed by thousands of researchers and experts from all over the world. This discipline-defining encyclopedia will serve research needs in numerous fields that are affected by the rapid pace and substantial impact of technological change. With an emphasis on modern issues and the presentation of potential opportunities, prospective solutions, and future directions in the field, it is a relevant and essential addition to any academic library's reference collection.

Take An Extra **30% Off**[1]

[1] 30% discount offer cannot be combined with any other discount and is only valid on purchases made directly through IGI Global's Online Bookstore (www.igi-global.com/books), not intended for use by distributors or wholesalers. Offer expires December 31, 2016.

Free Lifetime E-Access with Print Purchase

Take 30% Off Retail Price:

Hardcover with Free E-Access:[2] **$2,765**
List Price: $3,950

E-Access with Free Hardcover:[2] **$2,765**
List Price: $3,950

Recommend this Title to Your Institution's Library: www.igi-global.com/books

[2] IGI Global now offers the exclusive opportunity to receive free lifetime e-access with the purchase of the publication in print, or purchase any e-access publication and receive a free print copy of the publication. You choose the format that best suits your needs. This offer is only valid on purchases made directly through IGI Global's Online Bookstore and not intended for use by book distributors or wholesalers. Shipping fees will be applied for hardcover purchases during checkout if this option is selected.

The lifetime of a publication refers to its status as the current edition. Should a new edition of any given publication become available, access will not be extended on the new edition and will only be available for the purchased publication. If a new edition becomes available, you will not lose access, but you would no longer receive new content for that publication (i.e. updates). Free Lifetime E-Access is only available to single institutions that purchase printed publications through IGI Global. Sharing the Free Lifetime E-Access is prohibited and will result in the termination of e-access.

CPSIA information can be obtained
at www.ICGtesting.com
Printed in the USA
BVOW09*1934180517
484205BV00005B/10/P